JJ,

I am happy to share this book with you on a gray fall day, Nov. 7th, 1998.

David P. Gustafson

FOURTH EDITION

Structural Design Guide to the ACI Building Code

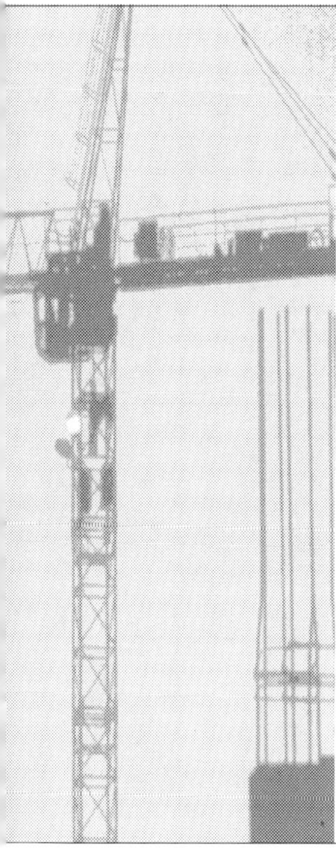

Structural Design Guide to the ACI Building Code

FOURTH EDITION

Edward S. Hoffman

President

Edward S. Hoffman, Ltd., Chicago, IL

David P. Gustafson

Vice President of Engineering

Concrete Reinforcing Steel Institute, Schaumburg, IL

Albert J. Gouwens

President

Gouwens Engineering Consultants, Inc., Elmhurst, IL

KLUWER ACADEMIC PUBLISHERS

Boston/Dordrecht/London

Distributors for North, Central and South America:
Kluwer Academic Publishers
101 Philip Drive
Assinippi Park
Norwell, Massachusetts 02061 USA

Distributors for all other countries:
Kluwer Academic Publishers
Distribution Centre
Post Office Box 322
3300 AH Dordrecht, THE NETHERLANDS

Library of Congress Cataloging-in-Publication Data

A C.I.P. Catalogue record for this book is available
from the Library of Congress.

Printed on acid-free paper.

Printed in the United States of America

Preface

This book is intended to guide practicing structural engineers familiar with earlier ACI building codes into more profitable routine designs with the ACI 1995 Building Code (ACI 318-95).

Each new ACI Building Code expresses the latest knowledge of reinforced concrete in legal language for safe design application. Beginning in 1956 with the introduction of ultimate strength design, each new code offered better utilization of high-strength reinforcement and the compressive strength of the concrete itself. Each new code thus permitted more economy as to construction material, but achieved it through more detailed and complicated design calculations. In addition to competition requiring independent structural engineers to follow the latest code for economy, it created a professional obligation to follow the latest code for accepted levels of structural safety.

The increasing complexity of codes has encouraged the use of computers for design and has stimulated the development of computer-based handbooks. Before computer software can be successfully used in the structural design of buildings, preliminary sizes of structural elements must be established from handbook tables, estimates, or experienced first guesses for input into the computer.

This book will guide the user to the various sections of the Code pertinent to design of common reinforced concrete structural elements. A brief explanation of the significance of these sections is presented, together with limits of applicability, the range in which results may control design; and, where possible, design short cuts to ensure automatic conformance to the Code without calculations.

This Guide does not duplicate nor replace the ACI Code, its Commentary, design handbooks, or use of computers. It complements the ACI Code and Commentary, shows how to take full advantage of available handbooks based on the Code, and should shorten time to develop computer design programs. It converts some code formulas from the review form (or trial designs) to direct design. It presents some simple appropriate formulas, tabulations, and charts for conservative longhand direct design.

Specifications for materials and special Code requirements superimposed upon the ASTM Specifications for materials are explained to aid the structural engineer to avoid difficulties with use of obsolete specifications.

The overall objective of this book is simply to save the engineer time in reinforced concrete design.

HOW TO USE THIS GUIDE

Code requirements applicable to the design of structural building elements, scattered through various Code chapters, have been assembled for the analysis and design of one-way slabs, one-way joists, beams, the various types of two-way slab systems with and without beams, prestressed flexural members, columns, walls, and footings.

Most of the numerical examples are based on normal weight concrete with f'_c = 4,000 psi for flexural members and 4,000, 5,000, or 6,000 psi for columns; and the standard Grade 60 reinforcement. For lightweight aggregate concrete, see Chapter 15. Other concrete strengths used are so indicated.

Provisions new to the Code are noted as such for especial attention. It is assumed that users of the guide are familiar with reinforced concrete design and structural analysis as well as the terms and symbols in common use.

This Guide is intended for use *with* the Code itself. Space limitations make it impracticable to include the Code. The Guide indicates the proper sections of the Code in the order that a designer would normally require their use for design of a particular building element. In the appropriate chapter for the element being designed, the engineer will find the applicable Code sections indicated in parenthesis thus: "(Section 00.00.00)" following the explanations of their application.

Explanations of requirements difficult to interpret are followed by numerical examples. Where several Code equations or requirements are applicable simultaneously and must be solved to determine which controls, computer solutions over the usual range have been included as convenient tables or curves. Where examination of computer solutions over a wide range show simple approximations by longhand possible, such short cuts, together with the limitations of range or accuracy are given. For especially difficult or unusual problems outside the scope of this guide, other references are cited to guide the engineer to a quick source of information for detailed study. References to the ACI Code Commentary are indicated thus: "(Commentary R00.00.00)."

No attempt has been made to explain each individual Section of the Code in this book. A large number of Code provisions which have provoked little or no question of interpretation in past codes have been repeated essentially without change in the 1995 Code.

Other exclusions are precast concrete and composite (precast with cast-in-place concrete) design procedures which involve no separate design theory but merely consideration of different load conditions due to construction sequences. An alternate (working stress) design method is permitted by Appendix A of the Code, but the working stress design method is uneconomical. The Code provisions for thin shell and folded-plate design serve only to include these structures within the scope of the Code. The variety of plates and shells precludes their inclusion here.

Two comprehensive indexes are provided, a subject index and a Code section reference index. The user wishing to locate all Code references to a particular subject, as well as the user interested in the interpretation of a particular Code section, should find this arrangement most convenient. Commentary sections referenced in the text are listed in a third index.

Inch-pound units are used throughout this Guide—consistent with the inch-pound version of the 1995 ACI Building Code. At press time, federal government-funded construction projects were required to be designed in metric units and built with metric materials. Construction projects in the private sector are not mandated to be metric. Some of the metric aspects of reinforced concrete design and construction are briefly discussed in Appendix A of this Guide.

Artwork for the book was prepared by Jesse Trevino. His skillful production of the drawings and charts is greatly appreciated by the authors.

The late Paul Rice originated the concept of this Guide. His vision, plus the participation and contributions of co-author Hoffman, culminated in the first edition of the Guide, which was published in 1973. The second edition under the stewardship of Rice and Hoffman followed in 1979. For the third edition in 1985, the senior authors were augmented with the addition of Gustafson and Gouwens. Although a quarter of a century has passed since the first edition was published, this fourth edition follows the style and format as envisioned and fostered by Rice. Paul Rice's legacy has inspired us to follow through with the preparation of this new edition, and we gratefully dedicate this book to his memory.

Edward S. Hoffman
David P. Gustafson
Albert J. Gouwens

Contents

1 Structural Materials, Specifications, and Testing

GENERAL

All structural materials for reinforced concrete and the standard acceptance tests for such materials are prescribed under the ACI Code as the standard specifications issued by the American Society for Testing and Materials (Section 3.8.1). The ASTM specifications, including the date of latest issue at the time each new ACI Code is itself adopted, are adopted by reference to the Code, as if set forth fully therein. Because the specifications are made an integral part of the Code, and the latter itself is intended to be incorporated into statutory general building codes as an integral part, it is legally necessary that the "outside" specifications be identified as to year of issuc.

Since the ASTM, like ACI, revises its specifications periodically to reflect new materials; improvements in production methods, materials, or testing; or new requirements, some of the materials specifications made part of a current ACI Code will always be obsolete. Many users of the Code simply specify material specifications of "latest issue" for their projects designed under the ACI Code. Usually, this procedure is not only satisfactory but desirable, since most revisions to material specifications upgrade the requirements. The specifying engineer, however, is responsible that materials used under statutory codes meet the legal requirements thereof, and so should be acquainted with any variations in later issues of the specifications.

STEEL REINFORCEMENT—ASTM SPECIFICATIONS

Applicable requirements are titled "Steel Reinforcement" (Section 3.5). For reinforced concrete, deformed reinforcing bars, bar mats, coated bars, wire, or welded wire fabric may be used (Section 3.5). Plain (smooth) bars or plain wire may be used only as spirals (Sections 3.5.4.1 and 3.5.4.2). Welded wire fabric may be composed of either plain or deformed wire (Sections 3.5.3.5 and

3.5.3.6). For composite steel-concrete construction, steel pipe or tubing (Section 3.5.6.2) or rolled shapes and built-up structural steel members may be used (Section 3.5.6.1). Wire, strand, or high-strength bars may be used for prestressing (Sections 3.5.5.1 and 3.5.5.2).

REINFORCING BARS

The ASTM Specifications, referenced in the Code, cover four different types of deformed steel reinforcing bars (hereafter termed "rebars" in this text). These specifications are identified as A615-94 for billet-steel, A706-92b for low-alloy steel, A616-93 for rail-steel, and A617-93 for axle-steel. Since the latter two types of rebars are not generally available except in a few areas of the country, they are not discussed in this text.

Bar sizes and Grades (minimum yield strength levels in ksi) covered by ASTM Specifications A615-94 and A706-92b are shown in Table 1-1. Mechanical requirements in the specifications are concerned with tensile properties and bending properties. The specifications' requirements for minimum tensile strength and percentage of elongation are included in Table 1-1, and bend test requirements are presented in Table 1-2. Also shown in Table 1-2 are the minimum finished bend diameters permitted by Section 7.2 on rebars used in construction. Finished bend diameters are measured on the inside of the bend. The

TABLE 1-1 ASTM Specifications—Bar Sizes, Grades, and Tensile Requirements

ASTM Specification and Type of Steel	Bar Sizes	Grade (Minimum Yield Strength, ksi)	Minimum Tensile Strength, ksi	Minimum Percentage of Elongation in 8-in. Gage Length
A615-94 Billet-Steel	#3 to #6	40	70	#3 . . . . 11 #4, #5, #6 12
	#3 to #11 #14, #18	60	90	#3 to #6 . . . . 9 #7, #8 . . . . 8 #9, #10, #11 . . . . 7 #14, #18 7
	#6 to #11 #14, #18	75	100	#6, #7, #8 . . . . 7 #9, #10, #11 . . . . 6 #14, #18 6
A706-92b Low-Alloy Steel	#3 to #11 #14, #18	60	80	#3 to #6 . . . . 14 #7, #8 . . . . 12 #9, #10, #11 . . . . 12 #14, #18 10

TABLE 1-2 ASTM Bend Tests and ACI Standard Bends

ASTM Specification and Type of Steel	Grade	Bar Sizes	ASTM Bend Test Pin Diameter*	ACI Finished Bend Diameters: Standard Bends	ACI Finished Bend Diameters: Stirrup and Tie Hooks†
A615-94 Billet-Steel	40	#3, #4, #5	3.5 d_b	6 d_b	4 d_b
		#6	5 d_b	6 d_b	6 d_b
	60	#3, #4, #5	3.5 d_b	6 d_b	4 d_b
		#6, #7, #8	5 d_b	6 d_b	6 d_b
		#9, #10,	7 d_b	8.5 d_b	—
		#11	7 d_b	8.5 d_b	—
		#14, #18	9 d_b (90°)	10.7 d_b	—
	75	#6, #7, #8	5 d_b	6 d_b	6 d_b
		#9, #10,	7 d_b	8.5 d_b	—
		#11	7 d_b	8.5 d_b	—
		#14, #18	9 d_b (90°)	10.7 d_b	—
A706-92b Low-Alloy Steel	60	#3, #4, #5	3 d_b	6 d_b	4 d_b
		#6, #7, #8	4 d_b	6 d_b	6 d_b
		#9, #10,	6 d_b	8.5 d_b	—
		#11	6 d_b	8.5 d_b	—
		#14, #18	8 d_b	10.7 d_b	—

* ASTM bend tests 180° except for #14, #18 Grade 60 bars (A615).
† 90° and 135°; industry practice limits stirrups and ties to bar size #8.
d_b = nominal bar diameter.

finished bend diameters for bar sizes #9 through #18 are slightly larger than the Code's minimum bend diameters to allow for "springback."

Grade 60 billet-steel (ASTM A615) bars are currently the most widely used type of rebar in reinforced concrete construction in the USA. An increase in usage of low-alloy steel rebars (ASTM A706) appears to be emerging. Low-alloy steel rebars are intended for applications where controlled tensile properties, or chemical composition restrictions to enhance weldability, or both, are required. The ASTM A706 Specification prescribes a minimum yield strength (60 ksi) and limits the maximum yield strength to 78 ksi. Minimum tensile strength is 80 ksi. In addition, the tensile strength must be 1.25 times actual yield strength. The latter requirement, a minimum tensile-yield ratio, will control when the actual yield strength of a bar exceeds 64 ksi. For example, if the actual f_y is 66 ksi (test value), the tensile strength would have to be at least 1.25 times 66 or 82.5 ksi. The controlled tensile properties are illustrated in Fig. 1-1. The requirements for tensile and bending properties in the A706 Specification are more restrictive than those in Specification A615, namely, larger values of minimum percentage of elongation (Table 1-1); and smaller pin diameters for

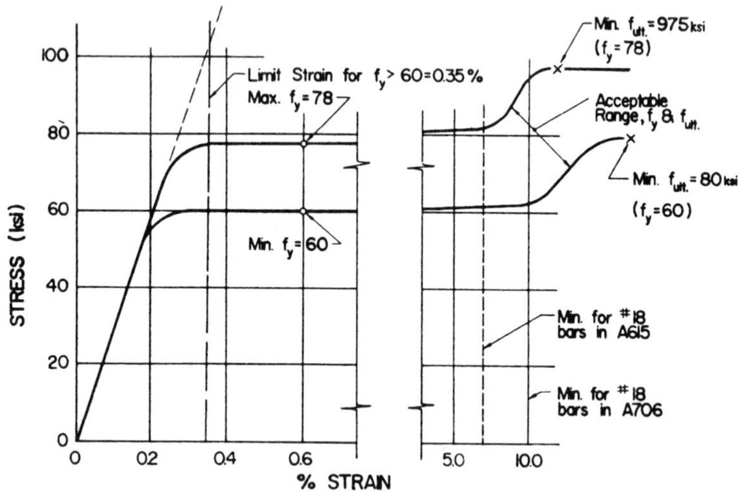

Figure 1-1 ASTM A706 tensile requirements for low-alloy steel bars.

bend tests (Table 1-2). Chemical composition restrictions in the ASTM A706 Specification are discussed under Weldability on page 5.

The Code permits the use of coated reinforcing bars, either epoxy-coated or zinc-coated (galvanized), as a corrosion-protection system (Section 3.5.3.7). Epoxy-coated rebars must conform to ASTM A775-94d or ASTM A934-95. The Code requires zinc-coated (galvanized) rebars to conform to ASTM A767-90. The bars that are to be coated must meet the requirements in Section 3.5.3.1. Proper use of the ASTM coated bar specifications requires inclusion of provisions in the project specifications for construction operations involving coated bars. The provisions should prescribe requirements for handling and placing (installation); storage; tying (fastening) and bar supports; limits on the amount of damaged coating; and the repair of damaged coating. Section 3 on Reinforcement and Reinforcement Supports in Specifications for Structural Concrete (ACI 301-96) includes such provisions. Guidelines for developing appropriate provisions for project specifications are included in a nonmandatory annex of A934-95.*

DEFORMED BAR MATS

The Code requires deformed bar mats to conform to ASTM A184-90 (Section 3.5.3.3), and the mats must be fabricated from rebars which meet the requirements in Section 3.5.3.1.

* At press time, the current edition of Specification A775 was dated 1996. The specification has been revised and now includes, similar to A934-95, a nonmandatory annex with guidelines for developing appropriate provisions for project specifications.

WELDABILITY OF REBARS

Weldability properties are specifically included in ASTM Specification A706-92b for low-alloy rebars. A section in its Scope states: "This material is intended for welding. ..." Weldability is accomplished in the specification by restrictions on chemical composition and on carbon equivalent. The ACI Code requires welding of rebars to conform to Structural Welding Code—Reinforcing Steel (ANSI/AWS D1.4-92). Weldability is at best a relative term. All steels are "weldable" somehow. The metallurgist defines weldability in terms of the chemical composition. The metallurgist's measure of weldability is the "carbon equivalent" content for which the maximum permitted under A706-92b is 0.55%. At or below this limit, the ANSI/AWS D1.4-92 prescribed welding procedures require little or no preheat at ordinary working temperatures. An important point here is that the same carbon equivalent (C.E.) formula is used in both A706-92b and in the ANSI/AWS Welding Code:

$$\text{C.E.} = \%\text{C} + \frac{\%\text{Mn}}{6} + \frac{\%\text{Cu}}{40} + \frac{\%\text{Ni}}{20} + \frac{\%\text{Cr}}{10} - \frac{\%\text{Mo}}{50} - \frac{\%\text{V}}{10}$$

ASTM A706-92b limits the more critical of these chemical elements affecting weldability: Carbon, max = 0.30%; and Manganese, max. = 1.50%. The A706-92b Specification requires the producer of the rebars to furnish a report of the chemical analysis and carbon equivalent.

In contrast to ASTM A706-92b, the Scope of Specification A615-94 for billet-steel rebars states: "Welding of the material in this specification should be approached with caution, since no specific provisions have been included to enhance its weldability. When the steel is to be welded, a welding procedure suitable for the chemical composition and ..." The ACI Code prescribes supplementing the A615-94 Specification to require a report of the chemical composition of the rebars (Section 3.5.2). For billet-steel bars, the ANSI/AWS Welding Code formula for determining the carbon equivalent is:

$$\text{C.E.} = \%\text{C} + \frac{\%\text{M}}{6}$$

As noted, the practical measure of "weldability" is mainly whether preheat is required and to a lesser extent exactly how much preheat. See Table 1-3, Minimum Preheat and Interpass Temperatures, for shielded metal arc welding with low-hydrogen electrodes.

The practical structural engineer defines weldability for rebars in terms of the strength achieved at a welded splice. The practical welder or contractor defines weldability in terms of cost—welding method required, amount of preheat, if any, and control of cooling, if any—required to deliver the performance normally specified by the Engineer. In this confusion, all parties agree that extremely rapid cooling such as quenching in cold liquid changes the ductility

TABLE 1-3 Minimum Preheat and Interpass Temperatures—ANSI/AWS DI.4-92

Carbon Equivalent %	Bar Sizes	Preheat and Interpass Temperatures °F
⩽0.40	#3–#11	None *
	#14 and #18	50
0.41–0.45	#3–#11	None *
	#14 and #18	100
0.46–0.55	#3–#6	None *
	#7–#11	50
	#14 and #18	200
	(Limit for bars meeting A706)	
0.56–0.65	#3–#6	100
	#7–#11	200
	#14 and #18	300
0.66–0.75	#3–#6	300
	#7–#18	400
⩾0.75	#3–#18	500

* When bars are below 32°F, preheat and maintain interpass temperature during welding to at least 70°F.

at the heated area of a weld in carbon steel bars. The steel in this area becomes embrittled.

The practical problem of securing dependable welding of rebars in the field is not resolvable by any single simple solution. Indeed, each essential step toward a solution requires practical consideration relating cost to quantity: (1) specifying the rebars, (2) assurance that materials as delivered meet the specification, (3) specifying the welding, (4) field inspection of welding, and (5) control testing of the results.

A. For large projects involving extensive welding over a long period of time on important elements:

1. Project specifications will be greatly simplified if suppliers are prepared to furnish low-alloy steel rebars meeting A706. (If not, see next paragraph, B.)
2. Mill tests reports, available upon request, show the results of the chemical analyses. The A706 specification requires the producer to furnish mill test reports.
3. Welding specified to follow ANSI/AWS D1.4-92 will seldom require any preheat which makes for less time and cost of the welding itself.
4. When preheat is not required, field inspection is far simpler and its cost is reduced also.

5. Welder qualification tests and occasional check tests are practicable for large projects, spreading the cost of same to a large amount of welding.

B. For ordinary projects involving some welding, off-and-on for important elements:

1. Project specifications should be open to include both A706 and A615 bars. If A706 bars are not available in a practical sense (at a cost resulting in over-all economy), the project specifications should require welding in accordance with ANSI/AWS D1.4-92 (rather than attempting to specify some custom-designed chemical limits to fit ANSI/AWS D1.4-92.)
2. Mill test reports of the chemical analyses must be secured and transmitted to the Contractor so that he can provide the preheat as required. The standard mill test report for A615 bars covers the principal chemical elements carbon and manganese, the elements required for determining the carbon equivalent in the ANSI/AWS Welding Code.
3. Since much or most of the welding will require preheat, heating equipment should be readily available at all times.
4. Field inspectors (and welders) will need temperature indicator "sticks" to mark the bars at the areas to be heated. Inspection should be available at all times when welding is undertaken on important elements.
5. More frequent check tests would be prudent.

C. For small projects involving very little welding, welding on few occasions, or welding at noncrucial areas:

1. Specify maximum preheat for an assumed C.E. > 0.75 on all bars welded.
2. Provide field inspection to establish full preheat as a regular part of the procedure.
3. Provide for field-welded tests to study the results before welding begins, and as a final precaution secure some welded specimens during the welding.

TACK WELDING (WELDING OF CROSSING REBARS FOR ASSEMBLY)

If a small portion of the cross section of a large cold rebar is rapidly heated to welding temperatures and allowed to cool equally rapidly (similar to the effect of quenching), a "metallurgical notch" effect is created (see Fig. 1-2). At this cross section, the original properties of the rebar such as elongation, bendability, dynamic load resistance, fatigue strength, ultimate strength, and yield strength will be reduced. This effect is most dramatically demonstrated commonly in reinforced concrete construction when a column cage is assembled by tack welding relatively small ties to large diameter vertical bars. When the assembly is performed under a combination of most adverse conditions (ambient temperature—cold; tie diameter to vertical bar diameter ratio—minimum),

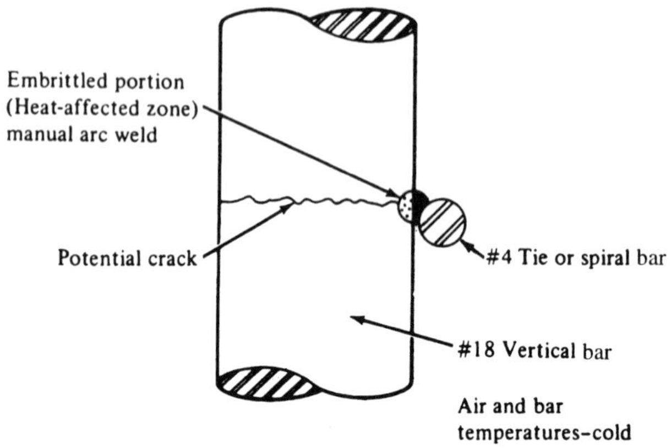

Figure 1-2 Tack welding under most adverse conditions.

the column cage is likely to shatter under any accidental impact. The individual bar failures will most likely occur in the large vertical bars, which snap straight across the cross section from the welds nearest the impact.

Laboratory tests on reinforced concrete beams containing beam cages with stirrups tack welded to the flexural bars (by manual arc welding under laboratory temperatures and conditions) have shown significant losses in the fatigue strength, etc., of the flexural bars. Again, the failure occurred in a brittle fashion at the welds.* The Code forbids field welding of crossing bars for assembly of reinforcement without permission of the Engineer (Section 7.5.4).

WELDED WIRE FABRIC (WWF)

Welded wire fabric, deformed and plain, is required to conform ASTM A497-94a and A185-94, respectively. WWF is fabricated from wires in diameters up to $\frac{5}{8}$ in.† The Code has two requirements in addition to those in the ASTM specifications: (1) for specified $f_y > 60{,}000$ psi, f_y, shall be measured at 0.35% strain, and (2) welded cross-wire spacings (welded intersections) shall not exceed 12 in. for plain wire fabric (Section 3.5.3.5.) or 16 in. for deformed wire fabric (Section 3.5.3.6). The limits on spacings of the welded intersections are not applicable when WWF is used as stirrups (Section 12.13.2).

The Code permits the use of epoxy-coated wire and welded wire fabric as a corrosion-protection system (Section 3.5.3.8). Epoxy-coated wire and WWF must conform to ASTM A884-94a. The wire which is to be coated must meet the requirements in Section 3.5.3.4, and the WWF to be coated has to meet Sec-

* *Reinforcement Anchorages and Splices*, 4th Edition, Concrete Reinforcing Steel Institute, 1997.
† The ASTM Specifications for plain wire (A82) and deformed wire (A496) have been revised, since ACI318-95 was adopted, to include a 0.75–in.–diameter wire size.

tions 3.5.3.5 and 3.5.3.6. As with the ASTM specifications for coated rebars, proper use of ASTM A884-94a requires inclusion of provisions in the project specifications for construction operations involving epoxy-coated wire and WWF. Such provisions are included in Specifications for Structural Concrete (ACI 301-96).[†]

EVALUATION OF TEST RESULTS ON REINFORCEMENT

It should be noted that the tension test requirements for reinforcing steel and prestressing steel are based upon the statistically impossible concept of an absolute minimum strength, except for the retest provisions of the ASTM specifications, which permit retesting when the specimen itself contains a flaw or when the testing equipment or testing procedure proves defective. Retests are also permitted by the ASTM rebar specifications if the results of an original tension specimen fail to meet the specified minimum requirements, but are within prescribed limits of the required tensile and yield strength, and required elongation. This minimum strength concept has been abandoned for evaluating the results of tests of concrete for which specific statistical methods and criteria have been provided. (See "Concrete," this chapter.)

In evaluating the test results for both the tensile and bending properties of rebars, some practical recognition of the statistical approach to variations is desirable. In practice the minimum strength concept of evaluating tests on rebars has probably developed and survived only because few projects require a sufficient number of tests for statistical evaluation; the coefficient of variation is low, usually about 5%; and normally rebars are produced to develop an average yield strength equal to 1.10 or 1.15 f_y. Steel is a factory-made material, and all lots of rebars are tested at the point of production regularly. Material failing to meet ASTM specifications in this control testing is rejected, and thus removed from the "population" tested by the user from field samples.

Periodically, the question of statistical evaluation of strength tests on reinforcement has been seriously considered. Quality assurance for large projects, e.g., dams, nuclear containment, and wastewater treatment plants, where construction may extend several years and involve hundreds of tension tests can only be achieved through statistical evaluation. Similarly, for numerous smaller projects under one agency, supplied over a period of years by a limited number of local producers, statistical evaluation offers the best method of quality assurance. Acceptance of tension testing on the basis of statistical evaluation is a

[†] At press time, the current edition of Specification A884-94a was dated 1996. Guidelines for developing appropriate provisions for project specification are included in a nonmandatory annex of A884-96.

common practice in the national standards for steel reinforcement in many foreign countries. The major advantage of statistical evaluation is in reducing construction costs overall, by avoiding last-minute rejections and costly job delays awaiting further testing, only to discover that the last single-field sample test reporting 99.5% f_y was *not* truly representative of the "lot" or "heat" furnished.

In bend tests, results are inherently even more variable due to the mechanical effect of the deformations. The stress-raising effect can vary not only as the fresh cut roll wears, giving smoother transitions, but also with the orientation of the bar in the bending equipment and the speed of the bending operation. The final test of satisfactory bending is field inspection of the bent bars. The bendability property requirements are somewhat self-enforcing due to the producers' concern with waste involved in breakage during bending.

The ASTM Specifications for rebars provide for retest in the event of a failure to pass the bend test. This provision is particularly enlightened in view of published basic data on bendability.* Bend tests are notoriously unrealistic in that the test result is very sensitive to surface defects, flaws, testing procedure, etc., to the extent that the bend test per se has little real significance to structural safety.

For small projects where too few tests are required to permit a statistical evaluation, the decision to reject material should not be made without careful retesting, and low tests close to the minimum specified should be considered within the overall framework established for safe tolerances in variables affecting the strength of the final product. In this consideration, the three philosophical criteria used for evaluating concrete tests might be of value: average (material) strength levels are intended to reduce the probability of (1) too many low tests, (2) too many consecutive low tests, and (3) individual tests being disturbingly low (for concrete, more than 500 psi below specified f_c').

CONCRETE MATERIALS

A number of provisions in the Code regarding concrete materials should be known to the structural engineer responsible for project specifications.

Chloride in concrete from any source has been found detrimental to prestressed reinforcement (stress corrosion) and to aluminum embedments (galvanic corrosion).† For prestressed concrete and concrete containing aluminum embedments of any type, the Code limits the accidental chloride ion content of mixing water (Section 3.4.2) and of admixtures (Section 3.6.3). The use of admixtures must be approved by the Engineer (Section 3.6.1).

---

* Bend Tests of Grade 60 Reinforcing Bars" by R.J. Kudder and D.P. Gustafson, *ACI Journal, Proceedings* V. 80, No. 3, May-June 1983, pp. 202–209.

† "Corrosion of Metals in Concrete (ACI 222R-89)", *ACI Manual of Concrete Practice*, Part 1, 1996.

Nonpotable water is to be permitted only after specified comparative cube tests show that it will produce at least 90% of the strength achieved with potable water (Section 3.4.3).

The Code does not permit the use of calcium chloride as an admixture in concrete which will be exposed to severe or very severe sulfate-containing solutions (Section 4.3.2). Limits on chloride ion concentrations in hardened concrete contributed from the ingredients are given in Section 4.4.1.

CEMENTS

Cements permitted under the Code are portland cement conforming to ASTM C150, blended hydraulic cements to ASTM C595, and expansive hydraulic cement to ASTM C845 (Section 3.2.1). One significant additional provision concerning cement is that the brand and type used in the work shall correspond to that upon which the selection of proportions is based (Sections 3.2.2 and 5.2.3). Other cementitious materials include fly ash and other pozzolans meeting ASTM C618, slag meeting ASTM C989, and silica fume meeting ASTM C1240 (Section 4.1.1).

With the refinements of statistical evaluation controlling concrete quality from establishment of original proportions and changes during the work, the importance of variations in the cement must not be minimized. The Code provides no specific limit on the minimum amounts of cement, but limits the water-cementitious materials ratios and compressive strength for normal weight concrete (compressive strength for lightweight aggregate concrete) for special conditions of exposure and severe concentrations of sulfate (Sections 4.2.2, 4.3.1, and 4.4.2). Special sulfate-resisting cements may be used for concretes which will be exposed to sea water, sulfate in soil, or sulfate in solution (Section 4.3.1.). For concrete exposed to deicing chemicals, Section 4.2.3 limits the amount of fly ash, other pozzolans, silica fume, slag, or blended cements used in the concrete.

Where the original proportions for the concrete strength specified, f_c' are based upon a statistical analysis of production from an established plant (field experience) over a long period, the resulting coefficient of variation will probably include cement variations, and so the requirement that the cement used shall correspond to that upon which proportions are based becomes simply a matter of using the same *type*.

Where laboratory trial mixtures are the basis for the proportions (Sections 5.3 and 5.6), the standard deviation used to establish the required excess of average strength above f_c' may be based on records of strength tests within 1,000 psi on similar proportions and materials (not necessarily identical) and similar controls of proportioning (Section 5.3). To eliminate the effect of cement variations in this situation, the specifier should require that a "corresponding" cement, usually the same type from the same source as used in the trial mixtures, be used for the work and provide for periodic tests on it, unless the trial mixtures were made with a blend of the several cements expected to be used. Where cement from several sources is anticipated, it is better practice to require trial mixtures separately, establishing separate mixture proportions for each to allow for the considerable coefficient of variation in strength-producing properties of cement from different sources.

Although of no structural significance (to safety) where a uniform color of the exposed concrete is desired, use of cement of the same type and from the same source will improve uniformity.

AGGREGATES

The Code requires that concrete aggregates conform to ASTM C33 or C330 for normal weight or lightweight aggregate, respectively. An exception provides that nonconforming aggregate with a satisfactory service record is acceptable (Section 3.3.1), where approved by the building official. The Code contains specific provisions parallel to those for cement requiring that the aggregate used in the structure be the same as in the trial mixtures. This obvious precaution is included in an overall provision (Section 5.2.3). Proportions are to be based on laboratory trial mixtures or field experience with the materials to be employed. The aggregate specifications are very broad and include materials within broad ranges of gradation, surface roughness, and particle shape. Conformance to C33 or C330 does not assure that all aggregates so conforming are to be used interchangeably in the same proportions under either specification.

CONCRETE QUALITY

The structural engineer is required to specify on the design drawings or in the project specifications the concrete compressive strength, f_c', for which each part of the structure is designed (Section 1.2.1-c). Unless otherwise specified, f_c' is based on the 28-day strength (Section 5.1.3). Field samples for acceptance tests must be laboratory-cured (Section 5.6.2.2). If additional tests are required to measure adequacy of curing in the structure or to determine when forms may be removed, parallel specimens from the same sampling as the acceptance tests should be cured in the field in accordance with ASTM C31.

The Code provides that the field curing procedures be improved when the field-cured strengths are less than 85% of companion laboratory-cured specimens or less than 500 psi above f_c' when 85% of the laboratory-cured test results are higher than $f_c' + 500$ psi (Section 5.6.3.4).

The Code emphasizes that concrete proportioning be based on the use of field experience or laboratory trial mixtures, except that the original proportions, where laboratory trial mixtures are not feasible and no experience record is available, may be based upon other experience or information if approved by the designer (Section 5.4.1). This alternate provision is limited for application to maximum $f_c' = 4{,}000$ psi; also applicable are Code Chapter 4 and Section 5.4.2.

For structural lightweight concrete, where the splitting strength is used as a design property, provision is made for tensile splitting tests (Section 5.1.4). No specific criteria are given for evaluation of the results, and the Code specifically notes that field tests are not to be used as a basis for acceptance. Unless the design engineer can be assured, by a well-documented series of splitting tests on the structural lightweight concretes available, that the average value will justify higher stresses, it should be conservative to use the values prescribed for use without tests in determining shear strength (Section 11.2) and the factors for tension development length (Section 12.2.4).

CONCRETE TESTS AND EVALUATION OF RESULTS

The primary interest of the structural engineer is that the concrete placed will develop the strength, f_c', used for design before the structural element concerned can receive its design load. Code provisions for initial proportioning (Sections 5.2 and 5.3) and for later adjustment of the initial proportions (Section 5.5) are intended to develop concrete with an average strength which is always greater than f_c'.

The procedures for evaluating the strength of the concrete produced (Section 5.6) are based upon probability concepts outside the scope of this book. The basic reference for this task is *Recommended Practice for Evaluation of Strength Test Results of Concrete* (ACI 214-77, Reapproved 1989); and *Simplified Version of the Recommended Practice for Evaluation of Strength Test Results of Concrete* (ACI 214.3R-88). Another reference is the Commentary to the Code.

When it has been determined that the strength of field samples is below acceptable limits, the Code provides for an orderly progression of further testing to evaluate the questionable concrete. If field-cured samples show deficient strengths and the companion laboratory-cured samples give acceptable results, further curing can be required. The Code provides for taking core samples that can be utilized to confirm the need for further curing or the results of further curing.

CORE TESTS

If the strength results on laboratory-cured specimens are unacceptable, core tests are permitted (Section 5.6.4.2). If parallel sets of core samples are taken, one set can be tested immediately and one after intensive further curing to evaluate the potential benefit of further curing. It will also be helpful to take numerous surface hardness (impact) readings in conjunction with any core testing. The results of the impact tests should be corrected by a calibration factor from impact readings on actual test cylinders and cores and the compression tests on same. These tests can be very helpful in locating areas of dubious strength and progress in strength gain under additional curing.

Note that core test criteria are different from criteria for the evaluation of the regular cylinder tests. There is an inherently larger scatter in core test results. Core test results are to be considered adequate if the average of three tests is equal to or greater than 0.85 f'_c, and no single core is less than 0.75 f'_c (Section 5.6.4.4).

A common difficulty in securing reliable core test results is the tendency toward using small diameter cores. For faster core drilling, less damaging holes, and the standard 2:1 ratio of length to diameter, core samples in practice are frequently of smaller diameter than desirable in relation to aggregate size. Since cores are cut through, and often loosen pieces of coarse aggregate lying partly within the core, the core samples are more sensitive to maximum aggregate size because of the resulting stress concentrations than are cast cylinders. As a rule, more reliable results will be achieved if the minimum core diameter is the same as that required for test cylinders, even if a correction factor for a smaller length to diameter ratio is required (ASTM C42-90).

STRENGTH EVALUATION

The last resort before structural strengthening, reduced load rating, or removal and replacement is a strength evaluation (Section 5.6.4.5 and Code Chapter 20). The Code provides for an analytical investigation or a load test (Section 20.1). (See Chapter 18, this book.)

2 Structural Analysis and Design—General

DESIGN METHODS

The Code is based upon design in two stages, strength and serviceability. Design for strength (safety) (Section 9.1.1) involves use of load magnification factors to obtain the required strength (Section 9.2) and strength-reduction (ϕ) factors (Section 9.3.1) to determine the design strength. Strength-reduction factors allow for various possible understrength effects and to reflect the required ductility, reliability, and relative importance of the member. The second stage of design, serviceability (Section 9.1.2), consists of controls on the computed deflection and crack widths under service loads (Sections 9.5 and 10.6). This design method is called "Strength Design." The design philosophy is referred to as "Load Resistance and Factor Design" by building codes for other construction materials.

An alternate design method using the actual service loads, in which all load magnification factors and ϕ-factors = 1.0, is permitted (Section 8.1.2). Under the alternate method, the straight-line theory of stress–strain relationship with reduced allowable stresses is used for flexure only; stresses due to combined flexure and axial load, shear, torsion, and bearing are computed with unit load factors by the same equations as in the principal (strength) design method, but with special capacity-reduction factors applied to computed capacities and allowable stresses (Appendix A: A.3, A.5, A.6, and A.7). Development lengths are the same by both methods of design (Appendix A.4).

The designer may wish to use an alternative method of strength design consisting of load magnification factors in ASCE 7 in conjunction with strength-reduction factors in Appendix C of the Code.

The Code is intended to become part of statutory general building codes by reference. Live loads with permissible reductions thereof as prescribed in the statutory code for application to all types of construction are to be used for reinforced concrete design under the Code (Section 8.2.2) See Fig. 2-1 for live load reductions permitted by the American Society of Civil Engineers Standard ASCE 7-95 Minimum Design Loads for Buildings and Other Structures. (Note:

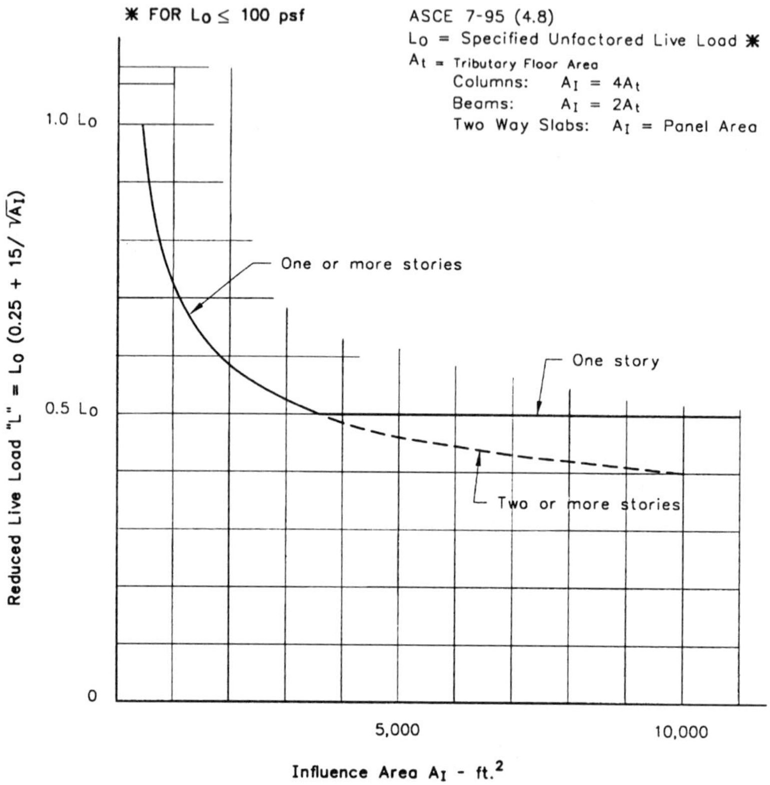

Influence Area A_I - ft.^{2}	Reduced Live Load "L" $L = L_o(0.25 + 15/\sqrt{A_I})$
400	1,000 L_o
1,000	0.724 L_o
2,000	0.585 L_o
3,000	0.524 L_o
4,000	0.487 L_o
5,000	0.462 L_o
6,000	0.444 L_o
7,000	0.429 L_o
8,000	0.418 L_o
9,000	0.408 L_o
10,000	0.400 L_o

Figure 2-1 Live load reduction—ASCE 7.

In the Code the term "load" means the actual dead weight of building elements or reduced minimum loads as prescribed for design by the statutory general code. "Factored load" means actual load times the load factor prescribed in the Code (Section 9.2). The Code, including (Chapter 21) "Special Provisions for Seismic Design," provides for design to resist the lateral loads of wind or earthquake, and requires consideration of the effects of forces due to prestressing, crane loads, vibration, impact, shrinkage, temperature changes, creep, expansion of shrinkage-compensating concrete, and unequal settlements of supports (Sections 8.2.3 and 8.2.4). It will be noted that, except for prestressed concrete (Section 18.19.3), provisions for the effect of fatigue have been omitted as not applicable to reinforced concrete in most building construction.

For both strength and alternate design methods, the modulus of elasticity in psi, E_c, for concrete is prescribed to vary with the concrete weight and compressive strength as:

$$E_c = w_c^{1.5} 33 \sqrt{f_c'} \text{ for concretes between 90 and 155 pcf}$$

or

$$E_c = 57{,}000 \sqrt{f_c'} \text{ for normal weight concrete (Section 8.5.1)}$$

Although "normal" is not defined, presumably concrete made with all natural aggregate would be normal weight, commonly varying between 140 and 150 pcf. Structural lightweight concrete is defined as concrete containing lightweight aggregate and not exceeding a weight of 115 pcf (Section 2.1).

DESIGN DRAWINGS AND PROJECT SPECIFICATIONS

For either design method, certain minimum data are required to be shown on the design drawings or project specifications (Section 1.2.1):

1. The *size* and *location* of all structural elements *and* reinforcing steel. "Size and location of reinforcing steel includes anchorage length, location and length of lap splices, and type and location of welded splices and mechanical connections; orientation of column verticals within the column; and where different lengths for staggered splices are used, the location of each within the column.
2. Provisions for dimensional changes. Strictly interpreted, such provisions include all joints. Construction joints are not only practical necessities, but the location of construction joints and the resulting sequence and size of elements cast are specific provisions to regulate shrinkage.
3. Contraction and isolation joints for structural plain concrete. Such joints divide structural plain concrete into flexurally discontinuous elements so that restraint due to creep, shrinkage, and temperature do not cause a buildup of excessive tensile stresses. Joints must reduce the thickness of members by 25% or more so that no tension or flexure can be developed across the joint after cracking. See Chapter 16 on Plain Concrete.
4. Specified compressive strength of concrete, f_c', and specified yield strength (f_y) or grade of reinforcing steel. If different strengths of concrete or grades of reinforcing steel are to be used for different elements on the same drawing, of course, the location of each should be shown or noted.

5. Magnitude, location, and, where applied in increments, sequence of prestressing forces. Magnitude must be completely described as "maximum jacking force," "initial prestress," "final prestress after losses," etc.
6. Live loads and other loads used in the design. (Although not specifically required, it is also desirable to show the design method used, since the Code permits use of two methods.)

Where design calculations are required (by the Building Official), computer design assumptions, input, and output data are acceptable. Results of model analyses are also recognized (Section 1.2.2).

FRAME ANALYSIS BY COEFFICIENTS (SECTION 8.3)

The approximate method of frame analysis is essentially unchanged from past codes (Section 8.3.3). Note that the uniformly distributed load used, w_u, is "factored load" (Section 8.0), which is the actual load times the load factors (Section 9.2), for use with the moment coefficients (Section 8.3.3). Note also that the limitations on the use of the approximate method (Section 8.3.3) include the limit of three on the ratio of live to dead load. Here the term "load" is the actual dead load and the actual specified live load with all reductions permitted by the general code. The limitations also restrict the use of the approximate method to nonprestressed concrete members with uniformly distributed loads.

The moment coefficients supply the maximum points of a moment envelope at the critical sections required for the design. The theoretical moment envelope consists of maximum moments within the limitations on adjacent unequal spans with live loads applied to all, alternate, or adjacent pairs of spans. In the approximate method of analysis, theoretical maximum negative moments, at centers of supports, have been reduced to allow for usual width of supports and some moment redistribution. Theoretical maximum positive moments have been increased for such redistribution. No further redistribution (limit design) is permitted with the moments calculated using the coefficients of this section (Section 8.4).

Coefficients are not prescribed for negative moments that can develop at the centers of spans, with L/D ratios equal to or greater than 1.0 in the shortest spans permitted under the limitations on the approximate method. See Fig. 2-2 for illustrations of the various coefficients of the approximate frame analysis method. Under the general "frame analysis" method (Section 8.3.1), some top reinforcement will often be required continuously across short spans. See Figs. 2-3 and 2-4 for a comparison of factored moments by the two methods of analysis for an example within the limitations of the approximate method.

An important advantage to using the approximate method of analysis when the loads and the frame dimensions are within the limitations prescribed, is that a "three-dimensional" analysis considering torsional stiffnesses of the members at right angles to the frame is not required. Such an analysis is required with the

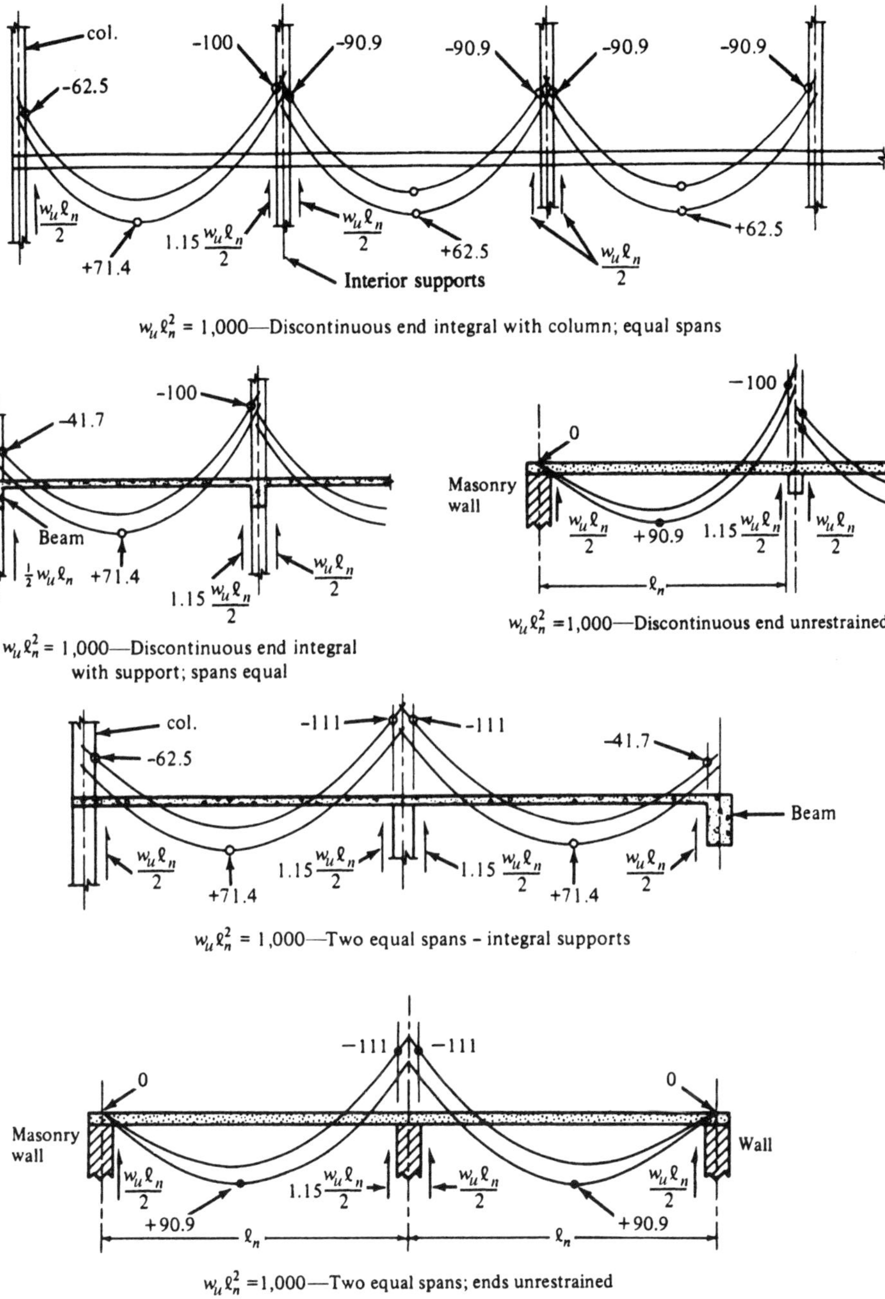

Figure 2-2 Approximate frame analysis coefficients.

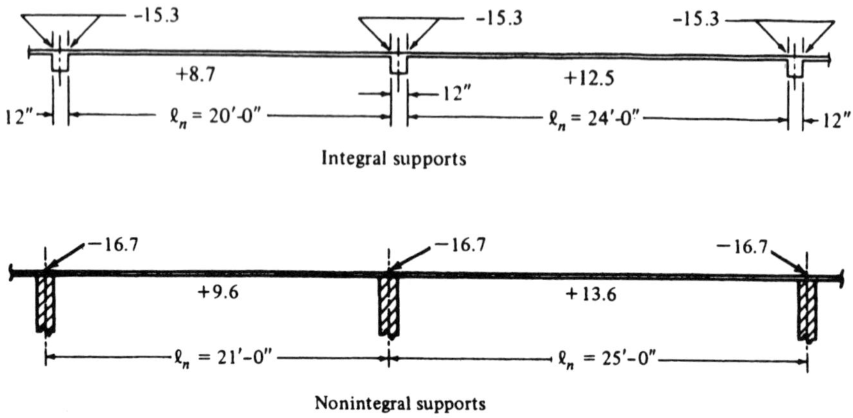

Figure 2-3 Approximate analysis; $-M_u$ at faces of support, ft-k.

general frame analysis whenever the flexural members are integral with their supports (Section 8.6.1) for beam-column frames, and (Section 13.7.5.1) two-way slab and column frames with or without beams).

EXAMPLE. To evaluate the approximate frame analysis (by coefficients) under typical, real conditions, consider a joist-slab continuous across alternate spans of 21′-0″ and 25′-0″ between centerlines of supports *not* integral with the joist-slab (worst condition for slab moments). The supports, if masonry walls, can be taken as 12 in. thick. Use a joist-slab with total depth of 13 in. (10 + 3) with 6 in. ribs @ 26″. This depth is less than the minimum (Section 9.5.2.1 and Table 9.5-a), and so computed deflections must be compared with those permitted for conditions (Table 9.5-b). Clear span ℓ_n = (21′-0″) − 12″ (support width) + 13″ (depth) ≦ 21′-0″ (center to center span) (Section 8.7.1). Use end span ℓ_n = 21′-0″; similarly for interior spans, ℓ_n = 25″-0″.

Let the factored live load, $w_\ell = 250$ psf. (Actual specified live load after reductions permitted, $L = 250/1.7 = 147$ psf.) The dead weight of the joist-slab, $D = 70$ psf for normal weight concrete. Factored dead load, $w_d = 1.4\,D = 98$ psf. Total factored load, $w_u = 348$ psf.

The approximate frame analysis provides moment coefficients to be applied to $w_u\ell_n^2$. Note that ℓ_n for the negative moments is the average of the adjacent clear spans at a support (Section 8.0).

For a 1-ft wide strip, the values of $w_u\ell_n^2$ are:

End, $\ell_n = 21'$; $w_u\ell_n^2 = (0.348)(21)^2 = 153$ ft-kips

Interior, $\ell_n = 25'$; $w_u\ell_n^2 = (0.348)(25)^2 = 217$ ft-kips

Average, $\ell_n = 23'$; $w_u\ell_n^2 = (0.348)(23)^2 = 184$ ft-kips

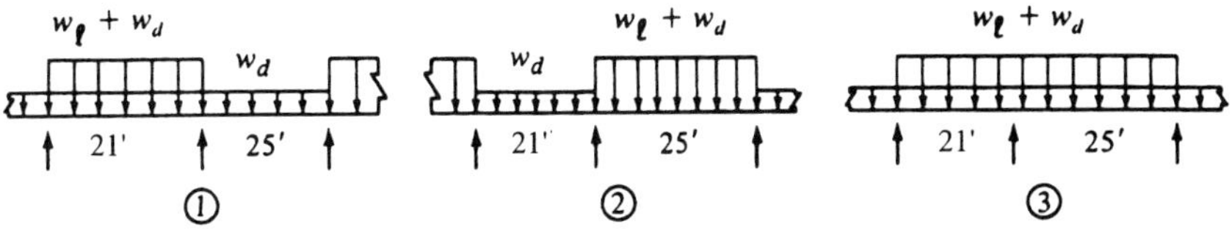

Loading patterns

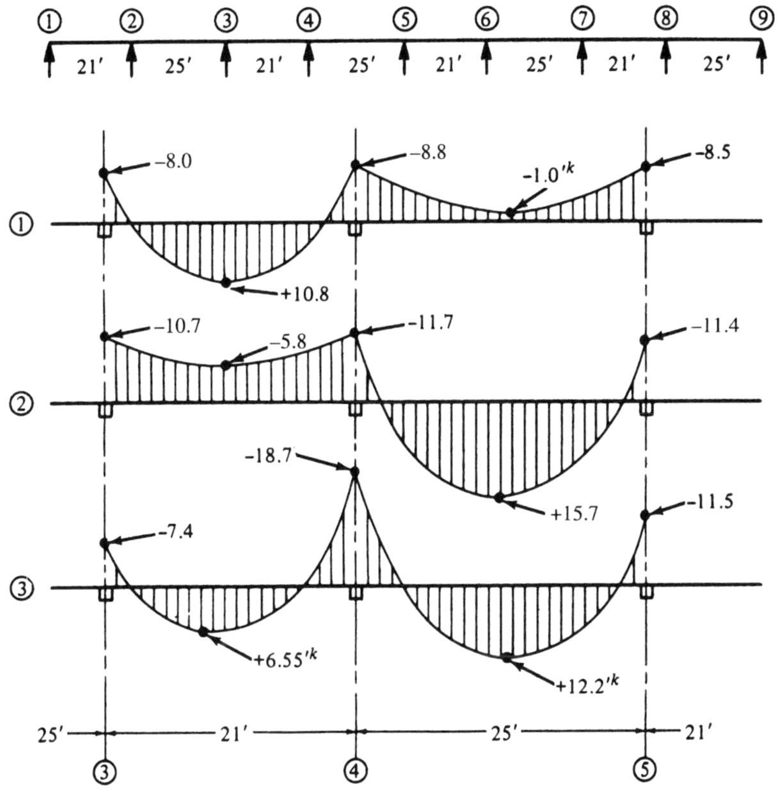

Figure 2-4 Frame analysis factored moments, ft-k, at midspan and centers of supports.

Maximum factored moments for interior spans per 1-ft wide strip, at the interior face of the first interior support and at the faces of all other interior supports,

$$-M_u = -\frac{1}{11}(184) = -16.7 \text{ ft-kips}$$

At the center of the 21-ft interior spans,

$$+M_u = +\frac{1}{16}(153) = +9.6 \text{ ft-kips}$$

At the center of the 25-ft interior spans,

$$+M_u = +\frac{1}{16}(217) = +13.6 \text{ ft-kips}$$

The preceding example was selected to show the application of the definition of clear span to nonintegral supports (Section 8.7.1). Although torsional stiffness of integral supports is not directly considered in the approximate frame analysis, clear span is measured from the face of each support. If integral supports were provided of the same width as in the preceding example, say 12-in.-wide beams monolithic with the joist-slab system, the clear spans would become 20 ft and 24 ft, respectively.

For a 1-ft-wide strip, the values of $w_u \ell_n^2$ become:

$$\text{End } \ell_n = 20 \text{ ft}; \qquad w_u \ell_n^2 = 139 \text{ ft-kips}$$

$$\text{Interior } \ell_n = 24 \text{ ft}; \qquad w_u \ell_n^2 = 200 \text{ ft-kips}$$

$$\text{Average } \ell_n = 22 \text{ ft}; \qquad w_u \ell_n^2 = 168 \text{ ft-kips}$$

Maximum factored moments for interior spans per 1-ft wide strip, at the interior face of the first interior support and at the faces of all other interior supports.

$$-M_u = -\frac{1}{11}(168) = -15.3 \text{ ft-kips}$$

At the center of the 20-ft interior spans,

$$+M_u = +\frac{1}{16}(139) = +8.7 \text{ ft-kips}$$

At the center of the 24-ft interior spans,

$$+M_u = +\frac{1}{16}(200) = +12.5 \text{ ft-kips}$$

See Fig. 2-3 for a summary of these factored moments.

APPROXIMATE METHOD VERSUS FRAME ANALYSIS

The maximum moment values for interior spans in Fig. 2-4 were derived from a frame analysis applied to the preceding example of approximate analysis (Section 8.3). Three loading patterns were considered as prescribed (Section 8.9.2):*

1. w_d on all spans with w_ℓ on alternate spans:
   (a) odd spans (Loading pattern (1)) and (b) even spans (Loading pattern (2))
2. w_d on all spans with w_ℓ on adjacent spans (Loading pattern (3))

Maximum values from these three analyses can be combined to form a moment envelope. Using the frame analysis method, if the torsional stiffness of integral supports can be neglected, the factored moments will be the same for integral or nonintegral supports where depth of the member is the same or more than the width of the nonintegral supports (Sections 8.7.1 and 8.7.2). Center-to-center spans are used for determination of the moments by frame analysis for both of the preceding examples.

One of the dangers of the commonly used general frame analysis computer programs is errors in the input data. A useful practice is to compare some of the maximum factored moments by the approximate analysis method. Negative moments at the center of supports in Fig. 2-4 must be reduced to the face of supports. Reduce all negative moments by (1/2) $(w_u \ell)$ 6/12 ft-kips using total w_u. Compare negative moments by the frame analysis, −16.5 and −16.9 ft-kips, to the average negative moment by approximate analysis, −16.7 ft-kips. Maximum positive moments in the long spans for frame analysis versus approximate analysis are +15.7 ft-kips versus +13.6 ft-kips. These values are well within the redistribution limit allowed with the general frame analysis (Section 8.4). See Fig. 2-5.

The comparison of the results from the frame analysis to the results from the approximate analysis for integral supports is also within the redistribution limit allowed for usual reinforcement ratios, though not as close as for nonintegral supports. The greatest difference, in positive moments, +10.8 versus +8.7 and +15.7 versus +12.5, would be reduced somewhat in the usual structure where the usual beam-column support stiffness would be significant (Section 8.6.1).

A significant difference is the minimum midspan moment for the short span which is −5.8 ft-kips by the frame analysis and 0 by the coefficients. Application of the frame analysis procedures (Section 8.4) would require top tensile reinforcement continuous across the short spans. Decades of experience with approximate analysis have shown such reinforcement, within the limitations of 20% difference in spans and L/D ratios not greater than 3.0, to be usually unnecessary. Some negative moment can be resisted by concrete tension,

---

* Note that *factored* loads are used for the various loading patterns. At the limiting ratio of live to dead load, $L/D = 3.0$, for approximate analysis, the ratio of *factored* loads $w_\ell/w_d = \frac{(1.7)}{(1.4)} 3 = 3.64$.

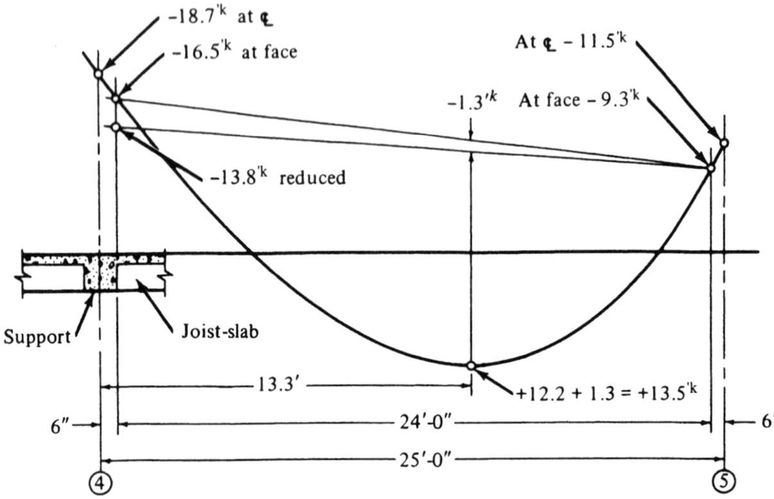

Figure 2-5 Moment redistribution (limit design).

approximately 8.5 ft-kips in this example. If tensile strength is exceeded, top cracking would occur and a significant redistribution of moments would result.

The general absence of such cracks in actual applications where $L/D \leqq 3.0$ indicates that the stress is less than the concrete tensile strength, perhaps because the rigorous pattern of full factored live load on alternate long spans and none on alternate short spans occurs very seldom.

Although most design programs use the general frame analysis method, it is also possible to automate the approximate method wherever the span and live/dead load limitations permit and concentrated loads are not anticipated. When either the live/dead load or the span ratios approach the limitation and full live load in alternate spans can be expected, the approximate method can be employed to evaluate negative moment at the center of short spans and possibly reduce top reinforcement.

For the preceding example, the approximate analysis can be extended to evaluate pattern live loading as follows:

$$\text{Total load in long spans, } -M_u = \frac{(0.35)(24)^2}{11} = -18.2 \text{ ft-kips}$$

$$\text{Dead load only in short spans, } -M_u = \frac{(0.10)(20)^2}{11} = -3.6 \text{ ft-kips}$$

$$\text{Average, } -M_u = \frac{18.2 + 3.6}{2} = -10.9 \text{ ft-kips}$$

$$\text{Simple span moment in short span, } +M_u = \frac{(0.10)(20)^2}{8} = +5.0 \text{ ft-kips}$$

$$\text{Negative moment at center of short span, } -M_u = -10.9 + 5.0 = -5.9 \text{ ft-kips}$$

It will be noted that the result by this approximate analysis is within 10% of the frame analysis result (Fig. 2-4).

The authors recommend further that, if the frame analysis is used for conditions *within* the limitations for approximate analysis, the probability of full live load in alternate spans be evaluated and any resulting negative moments at midspan be adjusted accordingly for design.

Where full live load in alternate spans is expected and an approximate analysis indicates negative moments, the authors recommend use of the frame analysis method, particularly for elements forming the primary framing system. It will, of course, frequently be desirable to employ a more accurate analysis, even where the code permits use of the coefficients, to determine unbalanced moments to columns, to consider effects of torsional stiffness or width of wide integral supports, etc. (see Chapter 9).

REDISTRIBUTION OF NEGATIVE FACTORED MOMENTS (LIMIT DESIGN)

In the previous example, frame analysis results for the negative factored moments for the 25-ft span and the reductions permitted by redistribution are shown in Fig. 2-5. Such redistribution is permitted when the net reinforcement ratio $\rho - \rho' \leqq 0.50\rho_b$ (Section 8.4.1). For Grade 60 bars and $f_c' = 4{,}000$ psi,

$$\rho_b = \frac{0.85\beta_1 f_c'}{f_y} \times \frac{87{,}000}{87{,}000 + f_y} \qquad \text{ACI Eq. (8-1)}$$

$$\rho_b = 0.0285$$

For the joists on a span of 25 ft, reinforcement could consist of four #4 straight top bars and one #5 + one #6 straight bottom bars, as shown in Fig. 2-6.

$$\text{Average width, } b = 6.5 \text{ in.}$$

$$\text{Effective depth, } d = 12 \text{ in.}$$

$$\text{At the face of the supports, } \rho = \frac{4 \times 0.20}{6.5 \times 12} = 0.010$$

Net ρ:

Assuming the two bottom bars are made the same length, and the minimum required embedment of 6 in. (Section 12.11.1) is provided for both, the bottom

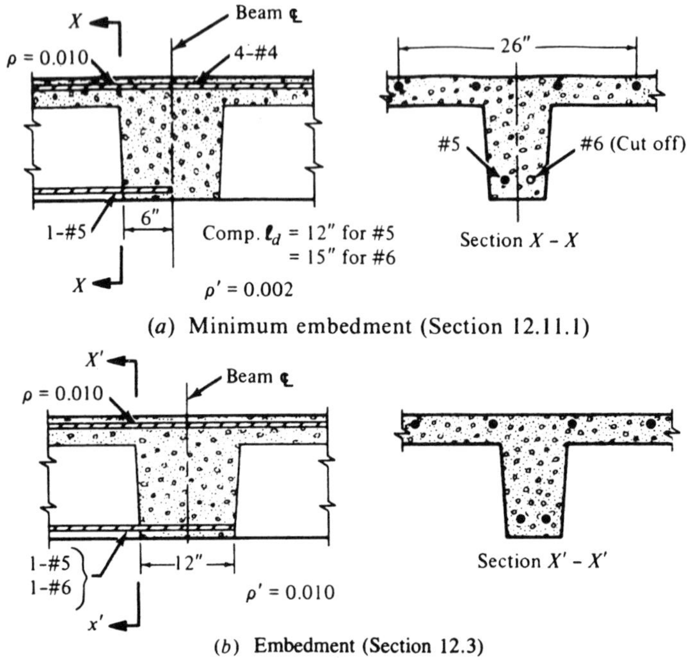

Embed

1 - #5 into support 6 in. (minimum allowed): $\rho - \rho' = 0.008$
Both bottom bars 6 in. (practical minimum): $\rho - \rho' = 0.005$
Both bottom bars 12 in. (practical maximum): $\rho - \rho' = 0.001$

Figure 2-6 Embedment and effectiveness of bottom bars as compressive reinforcement.

bars can be assumed 50% effective for compression at the face of the support (see Table 13-1).

$$\rho' = \frac{(0.44 + 0.31)}{6.5 \times 12}(0.50) = 0.005$$

$$\rho - \rho' = 0.010 - 0.005 = 0.005 \text{ at the face of the support}$$

Negative factored moment may be decreased* for design,

---

* $-M_u$ may also be increased in a like amount if desired (Section 8.4.1). The authors do not recommend increasing the negative moment for design. For practical reasons it is not desirable to increase congestion of the reinforcing steel at supports; also any error in placing steel is likely to reduce effectiveness of top steel. One exception occurs where a structural fire resistance analysis will be made. More negative and less positive steel will increase fire resistance. Reference: *Reinforced Concrete Fire Resistance*, CRSI, 1980.

$$(20)\left(1 - \frac{\rho - \rho'}{\rho_b}\right)\% \text{ (Section 8.4.1)}$$

$$(20)\left(1 - \frac{0.005}{0.0285}\right) = 16.5\%$$

The adjusted maximum negative moment for design becomes $(16.5)(1 - 0.165) = -13.8$ ft-kips. No adjustment should be made to the less-than-maximum negative moment at the other support. This adjustment, $16.5 - 13.7 = 2.8$ ft-kips, is compatible with an increase in the positive factored moment to $+12.2 + (1/2)(2.8) = +13.5$ ft-kips for this loading condition. No increase in the bottom reinforcement is required since the maximum positive factored moment is $+16.0$ ft-kips (Fig. 2-4).

PRACTICAL USES OF MOMENT REDISTRIBUTION

Note that the maximum reduction of negative reinforcement permitted, 20%, is reached when $\rho - \rho' = 0$. The provision offers an opportunity for savings in the total weight of reinforcement for continuous construction, particularly for heavily reinforced long span beams. A very short extension of the bottom bars into the supports, usually 19 bar diameters (for $f'_c = 4{,}000$ psi and $f_y = 60{,}000$ psi) for full compression embedment (Section 12.3), will often permit substantial further reduction of the top bar area over a length equal to more than 0.6 span.

In this short span example, a further embedment of 7 in. for full compression would permit the full reduction of 20% in top steel. Added weight

$$= \frac{7}{12}(1.5 + 1.0) = 1.5 \text{ lb per joist}$$

$$\text{Weight reduced} = (0.035)[1.0 + 0.3(20 + 24)](4)(0.668) = 1.4 \text{ lb per joist}$$

If all or some of the joists in this example are part of the "primary lateral load resisting system," full tension embedment is required (Section 12.11.2). Note that the redistribution of negative moment procedure will often enable the designer to provide this additional bottom steel with no increase in total steel weight.

TWO-WAY SLABS—ANALYSIS BY MOMENT COEFFICIENTS

Since the 1983 Code, the Code has provided a table of moment coefficients for end spans in two-way slabs (Section 13.6.3.3). (See Table 2-1.) This table provides designers with a simple procedure for distributing the total factored moment, M_0, in the end spans of two-way slabs. The "direct design" method applies to two-

TABLE 2-1 Distribution of End Span Total Factored Moment, M_0

	(1)	(2)	(3)	(4)	(5)
	Exterior Edge Unrestrained	Slab with Beam Between All Supports	Slab Without Beams Between Interior Supports: without edge beam	Slab Without Beams Between Interior Supports: with edge beam	Exterior Edge Fully Restrained
Interior negative factored moment	0.75	0.70	0.70	0.70	0.65
Positive factored moment	0.63	0.57	0.52	0.50	0.35
Exterior negative factored moment	0	0.16	0.26	0.30	0.65

way slabs within specified limitations (Section 13.6.1), one of which limits use to uniformly distributed gravity loadings with a L/D ratio not to exceed 2.

The direct design moment coefficients offer an excellent approach for quick preliminary designs. Factored moments may be increased or decreased by 10%, provided the total static moment for the panel is not less than M_0 (Section 13.6.7). For unusually stiff columns (rectangular with longside parallel to moments, very large to support many floors above, or wall-like extending across the strip), the stiffness ratio α_{ec} or α_c formulas for the equivalent column as given in the 1977 Code for the modified stiffness method may be used in place of the coefficients in Section 13.6.3.3 (Commentary R13.6.3.3). See Table 2-2.

$$\textit{Design \#1:}\quad h = 8''; c_1 = c_2 = 36''; \ell_1 = \ell_2 = 21'\text{-}0''$$

$$w_\ell = 200 \text{ psf; story height} = 12'\text{-}0''; M_0 = 289 \text{ ft-k}$$

$$\textit{Design \#2:}\quad h = 10''; c_1 = c_2 = 42''; \ell_1 = \ell_2 = 24'\text{-}0''$$

$$w_\ell = 250 \text{ psf; story height} = 12'\text{-}0''; M_0 = 589 \text{ ft-k}$$

Both designs #1 and #2, are within Direct Design limitations (Section 13.6.1).

It will be noted that the ACI 318 direct design method coefficients show close comparison except in the range 25% to 65% for exterior edge moments. The charts in Fig. 5-3 may be helpful to quick adjustments of the moment design

TABLE 2-2 Comparison—Coefficients vs. Equivalent Column (*M*, ft-kips)

Design	Method	Ext. (Edge) Col. $-M$	Int. Span $+M$	First Int. Col. $-M$
#1	Coeff.	-75 (= 0.26 M_0)	150 (= 0.52 M_0)	-202 (= 0.70 M_0)
	Equiv. Column	-139 (= 0.48 M_0)	118 (= 0.41 M_0)	-191 (= 0.66 M_0)
#2	Coeff.	-153 (= 0.26 M_0)	306 (= 0.52 M_0)	-412 (= 0.70 M_0)
	Equiv. Column	-283 (= 0.48 M_0)	241 (= 0.41 M_0)	-389 (= 0.66 M_0)

coefficients in this range—roughly between "ordinary" and "fully restrained" edge columns.

TWO-WAY SLAB AND COLUMN FRAMES—LATERAL LOADS

For this problem, the reader must be alert to know what *not* to read into the Code. A natural but mistaken interpretation of relevant sections in Chapter 13 dealing with analyses would be that the "equivalent frame" prescribed in great detail for both the "Direct Method Design Method" and the "Equivalent Frame Method" applies for lateral loadings, as well as gravity loadings. Not so. Commentary Section R13.5.1.1 limits the application of both of the analysis methods to gravity loads only.

The "equivalent column concept" is merely a mathematical device to account for the inefficiency of the slab-column connection for the transfer of moments generated in the slab (gravity load moments) to the column. In the common case where the column width, c_2, is much less than the transverse frame width, ℓ_2, the unbalanced negative slab moments midway between columns, cannot be directly transferred into the columns, but must be transferred by torsion in the slab and transverse beam. Only monolithic bearing wall supports extending the full width of the frame, ℓ_2, can be regarded as 100 percent efficient moment connections. In this sense, efficiency of a connection is defined as the ability to transfer unbalanced moments in proportion to the ratio $\alpha_c = \Sigma K_c/\Sigma(K_s + K_b)$. Thus, the terms, K_{ec}, and $\alpha_{ec} = K_{ec}/\Sigma(K_s + K_b)$ were developed to reduce the apparent column stiffness from K_c to K_{ec} and thus account for the "leakage" of a portion of the unbalanced gravity load moments around the column instead of into it (Commentary Fig. R13.7.2).

The casual reader or infrequent user of the Code must be very alert indeed to catch the significance of the word "only," which limits the use of the equivalent frame method to analysis of gravity loads *only*. See the first sentence in Commentary R13.5.1.1.

Under lateral loads, an unbraced frame develops the unbalanced moments in the columns. Thus, connections joining columns to the slabs at less than 100 percent efficiency permit "leakage" in the opposite direction from those generated in the slab system. This "leakage" in effect leaves more of the unbalanced moment in the columns as it represents an inefficient transfer of moment from column to slab system. To account for this effect properly in the lateral load analysis then, the apparent stiffness of the slab system must be reduced. Traditionally, practical structural engineers have done this by considering a reduced *width* of the slab effective for the lateral moment resistance. Practical rules-of-thumb vary among designers. Usually simple ratios are used such as: 0.5 ℓ_2, 0.25 ℓ_2, "column strip only," or column width plus some multiple of the slab thickness. Theoretical elastic analyses have been developed for "effective width" considering square and rectangular panels, column width-to-transverse span ratio (c_2/ℓ_2), etc., for simple cases without beams.*

The most comprehensive and elegant approach which includes the effects of column width, slab depth, beam torsion, and beams in flexure in the slab system is that by Parme.† Dr. Parme simply applies the same Code procedure for an "equivalent column" to define an "equivalent slab" stiffness. His method is simply stated:

K_c, K_s, K_b, and K_t = same as notation in the preceding paragraphs

K_{es} = equivalent slab stiffness for lateral load analysis

$$= \frac{\Sigma(K_s + K_b)}{1 + \Sigma(K_s + K_b)/K_t}$$

α_{es} = ratio of column stiffness to equivalent stiffness of the slab system.

$$= \Sigma K_c/K_{es}$$

The advice (Commentary Section R13.5.1.2) that the slab stiffness should include consideration of the effects of cracking and reinforcement in compression should be noted. For a lateral load analysis, a reasonable or an overestimated evaluation of the lateral displacement is desired. Reduction of stiffness due to cracking should not be based on the most severely cracked cross section of a member. Rather the stiffness should reflect the moment-end rotation relationship of the complete member. For flat plates and flat slabs, the Code permits the use of the stiffness and moment of inertia as:

* "Interaction of Shear Walls and Frames," Khan and Sbarounis, ASCE, ST-3, June 1964, pp. 307–312.

† Presented at ACI Convention, Mexico City, 1977.

Modulus of elasticity . . . . E_c (Section 8.5.1)

Moment of inertia (Section 10.11.1)
- Beams . . . . 0.35 I_g
- Columns . . . . 0.70 I_g
- Walls—uncracked . . . . 0.70 I_g
- Walls—cracked . . . . 0.35 I_g
- Flat plates and flat slabs . . . . 0.25 I_g

Correct *relative* stiffness ratios suffice for gravity load analyses, but may be misleading if used for displacement due to lateral loads. It is usually desirable to ensure that lateral displacement actually occurring is tolerable, and further a realistic value is required for safety to calculate moment magnification, "$P\Delta$" effect (Code and Commentary Sections 10.10, and 10.11).

STRENGTH DESIGN

The first and primary stage of design is for strength (safety). The Code requires that all structural members and structures have a minimum computed design strength at all sections equal to the effects of prescribed factored loads and forces upon the section (Section 9.1.1). Perhaps the two most important requirements in the entire ACI Code are those in Section 9.1, each succinctly stated in single sentences. Too many occasional users of the Code overlook or forget Section 9.1 simply because it is so succinct. Those familiar with the Code in day-to-day use concentrate upon and remember specific "critical sections" defined everywhere in the Code, often forgetting that Section 9.1.1 requires that *all* sections be investigated for strength. Similarly, Section 9.1.2 requires that (all) "members" shall meet *all* requirements for performance at service load levels. Load factors and strength-reduction (ϕ) factors are separately prescribed, and are intended together to provide against *overloads* above the minimum live load, wind load etc., prescribed by the general code, *overrun* of estimated dead weight of materials, and *understrength* resulting from normal variations in the strength of reinforced concrete. These factors do not, and are not intended to, protect against gross errors or negligence in analysis, design, or construction.

The Code prescribes in general an elastic analysis of the structure (exception, "Redistribution," Section 8.4.1 or Appendix B8.4.1). Except for simple span construction, in all monolithic reinforced concrete structures or where the connections of various elements are made integral, some of the sections can yield without causing failure of the structure and thus some additional load capacity is available. With the one exception noted, this additional capacity is conservatively neglected under the Code. Critical sections are designed on the basis of an elastic analysis of the structure, under the factored loads, for the nominal strength of cross sections of members, reduced by specified ϕ-factors (Sections 9.2 and 9.3 or Appendix B9.2 and B9.3). Examples showing applications for strength design to various common structural elements and systems are offered in Chapters 3 through 14 of this book.

General Assumptions for Strength Design—Flexure with No Axial Load

Strength design is based on the assumptions summarized in Fig. 2-7 and on the satisfaction of applicable conditions of equilibrium and compatibility of strains. The Code permits the use of any shape of compressive stress block that will

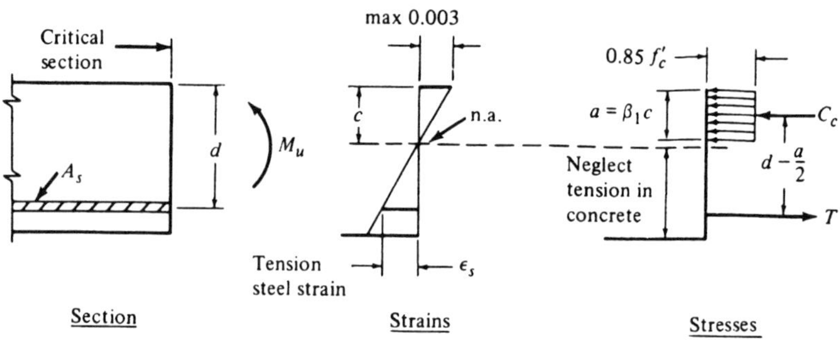

Note: Strains are directly proportional to the distance from the neutral axis, n.a.; 0.003 is the assumed maximum strain usable for the concrete in compression (Section 10.2.3).

For $f'_c \leq 4{,}000$ psi, $\beta_1 = 0.85$

For $f'_c \geq 4{,}000$ psi

$$\beta_1 = 0.85 - 0.05\,\frac{(f'_c - 4{,}000)}{1{,}000} \geq 0.65$$

(*a*) Singly-Reinforced Sections: Design moment strength, $\phi M_n = \phi A_s f_y \left(d - \frac{a}{2}\right)$

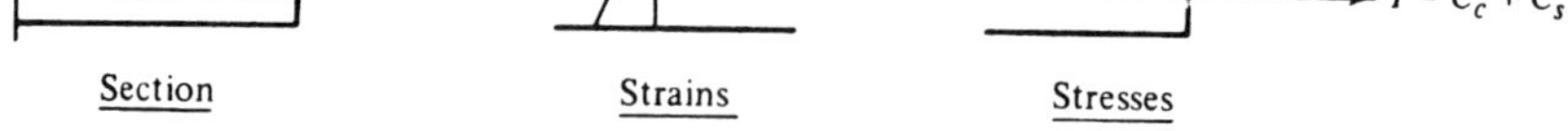

Other strains and stresses same as in (*a*). The stress in the compression steel $f'_s = (29 \times 10^6)(\epsilon_s) \leq f_y$. For all ϵ_s values greater than that corresponding to a stress, f_y, the compression stress is taken as f_y.

(*b*) Doubly-Reinforced Sections:

Design moment strength, $\phi M_n = \phi C_c(d - a/2) + \phi C_s(d - d')$

$$= \phi\left(A_s - A'_s \frac{f'_s}{f_y}\right)(f_y)(d - a/2) + \phi A'_s f'_s (d - d')$$

Figure 2-7 General assumptions for strength design (Section 10.2).

result in strength predictions that are in agreement with available tests, but only the rectangular stress block is prescribed in detail (Sections 10.2.6 and 10.2.7). The authors recommend use of the rectangular stress block for manual design, and for simplicity of programming computer design. See also Fig. 10-5 for applications of the strength design assumptions to column design. The strength-reduction factor, ϕ, in the Fig. 2-7 expressions for flexure is 0.90 (Section 9.3.2). The effects of axial loads, compression or tension, must be included with the effects of flexure. See Chapter 10 for examples.

The most important requirement for strength design involves the definition of "balanced conditions," and the limitation of tensile reinforement based thereon. The entire concept of strength design for flexure utilizes the strength-reduction factor, $\phi = 0.90$, and is based on a gradual yielding tensile failure. Lower ϕ-factors are prescribed where a brittle compression failure is to be expected, as in columns.

Balanced conditions are defined as simultaneously reaching a compression strain in the outer fiber of the concrete of 0.003 (crushing) and tensile strain in the tensile reinforcement corresponding to that for the specified yield strength, f_y (Section 10.3.2) (see Fig. 2-8). The Code limits tensile reinforcement in flexure to produce ductile behavior of the structure in general and to preclude brittle failure at any point under factored loads. To achieve ductile behavior the tension steel must yield before the concrete in compression crushes. The limit prescribed is 0.75 times the ratio ρ_b for singly-reinforced members, Fig. 2-8 (a), plus additional tensile reinforcement sufficient to balance the force in compressive reinforcement, if any, Fig. 2-8 (b) (Section 10.3.3). With the prescribed limit of tensile reinforcement, a yielding failure in singly-reinforced members will occur under any of the following combinations of material strengths assuming construction to the design dimensions:

Let f_{ya} = actual yield strength of steel; f_y = specified
f'_{ca} = actual concrete strength; f'_c = specified

1. $f_{ya} \leqq f_y$ and $f'_{ca} \geqq 0.75\, f'_c$
2. $f_{ya} \leqq 1.33\, f_y$ and $f'_{ca} \geqq f'_c$
3. $f_{ya} \geqq 1.33\, f_y$ and $f'_{ca} \geqq (f_{ya}/f_y - 0.33)(f'_c)$

When condition (3) is exceeded, the failure will be brittle, but will not occur until the actual loading is at least one-third larger than the factored loading. For doubly-reinforced members, these same limits also apply, if the steel in compression has the same stress-strain relationship as that in tension. This assumption is commonly accepted. In doubly-reinforced members, brittle failure at the above three "balancing" conditions is unlikely. Tests show ultimate strains up to 0.006 on test cylinders, and over 0.01 in the (doubly-reinforced) columns of concrete frames with strain gradients (double curvature).*

* Research at the University of Texas, Austin. See Discussion page 143, *ACI Jour.*, March 1977.

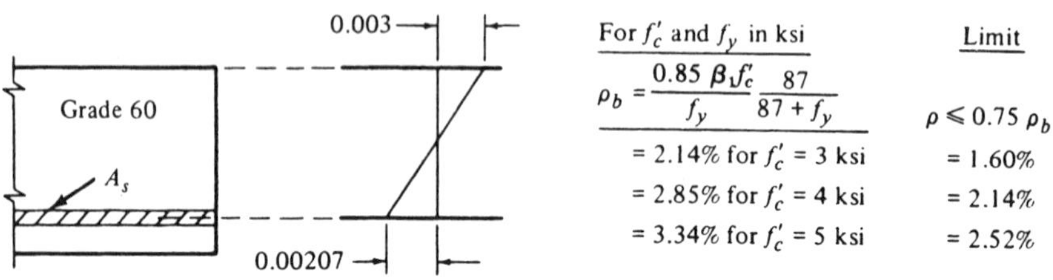

(*a*) Limits on tensile steel -- singly-reinforced elements

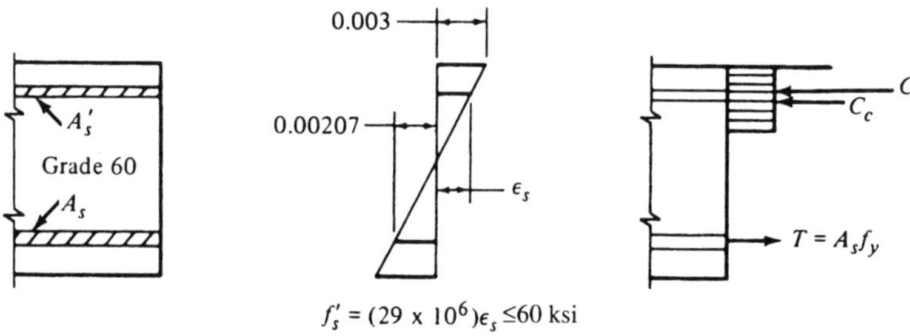

$C_c = (0.85 f'_c)$ ba $\quad C_s = A'_s f'_s \quad$ Max $T = 0.75\,C_c + C_s$

(*b*) Limits on tensile steel – doubly-reinforced elements

Figure 2-8 Limiting reinforcement in flexure (Section 10.3.2).

The limit on the amount of tensile reinforcement applies also to columns under small axial load where the ϕ-factor is gradually reduced from $\phi = 0.90$ for zero axial load to $\phi = 0.70$ (tied columns) for axial load equal to $0.10\, f'_c A_g$ (Section 10.3.3). See interaction diagrams for columns from axial load zero to maximum, Chapter 10.

UNIFIED DESIGN PROVISIONS

Structural design of reinforced concrete or prestressed concrete flexural and compression members to resist factored dead loads and live loads may be designed by Appendix B of the Code (Section 8.1.3). Alternate provisions are given for redistribution of moments (B8.8.4) based on the magnitude of the strain in the tensile reinforcement. Criteria for selection of the ϕ-factors are taken as 0.90 for sections when the net tensile strain is $\geqslant$0.005. For compression-controlled members, $\phi = 0.70$ for tied columns and ϕ varies linearly in a transition zone between tension-controlled and compression-controlled sections.

SERVICEABILITY

Satisfactory levels of safety (strength) do not alone ensure serviceability of a structure under its ordinary conditions of "service" loads, exposures, special requirements, etc. In this sense, serviceability is defined as acceptable conditions for the intended use of the structure, including the deflection of floors and roofs from level; crack widths for exposure conditions; stiffness against vibration, sway, or other movement; and any other structural property required.

Unnecessary additional computations for serviceability can be avoided in the design of most routine elements. For both one-way reinforced and two-way reinforced nonprestressed construction, the Code prescribes minimum thickness limits above which deflections can be assumed to be tolerable (Section 9.5.2, one-way; Section 9.5.3, two-way).

DEFLECTION. If lesser thicknesses must be used or if deflection must be computed for any reason, standard procedures for such computations are given (Section 9.5.2) and acceptable limits on deflection so computed are given (Table 9.5-b). The simplest illustration of possible variation between design for strength only and design for the serviceability requirement that elastic deflection under live load be less than $\ell/360$ (Table 9.5-b for floors not supporting nor attached to nonstructural elements likely to be damaged by large deflections) is a one-way slab (See Fig. 3-8).

CRACK CONTROL. Limiting computed crack widths were established for various conditions of exposure. These limits have been converted into limiting parameters "z" (Section 10.6.4). An equation for the computation of z-values based on laboratory research is given (Section 10.6.4, Eq. 10-5). If reinforcement is distributed as required (Section 10.6) to result in computed z-values less than the limits prescribed for various exposures, the crack widths in one-way construction should be satisfactory.

Maximum values of $z = 175$ and 145 kips/in. are specified (Section 10.6.4) for interior and exterior exposure, respectively. These values may be reduced by the factor 1.2/1.35 for slabs (Commentary R10.6.4). Lower maximum z values should be used for structures subjected to very aggressive exposure or designed to be watertight. ACI Committee 350 "Environmental Engineering Concrete Structures" recommends limiting service load design stresses by use of z values equal to or less than 115 for crack control in liquid-retention structures under "normal sanitary exposures." For "severe sanitary exposures,*" severe exposure to deicing salts or seawater, additional protection such as coated reinforcing bars or protective linings are suggested.

Calculation of z-values for each element of construction would be tedious. For practical application, the authors recommend that users convert the equation for z to solve the maximum bar spacings. The maximum bar spacings can be tabulated for each size bar, each standard cover specified, and for each z-value. Such tabulations can be prepared for any special conditions of exposure and cover desired. See Chapter 3 for bar spacings in one-way slabs, common cover, interior and exterior. See Fig. 2-9 for design—service load tension stresses in reinforcement vs. bar spacings for different size bars, normal sanitary structure exposure conditions, and cover $\geqslant$2 in.

* Severe sanitary exposures exist with liquids containing sulfate in excess of 1500 ppm or pH less than 5.

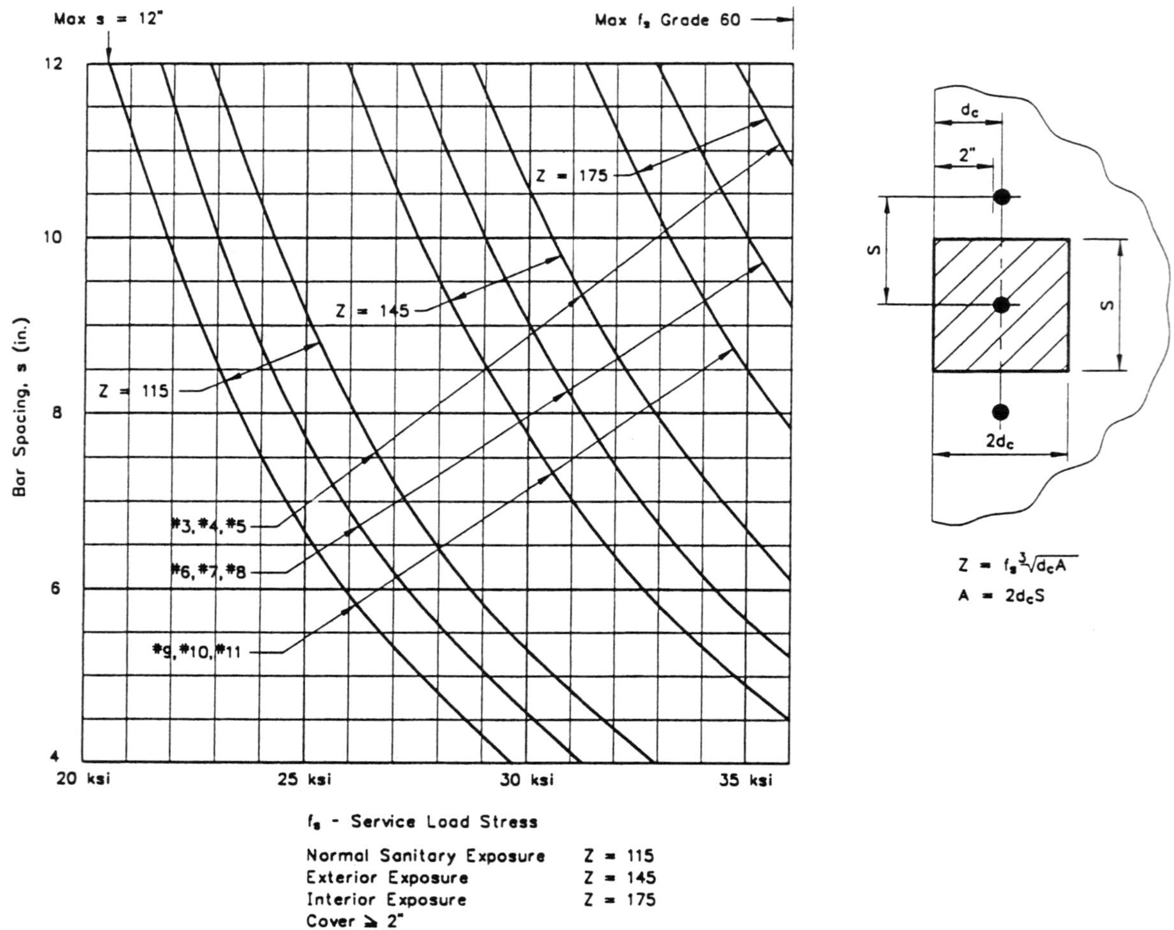

Figure 2-9 Service load stresses for crack control.

3 One-Way Reinforced Concrete Slabs

GENERAL

The simplest routine structural element for illustration of design provisions in the Code is the one-way slab. A one-way slab is defined for purposes of this book as a flexural member with depth small relative to other dimensions, supporting (gravity) loads applied normal to and directly over its surface, spanning in one direction between parallel supports, and reinforced for flexure in that direction only. For purposes of analysis, one-way slabs may be restrained to any degree at the supports or may be unrestrained. A number of Code provisions refer to "flexural members," which include one- and two-way slabs, beams, girders, footings, and, where bending exists with the axial load, walls and columns. In general, when the Code provision is intended to apply to one-way slabs, the term will be used in the sense of the definition herein. For example, the Code requires stirrups in most beams, but specifically exempts slabs and footings from the requirement (Section 11.5.5.1). Where the Code term "structural slabs" is used, as in the provision that the minimum reinforcement in the direction of span shall be that required for temperature and shrinkage (Sections 7.12 and 10.5.4), the requirement applies to both one- and two-way slabs as herein defined. In this book, one-way slabs and beams are discussed separately. Beams are defined herein as one-way flexural members designed to support concentrated line loads such as slabs, girders as one-way flexural members designed to support concentrated loads such as beams. Either beams or girders or two-way slabs may be utilized as primary members of a frame. For design of beams, see Chapter 9. For design of two-way slabs, see Chapters, 5, 6, and 7.

TYPES OF ONE-WAY SLABS

One-way slabs may be solid, hollow, or ribbed. When ribbed slab dimensions conform to the standard dimensional limitations for joist-slab construction (Section 8.11) special Code provisions apply. (For design of one-way joist-slab systems, see Chapter 4.) Ribbed slabs larger than the minimum limits are designed as slabs and beams (Section 8.11.4). Various shapes of totally enclosed void volumes can be incorporated into one-way slabs. Special hollow sections (precast blocks) are used in various proprietary precast planks to form one-way slabs. Proprietary longitudinal void core forms can be employed for cast-in-place construction of one-way slabs. Some of these forms are inflatable and may be reused; others, such as fiber or cardboard cylinders, are used once and left in place. Since most of the hollow slab designs are special, either proprietary or requiring unique computations, they are outside the scope of this book. One-way slabs need not be of uniform depth; they may be haunched. Where haunches are used, the effect of the haunches on stiffness, moment, flexural and shear strength, and deflection must be considered (Section 8.6.2). The discussion and examples following in this chapter will be limited to the common one-way slab, solid, of uniform depth, and designed for uniformly distributed loads.

SLAB REINFORCEMENT

Reinforcement for one-way slabs commonly consists of straight bars. Welded wire fabric is used where some special feature of the structure makes it advantageous. The maximum reinforcement permitted for slabs can be easily computed (see Fig. 2-8). The minimum reinforcement permitted (Section 10.5.4) is that required for temperature and shrinkage (Section 7.12). See Table 3-1 for a summary of both maximum and minimum limits on reinforcement for common slab thicknesses.

Section 7.12.1.2 was adopted in the 1995 Code to alert the designer to cases where the minimum reinforcement for shrinkage and temperature required by

TABLE 3-1 Minimum and Maximum Areas of Steel in One-Way Slabs

LIMITING AREAS OF GRADE 60 REINFORCEMENT (IN.²/12 IN. WIDTH)					
	ONE-WAY SLAB THICKNESS, *h*				
Limit	**4 in.**	**5 in.**	**6 in.**	**7 in.**	**8 in.**
Minimum $A_s = 0.0018\ bh$ [(a)]	0.0865	0.108	0.130	0.152	0.173
Maximum $\rho = 0.75\ \rho_b$ [(b)] where $\rho = A_s/bd$					
For $f_c' = 3{,}000$ psi [(c)] $0.75\ \rho_b = 0.0161$	0.578	0.761	0.955	1.15	1.34
For $f_c' = 4{,}000$ psi [(c)] $0.75\ \rho_b = 0.0213$	0.768	1.01	1.26	1.52	1.77

(a) As limited by Sections 10.5.4 and 7.12.

(b) $\text{Max}\ \dfrac{A_s}{bd} = 0.75\left[\dfrac{(0.85)(\beta_1)(f_c')}{f_y}\ \dfrac{(87)}{(87 + f_y)}\right].$

(c) Applicable to bar sizes #3, #4 and #5 with $^3/_4$ in. clear cover for interior exposures.

Section 7.12.2.1 may not be adequate. The cases according to Commentary Section R7.12.1.2 are where other structural elements such as large columns and structural walls provide significant restraint to shrinkage and temperature movements. For these cases, the amount of shrinkage and temperature reinforcement should be increased beyond that required by Section 7.12.2.1.

SERVICEABILITY

Serviceability requirements of the Code for one-way slabs involve control of flexural cracking and control of deflections (Sections 10.6 and 9.5, respectively). For one-way solid slabs these limitations can be expressed most easily as maximum bar spacings and minimum thicknesses to avoid need for any added calculations. See Table 3-2 for bar spacings and Table 3-3 for minimum thicknesses.

TABLE 3-2 Maximum Bar Spacings for Crack Control in One-Way Slabs

Grade 60 Bar	MAXIMUM BAR SPACINGS FOR CRACK CONTROL[a]		
	INTERIOR EXPOSURES		EXTERIOR EXPOSURES
	Clear cover[b] = 3/4 in.	Clear cover[d] = 1 in.	Clear cover[b] = $1\frac{1}{2}$ in. for #3, #4, #5; = 2 in. for #6
#3	(46″) ≦ $3h$ ≦ 18″[c]	(29″) ≦ $3h$ ≦ 18″[c]	8″
#4	(41″) ≦ $3h$ ≦ 18″	(26″) ≦ $3h$ ≦ 18″	$7\frac{1}{2}$″
#5	(36″) ≦ $3h$ ≦ 18″	(23″) ≦ $3h$ ≦ 18″	7″
#6	(32″) ≦ $3h$ ≦ 18″	(21″) ≦ $3h$ ≦ 18″	4″

[a] For Grade 60 bars, use f_s = 36 ksi (Section 10.6.4); spacings to satisfy $z = f_s\sqrt[3]{d_cA}$ (Eq. 10-5) for values of z = 156 for interior exposure; z = 129 for exterior exposure (Section 10.6.4, maximum z = 175 and 145, respectively, for beams; Commentary, z for slabs = (1.2/1.35) (z) for beams).
[b] To satisfy cover requirements for cast-in-place concrete (Section 7.7.1).
[c] Spacings are limited to $3h$ but not more than 18 in. (Section 7.6.5).
[d] Required under some general codes for maximum fire rating.

ANALYSIS

One-way slabs may be analyzed by the approximate method (coefficients for maximum shears and moments at critical sections) (Section 8.3.3) or by the frame analysis method (Sections 8.6, 8.7 and 8.9) (see Figs. 2-1, 2-2, 2-3). For manual calculations, the authors recommend the use of approximate analysis where conditions of design permit, for maximum efficiency of design time and use of material. See Fig. 3-1.

EXAMPLE. ANALYSIS AND DESIGN. Design a one-way slab system for a nominally flat roof (where ponding is prevented by positive drainage). All spans are 16′-0″ from center-to-center of 12-in.-wide integral supports. Clear spans are 15′-0″. Superimposed service loads consist of 30 psf live and 16 psf dead. Use Grade 60 bars and a concrete strength, $f_c' = 4{,}000$ psi.

TABLE 3-3 Maximum Spans for One-Way Slabs Without Computation of Deflections (Normal Weight Concrete)

MAXIMUM CLEAR SPANS ℓ_n[a]				
Slab Thickness (inches)	Simple Span ($\ell_n/20$)	One End Continuous ($\ell_n/24$)	Both Ends Continuous ($\ell_n/28$)	Cantilever Span ($\ell_n/10$)
4	6′–8″	8′–0″	9′–4″	3′–4″
$4\frac{1}{2}$	7′–6″	9′–0″	10′–6″	3′–9″
5	8′–4″	10′–0″	11′–8″	4′–2″
$5\frac{1}{2}$	9′–2″	11′–0″	12′–10″	4′–7″
6	10′–0″	12′–0″	14′–0″	5′–0″
$6\frac{1}{2}$	10′–10″	13′–0″	15′–2″	5′–5″
7	11′–8″	14′–0″	16′–4″	5′–10″
$7\frac{1}{2}$	12′–6″	15′–0″	17′–6″	6′–3″
8	13′–4″	16′–0″	18′–8″	6′–8″
$8\frac{1}{2}$	14′–2″	17′–0″	19′–10″	7′–1″
9	15′–0″	18′–0″	21′–0″	7′–6″
$9\frac{1}{2}$	15′–10″	19′–0″	22′–2″	7′–11″
10	16′–8″	20′–0″	23′–4″	8′–4″

[a] (Section 9.5.2.1 and Table 9.5-a).

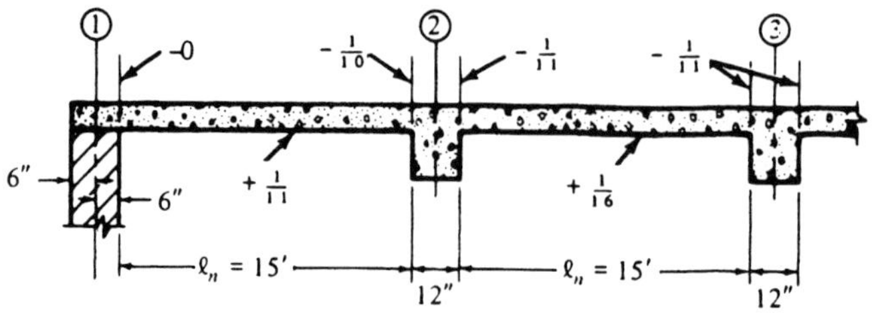

Figure 3-1 Moment coefficients of factor, $w_u\ell_n^2$.

Step 1: Thickness, h

See Table 3-3 for $\ell_n = 15'\text{-}0''$.

For end span, min $h = 7\frac{1}{2}$ in., and for interior span, min $h = 6\frac{1}{2}$ in. unless computed deflections are within allowable limits.

Since the superimposed loads are very light and the maximum immediate deflection allowable under live load, L, is $\ell_n/180 = 1$ in., a 6 in. thickness will be assumed. Immediate deflection must be computed to verify that it is less than 1 in. (Section 9.5.2.1 and Table 9.5-b).

Step 2: Analysis by Coefficients (Section 8.3.3)

Required strength, $U = 1.4\ D + 1.7\ L$ (Section 9.2.1, Eq. 9-1).

$$w_d = 1.4(75 + 16) = 127 \text{ psf}; w_\ell = 1.7 \times 30 = 51 \text{ psf};$$

$$w_u = 127 + 51 = 178 \text{ psf}$$

Factored shear at the first interior support, $V_u = (1/2)(15)(0.178)(1.15) = 1.53$ k.
Factored shear at all other supports, $V_u = 1/2(15)(0.178) = 1.33$ k.
Coefficients for factored moments are shown in Fig. 3-1.
Factored moment(s) = Coefficient(s) × $w_u \ell_n^2$ (Fig. 3-2)
where $w_u \ell_n^2 = (0.178)(15)^2 = 40.0$ ft-k per 12 in. width of slab
(Critical section for factored negative moment is at the face of the supports; for shear, at a distance, d, from the face of supports.)

Step 3: Determination of Steel Areas Required and Bar Sizes for Use

The minimum steel (Section 7.12.2) permitted, $A_s = 0.0018\ bh$.

$$\text{Min } A_s = (0.0018)(6 \times 12) = 0.130 \text{ in.}^2/\text{ft (See Table 3-1.)}$$

Cover required = 3/4 in. (Section 7.7.1-c) for condition of not being exposed to weather—assuming roofing material protects top surface of slab. For #4 bars, the effective depth, $d = 6$ in. − 0.75 in. − 1/2(0.5 in.) = 5 in.
Required areas of reinforcement using the expression from page 258:

$$A_s = 0.85\left(1 - \sqrt{1 - \frac{2.35\ M_u}{\phi f'_c bd^2}}\right)\frac{bdf'_c}{f_y}$$

Exterior span, bottom bars: from Fig. 3-2 using maximum $+M_u = 3.64$ ft-k,

$$A_s = 0.85\left(1 - \sqrt{1 - \frac{2.35(3.64 \times 12)}{0.90(4)(12)(5)^2}}\right)\frac{12(5)(4)}{60} = 0.17 \text{ in.}^2/\text{ft} \quad \text{Use\#4@14}$$

Exterior span, first interior support, top bars; from Fig. 3-2 using $-M_u = 4.00$ ft-k,

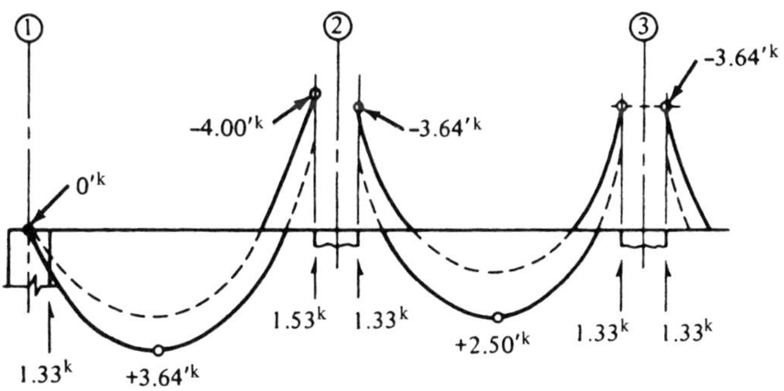

Figure 3-2 Factored moments and shears at critical sections.

$$A_s = \left(\frac{4.00}{3.64}\right)(0.17) = 0.19 \text{ in.}^2/\text{ft.} \qquad \text{Use \#4@12}$$

Other interior supports, top bars: the factored moment is the same as for the exterior span bottom bars, use #4@14. Interior spans, bottom bars:

$$A_s = \frac{2.50}{3.64}(0.17) = 0.12 \text{ in.}^2/\text{ft} < 0.130 \text{ in.}^2/\text{ft min}$$

The maximum spacing for bars (Section 7.6.5) is $3h = 3h \times 6$ in. = 18 in., but not more than 18 in. Since #4 bars have been selected for use elsewhere, use #4@18 for the bottom bars in the interior spans and also for the temperature and shrinkage steel in the other direction. The requirement for minimum temperature and shrinkage reinforcement is the same as for minimum flexural reinforcement (Sections 10.5.4 and 7.12.2).

Step 4: Check Required Tension Development Lengths for the #4 Bars

Since all of the bottom bars are #4, tension development length need only be checked at the most critical point, which is the discontinuous end (Commentary Section R12.11.3). The end embedment available is 12 in. − 2 in. (cover) = 10 in. for bottom bars at the discontinuous end; at interior supports an end embedment of 6 in. (Section 12.11.1) for all bars should be ample. Using only one length of bottom bar will simplify detailing, placing, and inspection, as well as design (see Fig. 3-3).

$$\ell_d \leq \ell_a + M_n/V_u$$

$$\ell_a = 12 - (2 + 6) = 4 \text{ in.}$$

Compute nominal moment strength, M_n, for bottom bars, #4@14, $A_s = 0.17$ in.^{2}/ft

$$a = \frac{A_s f_y}{0.85 f'_c b} = \frac{0.17(60)}{0.85(4)(12)} = 0.25 \text{ in.}$$

$$M_n = A_s f_y(d - a/2) = 0.17(60)[5 - 0.25/2] = 49.7 \text{ in.-k/ft}$$

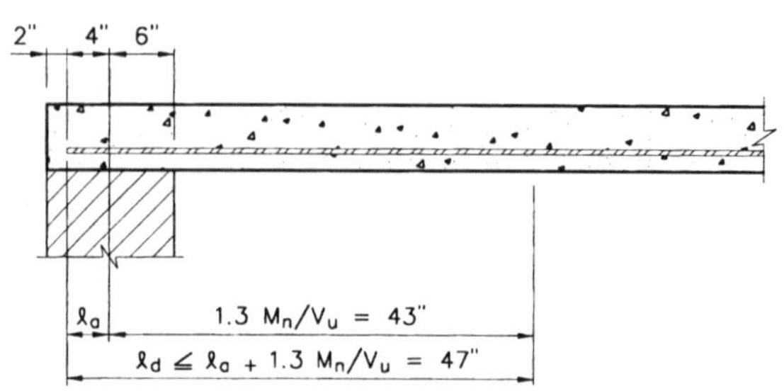

Figure 3-3 Development length at the discontinuous end.

$$\frac{M_n}{V_u} = \frac{49.7}{1.33} = 33 \text{ in.}$$

The value of M_n/V_u can be increased by 30% when the ends of the bars are confined by a compressive reaction (Section 12.11.3).

$$\ell_d \leqq \ell_a + 1.3\, M_n/V_u = 4 + 43 = 47 \text{ in.}$$

For a #4 bar, $\ell_d = 19$ in. (See Table 13-4)

The Code will permit two-thirds of the bottom bars at discontinuous ends and three-quarters at continuous ends to be cut short of the support (Section 12.11.1). Similarly, several lengths of top bars could be used for a small saving in material. For this example, assume the saving in material is less than the added expenses for design, detailing, fabricating, placing, and inspection. In this case, then, no further check of development lengths is necessary.

Step 5: Check Design Shear Strength

In any one-way flexural member where shear may be critical, the final investigation of shear cannot be made until factored moments have been determined and flexural steel selected and detailed, including cutoff points. The more exact formula for nominal shear strength of the concrete, $V_c = [1.9\sqrt{f'_c} + 2{,}500\, \rho_w V_u d/M_u] b_w d \ldots$ (Eq. 11-5), but $V_c \leqslant 3.5\sqrt{f'_c}\, b_w d$, involves the tensile reinforcing steel ratio, ρ_w, at any section considered for shear, (Section 11.3.2.1). It may save time where shear is expected to be critical to make a preliminary shear investigation using the approximate formula for nominal shear strength, $V_c = 2\sqrt{f'_c}\, b_w d$ (Section 11.3.1.1).

In the example here, a one-way solid slab even less than minimum thickness/span ratio, the shear check is a waste of design time. The Code requires, however, that shear be investigated. To demonstrate for completion of this example: the maximum factored shear at the face of the first interior support is 1.53 k per 12 in. width. The critical section, $d = 5$ in. from the face of the support (Fig. 3-4) must resist a factored shear,

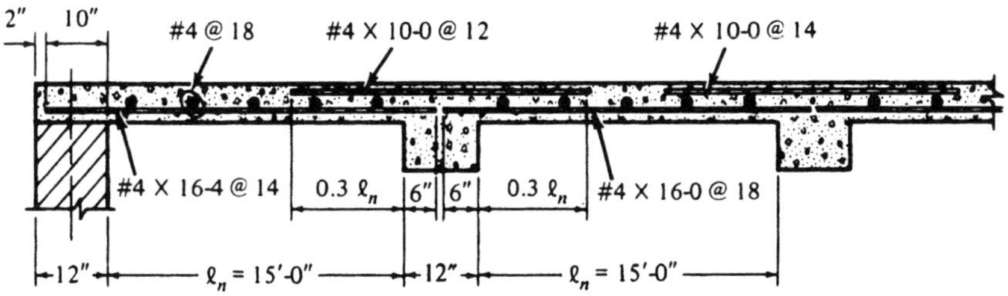

Figure 3-4 Details of reinforcement.

$$V_u = 1.53 - (0.178)(5/12) = 1.46 \text{ k (Section 11.1.3.1)}$$

$$\phi V_n = \phi V_c = \phi 2\sqrt{f_c'}\, b_w d$$

$$= (0.85)(2)\sqrt{4000}\,(12)(5)/1000$$

$$= 6.45 \text{ kips}$$

$$V_u < \phi V_c \text{ (Section 11.1.1, Eq. 11-1) O.K.}$$

Step 6: Deflection Check

The coefficients of the approximate frame analysis give maximum moments for service loads which result in the following maximum deflections:

$$\text{End span } \Delta = \frac{3.25\, w\ell_n^4}{384\, EI} \text{ (= 65\% of simple span deflection)}$$

$$\text{Interior span } \Delta = \frac{2\, w\ell_n^4}{384\, EI} \text{ (= 40\% of simple span deflection)}$$

In this case the critical deflection occurs in the end span.

For deflection computations, the Code prescribes an effective moment of inertia (Section 9.5.2.3, Eq. 9-7)

$$I_e = \left(\frac{M_{cr}}{M_a}\right)^3 I_g + \left[1 - \left(\frac{M_{cr}}{M_a}\right)^3\right] I_{cr} \leqslant I_g$$

where

$$M_{cr} = \frac{f_r I_g}{y_t} \text{ and } f_r = 7.5\sqrt{f_c'} = 475 \text{ psi for } f_c' = 4{,}000 \text{ psi} \quad \text{(Section 9.5.2.3)}$$

$$M_{cr} = \frac{(475)(216)}{3} = 34{,}200 \text{ in.-lbs}; \; I_g = \frac{(12)(6)^3}{12} = 216 \text{ in.}^4$$

$$y_t = 3 \text{ in.}$$

Service load, w_s, = 75 + 16 + 30 = 121 psf
Factored load, w_u = 178 psf
Compute service load moments, M_a, from $w_s/w_u \times$ factored load moments
For maximum $+M = 3.64$ ft-k, and $-M = -2.94$ ft-k

$$+M_a = (121/178)3.64 = 2.47 \text{ ft-kips/12 in. width}$$

$$-M_a = (121/178)2.94 = 2.00 \text{ ft-kips/12 in. width}$$

$$\text{Average } M_a = 2.23 \times 12 = 26.8 \text{ kip-in./12 in. width}$$

$$\text{Average } I_{cr}\text{: average } \rho = \frac{0.18}{5 \times 12} = 0.003.$$

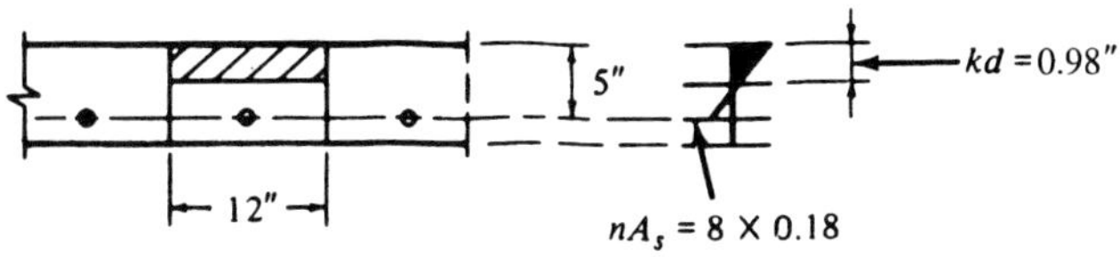

Figure 3-5 Cracked transformed section.

$$k = \sqrt{2\,n\rho + (n\rho)^2} - n\rho$$

$$k = 0.196 \quad \text{See Fig. 3-5.}$$

Average I_{cr} for negative and positive moment sections, neglecting compression steel at the supports:

$$I_{cr} = d^3[4\,k^3 + 12\,n\rho(1 - 2k + k^2)] = 27 \text{ in.}^4 \text{ per 12 in. strip}$$

$$I_e = \left(\frac{34.2}{26.8}\right)^3 (216) + \left[1 - \left(\frac{34.2}{26.8}\right)^3\right](27) > I_g;\ \text{use } I_g = 216 \text{ in.}^4$$

$$E_c = 57{,}000\sqrt{f_c'} \text{ (Section 8.5); } E_c = 57{,}000\sqrt{4{,}000} = 3{,}600{,}000 \text{ psi}$$

Instantaneous live load, $\Delta = \dfrac{(3.25)}{384}\,\dfrac{(30)(15)^4(1728)}{(3{,}600{,}000)(216)} = 0.029 \text{ in.} < 1.00 \text{ in.}$ allowable (Table 9.5-b).

Conservatively neglecting the compression steel at the support sections, the long-term deflection factor, λ, is equal to 2.0 for a duration of five years or more (Section 9.5.2.5).

$$\text{Instantaneous dead load, } \Delta = (0.029)\left(\frac{91}{30}\right) = 0.086 \text{ in.}$$

$$\text{Additional long-term deflection, } \Delta = (0.086)(2.0) = \underline{0.172 \text{ in.}}$$

$$\text{For camber, total long-term deflection} = 0.258 \text{ in.}$$

If this roof is not supporting nor attached to nonstructural elements likely to be damaged by large deflections, these calculations would satisfy both Code and practical requirements (Section 9.5, Table 9.5-b). If the roof *were* attached to or supporting nonstructural elements likely to be damaged by large deflections, the sum of the live load, plus additional long-term load deflections computed as above, must be no greater than $\ell_n/480$ (Section 9.5, Table 9.5-b). Additional long-term, plus live load $\Delta = 0.172 + 0.029 = 0.201$ in.; $\ell_n/480 = \dfrac{180}{480} =$ 0.375 in. $>$ 0.201 in. O.K.

For a more complete discussion of deflection computations and limits thereon, see Chapter 4.

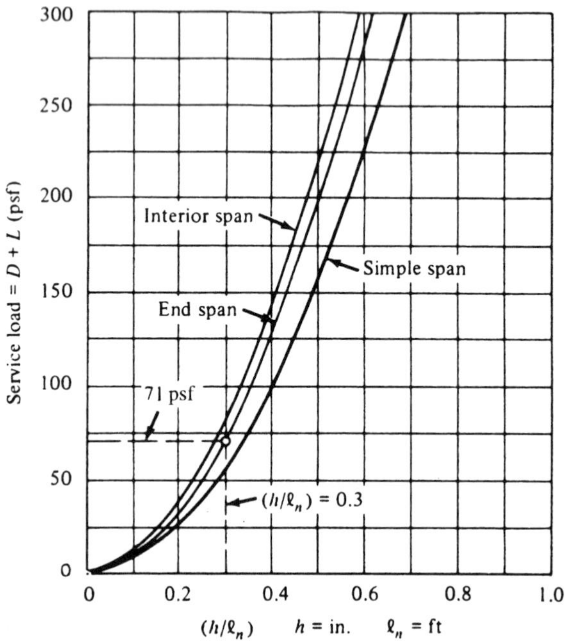

Figure 3-6 Load at which $M_{cr} = M_a$.

Note that whenever the cracking moment, M_{cr}, is larger than the positive or negative actual service load moments, the gross moment of inertia, I_g, can be used as the effective moment of inertia, I_c, saving computations for the cracked section moment of inertia (See Fig. 3-6).

Step 7: Bar Spacings for Crack Control (Section 10.6).

When the specified yield strength, f_y, for design exceeds 40,000 psi, bars must be spaced such that the quantity, $z = f_s \sqrt[3]{d_c A}$ ... (Section 10.6.4, Eq. 10-4) does not exceed 175 kips/in. for interior exposures and 145 kips/in. for exterior exposures for beams and (1.2/1.35) times these values for one-way slabs (Section 10.6.4 and Commentary).

f_s = flexural tensile steel stress at service load; use $0.60 f_y$

= 36 ksi unless calculated.*

$$* f_s = \frac{\text{Area of steel required}}{\text{Area of steel furnished}} \times \frac{D + L}{w_u} \times 0.90 f_y$$

where $w_u = (1.4D + 1.7L)$.

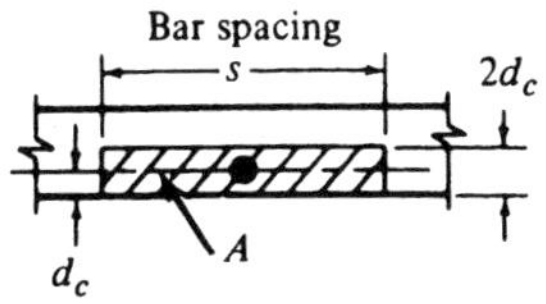

Figure 3-7 Tributary tensile area per bar = A.

d_c = cover from nearest surface to the center of the bar

= clear cover + 1/2 bar diameter

$A = (2)(d_c)(s)$, where s = bar spacing (See Fig. 3-7.)

For #4 bars with standard cover = 3/4 in., d_c = 1 in. Maximum $z = 175 \times \frac{1.2}{1.35} = 156$ kips/in. for interior exposures. Solving for the maximum allowable spacing of bars, s:

$$\sqrt[3]{d_c A} = \frac{z}{f_s} = \frac{156}{36} = 4.33$$

When $d_c = 1$, $\sqrt[3]{d_c A} = \sqrt[3]{2s}$; solving, maximum spacing, $s = 40$ in. > limit $3h = 18$ in. (Section 7.6.5).

DESIGN AIDS

In routine design, the last three steps and part of Step 3 in the preceding example can be avoided by the use of design aids to ensure automatic conformance with all applicable code requirements. Such design aids are published in handbooks, including those by CRSI and ACI, and any special aids required for particular job conditions can be easily prepared.

For Step 3, selection of bar sizes, the minimum and maximum (singly-reinforced) steel areas permitted for each common slab thickness can be tabulated for Grade 60 reinforcement and concrete strengths generally used (see Table 3-1).

For Step 4, tension development lengths for each size of bar can be calculated for any concrete strength and grade of steel. Thereafter, the design step will consist merely of comparing the embedment length available, from a sketch of design dimensions, to that required for development of the various bar sizes. Ordinarily, a designer then simply selects the largest bar which can be developed. See development length and lap splice length tables in Chapter 13.

The deflection computation as noted in the text can be avoided entirely by conformance within the minimum thickness/span ratios prescribed (see Table 3-3). Where it is desirable to use lesser thicknesses of slab, for light loads and/or

TABLE 3-4 Moments of Inertia and Cracking Moments for Normal Weight Concrete One-Way Slabs

h for I_g and M_{cr} / d for I_{cr}	I_g in.^{4} per 12 in. strip	M_{cr}^a in.-k per 12 in. strip	I_{cr} (IN.^{4} PER 12 IN. WIDTH)[a] Tensile Reinforcement Ratio, $\rho = A_s/bd$ (for continuous members, use average top and bot. ρ)									
			0.0018	0.002	0.003	0.004	0.005	0.006	0.008	0.010	0.012	0.014
3″	27	8.54	3.7	4.1	5.8	7.5	9.0	10.4	13.1	15.5	17.8	19.9
$3\frac{1}{2}$″	43	11.65	5.9	6.5	9.3	11.8	14.3	16.5	20.8	24.6	28.2	31.6
4″	64	15.2	8.8	9.7	13.8	17.7	21.3	24.7	31.0	36.8	42.1	47.1
$4\frac{1}{2}$″	91	19.2	12.6	13.8	19.7	25.2	30.3	35.1	44.1	52.4	60.0	67.1
5″	125	23.7	17.3	19.0	27.0	34.5	41.6	48.2	60.5	71.8	82.3	92.0
$5\frac{1}{2}$″	166	28.8	23.0	25.3	36.0	46.0	55.3	64.2	80.6	95.6	110	122
6″	216	34.2	29.9	32.8	46.7	60.0	71.8	83.3	105	124	142	159
$6\frac{1}{2}$″	275	40.0	38.0	41.7	59.4	75.8	91.3	106	133	158	181	202
7″	343	46.5	47.4	52.1	74.2	94.7	114	132	166	197	226	252
$7\frac{1}{2}$″	422	53.2	58.3	64.1	91.2	117	140	163	204	242	278	310
8″	512	60.8	70.8	77.7	111	141	170	197	248	294	337	377
$8\frac{1}{2}$″	614	68.7	84.9	93.2	133	170	204	237	297	353	404	452
9″	729	77.0	101	111	158	201	242	281	353	419	480	536
$9\frac{1}{2}$″	857	85.5	119	130	185	237	285	331	415	493	564	631
10″	1000	95.0	138	152	216	276	332	386	484	575	658	756

[a] For $f'_c = 4{,}000$ psi; $f_r = 7.5\sqrt{f'_c} = 475$ psi; $n = 8$

$M_{cr} = 2 f_r h^2$; $k = \sqrt{2\rho n + (\rho n)^2} - \rho n$

$I_{cr} = d^3[4k^3 + 3\rho n(1 - 2k + k^2)]$

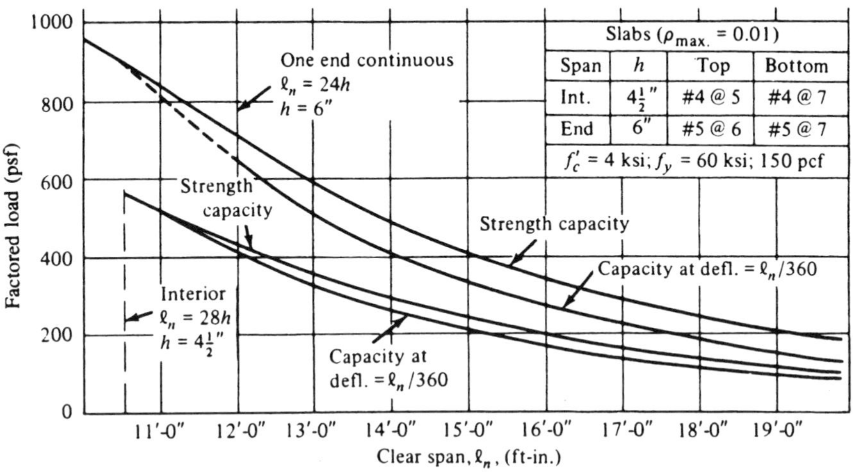

Figure 3-8 One-way slabs, strength and deflection capacity.

areas where immediate deflection is not likely to be harmful, design aids consisting of tabulated values for M_{cr} and I_{cr} for the various slab thicknesses and steel percentages, are helpful (see sample in Table 3-4). Also see Fig. 3-8 for an example of load capacities limited by strength requirements only versus loads limited by the immediate service live load deflection requirement, $\Delta < \ell_n/360$. (Fig. 3-8 was prepared from load tables in the *CRSI Handbook*, 1996.)

Finally, automatic conformance to the bar spacing limits (for crack control) can be assured with no computations at all in design for routine elements subject to the standard criteria for ordinary interior or exterior exposures. In selecting the bar sizes, the designer merely has to keep spacings within the tabulated limits for that bar size and exposure, as in sample Table 3-2. Reinforcement in structures such as wastewater treatment plants usually is provided with larger cover (commonly 2 in. minimum clear cover) and the resulting deeper tributary areas of concrete in tension per bar will require closer spacings of bars. Similarly, any structure intended to be watertight may utilize a lower service load stress or a lower limiting z value or both. Special tables similar to Table 3-2 can be prepared as a preliminary design step to avoid any further crack control computations in the design of separate elements.

LONGER SPANS

For spans longer than those of Table 3-3, complete deflection calculations can often be avoided if the slab is lightly loaded. There is an inherent discontinuity in the Code-prescribed procedure for computing deflections. The expression for effective moment of inertia,

$$I_e = \left(\frac{M_{cr}}{M_a}\right)^3 I_g + \left[1 - \left(\frac{M_{cr}}{M_a}\right)^3 I_{cr}\right]$$

reduces to $I_e = I_g$ when the cracking moment, M_{cr}, is greater than the applied service load moment for which deflection is being computed, M_a. Immediate deflection under service live load is limited to $\ell_n/360$ for all floor construction (Table 9.5-b) and other deflection conditions can be computed proportionately. For immediate deflection, M_a is the moment of the service dead and live loads.

$$M_{cr} = \frac{f_r I_g}{(1/2)(h)} \quad \text{where } f_r = 7.5\sqrt{f_c'} \text{ (Section 9.5.2.3.)}$$

$$M_{cr} = 15\sqrt{f_c'}\,(h)^2 \text{ in.-lbs} = 1.25\sqrt{f_c'}\,(h)^2 \text{ ft-lbs}$$

Equate the cracking moment, M_{cr}, to the maximum applied service load moment, M_a, and solve for the maximum applied service load when $M_{cr} = M_a$: Use h in inches: ℓ_n in feet: for M in ft-lbs.

Simple spans:

$$\frac{w(\ell_n)^2}{8} = 1.25\sqrt{f_c'}\,(h)^2 \tag{3-1}$$

$$w_{cr} = 10\sqrt{f_c'}\,(h/\ell_n)^2$$

Similarly,

End spans:

$$w_{cr} = 12.5\sqrt{f_c'}\,(h/\ell_n)^2 \tag{3-2}$$

Interior spans:

$$w_{cr} = 13.75\sqrt{f_c'}\,(h/\ell_n)^2 \tag{3-3}$$

At loads equal or less than the maximum from Eqs. (3-1), (3-2), or (3-3), $M_{cr} \geqslant M_a$ and so I_g is used in the computation for deflection. When I_g is used in the equations for deflection (see Example Step 6), and equated to a total service load deflection $\Delta = \ell_n/360$, the following equations result:

$$\text{Simple span: } w_{max} = 5{,}250\,(h/\ell_n)^3$$

$$\text{End span: } w_{max} = 8{,}050\,(h/\ell_n)^3$$

$$\text{Interior span: } w_{max} = 13{,}050\,(h/\ell_n)^3$$

EXAMPLE. For a 6-in.-thick slab, end span = 20′-0″, $f_c' = 4{,}000$ psi, find the maximum load where $M_{cr} = M_a$ and check if deflection exceeds $\ell_n/360$. Use h in inches; ℓ_n in feet. $(h/\ell_n) = 0.30$ (see Fig. 3-6).

$$w_{cr} = (12.5)(63.2)(6/20)^2 = 71 \text{ psf for } M_{cr} = M_a$$

Compare to load causing $\Delta = \ell_n/360$ computed with I_g:

$$w_{max} = (8{,}050)(0.3)^3 = 217 \text{ psf} > 71 \text{ psf, and so } \Delta < \ell_n/360$$

Note that the total actual service load (71 psf) = $D + L$. Note also that the last step above will seldom be necessary. When load is limited to less than the cracking moment, deflections will seldom be large. See Fig. 3-8 for the effect of the deflection limitations on load capacity of more heavily loaded (and more heavily reinforced) slabs. See Table 3-4 for a convenient tabulation of values of I_g, M_{cr}, and I_{cr} for common slab thicknesses and steel ratios.

WELDED WIRE FABRIC

Welded wire fabric (WWF) is commonly used for reinforcement in one-way slabs with short spans. Two types of wire are used to form WWF deformed reinforcement—plain and deformed (plain, Section 3.5.3.5 and deformed, Section 3.5.3.6). Under the ASTM specifications, plain wire sizes are designated by the letter, *W*-, followed by the cross-sectional area in hundredths of square inch units from W-45 (3/4 in. in diameter) to W-0.5 (0.08 in. in diameter); in half sizes to W-6; and multiples of two (hundredths) for sizes above W-8. Deformed wire sizes are designated similarly in sizes from D-45 to D-1. Welded intersections comprise the bond value for plain WWF; the Code limits the maximum spacing of welded plain wire intersections to 12 in. (Section 3.5.3.5). For deformed WWF, welded intersections are limited to a maximum spacing of 16 in. (Section 3.5.3.6).

Since a great variety of wire size and spacing combinations are possible, it is best to determine those readily available in the geographical area or region for proposed use, and to use as few combinations as possible for practical applications. Even combinations of plain and deformed wire fabric can be obtained as special orders.

For one-way slabs with spans less than 10-0″ (center-to-center) a single layer of draped fabric may be used for both top and bottom reinforcement (Section 7.5.3). (Wire sizes for draped reinforcement should be kept to about 1/4 in. diameter (W-5 or D-5) or less to permit draping in short distances.) The Code also permits the use of equal top and bottom factored moments for design in slabs with spans 10′-0″ or less, $\pm M_u = 1/12\ w_u \ell_n^2$ (Section 8.3.3). For one-way spans within these limits, WWF is capable of being placed in a minimum of time. Cross-wires are designed to satisfy the requirements for minimum temperaure and shrinkage steel (Section 7.12) (see Fig. 3-9).

For use in one-way spans 10′-0″ or less, the design can be quite simple. In addition to flexure, shear and minimum steel areas, only the lap splice and end embedment requirements need to be computed (Sections 12.7, 12.8, 12.18 and 12.19). The splice arrangements and end embedments desired can usually be shown most simply as typical details. End embedment for plain WWF consists of two cross wires with the nearest one embedded at least 2 in. past the point of maximum factored moment (Section 12.8). For deformed WWF, the required

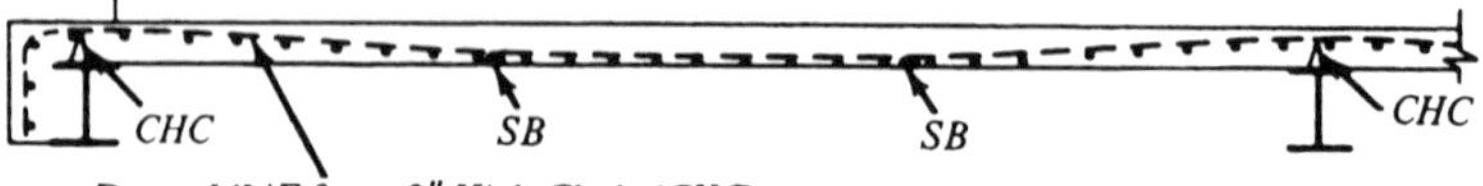

Figure 3-9 Welded wire fabric—end spans (interior spans similar).

embedment is slightly less (Section 12.7). For quick, safe design, the authors recommend simply specifying the same end anchorage for both. Similarly, splices for plain WWF are very simple—two cross-wires overlapped a minimum of 2 in. for stresses in excess of $1/2\,f_y$ and one overlapped a minimum of 2 in. for stress of $1/2\,f_y$ or less (Section 12.19). Two factors, are required to compute minimum lap splice lengths, whichever is greater, for deformed WWF (Section 12.7.2). Again, to avoid field confusion and a waste of design and inspection time, the authors recommend use of the same typical details for both plain and deformed WWF lap splices. See Figs. 13-18 through 13-23 for WWF splice and embedment typical details.

Flat sheets of heavier wire fabric usually with the same wire sizes and same spacings in each direction ("two-way" fabric) are used in two-way construction for somewhat longer spans.

EXAMPLE. Use welded wire fabric as the top and bottom reinforcement in a floor slab to span 10′-0″ from center-to-center of supporting steel beams. The service live load specified is 80 psf. Use concrete with $f'_c = 4{,}000$ psi weighing 150 pcf (see Fig. 3-8).

Try thickness, $h = 4$ in.; $d = 4.00 - 0.75$ (cover) $- 0.15$ in. (1/2 wire dia.)

$$= 3.1 \text{ in.} \qquad D = \frac{4}{12}(150) = 50 \text{ psf.}$$

$$U = 1.4D + 1.7L \text{ (Section 9.2.1)}$$

$$U = (1.4)(50) + (1.7)(80) = 206 \text{ psf}$$

$$\pm M_u = 1/12\, w_u \ell_n^2 \text{ (Section 8.3.3)}$$

$$\pm M_u = (1/12)(0.206)(10)^2(12) = 20.6 \text{ in.-kips/ft}$$

Using the expression from page 258 for calculating the required area of reinforcement,

$$A_s = 0.85\left(1 - \sqrt{1 - \frac{2.35M_u}{\phi f'_c bd^2}}\right)\frac{bdf'_c}{f_y}$$

$$A_s = 0.85\left(1 - \sqrt{1 - \frac{(2.35 \times 20.6)}{0.90(4)(12)(3.1)^2}}\right)\frac{(12)(3.1)(4)}{60} = 0.127 \text{ in.}^2/\text{ft}$$

Area of steel (minimum): $A_s = 0.0018\ bh$ (Section 7.12). Minimum $A_s = (0.0018)(12)(4) = 0.0864$ in.^{2}/ft (see Table 3-5 for wire size). Use: 6 × 6-W7 × W4.5 unless large quantities are required, in which case 6 × 6-W6.35 × W4.32 welded wire fabric meeting exact areas required could be specified.

Check design shear strength.

$$V_u = 1.15 w_u (1/2)\ell_n \text{ (Section 8.3.3)}$$

$$V_u = 1.15(206)(1/2)(10) = 1{,}180 \text{ lb}$$

$$\phi V_n = \phi V_c = \phi 2\sqrt{f'_c}\, b_w d$$

$$= (0.85)(2)\sqrt{4000}\,(12)(3.1)$$

$$= 4000 \text{ lb}$$

$$\phi V_n > V_u \text{ (Section 11.1.1, Eq. 11-1) O.K.}$$

For end embedment and details for supporting the welded wire fabric, see Fig. 3-9. The design would be completed with typical splice details selected from those illustrated in Chapter 13.

TABLE 3-5 Standard Wire Sizes for Reinforcement

STANDARD STEEL WIRE GAUGES				DEFORMED WIRE (A496)	PLAIN WIRE (A82)	NOMINAL	
Diameter (in.)	AS&W Gauge	Diameter (in.)	Area (sq. in.)	Size	Size	Diameter (in.)	Area (sq. in.)
				D-45	W-45	0.757	0.45
1/2	—	.5000	.19635	D-31	W-31	0.628	0.310
—	7/0	.4900	.18857	D-30	W-30	0.618	0.300
15/32	—	.46875	.17257	D-29	—	0.608	0.29
—	6/0	.4615	.16728	D-28	W-28	0.597	0.280
7/16	—	.4375	.15033	D-27	—	0.586	0.27
—	5/0	.4305	.14556	D-26	W-26	0.575	0.260
13/32	—	.40625	.12962	D-25	—	0.564	0.25
—	4/0	.3938	.12180	D-24	W-24	0.553	0.240
3/8	—	.3750	.11045	D-23	—	0.541	0.23
—	3/0	.3625	.10321	D-22	W-22	0.529	0.220
11/32	—	.34375	.092806	D-21	—	0.517	0.21
—	2/0	.3310	.086049	D-20	W-20	0.504	0.200
5/16	—	.3125	.076699	D-19	—	0.491	0.19
—	0	.3065	.073782	D-18	W-18	0.478	0.180
—	1	.2830	.062902	D-17	—	0.465	0.17
9/32	—	.28125	.062126	D-16	W-16	0.451	0.160
—	2	.2625	.054119	D-15	—	0.437	0.15
1/4	—	.2500	.049087	D-14	W-14	0.422	0.140
—	3	.2437	.046645	D-13	—	0.406	0.13
—	4	.2253	.039867	D-12	W-12	0.390	0.120
7/32	—	.21875	.037583	D-11	—	0.374	0.11
—	5	.2070	.033654	D-10	W-10	0.356	0.100
—	6	.1920	.028953	D-9	—	0.338	0.09
3/16	—	.1875	.027612	D-8	W-8	0.319	0.080
—	7	.1770	.024606	D-7	W-7	0.298	0.070
—	8	.1620	.020612	D-6	W-6	0.276	0.060
5/32	—	.15625	.019175	—	W-5.5	0.264	0.055
—	9	.1483	.017273	D-5	W-5	0.252	0.050
—	10	.1350	.014314	—	W4.5	0.240	0.045
1/8	—	.1250	.012272	D4	W4	0.225	0.040
—	11	.1205	.011404	—	W-3.5	0.211	0.035
—	12	.1055	.0087147	D-3	W-3	0.195	0.030
3/32	—	.09375	.0069029	—	W-2.5	0.178	0.025
—	13	.0915	.0065755	D-2	W-2	0.159	0.020
—	14	.0800	.0050266	—	W-1.5	0.138	0.015
—	15	.0720	.0040715	D-1	W-1	0.113	0.010
—	16	.0625	.0030680	—	W-0.5	0.080	0.005

4 One-Way Joist Systems

GENERAL

Concrete joist construction consists of a monolithic combination of regularly spaced ribs and a top slab. It may be constructed with permanent or removable fillers between ribs (joists) arranged to span in one direction or in two orthogonal directions (Section 8.11.1). Two-way joist construction which spans in two orthogonal directions, commonly referred to as waffle slab construction, must conform to the requirements of Chapter 13 of the 1995 Code for slab systems reinforced in flexure in more than one direction. Chapter 7 of this book gives a detailed explanation of these requirements. One-way joist construction with joists parallel to one another must conform to the requirements for analysis and design of the 1995 Code which will be discussed in this chapter.

FORMS

Removable form fillers can be made with ready-made steel "pans" of standard size or from hardboard, fiberboard, glass reinforced plastic, or corrugated cardboard fillers. Some firms that supply such removable fillers also subcontract to erect and remove them. Other firms only rent or sell such fillers.

Lightweight and normal weight concrete or clay tile filler blocks that remain in place as permanent fillers are also used in some proprietary designs.

Standard Reusable Forms

Standard types and sizes of removable forms for one-way and two-way joist construction are listed in ANSI/CRSI Standards A48.1-1986 and A48.2-1986, respectively. Sizes of standard end taper forms and special filler forms are also included in the ANSI Standards.

JOIST-SLAB CONSTRUCTION LIMITATIONS

Concrete joist construction that does not conform to the dimensional limitations of the Code (Sections 8.11.1, 8.11.2, and 8.11.3) must be designed as slabs and

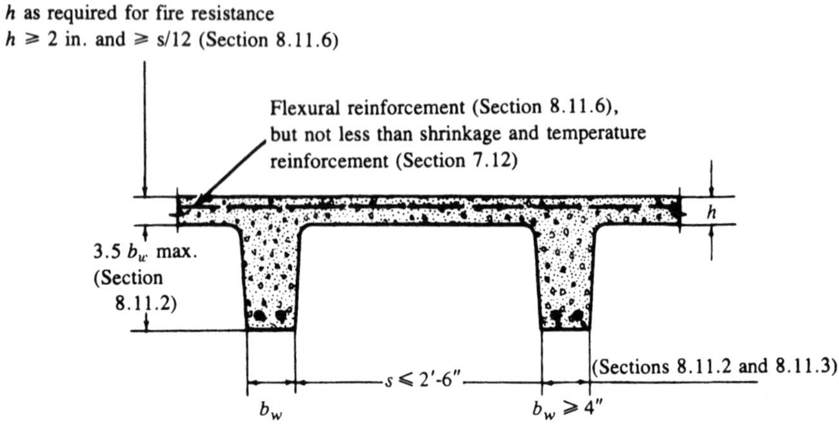

Figure 4-1 Limiting dimensions—concrete joist construction, removable fillers.

beams (Section 8.11.4). Distribution (bridging) joists are not required by the Code.*

Limitations on the width and spacing of joist ribs, depth of fillers, and thickness of top slabs are shown in Fig. 4-1 for joists with removable fillers.

Top slabs that contain horizontal conduits or pipes that are allowed by the Code (Section 6.3) must meet the thickness limitations shown in Fig. 4-1 and, in addition, must have a thickness not less than 1 in. plus the total overall depth of the conduit or pipe (Section 8.11.7).

MINIMUM AND MAXIMUM REINFORCEMENT

The positive and negative moment flexural reinforcement ratio, ρ, for one-way joist construction, must be equal to or greater than $3\sqrt{f_c'}/f_y$ and not less than $200/f_y$, unless the area of both the positive and negative reinforcement at every section is equal to one-third greater than the amount required by analysis (Section 10.5.1). The ratio, $200/f_y$, will govern, except when $f_c' > 4{,}440$ psi.

The positive and negative moment reinforcement ratio, ρ, must be equal to or less than 0.75 of the ratio which produces balanced strain conditions in pure flexure (Section 10.3.3). In determining the maximum reinforcement ratio for positive moment, the top slab or compression flange is included in the concrete

---

* Industry practice as recommended by CRSI is to furnish distributing ribs (bridging) with one continuous #4 bar top and bottom; one rib for spans 20–30 ft; two for spans 30–40 ft; and three for spans more than 40 ft. CRSI Professional Members' Structural Bulletin #8, "Effective Width of One-Way Monolithic Joist Construction as a Two-Way System," presents computer-simulated load test studies of load transfer developed by this industry practice. The conclusion is that use of a minimum effective width of three joist spacings for two-way action is a conservative approach to use of the usual live load reductions (Fig. 2-1, in this book). See further discussion at the end of this chapter.

area *bd*. For negative moment the maximum reinforcement ratio is based on the area of the joist rib $b_w d$.

Negative moment reinforcement should be distributed in zones of maximum concrete tension and a part of the main tension reinforcement must be placed in the effective flange width of the joist (Section 10.6.6). The entire top slab will be within the effective flange width (Section 8.10.2) for joist construction ($s \leqslant 16h$ in Fig. 4-1), and the authors recommend that the negative moment bars should be uniformly spaced throughout the entire top flange for effective crack control.

For cantilevers and other statically determinate situations, with the flange in tension, another provision for minimum reinforcement is imposed by the Code (Section 10.5.2). The minimum area of reinforcement must be equal to or greater than the smaller of $6\sqrt{f'_c}\, b_w d/f_y$ (Eq. 10-4) or $3\sqrt{f'_c}\, b_w d/f_y$ (Eq. 10-3) with b_w in Eq. 10-3 taken as the width of the flange.

MAXIMUM SPACING OF TOP BARS (CRACK CONTROL)

The maximum spacing of negative moment reinforcing bars in the top slab is limited by $z = f_s \sqrt[3]{d_c A}$ (Section 10.6.4, Eq. 10-5). This equation can be expressed in terms of bar spacing, $s_b = (z/f_s)^3/2d_c^2$, as shown in Fig. 4-2.

Table 4-1 gives maximum bar spacings, s_b, in one-way slabs for crack control with $f_s = 0.6f_y$, Grade 60 bar sizes, the usual range of concrete cover, and for interior and exterior exposures. In one-way slabs for interior exposure, z (Section 10.6.4) is $(175)(1.2/1.35) = 156$ kips per inch; and for exterior exposure, $(145)(1.2/1.35) = 129$ kips per inch (Commentary R10.6.4). For exterior exposure, Code Eq. (10-5) does not apply to structures subjected to very aggressive exposure or to structures which are to be watertight (Section 10.6.5). Special precautions are required for each exposure.

The values of $z = 175$ and 145 kips per inch for interior and exterior exposure, respectively, were selected to limit crack widths in beams to 0.016 and 0.013 in. and assure many fine hairline cracks rather than a few wide ones (Commentary R10.6.4).

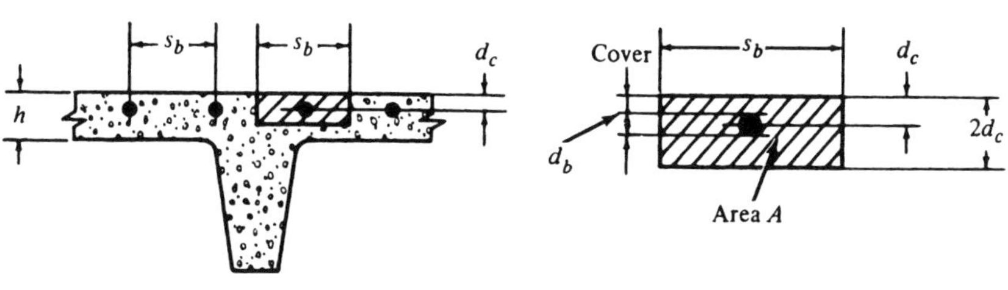

Figure 4-2 Crack control.

TABLE 4-1 Joist Construction—Maximum Spacing of Negative Moment Reinforcing Bars in Top Slab for Crack Control (inches)*

Bar Size	INTERIOR EXPOSURE COVER IN INCHES			EXTERIOR EXPOSURE COVER IN INCHES		
	5/8″	3/4″	1″	1-1/4″	1-1/2″	2″
#3	18	18	18	11	8	—
#4	18	18	18	10	$7\frac{1}{2}$	—
#5	18	18	18	$9\frac{1}{2}$	7	—
#6	18	18	18	—	$6\frac{1}{2}$	4
#7	18	18	18	—	6	$3\frac{1}{2}$
#8	18	18	18	—	$5\frac{1}{2}$	$3\frac{1}{2}$
#9	18	18	$16\frac{1}{2}$	—	$5\frac{1}{2}$	$3\frac{1}{2}$
#10	18	18	15	—	5	3
#11	18	18	14	—	$4\frac{1}{2}$	3

*Spacing must be equal to or less than $5h$ and 18 in. (Sections 10.5.4, 7.12.2.2, and 8.10.5.2).

It is obvious from Table 4-1 that the Code (Section 10.6) does not limit the maximum spacing of joist construction negative moment top bars for interior conditions of exposure except for reinforcing bar sizes #9, #10, #11 with 1 in. concrete cover unless, of course, the spacing is further limited by a maximum of five times the thickness of the top slab. For exterior conditions of exposure, the Code (Section 10.6) does limit the maximum spacing of bar sizes #3 through #11 with 1-1/4, 1-1/2, and 2 in. of concrete cover. #14 and #18 bars are not practical for use as negative moment bars in top slabs of joist construction and were, therefore, omitted from Table 4-1.

MINIMUM CONCRETE COVER

Required concrete cover for protection of joist reinforcement in cast-in-place construction and in precast concrete (Sections 7.7.1 and 7.7.2) is shown in Fig. 4-3. When corrosive environments require greater concrete protection for more severe conditions of exposure (Section 7.7.5), or when the "general code" requires more cover for fire resistive ratings, it should be provided (Section 7.7.7). Such cover should be regarded as "sacrificial", as it is considered in other reinforced concrete structures or systems, e.g., to extend life of pavement and bridge decks against abrasion, or environmental engineering structures against concrete surface deterioration.[†]

[†] For cover greater than 2 in., use 2 in. in Eq. 10-5. For conditions of extreme exposure, tests and performance show additional cover extends life of the structure. Eq. 10-5 for surface crack width is significant only for the range of cover less than 2 in. See report, "Environmental Engineering Concrete Structures (ACI 350R-89)", *ACI Manual of Concrete Practice*, Part 4, 1997.

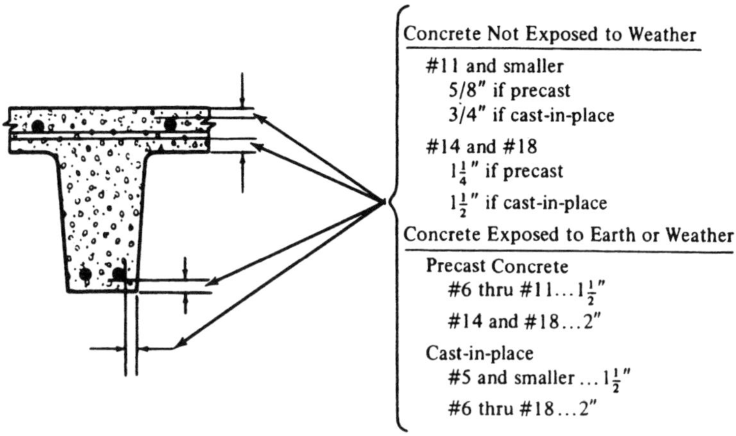

Figure 4-3 Concrete cover for protection of reinforcement.

MINIMUM CONCRETE COVER AND TOP SLAB THICKNESS FOR FIRE RESISTANCE

The required thickness of concrete top slabs and the required concrete cover for protection of reinforcement for different fire resistive ratings depends on the "general code" that is being followed. Table 4-2 lists the top slab thickness required for fire resistance by the *Uniform Building Code*, 1994 Edition, and *Standard Building Code*, 1994 Edition, for joist construction made from normal weight and structural lightweight concrete with aggregates such as limestone, calcareous gravel, trap rock, slag, expanded clay, shale, slate or any other aggregates possessing equivalent fire-resistive properties. The practical structural engineer will most likely round the decimal values of minimum top slab thickness in Table 4-2 to the nearest $^1\!/_4$ in. or $^1\!/_2$ in. Providing a minimum $^3\!/_4$ in. concrete cover of reinforcement in restrained construction is adequate under the *UBC* and *SBC* for fire-resistive ratings up to 4 hours.

In addition to the fire resistive requirement, the thickness of the top slab of concrete joist construction must be sufficient to meet the dimensional requirements shown in Fig. 4-1 and to provide adequate cover for protection of reinforcement as shown in Fig. 4-3. Table 4-3 shows the minimum concrete top slab thickness necessary for cover of negative moment reinforcement with #3 shrinkage and temperature bars perpendicular to the joists.

Shrinkage and temperature steel must be provided for one-way joist construction (Section 8.11.5.3). Such reinforcement, placed perpendicular to the ribs, must not be spaced further apart than five times the slab thickness nor more than 18 in. The amount of such reinforcement, expressed as a ratio of reinforcement area to concrete area, varies with the yield strength of the steel (Section 7.12).

TABLE 4-2 Joist Construction—Minimum Top Slab Thickness for Various Fire-Resistive Ratings

Type of Concrete	FIRE-RESISTIVE RATING (HOURS)		
	1 Hour	2 Hours	3 Hours
Normal Weight Concrete			
Top slab thickness[a]			
Siliceous aggregate	3.5″	5.0″	6.2″
Carbonate aggregate	3.2″	4.6″	5.7″
Structural Lightweight Concrete			
Top slab thickness[a]			
Sand-lightweight	2.7″	3.8″	4.6″
Lightweight	2.5″	3.6″	4.4″

[a] From Table 7-7-C-C in UBC Std. 7-7 or Table 709.2.2.1 in *Standard Building Code.*

TABLE 4-3 One-Way Joist Construction—Minimum Top Slab Thickness for Various Thicknesses of Concrete Cover of Reinforcement with #3 Shrinkage and Temperature Reinforcement Perpendicular to Joists

Bar Size	CONCRETE COVER IN INCHES					
	5/8″	3/4″	1″	1-1/4″	1-1/2″	3″
#3	2″	$2\frac{1}{2}$″	3″	$3\frac{1}{2}$″	4″	5″
#4	$2\frac{1}{2}$″	$2\frac{1}{2}$″	3″	$3\frac{1}{2}$″	4″	5″
#5	$2\frac{1}{2}$″	$2\frac{1}{2}$″	3″	$3\frac{1}{2}$″	4″	5″
#6	$2\frac{1}{2}$″	3″	$3\frac{1}{2}$″	4″	$4\frac{1}{2}$″	$5\frac{1}{2}$″
#7	$2\frac{1}{2}$″	3″	$3\frac{1}{2}$″	4″	$4\frac{1}{2}$″	$5\frac{1}{2}$″
#8	3″	3″	$3\frac{1}{2}$″	4″	$4\frac{1}{2}$″	$5\frac{1}{2}$″
#9	3″	3″	$3\frac{1}{2}$″	4″	$4\frac{1}{2}$″	$5\frac{1}{2}$″
#10	3″	$3\frac{1}{2}$″	4″	$4\frac{1}{2}$″	5″	6″
#11	$3\frac{1}{2}$″	$3\frac{1}{2}$″	4″	$4\frac{1}{2}$″	5″	6″

The required shrinkage and temperature steel of Grade 60 bars and WWF for the top slab of one-way joist construction, placed perpendicular to the joists, is shown in Table 4-4.

SHEAR

The basic requirement for shear as a measure of diagonal tension is computed from the Code formula

$$\phi V_n \geqslant V_u \quad \text{(Section 11.1.1, Eq. 11-1).}$$

TABLE 4-4 Shrinkage and Temperature Reinforcement of Grade 60 Bars and WWF Per Foot of Slab Width for One-Way Joist Construction

Thickness of Top Slab (Inches)	SHRINKAGE AND TEMPERATURE REINFORCEMENT	
	Required Area (square inches)	Size and Spacing
2	0.043	4 × 12-W1.5/W1
$2\frac{1}{2}$	0.054	4 × 12-W2/W1
3	0.065	4 × 12-W2.5/W1
$3\frac{1}{2}$	0.076	#3 @ $17\frac{1}{2}''$
4	0.086	#3 @ 15″
$4\frac{1}{2}$	0.097	#3 @ $13\frac{1}{2}''$
5	0.108	#3 @ $12\frac{1}{2}''$
$5\frac{1}{2}$	0.119	#3 @ 11″
6	0.130	#3 @ 10″

This equation is set forth in terms of total forces instead of stresses and says that the design shear strength, ϕV_n, must be equal to or greater than the factored shear force, V_u, at the cross section being considered. The design shear strength is the product of the strength-reduction factor, $\phi = 0.85$ (Section 9.3), and the nominal shear strength, V_n. The nominal shear strength is computed by

$$V_n = V_c + V_s \text{ (Section 11.1.1, Eq. 11-2)}$$

where V_c is nominal shear strength provided by concrete (Section 11.3) and V_s is nominal shear strength provided by reinforcement (Section 11.5.6).

The nominal shear strength provided by the concrete for one-way and two-way concrete joist construction can be taken as 10% greater than the values used for other structural elements (Section 8.11.8). The nominal shear strength provided by the concrete of joist construction, not subjected to axial load or torsion, is equal to the following amounts unless a detailed analysis is made that considers the tension reinforcement ratio, and the factored moment and shear at the section of the joist considered.

Normal Weight Concrete (Section 11.3.1.1)

$$V_c = 2.2\sqrt{f'_c}\, b_w d$$

Lightweight Aggregate Concrete—f_{ct} average splitting tensile strength specified (Section 11.2.1.1)

$$V_c = 2.2 b_w d f_{ct}/6.7 \text{ where } f_{ct}/6.7 \leq \sqrt{f'_c}$$

Sand-Lightweight Concrete—f_{ct} not specified (Section 11.2.1.2)

$$V_c = 1.87\sqrt{f'_c}\, b_w d$$

All-Lightweight Concrete—f_{ct} not specified (Section 11.2.1.2)

$$V_c = 1.65\sqrt{f'_c}\, b_w d$$

When a detailed analysis is made that considers the tension reinforcement ratio and the factored moment and shear, then V_c can be increased to the following amounts, providing $V_u d/M_u$ is taken not greater than 1.0.

Normal Weight Concrete (Section 11.3.2.1)

$$V_c = [2.09\sqrt{f'_c} + 2750\rho_w V_u d/M_u]b_w d \leq 3.85\sqrt{f'_c}\, b_w d$$

Lightweight Aggregate Concrete—f_{ct} specified

$$V_c = [2.09 f_{ct}/6.7 + 2750\rho_w V_u d/M_u]b_w d \leq 3.85\sqrt{f'_c}\, b_w d$$

Sand-Lightweight Concrete—f_{ct} not specified

$$V_c = [1.78\sqrt{f'_c} + 2338\rho_w V_u d/M_u]b_w d \leq 3.27\sqrt{f'_c}\, b_w d$$

All-Lightweight Concrete—f_{ct} not specified

$$V_c = [1.57\sqrt{f'_c} + 2062\rho_w V_u d/M_u]b_w d \leq 2.89\sqrt{f'_c}\, b_w d$$

When factored shear exceeds the design shear strength of the concrete, stirrups are required and should be provided (Section 11.5.6.1).

DESIGN

Code requirements for joist construction will be demonstrated by use of a design example of one-way joist construction as shown in Fig. 4-4 for an interior clear span of 30′-0″, constructed from normal weight concrete, with $f'_c = 4{,}000$ psi and Grade 60 reinforcement. A 3-1/2 in. top slab for a 1-hr fire rating will be used with 6 in. wide joist ribs and 14 in. deep by 30 in. wide removable fillers with 3 ft. long tapers at each end of the span. The tapered end fillers increase the shear capacity by increasing the width of the joist from 6 in. to 11 in. at the support.

EXAMPLE

Loads. The service and factored loads selected for the example shown in Fig. 4-4 are as follows:

Joist and bridging	77 psf	
Ceiling	10 psf	
Piping.	17 psf	
	(104)(1.4)	= 145 psf
Live load	(150)(1.7)	= 255 psf
w_u.		400 psf
w_u per linear foot of joist rib =	(0.40)(3)	= 1.2 klf

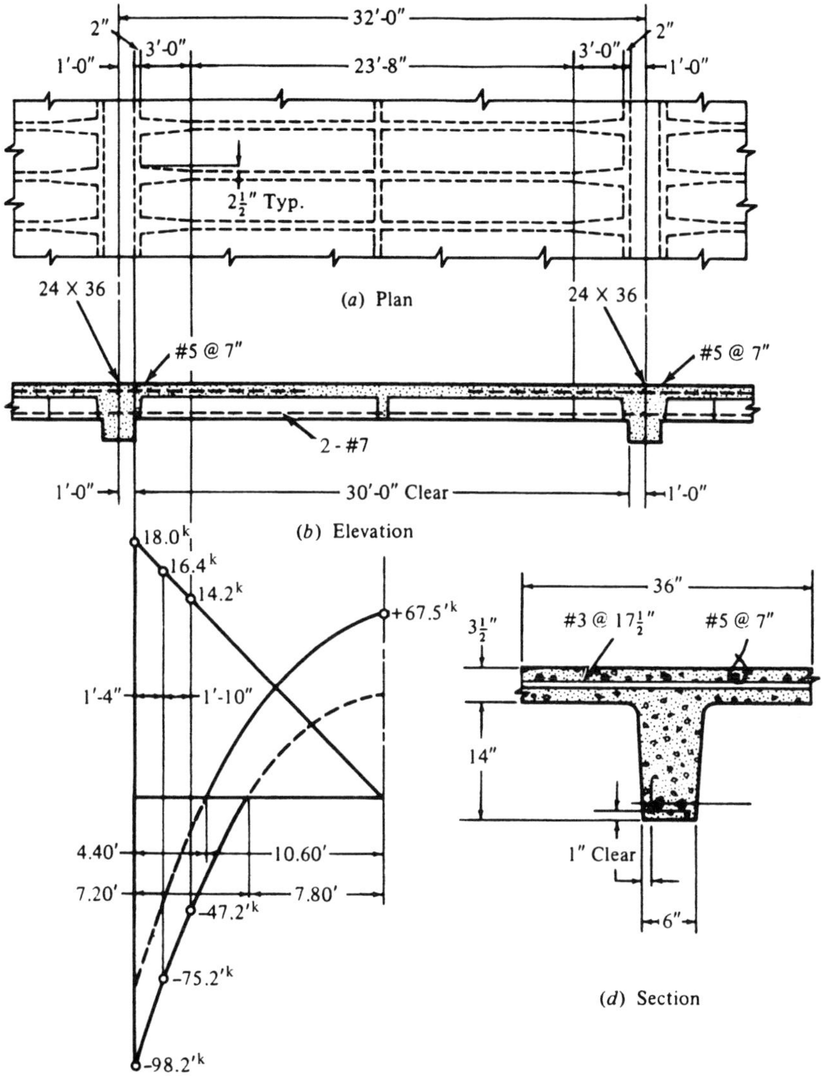

(c) Factored shear and moment
$w_u = 1.20$ k/ft

Figure 4-4 One-way joist construction—example.

Analysis

Factored shears and moments were calculated using the simplified assumptions of frame analysis and coefficients (Section 8.3.3).

Factored negative moment at the face of supports:

$$-M_u = (1/11)(1.2)(30)^2 = 98.2 \text{ ft-k per rib}$$

Factored positive moment at center of span:

$$M_u = (1/16)(1.2)(30)^2 = 67.5 \text{ ft-k per rib}$$

Flexural Reinforcement Required for Factored Moments

Using the expression from page 258 for calculating the required areas of reinforcement, and checking the required A_s for minimum and maximum limits:

$$A_s = 0.85\left(1 - \sqrt{1 - \frac{2.35M_u}{\phi f_c' bd^2}}\right)\frac{bdf_c'}{f_y}$$

Negative moment reinforcement at face of supports ($b \approx 6$ in.):

$$A_s = 0.85\left(1 - \sqrt{1 - \frac{2.35(98.2 \times 12)}{0.90(4)(6)(16)^2}}\right)\frac{6(16)(4)}{60} = 1.60 \text{ in.}^2$$

$$\text{Max } \rho = 0.75\rho_b = 0.0213 \text{ (Section 10.3.3)}$$

$$\text{Required } \rho = A_s/b_w d = 1.60/(6)(16) = 0.0167 < 0.75\rho_b, \text{ okay}$$

$$\text{Min } A_s = 0.0033b_w d = 0.0033(6)(16) = 0.32 \text{ in.}^2 \text{ (Section 10.5.1)}$$

Use #5 @ 7″ top bars. $\quad A_s = (0.31)(36)/7 = 1.59 \text{ in.}^2$

Positive moment reinforcement at center of span ($b = 36$ in.):

$$A_s = 0.85\left(1 - \sqrt{1 - \frac{2.35(67.5 \times 12)}{0.90(4)(6)(16)^2}}\right)\frac{36(16)(4)}{60} = 0.95 \text{ in.}^2$$

$$\text{Min } A_s = 0.0033bd = 0.0033(36)(16) = 1.90 \text{ in.}^2 > 0.95 \text{ in.}^2 \text{ (Section 10.5.1)}$$

$$\text{Use } A_s = 4/3(0.95) = 1.27 \text{ in.}^2 \text{ (Section 10.5.3)}$$

Use 2-#7 bottom bars. $\quad A_s = 1.20 \text{ in.}^2$ (within 5% of required A_s)

Structural Integrity

The Code requires at least one of the bottom bars to be lap spliced over the support (Section 7.13.2.1). A Class A tension lap splice is required. For a #7 bottom bar, the required lap splice length is 42 in. from Table 13-6.

Shear

Check the shear at a distance d from the support where the minimum joist width is $6 + (5)(22)/(36) = 9.05$ in. and at the end of the taper where the minimum joist width is 6 in. An average joist width is used in the shear calculations and is based on an assumed slope on the sides of the removable fillers of 1 in 12.

Factored shear at d from support (see Fig. 4-4)

$$V_u = 16{,}400 \text{ lb}$$

$$\phi V_c = \phi 2.2 \sqrt{f_c'}\, b_w d$$

$$= (0.85)(2.2)\sqrt{4000}\,(9.05 + 1.17)(16)$$

$$= 19{,}339 \text{ lb} > 16{,}400 \text{ lb}$$

$\therefore$ Stirrups not required

Factored shear at end of 3′-0″ taper (see Fig. 4-4)

$$V_u = 14{,}200 \text{ lb}$$

$$\phi V_c = (0.85)(2.2)\sqrt{4000}\,(6 + 1.17)(16)$$

$$= 13{,}568 \text{ lb} < 14{,}200 \text{ lb}$$

$$V_u > \phi V_c$$

Because the factored shear at the end of the taper is greater than ϕV_c, based on $2.2\sqrt{f_c'} = 139$ psi, a more detailed calculation of the design shear strength of the concrete will be made as follows:

$$\phi V_c = 0.85[2.09\sqrt{f_c'} + 2750\rho_w V_u d/M_u]b_w d$$

$$= 0.85\left[2.09\sqrt{4000} + \frac{(2750)(1.49)(14{,}200)(16)}{(7.17)(16)(47{,}200)(12)}\right](7.17)(16)$$

$$= 0.85(132 + 14)(7.17)(16)$$

$= 14{,}286$ lb which exceeds the factored shear, V_u, at end of 3′-0″ taper

A graphical presentation of factored shear and design shear strength of the concrete is shown in Fig. 4-5.

Figure 4-5 shows that for one-way joist construction without stirrups the use of tapered fillers increases the clear span range from 24′-0″ to 30′-0″ or approximately 25%. The critical section for shear occurs at the end of the tapered filler. Between the end of the tapered filler and a distance d from the support, the factored shear, V_u, decreases at a faster rate than the design shear strength of the concrete, ϕV_c (Sections 8.11.8, 11.3.1.1, and 11.3.2.1).

The use of calculations based upon *total force* for shear design as prescribed in the Code is illustrated here. Where design aids are used to reduce design time in practical work, they are conveniently expressed in terms of unit *stresses*. See Chapter 9 for such aids and terminology to adapt Code formulas to their use.

The depth of the joist construction shown in Fig. 4-4 barely meets the minimum thickness limitation of $\ell_n/21$ for interior span ribbed one-way slabs in the Code (Table 9.5a):

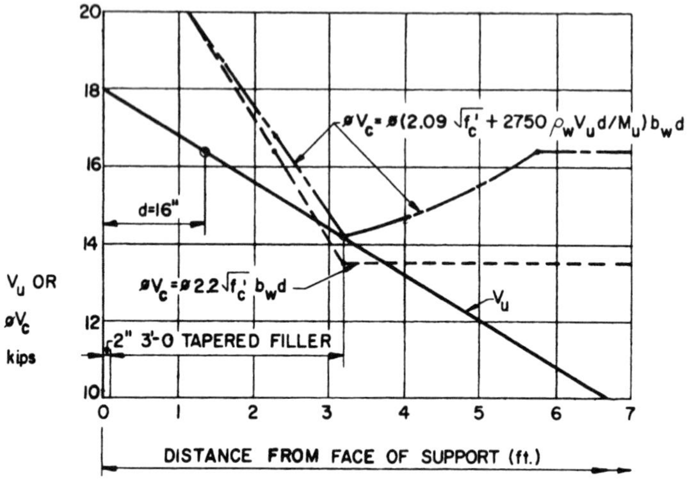

Figure 4-5 Shear strength example—tapered end joist.

$$17\text{-}1/2'' > (30)(12)/(21) = 17.10''$$

and computations for deflection are not required (Section 9.5.2.1). The deflection is calculated here to demonstrate the application of deflection criteria in the Code. In this computation ℓ_n is equal to the clear span (Section 8.7.3).

SECTION PROPERTIES FOR DEFLECTION CALCULATIONS

Determine the gross moment of inertia of the concrete section for positive moment ($+I_g$).

$$\begin{aligned}(30)(3.5) &= 105 \times 1.75 = 184\\(17.5)(6) &= \underline{105} \times 8.75 = \underline{919}\\ &\quad 210 \qquad\qquad\quad 1093\end{aligned}$$

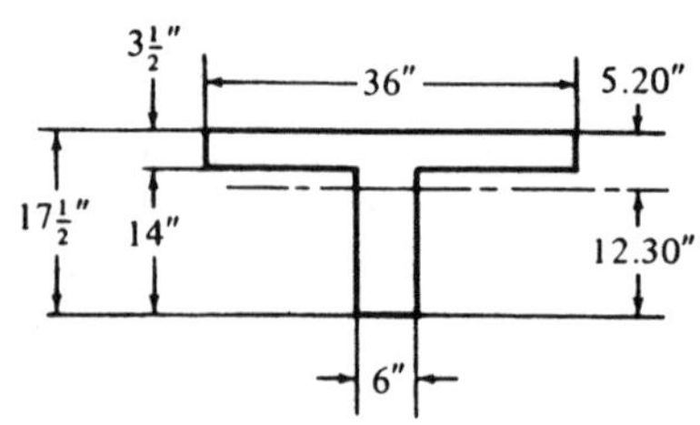

Figure 4-6 I_g for + Moment.

$$y_t(\text{top}) = 1093/210 = 5.20 \text{ in.}$$

$$y_b(\text{bottom}) = 17.5 - 5.20 = 12.3 \text{ in.}$$

$$\begin{aligned}(12.30)^3(6)/(3) &= 3721\\ (5.20)^3(36)/(3) &= \underline{1687}\\ &\ 5408\\ (-1.70)^3(30)/(3) &= \underline{-49}\\ +I_g &= 5359 \text{ in.}^4\end{aligned}$$

Determine the gross moment of inertia of the concrete section for negative moment ($-I_g$).

$$\begin{aligned}(25)(3.5) &= 87.5 \times 1.75 = 153\\ (17.5)(11) &= \underline{192.5} \times 8.75 = \underline{1684}\\ &\ 280.0 \qquad\qquad\ \ 1837\end{aligned}$$

$$y_t(\text{top}) = 1837/280 = 6.56 \text{ in.}$$

$$y_b(\text{bottom}) = 17.5 - 6.56 = 10.94 \text{ in.}$$

$$\begin{aligned}(10.94)^3(11)/(3) &= 4820\\ (6.56)^3(36)/(3) &= \underline{3390}\\ &\ 8210\\ (-3.06)^3(25)/(3) &= \underline{-239}\\ -I_g &= 7971 \text{ in.}^4\end{aligned}$$

Determine the cracked transformed moment of inertia for positive moment ($+I_{cr}$).

$$(36)(kd)^2/(2) = (8.32)(16 - kd)$$

$$kd = 2.50 \text{ in.}$$

$$\begin{aligned}(36)(2.5)^3/(3) &= 188\\ (8.32)(13.5)^2 &= \underline{1516}\\ +I_{cr} &= 1704 \text{ in.}^4\end{aligned}$$

Determine the cracked transformed moment of inertia for negative moment ($-I_{cr}$).

$$(11)(kd)^2/(2) + (16.6)(kd - 1.5) = (12.7)(16 - kd)$$

$$kd = 4.30 \text{ in.}$$

$$\begin{aligned}(11)(4.30)^3/3 &= 291\\ (12.7)(11.70)^2 &= 1738\\ (16.6)(2.80)^2 &= \underline{130}\\ -I_{cr} &= 2159 \text{ in.}^4\end{aligned}$$

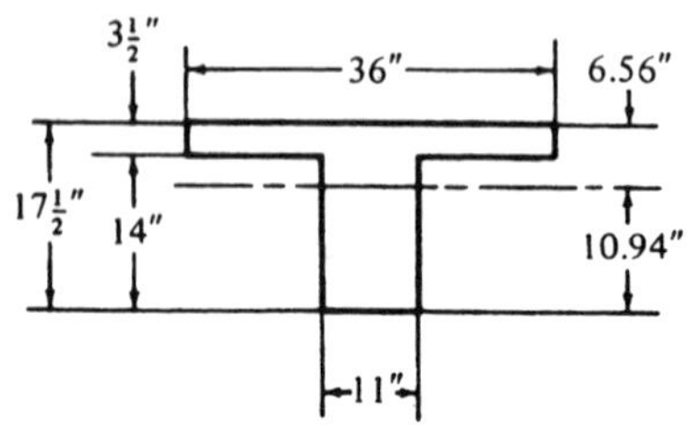

Figure 4-7 I_g for − Moment.

Note that $2n$ is used for the modular ratio of the compression steel (Appendix A.5.5).

Determine the modulus of rupture of the concrete (f_r) (Section 9.5.2.3).

$$f_r = 7.5\sqrt{f_c'} = (7.5)(\sqrt{4000}) = 474 \text{ psi}$$

Determine cracking moments (M_{cr}) (Section 9.5.2.3).

$$\begin{aligned} -M_{cr} &= f_r I_g / y_t \\ &= (474)(7971)/(6.56) = 576{,}000 \text{ in.-lbs} \\ &= 48.0 \text{ ft-kips} \\ +M_{cr} &= (474)(5359)/(12.30) = 206{,}000 \text{ in.-lbs} \\ &= 17.2 \text{ ft-kips.} \end{aligned}$$

The cracking load based on equal spans and a loading pattern that results in a maximum positive moment of $w\ell_n^2/16$ is

$$\begin{aligned} w_{cr} &= (17{,}200)(16)/(3)(30)^2 \\ &= 102 \text{ psf} \end{aligned}$$

Determine the maximum moment for service dead and service live load (M_a) for the stage of loading that produces maximum deflection.

$$\begin{aligned} M_a &= -M = +M = w\ell_n^2/16 \\ &= (0.254)(3)(30)^2/(16) \\ &= 42.9 \text{ ft-kips} \end{aligned}$$

Determine the effective moment of inertia for positive moment ($+I_{eff}$) (Section 9.5.2.3).

$$\begin{aligned} +I_{eff} &= (M_{cr}/M_a)^3 I_g + [1 - (M_{cr}/M_a)^3] I_{cr} \leq I_g \\ &= (17.2/42.9)^3(5359) + [1 - (17.2/42.9)^3](1704) \\ &= 344 + 1594 \\ &= 1938 \text{ in.}^4 \end{aligned}$$

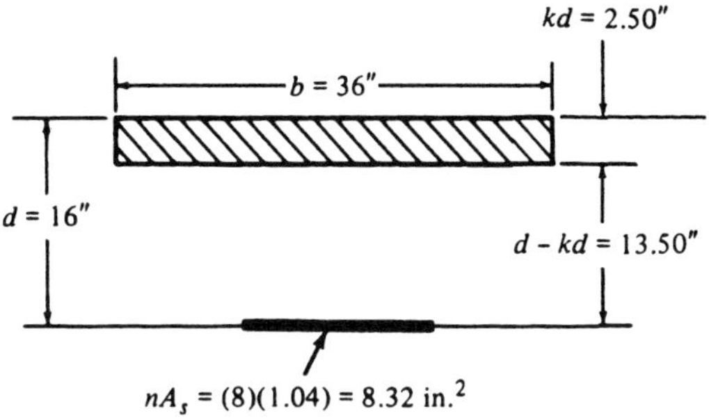

Figure 4-8 I_{cr} for + Moment.

Determine the effective moment of inertia for negative moment $(-I_{\text{eff}})$ *(Section 9.5.2.3).*

$$M_{\text{cr}}/M_a = 48.0/42.9 > 1$$

$$\therefore\ I_{\text{eff}} = I_g = 7971 \text{ in.}^4$$

The necessary parameters essential to calculate the deflection of the interior span joist are now available (Section 9.5.2).

DEFLECTION COMPUTATIONS

The instantaneous deflection under service dead load and service live load is based on the clear span, ℓ_n, as allowed for two-way construction, because the negative gross moment of inertia of the joist increases about 20 times in the width of the beams. The loading condition for deflection under service dead and live load is that which produces maximum positive moment $(w\ell_n^2/16)$ and maximum deflection (2) $w\ell_n^4/384E_cI$. The effective moment of inertia is used in deflection calculations and may be taken equal to the average of the values for

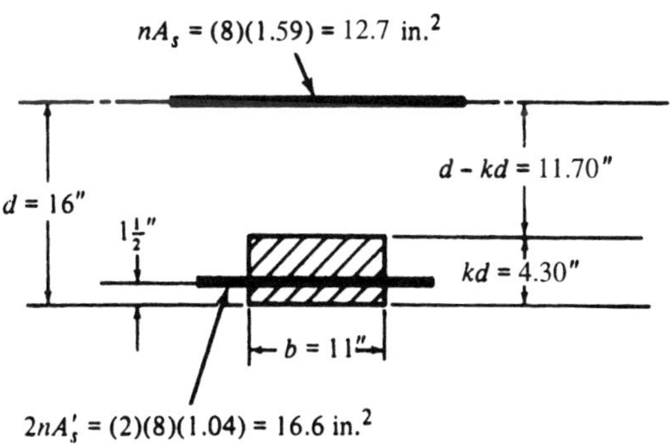

Figure 4-9 I_{cr} for − Moment.

the critical positive and negative moment sections (Section 9.5.2.4) as recommended by ACI Committee 435. Note that $I_{cr} \leqslant I_{eff} \leqslant I_g$.

$$\Delta_{(D+L)} = \frac{(2)(D+L)\ell_n^4}{384E_cI_{eff}}$$

$$= \frac{(2)(254)(3)(30)(360)^3}{(384)(3{,}600{,}000)(1/2)(1938+7971)}$$

$$= \frac{(2)(254)(3)(30)(360)^3}{(384)(3{,}600{,}000)(4954)}$$

$$= 0.311 \text{ in.}$$

Computed Immediate Live Load Deflection

The instantaneous deflection for service live load, based on the effective moment of inertia for service dead and live load, is

$$\Delta_L = (0.311)(150)/(254) = 0.184 \text{ in.}$$

Computed Long-Term Deflections

It is recommended, however, that calculations for the effect of the sustained service load ($D = 104$ psf) be based on the moment of inertia of the cracked section in lieu of I_{eff} because "tests seem to suggest that the tension in the concrete diminishes or even disappears with time for moments above the cracking moment."* Actually, the cracking moment per se becomes unimportant once cracking has occurred. Under *any* sustained load thereafter the tension carried by the concrete between cracks diminishes and additional deflection occurs with time.

The instantaneous deflection for equal spans with all spans loaded with a service dead load of 104 psf using the average cracked moment of inertia is

$$\Delta_{(D=104)} = \frac{D\ell_n^4}{384E_cI_{cr}} = \frac{(104)(3)(30)(360)^3}{(384)(3{,}600{,}000)(1/2)(1704+2041)}$$

$$= \frac{(104)(3)(30)(360)^3}{(384)(3{,}600{,}000)(1872)}$$

$$= 0.169 \text{ in.}$$

The multiplier, λ, to be applied to the instantaneous deflection of 0.169 in. for a dead load of 104 psf to calculate the additional long-term deflection (Sec-

* "Allowable Deflections," *Journal of the American Concrete Institute*, June 1968, Chapter 2.

tion 9.5.2.5) is equal to 2 (for a duration of five years or more, and no compression steel with positive moment).

The additional long-term deflection for a sustained service load of 104 psf is

$$\Delta_{\text{(long-term)}} = (0.169)(2) = 0.338 \text{ in.}$$

Deflection After Installation of Non-Structural Items

The floor of the example shown in Fig.4-4 is attached to a plaster ceiling and piping that is likely to be damaged by large deflections. For such a floor the Code recommends that the sum of the additional long-term deflection due to all sustained loads and the immediate deflection due to any additional live load be not more than 1/480 of the span unless adequate measures are taken to prevent damage to these attached elements (Code Table 9.5b). The sum of additional long-term deflection due to sustained service load ($D = 104$ psf) and immediate live load ($L = 150$ psf) is equal to

$$\Delta_{\substack{\text{(additional long-term plus}\\ \text{instantaneous live load)}}} = 0.338 + 0.184$$

$$= 0.522 \text{ in.}$$

This deflection is equal to 1/690 of the clear span and is less than the maximum value of 1/480 of the span allowed by Table 9.5b of the Code. A joist depth of 17-1/2 in. therefore satisfies the Code deflection criteria (Section 9.5.2.1).

SUMMARY DEFLECTION COMPUTATIONS

A graphical presentation of the deflection of the interior span joist of Fig. 4-4 is shown in Fig. 4-10. The deflection for a cracking load of 102 psf was calculated for equal spans with a maximum positive moment of $w\ell_n^2/16$ and is

$$\Delta_{\text{cr}} = \frac{2w\ell_n^4}{384E_cI_g} = \frac{(2)(102)(3)(30)(360)^3}{(384)(3{,}600{,}000)(5359)}$$

$$= 0.116 \text{ in.}$$

In order to study the shape of the load-deflection curve from cracking load to factored load, two additional instantaneous deflections were calculated, one for $D + L = 200$ psf and another for $1.4D + 1.7L = 400$ psf. Both of these calculations were based on equal spans with alternate panels loaded for an assumed maximum positive moment equal to $w\ell_n^2/16$ and an assumed maximum deflection equal to (2) $w\ell_n^4/384E_cI$. The average of the effective moment of inertia (I_{eff}) of the critical positive and negative moment sections was used in these calculations, which are as follows:

$$\Delta_{(D+L=200\text{ psf})} = \frac{2w\ell_n^4}{384E_cI_{\text{eff}}}$$

$$= \frac{(2)(200)(3)(30)(360)^3}{(384)(3{,}600{,}000)(1/2)(1938 + 7971)}$$

$$= 0.245 \text{ in.}$$

$$\Delta_{(1.4D+1.7L=400\text{ psf})} = \frac{2w\ell_n^4}{384E_cI_{\text{eff}}}$$

$$= \frac{(2)(145 + 255)(3)(30)(360)^3}{(384)(3{,}600{,}000)(1/2)(1938 + 7971)}$$

$$= 0.491 \text{ in.}$$

There is a sudden change in the slope of the load-deflection curve for loads greater than the cracking load. An effective moment of inertia (I_{eff}) for deflection calculations is prescribed between the gross moment of inertia (I_g) and the cracked transformed moment of inertia (I_{cr}) (Section 9.5.2.3, Eq. 9-7). Note the transition in Fig. 4-10 for the joist shown in Fig. 4-4.

WIDE-MODULE JOIST CONSTRUCTION

Wide-module joist construction, which is also referred to as "skip-joist" construction, is an approach to reduce form costs and develop longer spans than standard one-way joist systems (Section 8.11). Where fire codes establish thickness of top slabs at or about 4.5 in., the flexural capacity of the slab is underutilized within limitations of standard joist dimensions with maximum clear spacing between joists of 30 in. (Section 8.11.3). The wide-module joist concept

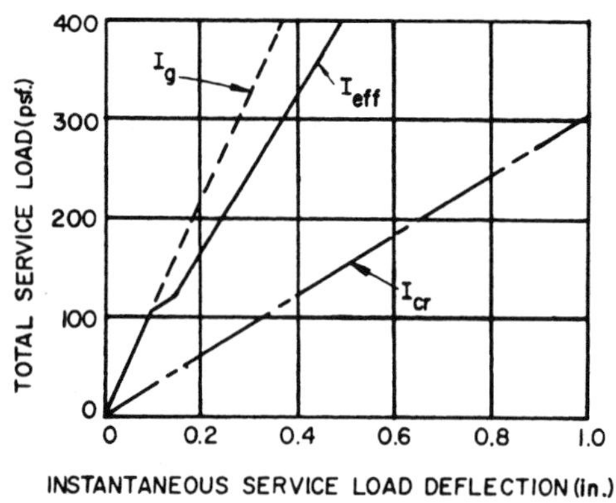

Figure 4-10 Deflection of interior span joist.

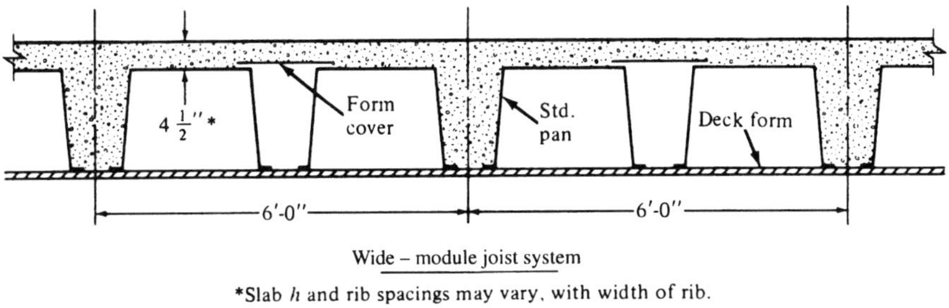

Figure 4-11 General arrangement of standard reusable forms for wide-module joist systems.

utilizes standard reusable joist forms with alternate ribs blocked off. See Fig. 4-11. It usually employs the deeper-size forms with rib depths 16 in. or 20 in. below the slab. Rib spacings may be 6′-0″ or more depending upon thickness of rib, or module established by architectural reasons.

The alternate name "skip-joist" is accurate only in that a potential rib is indeed omitted or "skipped." The ribs of course are designed as beams. Minimum cover (Section 7.7.1c) is 1.5 in. (instead of 0.75 in.) as in joists. Minimum shear reinforcement (Section 11.5.5) is required. Shear carried by the concrete is 10 percent less than that allowed for joists (Sections 8.11.8 and 11.3). Draped two-way reinforcement in the slab is permitted (Section 7.5.3).

The principal practical problem is providing shear reinforcement in amounts required (Section 11.5) detailed for practicable placing in narrow sections. Vertical U-stirrups are of course acceptable (Section 12.13.2), although practicable bending limitations may require that they be set at an angle to the longitudinal reinforcement. See Fig. 4-12. Single leg stirrups alternating direction of the hooked ends have been considered (Section 12.13.2.1). The Code also permits single leg, deformed or plain welded wire fabric (WWF) meeting special requirements (Section 12.13.2.4). Fig. 4-12 shows several possible details. Minimum shear reinforcement requirements will control in most cases, either throughout the span or at a short distance from supports (Section 11.5.5).

EFFECTIVE WIDTH OF ONE-WAY MONOLITHIC JOIST CONSTRUCTION AS A TWO-WAY SYSTEM

The term "one-way" applied to flexural elements usually refers to walls, slabs, or joists reinforced for flexure in only one principal direction. For convenience one-way slabs or walls are often designed for flexure as if they consisted of separate one foot wide strips. "One-way"joists similarly are designed as if they were single separate elements. "Joist construction" as defined (Section 8.11) is (1) required to be monolithic and (2) consists of ribs and top slab arranged to span in one *or two* orthogonal directions. Long-standing industry recommendations for one-way monolithic joists require distribution ribs for overloads on a

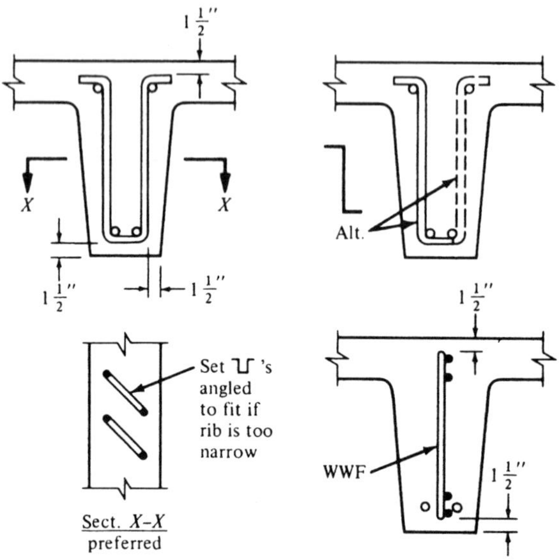

Figure 4-12 Various arrangements of shear reinforcement.

single joist as span lengths increase and joists become too deep for lateral distribution by the top slab only:

> *Distributing Ribs. Load distribution ribs are ribs normal to direction of span. As the name implies, load distribution ribs are provided to equalize deflection and thus ensure that the effect of concentrated or unequal loads is often omitted for roofs designed for uniform load only. For floor construction, use distributing ribs at least 4 inches wide, reinforced with at least one #4 bar continuous in both top and bottom, and spaced as follows:*
>
> *None in spans less than 20 feet*
> *One near the center of spans 20 feet to 30 feet, and*
> *Two near the third points of spans over 30 feet.**

For convenience in analysis, monolithic one-way joist construction is designed as separate one-way elements. It is, of course, well known that the connecting slabs and distribution ribs ordinarily provided effect a considerable lateral transfer of any load applied to a single joist. Special analyses are sometimes necessary for determination of the effect of unusual design load concentrations, changing original design loads or cutting large openings for new occupancy requirements. The routine analysis of this lateral distribution for

* *CRSI Handbook*, 1996.

determination of the "minimum effective width" for purposes of live load reductions, permitted under most codes, is too complicated for practical applications. Studies by CRSI provide simple rules for lateral transfer of concentrated load or minimum effective width of monolithic joist construction within the practical range of spans.*

The studies consisted of computer-simulated load tests of the center joist in an eleven-joist group. In addition to dead load on eleven joists, six test load increments were applied, each equal to one design live load, on the center joist.

At load test increment "Six" ($D + 5L$), critical conditions were reached for:

1. End shear in joist and
2. Positive moment in distribution ribs (requiring 1-#5 top and bottom).

Total transfer of live load (both ways from center loaded joist) at failure was approximately:

Spans less than 20′-0″ (monolithic slabs only):	30% transfer
Spans 20′-0″ $\leqslant L \leqslant$ 30′-0″ (monolithic slabs plus center rib):	40% transfer
Spans > 30′ (monolithic slabs plus two ribs at third points):	55% transfer

* *CRSI* Structural Bulletin N. 8.

5 Two-Way Solid Flat Plate Design

GENERAL

Following accepted terminology, a flat plate is considered here as a two-way reinforced flat slab of uniform depth without interior beams, drop panels, brackets, or capitals that is supported directly by columns spaced to form a square or rectangular grid. Spandrel beams at edges and large openings will usually be required for heavy loads, except where large columns are available, or where special provisions are made for shear transfer of loads to edge columns. Cantilever edges can be structually advantageous particularly where spandrel beams are undesirable for architectural reasons.

The 1995 ACI Building Code is intended to become part of general statutory building codes. Live load, as distinguished from factored live load (Section 2.1), is defined as the live load specified by the general code. Most building codes allow a reduction of the specified live load based on the tributary area supported by the member, the influence area affected by the loads, and the magnitude of the live load. The Uniform Building Code, the BOCA National Building Code, and others allow such live load reductions for flat slabs considering the area of the panel as the tributary area supported by the member and the influence area. (Refer to Minimum Design Loads for Buildings and Other Structures ASCE 7-95 and Fig. 2-1 live load reduction.)

1995 REVISIONS TO THE CODE ON TWO-WAY SLABS

The basic terminology from the 1989 Code that was continued in 1995 is as follows:

Nominal strength. Strength of a member or cross section calculated by the strength design method before application of strength-reduction ("ϕ") factors. Identified by the subscript "n", e.g., P_n, M_n, *etc.*

Design strength. Nominal strength reduced by appropriate ϕ-factor. No special notation. Identified simply as: ϕP_n, ϕM_n, ϕV_n, etc.

Required strength. Factored axial loads, moments, shears, or combinations thereof used to proportion members. Identified as P_u, M_u, V_u, etc.

Design procedure. $P_u \leqslant \phi P_n$; $M_u \leqslant \phi M_n$; etc.

Six significant revisions to the provisions for two-way flat slabs have been made in the 1995 Code.

The first revision requires that all bottom bars within column strips be continuous or spliced with Class A tension lap splices as shown in Fig. 13.8 of the Code (Section 13.8.5). This requirement provides some residual ability to span to adjacent supports in case a single support is damaged (Commentary R13.3.8.5).

The second revision permits an adjustment to the amount of unbalanced moment transferred between slab and column by shear or by flexure (Section 13.5.3.3). For an unbalanced moment about an axis parallel to the edge of an exterior support, the amount of the unbalanced moment transferred by flexure can be increased to 100% ($\gamma_f = 1.0$) provided that $V_u \leqslant 0.75\,\phi V_c$ and at a corner support when $V_u \leqslant 0.5\,\phi V_c$. For unbalanced moments at interior supports, and for unbalanced moments about an axis transverse to the edge at exterior supports, the value of γ_f is permitted to be increased up to 25% provided that $V_u \leqslant 0.4\,\phi V_c$. See Fig. 5-1.

$$\gamma_f = 1/[1 + (\tfrac{2}{3})\sqrt{b_1/b_2}] \quad \text{(Section 13.5.3.3)}$$

b_1 = width of the critical shear section measured in the direction of the span for which moments are determined.

b_2 = width of the critical shear section measured in the direction perpendicular to b_1.

It is important to note that when $\gamma_f M_u$ is increased, $\gamma_v M_u$ can be decreased. The flexural reinforcement ratio, ρ, placed within an effective slab width between lines that are $1\,\frac{1}{2}$ times the slab thickness ($1.5h$) outside opposite faces of the column, is limited to a maximum of $0.375\ \rho_b$ (Section 13.5.3.3).

The third revision reduces the maximum amount of live load (unfactored) permitted by the Direct Design Method from three times the dead load (unfactored) to two times the dead load (Section 13.6.1.5). However, variations from this limitation are permitted if it can be shown by analysis that this limitation does not apply, as for a slab system carrying a nonmovable load such as a water reservoir (Commentary R13.6.1.8).

The fourth revision requires that the direct design gravity load moment transferred between the slab and the edge column be set equal to $0.3\ M_o$ (Section 13.6.3.6).

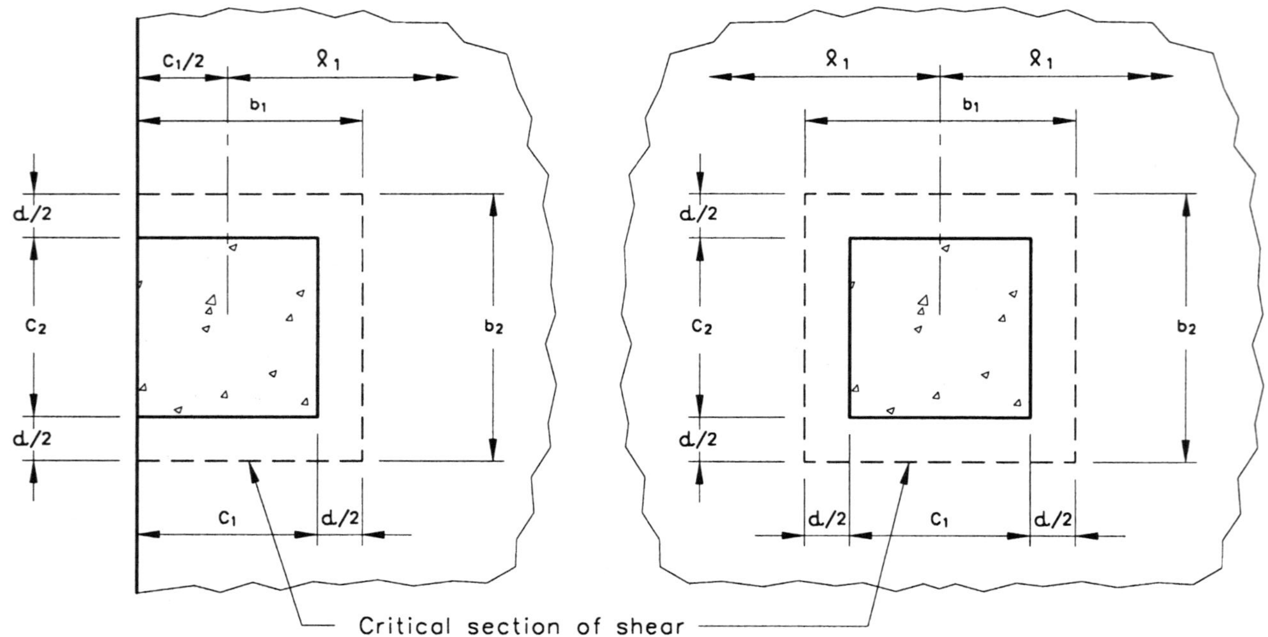

Figure 5-1 Plan at exterior and interior columns.

The fifth revision removes the stiffness, K_t, for torsional members perpendicular to the direction of the span for moments, from the Code to the Commentary (Commentary R13.7.5) and the definition of the cross-sectional torsional constant to the notation (Section 13.0).

The sixth revision removes the 1989 direct design code provisions for effects of pattern loading, because the ratio of live load to dead load (unfactored loads) was reduced from 3 to 2.

GENERAL DESIGN PROCEDURES

Shear at exterior columns is the "Achilles" heel of two-way slab to column systems. The unbalanced negative slab factored moments at the columns must be transferred to the columns through flexure and shear in accordance with a formula involving column width, depth, and effective depth of slab (Sections 11.12.6 and 13.6.3.6). Shear due to transfer of unbalanced moment is combined with uniform vertical shear due to gravity load and becomes a critical value, particularly at edge columns without spandrels, beams, or cantilevers.

The equivalent frame method must be used when the structure does not fall within the limitations on the use of the direct design method. A limit on live-to-dead load ratio of 2 has been added.

Those engineers frequently undertaking flat slab design will find use of a design handbook, a computer design program, or at least design aids for determining various element stiffnesses essential to profitable operations. In the remainder of this chapter, step-by-step procedures with applicable Code section references will be given. Some typical design aids for manual calculation of stiffnesses are presented. This material should aid the user in developing a computer design program or a full set of large-scale design aids for precise manual calculation. Finally, some simple approximations for quick longhand calculation are offered. These approximations should give acceptable preliminary designs for estimates, and may often serve as final designs, since the Code permits a 10% adjustment of all moments within the required total panel moment (Section 13.6.7).

DIRECT DESIGN METHOD (SECTION 13.6)

In addition to the principal reference section, the Code user will find it necessary to refer to new provisions for transfer of moment and shear applicable to the Direct Design method (Section 13.6.3.6). Sections 13.5.3 and 13.5.4 refer the user to Chapter 11, "Shear and Torsion." Finally, a number of the formulas necessary for shear computations are located in the Code Commentary Chapter 11. The best way to illustrate the sequence of references applicable to direct design is by an example. (All of the operations illustrated here apply also to flat slabs, waffle flat slabs, and two-way slabs. In subsequent chapters only the additional operations required for these designs are shown, with references to the simpler example here.)

Sample Problem—Direct Design Method

The most straightforward example is a flat plate without spandrels, with the outer edge flush with the outer face of the exterior columns. This location of the edge columns is particularly sensitive to analyze as well as to design for shear. As a simple case which might serve for a low-rise motel or dormitory, consider a superimposed factored load of 100 psf on square panels with spans of 20 ft between column centerlines. The story height is 12 ft. The same square column cross section will be maintained from footings to roof for proven overall economy. Lateral forces are to be resisted principally by other stiffer elements. Assume normal weight concrete, $f'_c = 4{,}000$ psi, is an economical choice in the area. Use Grade 60 bars for the maximum economy in reinforcement.

SOLUTION. Note that the direct design method is a review, requiring the designer to first select trial design dimensions and then check same for strength and serviceability. A step-by-step procedure follows:

Step 1. Plate thickness (h)

Even if the designer intends to compute the deflection and to reduce the first trial thickness to the absolute minimum without exceeding the maximum allowable computed deflection (Section 9.5.3.4), he must first complete a trial design through selection of reinforcement, since both the cracking moment and the cracked transformed moment of inertia are required for the deflection computations (Section 9.5.2.2).

If, as usual, the designer wishes to avoid deflection calculations entirely, the minimum thickness must satisfy Code Table 9.5(c) (Section 9.5.3.2); $h = \ell_n/30 = (240 - 16)/30 = 7.47$ in. Select 8 in. as the first trial. Compute various floor loads: factored live load, $w_\ell = 100$ psf; dead load $= 8/12 \times 150$ psf; $w_d = 100 \times 1.4 = 140$ psf; total factored load $= 240$ psf. (Note term "factored load" means specified load times the appropriate load factor in (Section 2.1).) Ratio of the live to dead loads $= \dfrac{100/1.7}{140/1.4} = \dfrac{59}{100} = 0.59$. This ratio applies whenever the term "factored" is omitted from Code provisions.

Step 2. Trial column dimensions.

For square columns, $c_1 = c_2$. There are no limits, minimum or maximum, on column stiffness. To select the square column dimensions, $c_1 = c_2$, a trial size must be assumed. Try $c_1 = c_2 = 16$ in. (see guide to selection of minimum column size, Fig. 5-38). Figure 5-38 shows minimum size for square exterior columns (17 in.) with minimum size square interior columns (12 in.). When both are to be made the same size, the exterior minimum may control, but the increased stiffness of the larger-than-minimum interior column will permit some reduction in the exterior column size shown.

Step 3. Exterior panel moments by Code coefficients. See Fig. 5-2.

$$M_o = \frac{w_u \ell_2 \ell_n^2}{8} \quad \text{(Section 13.6.2.2, Eq. 13-3)}$$

$\ell_n = 20$ ft $- 16/12 = 18.67$ ft, assuming 16 in. square first interior column

$$M_o = \frac{(0.240)(20)(18.67)^2}{8} = 209 \text{ ft-kips total panel moment}$$

Distribution (Section 13.6.3.3—Column 3) for slab without beams between interior supports and without edge beams.

1st interior column	$-M_u = -(0.70)(209)$	$= -146$ ft-kips
Positive moment	$+M_u = +(0.52)(209)$	$= 109$ ft-kips
Exterior column	$-M_u = -(0.26)(209)$	$= -54$ ft-kips
See Fig. 5-2(b)		

Considering the 10% redistribution factor permitted (Section 13.6.7), the experienced designer will immediately recognize several short cuts within any given or usual range of column sizes that he may encounter. For practical placing, congestion of the top slab bars passing through the column verticals can be a problem. With no edge beam, all the top bars must be anchored at the edge of the slab, and must be in the column strip. For $\beta_t = 0$, use 100% of panel moment for the column strip (Section 13.6.4.2). Furthermore, the Code requires placing up to 60% of these bars within the column width plus 1.5 times h on each side, $c_2 + 3h$ (Section 13.5.3.2). When these requirements get too restrictive for practical placing, decrease the top bar area at the edge by 10%, and add approximately 5% to the area of the bottom bars. For the example, one could reduce $-M_{ext}$, $0.10 \times 54 = 5.4$ ft-kips and add half, 2.7 ft-kips at the center of the span which equals 2.5% times $+M_u$. Negative moment change at the first interior column can be disregarded as negligible; see Fig. 5-3.

Step 5. Shear.

Unfortunately, the designer must proceed through the first four steps before a close trial of the assumed column size for shear can be attempted, since the Code provisions for investigation of shear at an exterior column are quite complicated.

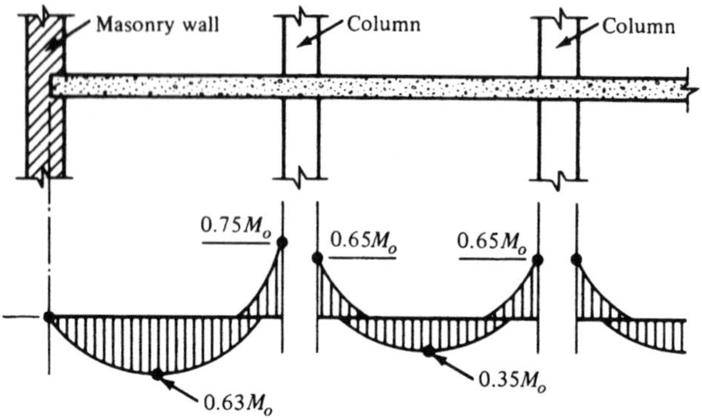

(a) Two-way slab with exterior edge unrestrained.

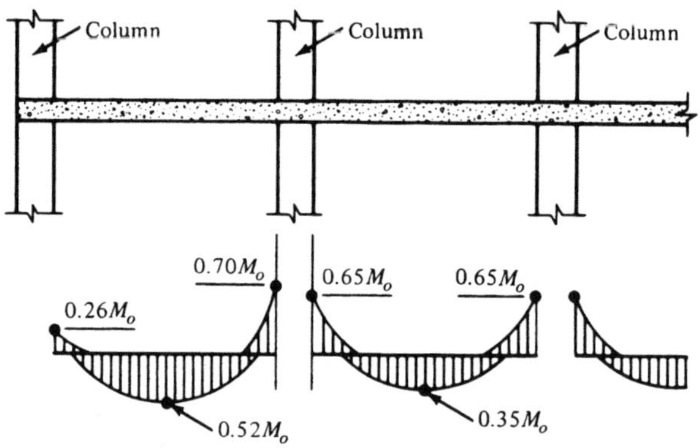

(b) Two-way slab with columns and without beams.

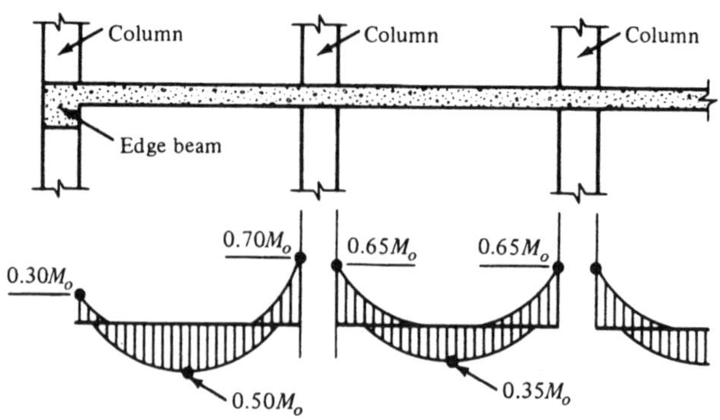

(c) Two-way slab with columns and edge beam.

Figure 5-2 Exterior panel moment coefficients.

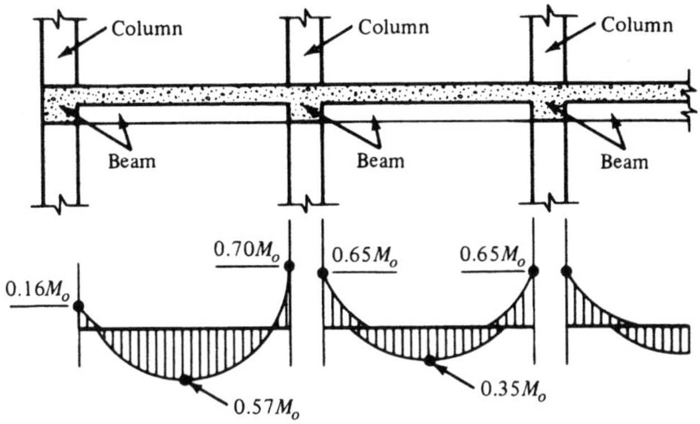

(d) Two-way slab with columns and beams between all supports.

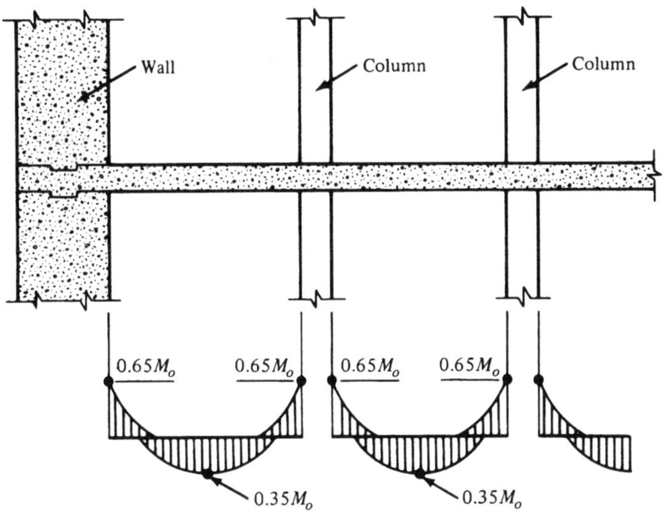

(e) Two-way slabs with exterior edge fully restrained.

Figure 5-2 (*Continued*)

It will be convenient to subdivide this step into three parts: computation of shear constants for the sections in the joint, adjustment of the exterior moment to the "center of shear" (c.s.), and calculations of the maximum unit shear combining a uniform vertical load shear with moment-induced shear in an assumed elastic distribution.

Shortcut. It will be helpful to avoid numerous trials if the designer can select a reasonably close first trial size column. For square edge columns flush with the edge, the authors find moment-induced shear about half the total allowable

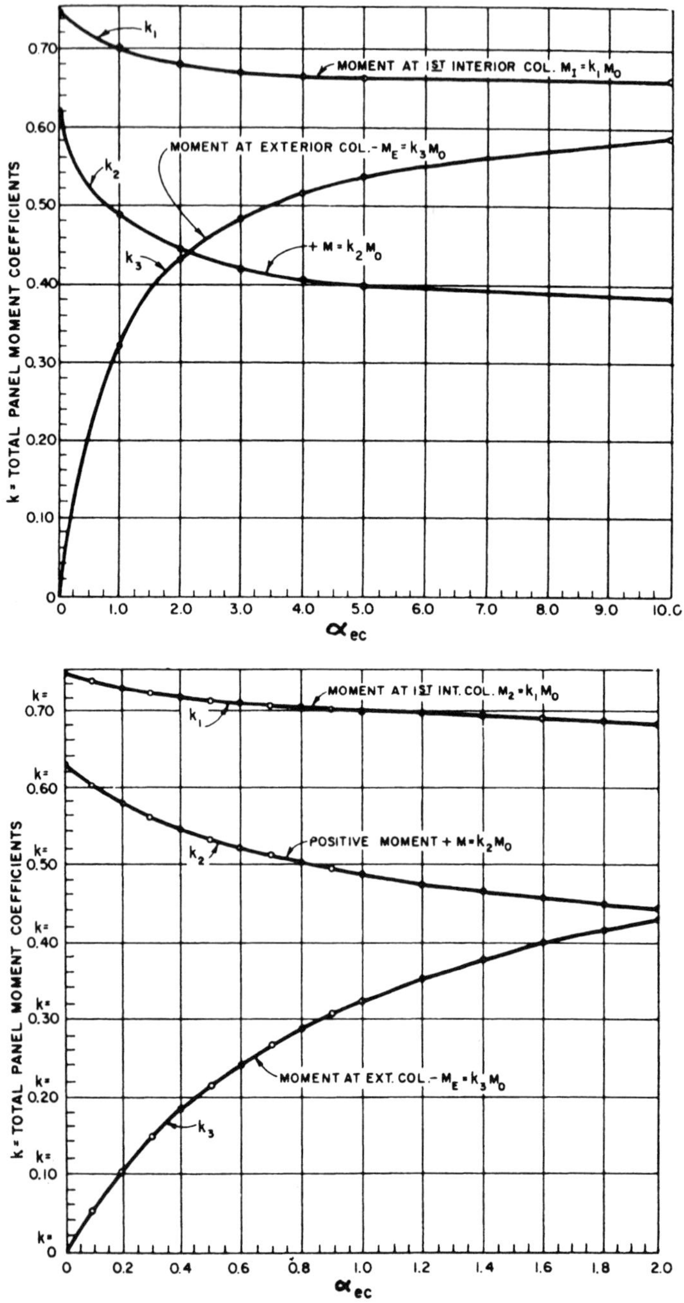

Figure 5-3 Total panel moment coefficients for edge panels. (*From the CRSI Handbook, 1992. Courtesy of the Concrete Reinforcing Steel Institute*)

value. For such columns, try a column size for which the gravity load shear uniformly distributed around the 3-sided critical section is 50% of the allowable shear ($2\sqrt{f'_c}$). For rectangular columns with long side parallel to the edge, try 55% ($2.2\sqrt{f'_c}$); with long side at right angles to the edge, 45% ($1.8\sqrt{f'_c}$). For square interior columns, neglect moment-induced shear in selecting the first trial size.

Constants. (See Code Section 11.12.6 and for formulas the Commentary, same section.) First consider the exterior column of our example as most likely to be critical. Average effective depth for two-way bars assuming maximum size #6 bars and 3/4 in. cover, $d = 8 - 3/4 - 3/4 = 6\ 1/2$ in. See Fig. 5-4 for designations identifying dimensions. Perimeter of the critical shear section $= b_o$ (Section 11.12.1.2). For derivation of shear equations, see Fig. 5-34.

$$b_o = 2(c_1 + d/2) + (d/2 + c_2 + d/2) = 61 \text{ in.}$$

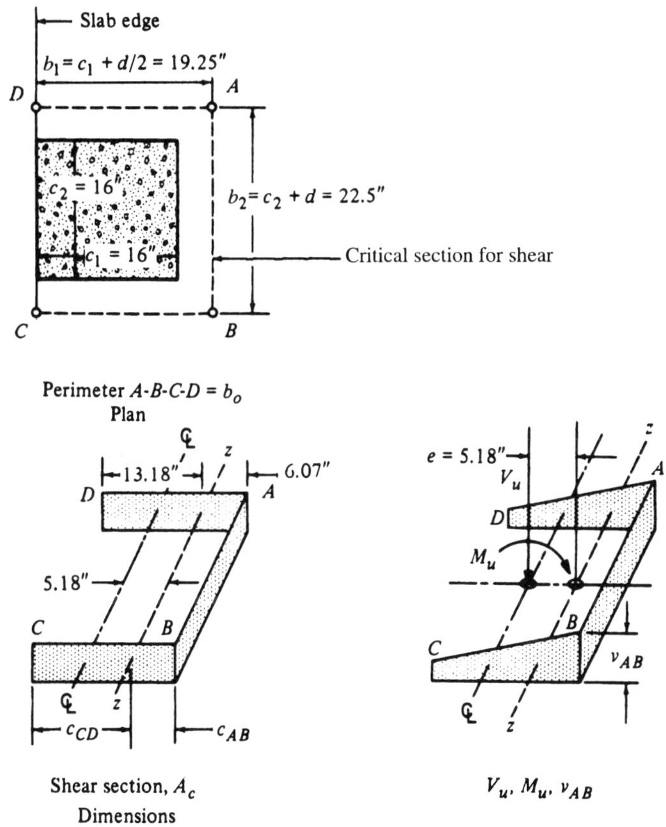

Figure 5-4 Shear constants.

Area of critical shear section, $A_c = b_o d = 61 \times 6.5 = 396.5$ in.

Axis z-z passes through the center of gravity of the shear area; point c.s. on the z-axis is the centroid of shear areas.

$$c_{AB} = \frac{(2)(6.5)(19.25)(1/2)(19.25)}{(396.5)} = 6.07 \text{ in.}$$

$$c_{CD} = 19.25 - 6.07 = 13.18 \text{ in.}$$

Center of column to the c.s. = 13.18 − 8.00 = 5.18 in.

Face of column to the c.s. = 8.00 − 5.18 = 2.82 in.

$$J_c = \frac{(19.25)(6.5)^3}{6} + \frac{(6.5)(19.25)^3}{6}$$

$$+ 2(6.5)(19.25)(19.25/2 - 6.07)^2 + (6.5)(22.5)(6.07)^2$$

$$J_c = 17{,}160 \text{ in.}^4$$

γ_v is the fraction of the unbalanced moment to be transferred to the column by shear about the centroid of the shear area (Section 13.5.3.1)*:

$$\gamma_v = 1 - \gamma_f = 1 - \frac{1}{1 + \frac{2}{3}\sqrt{b_1/b_2}}$$

$$= 1 - \frac{1}{1 + \frac{2}{3}\sqrt{19.25/22.5}} = 0.381$$

The reduction in shear at the face of the exterior column due to the difference in end moments at the exterior and first interior columns is:

$$\frac{(M_{\text{ext}} - M_{int})}{\text{clear span}} = \frac{(54 - 148)}{18.67} = -5.0 \text{ kips}$$

Shear at the face of the column: $V_f = (0.5)(0.240)(20)(18.67) - 5 = 39.8$ kips. See Fig. 5-5.

Reaction at exterior column: $V_{\text{c.s.}} = 39.8 + (0.240)(20.00)(16/12) = 46.2$ kips.

Moment at the centerline of the column:

$$M_{CL} = -54.0 - \frac{(39.8 + 43.0)(8)}{2(12)} = -81.6 \text{ ft-kips}$$

---

* Note that $b_1 = c_1 + \frac{d}{2}$ and $b_2 = c_2 + d$.

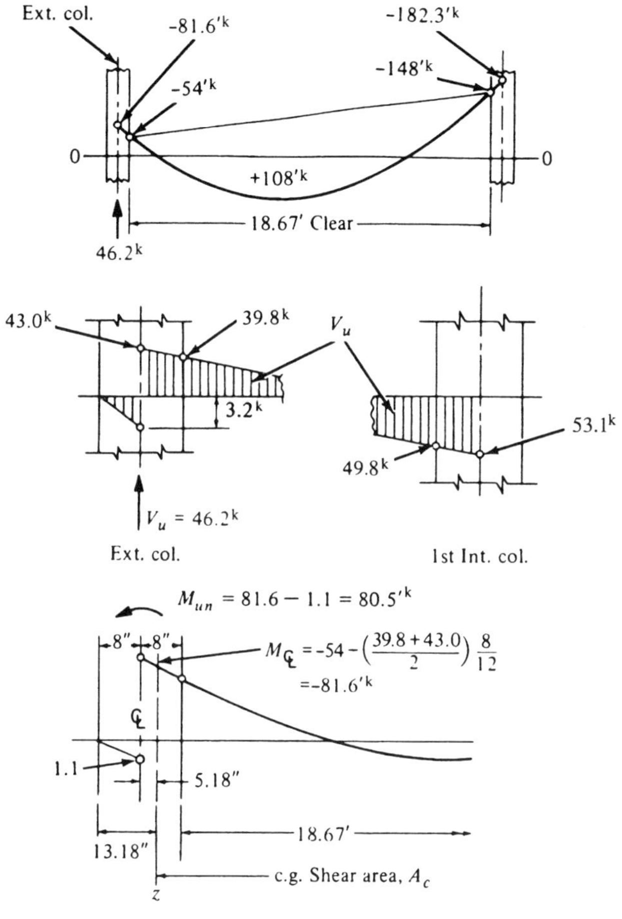

Figure 5-5 Factored shear and moment at column centerlines.

M_{un} (unbalanced moment to be transferred between slab and edge column) $= 0.30M_o$ (Section 13.6.3.6) $= (0.30)(209) = 62.7$ ft-kips

Combined shear due to gravity loads and to moment transfer:

$$v_{u(AB)} = \frac{V_{\text{c.s.}}}{A_c} + \frac{\gamma_v M_{un} c_{AB}}{J_c} \text{ (Commentary R11.12.6.2)}$$

$$= \frac{(46{,}200)}{(396.5)} + \frac{(0.381)(62{,}700)(12)(6.07)}{(17{,}160)}$$

$$v_{u(AB)} = 116 \text{ psi} + 101$$

$$= 217 \text{ psi}$$

Allowable shear when shear reinforcement is not provided, v_c:

$$v_c = \phi 4\sqrt{f_c'} = 0.85(4)\sqrt{4000} = 215 \text{ psi} \approx v_{u(AB)} \text{O.K.}$$

Consider the column size selected satisfactory for shear provided that it is not necessary to use bars larger than #6 both ways, and conservative if the average effective depth is more than that assumed, 6.5 in., due to use of smaller bars. No shear reinforcement will be required.

Step 6. Shear at the first interior column.

Most of the calculations required are the same as in Step 5, and will not be repeated in detail here. Even with the same size of column, the shear constants will change because the four sides of the interior column form a critical shear section symmetrical about both axes. The column centerline becomes the center of the shear area. Also, since only the prescribed unbalanced moment need be transferred to the column (Section 13.6.9.2) the combined shear calculation is much less critical. See Fig. 5-6 for dimensions of the critical shear section. For our example, try a 16 in. square column here also. $c_1 = c_2 = 16$ in. $A_c = 585$ in.^{2} (four sides). $J_c = 50{,}400$ in.^{4} The moment transfer fraction, $\gamma_v = 0.4$, since $b_1/b_2 = 1$. $c_{AB} = 0.5(16 + 6.5) = 11.25$ in. The unbalanced moment to be transferred, half to the column above and half to the column below, M_{un} for square panels, ℓ_1 and ℓ_2, reduces to

$$\text{Code } M_{un} = (0.07)[(w_d + .5w_\ell)(\ell_2)(\ell_n)^2 - w_d'\ell_2'(\ell_n')^2] \text{ (Section 13.6.9.2)}$$
$$= (0.07)\{[140 + (0.5)(100)](20)(18.67)^2 - (140)(20)(18.67)^2\}$$
$$= 24{,}400 \text{ ft-lb}$$

Factored shear at interior columns other than first line = panel load. Panel load = 0.240 $(20)^2$ = 96 kips. Also, for the first interior panel, add the shear deducted from exterior column due to difference in negative factored moments in the exterior panel. Factored shear at interior column $V_u = 96.0 + 5.0 = 101.0$ kips, all panels loaded. Design shear strength $\phi V_c = \phi 4\sqrt{f_c'}\, b_o d = (0.85)4\sqrt{4000}\,(4)(22.5)(6.5) = 125{,}800$ lb, O.K. $> V_u = 101{,}000$ lb. Com-

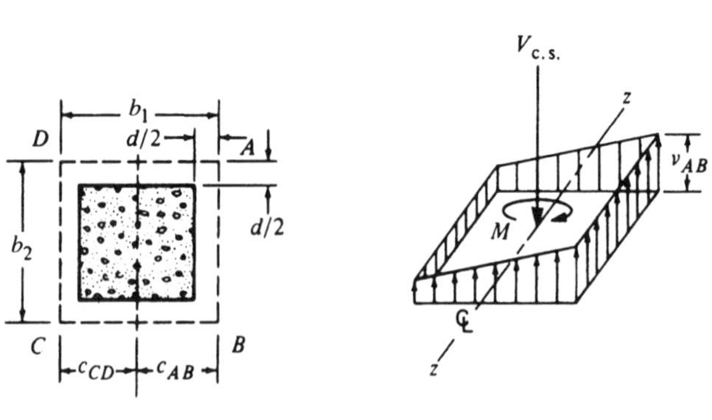

Figure 5-6 Shear at interior column.

bined shear plus moment transfer for panel with one-half live load on one side only $V_u = 5{,}000 + (140)(20)^2 + 0.5(100)(20)^2 = 81{,}000$ lb.

$$v_{u(AB)} = \frac{81{,}000}{(585)} + \frac{(0.4)(24{,}400)(12)(11.25)}{(50{,}400)} = 138 + 26$$
$$= 164 \text{ psi} < 215 \text{ psi allowable}$$

Beam shear. An investigation of beam shear is required (Section 11.12.1.1). Although it might be critical for long rectangular panels, it is unnecessary in square flat panels. In this example, $b_o = 240$ in.; $A_c = b_o d = 240 \times 6.5 = 1560$ in.^{2} Factored shear is a maximum at a distance d from the face of the first interior column, $V_u = 1/2$ exterior panel load + the increase from the exterior column moment adjustment (See Figs. 5-6 and 5-7).

$$V_u = 49{,}900 - (0.240)(20)(6.5/12) = 47{,}300 \text{ lb}$$

$$\text{Design shear strength } \phi V_c = \phi 2\sqrt{f_c'}\, b_w d = (0.85)(2)\sqrt{4000}\,(1560)$$
$$= 167{,}700 \text{ lb} > V_u = 47{,}300 \text{ lb, O.K.}$$

Step 7. Distribution of panel moments to column and middle strips

The distribution factors are based upon elastic conditions of distribution measured in extensive load tests. With no beams, $\alpha_1 = 0$ (Section 13.0), and $\beta_t = 0$ if the attached torsional member is a small strip of slab as in the example. For square panels, $c_1 = c_2$, the moments are distributed in the following percentages for the example used:

Location	Panel Moment	Column Strip	Middle Strip	Reference Sections
Face of ext. column	−54	100%	0%	13.6.4.2
+ Moment ext. panel	+108	60%	40%	13.6.4.4
Face of 1st int. col.	−148	75%	25%	13.6.4.1
+ Moment int. panels	$0.35\ M_o$	60%	40%	13.6.4.4
Face other int. cols.	$0.65\ M_o$	75%	25%	13.6.4.1

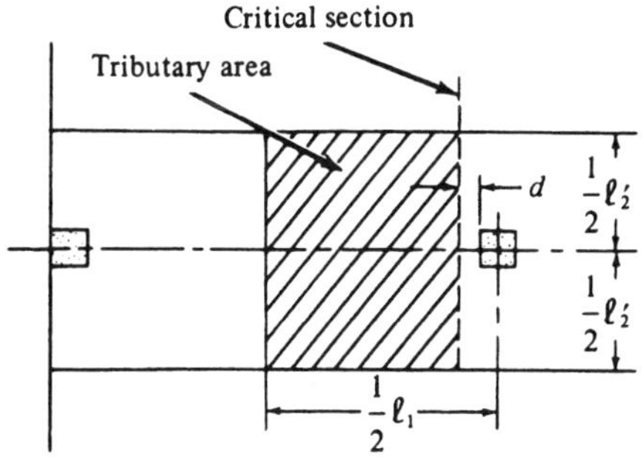

Figure 5-7 Beam shear.

Step 8. Selection of rebars (Section 13.3)

Computation of the steel areas required follows principles familiar to all users of the previous Codes. This final step in design is identical whether the factored moments have been determined by direct design or by equivalent frame analysis. Where there are two approximately identical crossing layers of bars required, as in square symmetrical panels at centers of the spans or at interior columns, use of the average depth is suggested for computations. Even if the two spans, ℓ_1 and ℓ_2, differ by a small amount, a conservative design can be based upon the longer span and both layers of bars can be made identical. This approach saves design computation time, time in specification of the steel placing sequence, and the time for field inspection. It will also simplify fabrication and reduce field placing time (and avoid possible costly errors) for cost savings to all concerned.

The top steel in the column strips must be placed such that the following fractions of the unbalanced moments given by

$$\gamma_f = \frac{1}{1 + 2/3\sqrt{b_1/b_2}} \qquad \text{(Code Equation 13-1)}$$

will be transferred by flexure in a slab width equal to $c_2 + 3h$ (Section 13.5.3.2). At an exterior column this can require nearly as much steel as required for the slab moment at the face of the column. For the direct design example, the steel requirements at the exterior column are as follows:

$$M_u(\text{Face of column}) = 54 \text{ ft-kips}$$

$$A_s = M_u/\phi f_y jd = (54)(12)/(0.9)(60)(0.98)(7)$$

$$= 1.75 \text{ in.}^2 \; 9\text{-\#4@}120/9 = 13.3 \text{ in. centers}$$

M_u (unbalanced moment to be transferred between slab and edge column)

$$= 0.30M_o \text{ (Section 13.6.3.6)} \quad \text{(see Fig. 5-5)}$$

$$= 0.30(209) = 62.7 \text{ ft-kips}$$

$$A_s = M_u\gamma_f/\phi f_y jd = (62.7)(12)(0.619)/[(0.9)(60)(0.94)(7)]$$

$$= 1.31 \text{ in.}^2 \; 7\text{-\#4@}40/7 = 5.7 \text{ in. centers}$$

At the exterior column place 7-#4 top bars in a 40 in. wide (16 + 8 × 3) width of column strip centered over the columns and provide minimum steel (#4@16 in. centers in each of the 40 in. wide remaining strips (Section 13.3.2)).

The typical detail drawing (Code Fig. 13.3.8) prescribes the minimum extensions, and cutoff points for flexural reinforcement in two-way slabs without beams. For greater ease in detailing, fabricating, and placing, simpler combinations are shown. Unfortunately one design requirement shown could increase time for design, detailing, and placing—the specification of all extensions into the span and cutoff points on the basis of the clear span. With columns of variable size and shape, some offset from the column centerlines is permitted, and perhaps being reduced in size for upper floors, this requirement literally followed could become a nightmare for all concerned. The authors would advise users to show a typical bar schedule on their structural drawings using all straight bars (end hooks excepted) with dimensions converted conservatively to percentages of center-to-center spans and conservatively referenced to same. See Fig. 5-8 for an example based on our sample problem. Conversion of the Code coefficients for bar extensions based on clear span and measured from the face of supports, to the basis of centerline span, ℓ_1, measured from centerline is easily accomplished. Let Code coefficient = k', new coefficient = k, and the average depth of column at each end of span = $ck\ell_1 = k'\ell_n + c/2 = k'(\ell_1 - c) + c/2$. $0.20 \leqslant k' \leqslant 0.33$, and $0.17\, c \leqslant (-k' + 1/2)c \leqslant 0.30\, c$. For 16 in. square columns, $c = 0.067\,\ell_1$ and the correction varies from a minimum of 0.006 to 0.011 times ℓ_1. Use $0.01\ \ell_1 k = k' + 0.01$.

In computing required steel areas and selecting bar sizes, the following controls will ensure conformance to the Code and a practical design:

1. Minimum $\rho = 0.0018\, bh$ for Grade 60 bars (Section 7.12) for either top or bottom steel, or where both are fully developed, the sum of the two. These minima apply separately in each direction.
2. Maximum bar spacing is $2h$ (Section 13.3.2), but not to exceed 18 in. (Section 7.6.5). There are no special Code requirements for crack control in two-way slabs. Note that in the crack control provisions for beams and one-way slabs (Section 10.6.2), the distribution of flexural reinforcement in two-way slabs must conform to the requirements in Section 13.3.
3. Maximum top bar spacing at all interior locations subject to construction traffic (no Code limit). The CRSI recommends not less than #4 @ 12 to provide adequate rigidity and to avoid displacement of top bars with standard bar support layouts under ordinary foot traffic.*
4. Maximum ρ is limited to $0.75\ \rho_b$ (Section 10.3.3); however, the authors recommend that $\rho_{\max} \leqq 0\ 50\ \rho_b$ to permit redistribution (Section 8.4), to provide ductility, to avoid too flexible systems subject to objectionable vibration or deflection, and for a practical balance to achieve overall economy of materials, construction, and design time.
5. Size of bars. Generally, the largest size of bars which will satisfy the maximum limits on spacing will provide overall economy. Critical dimensions which limit size are the thickness of plate available for hooks

---

* *CRSI Manual of Standard Practice.*

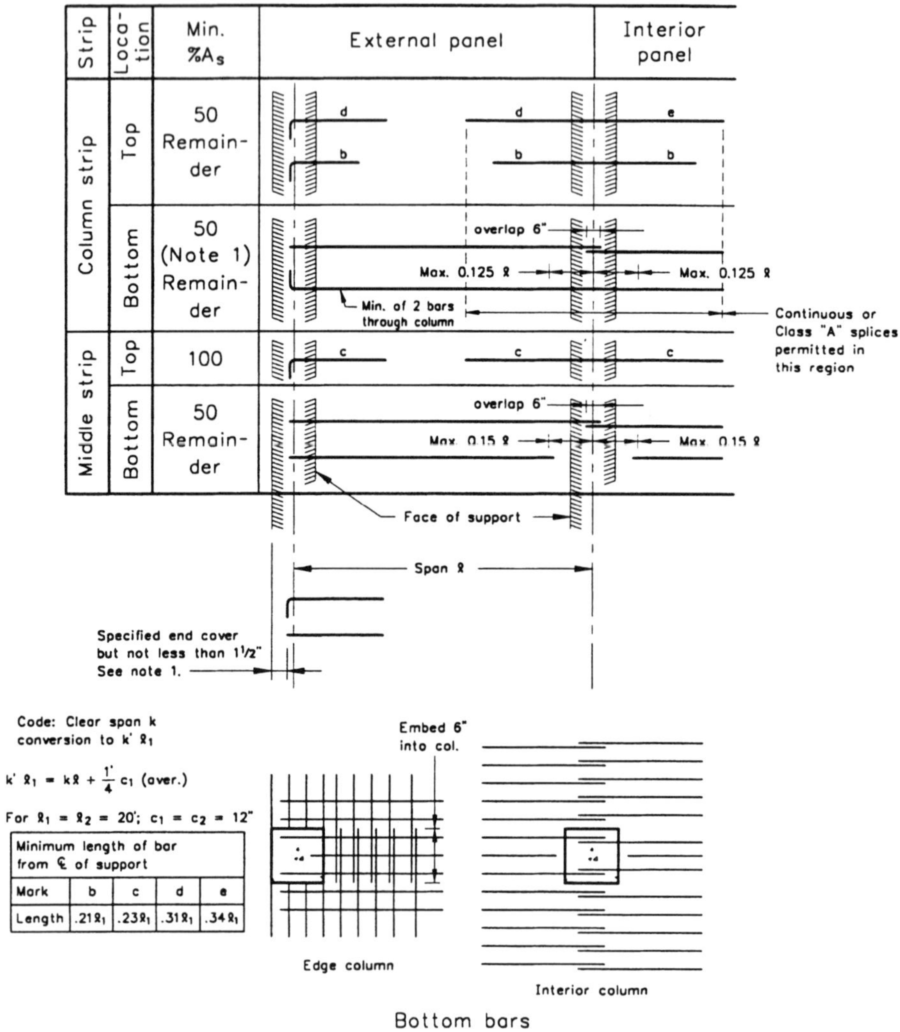

Notes:

1. Place alternate lengths of column strip bottom bars except within the width of the column verticals. Use the shorter length bars in this width to clear the column verticals as shown, except at edge columns, embed 6″ into edge columns.
2. Place half of the column strip top bars within the middle third of the column strip.
3. Place alternate lengths of middle strip bottom bars.

Figure 5-8 Flat plate reinforcement—recommended details for laterally braced frames only.

and the distance from the critical section to edge of plate. In the example, this distance for embedment of top bars at the edge is equal to the column side dimensions, *c*, minus the end cover. For full efficiency, the top bars must be fully developed at the critical section for moment, the inner face of exterior columns (Section 13.6.3.1). When excess steel area must be provided to satisfy minimum area or maximum spacing requirements, the required development length based on full f_y may be reduced in the ratio of A_s req'd/A_s used (Section 12.2.4). See Chapter 13, tabulation of end hook embedments, for full f_y. Cantilevered edges afford the designer great flexibility here in selecting the bar size and providing space for full development.

Bars with end hooks that will not fit within the slab depth have either caused job delays or resulted in loss of depth assumed for design when tilted. The practice by some designers of specifying hooks bent sideways for such conditions increases fabricating costs disproportionately to the benefit obtained.

ROOF CONSTRUCTION

The principal change in design from floor slabs to roof slabs results from the loss of the column above the flat plate. Another problem is the connection of column to slab for transmission of moment by flexure at the exterior column particularly (see suggestions, Fig. 5-9). Note that a spandrel beam may be used to advantage in the roof slab even if not required on floors below, particularly with flush edges (no cantilever), or that a cantilever at the roof level has especial structural utility.

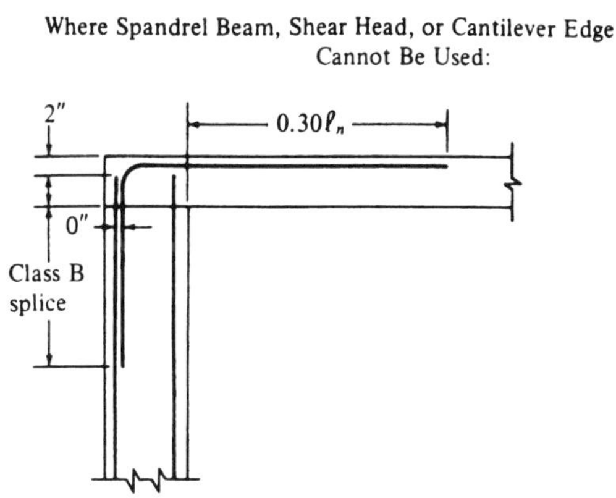

Figure 5-9 Roof plate design adjustment.

EQUIVALENT FRAME METHOD

References

In addition to the principal reference (Section 13.7), the user will find it necessary to refer to notations (Section 13.0), definitions (Section 13.2), and provisions for transfer of unbalanced bending moment and shear between slab and column (Sections 13.5.3 and 13.5.4). All of these provisions are applicable to both direct design and equivalent frame methods. Sections 13.5.3 and 13.5.4 refer the user to Chapter 11, "Shear and Torsion." Formulas for shear computations are also located in the Code Commentary on Chapter 11.

The equivalent frame method of elastic analysis for two-way slabs is based on studies reported in Commentary References 13.18, 13.19, and 13.20.

Details of the equivalent frame method that were given in the ACI 318-89 Commentary were removed from ACI 318-95 because of the number of computer programs available for the elastic analysis of two-way slabs utilizing the equivalent frame method. However, equivalent frame manual design examples are given in this book to familarize the designer with the fundamentals of the equivalent frame method of analysis that is explained in the above references.

Equivalent Frame Concept

The equivalent frame method of a flat plate design is an elastic analysis of a structural frame consisting of a row of column members and of horizontal slab members that are one panel in width and extend transversely each side of the centerline of the row of equivalent columns halfway to the next adjacent row of columns. An equivalent column can be used for vertical load analysis (Commentary R13.7.4). The equivalent column concept is necessary because of the fact that moments "leak" around the column which was not recognized in Codes before 1971. This concept can best be explained as follows.

If a beam 2 ft wide framed into an exterior column also 2 ft wide and of infinite flexural stiffness in the plane of the beam, there would be no rotation of the joint within the column. The negative moment in the beam at the face of the column would equal the fixed end moment. Under this concept an equivalent column is defined as the column and any attached transverse members which contribute torsional resistance to the joint. In this case there are no torsional members and the joint is complete within the width of the column. For this case then, the stiffness of the equivalent column is equal to the stiffness of the column alone.

If a two-way slab 20 ft wide framed into the same column, however, the condition of no rotation of the joint would exist for only the central portion of the slab 2 ft wide at the column (see Fig. 5-10). For the remaining two portions of slab extending 9 ft transversely to the column on each side, rotation of the slab would occur and increase from zero at the side face of the column to a maxi-

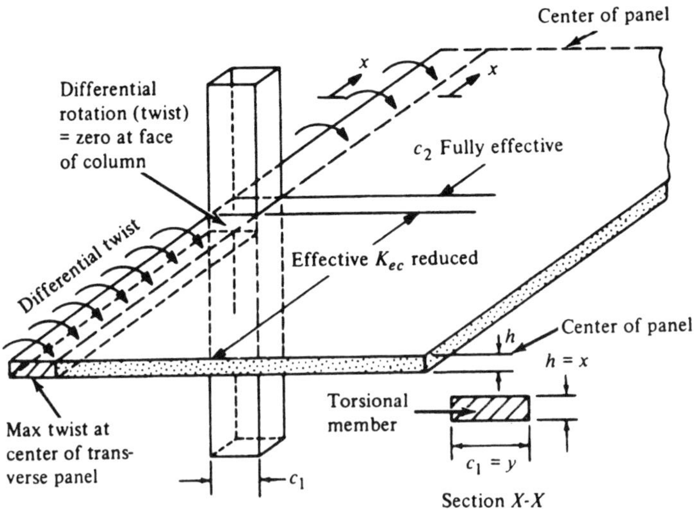

Figure 5-10 Equivalent column concept.

mum at the end of each 9 ft portion of slab. Because of this rotation, negative moment at the face of the column would be less than the full fixed end moment. Obviously, the equivalent (or effective) stiffness of the exterior column is decreased by the rotation of the attached transverse members.

The equivalent column stiffness, K_{ec}, is, therefore, always less at a joint with a two-way slab than the stiffness of the column itself, K_c, since the differential torsional rotation between slab and column will occur unless the torsional members possess infinite torsional stiffness. The flexibility (reciprocal of stiffness) of the equivalent column is increased by the flexibility of the torsional member (see Fig. 5-11 and 5-12). Note that the ratio, K_{ec}/K_c approaches unity only when the torsional stiffness, K_t, is much greater than the column stiffness, K_c; when $K_t \geqslant 50\ K_c$, the chart shows that $K_{ec} \approx K_c$.

Scope of the Equivalent Frame Method

There are no limitations on the maximum ratio of live load to dead load, of longitudinal to transverse span, or of the amount that successive span lengths can differ one from the other. The analysis of the equivalent frame is made for factored live load on all spans when the (actual) live load is variable, but does not exceed 0.75 times the (actual) dead load (Section 13.7.6.2). A pattern loading sequence with three-fourths of factored live load must be used for determining maximum moments and shears when the live load specified (in the General Code) exceeds 0.75 times the dead. "Factored live load" (w_ℓ) is the specified live load multiplied by the live load factor. "Live load" and "dead load" mean specified or actual loads (see Chapter 2 and Fig. 2-1).

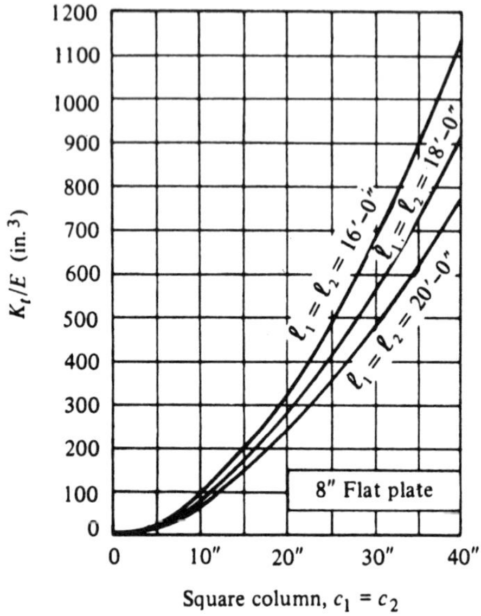

Figure 5-11 Torsional stiffness, K_t.

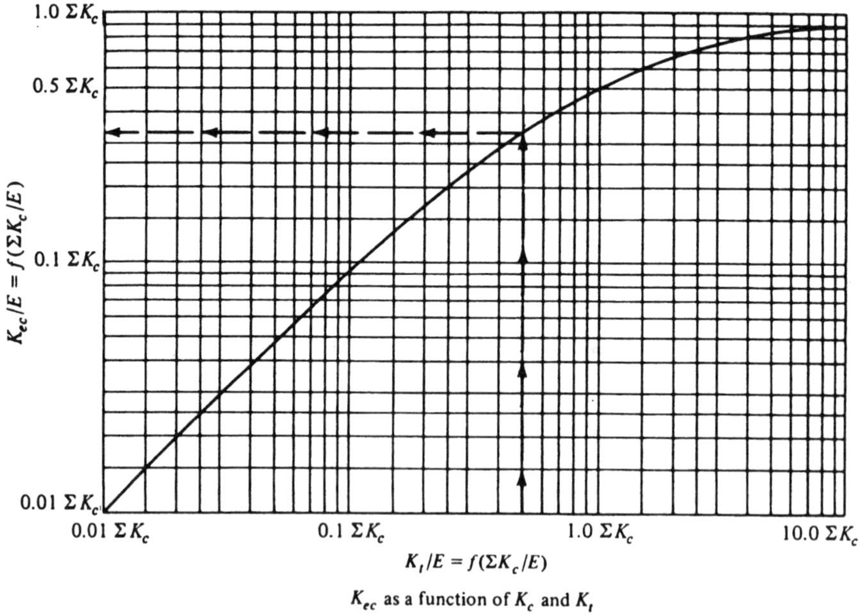

Figure 5-12 Flexural stiffness, K_{ec}, of equivalent columns. (*From the CRSI Handbook, 1992. Courtesy of the Concrete Reinforcing Steel Institute*)

Shear reinforcement consisting of bars, wires, or of steel I—or channel shapes can be used with flat plate construction to minimize column sizes (Section 11.12). For usual flat plate loads and spans, shear reinforcement is not required. If column loads are small, the minimum column size is determined by limiting the shear stress on the critical shear section, located a distance $d/2$ from the column, caused by gravity loads and unbalanced gravity load moments to the value, $\phi 4\sqrt{f'_c}$ allowed on concrete (Section 11.12.6).

The sequence of references applicable to the equivalent frame analysis is shown by use of the same example used with the direct design method—which makes possible a comparison of the two methods. (All of the operations illustrated for the equivalent frame method also apply to two-way slabs supported on beams, flat slabs with drop panels and/or capitals, which utilize two-way joists (see Chapters 6, 7, and 8). For these designs, only the additional operations required by the modifications therein are shown.)

SAMPLE PROBLEM. Superimposed factored live load, w_ℓ, of 100 psf on square panels with spans of 20 ft between column centerlines. The story height is 12 ft. The same square column cross section will be maintained from footings to roof for proven economy. Lateral forces are to be resisted principally by other stiffer elements. Assume that normal weight concrete $f'_c = 4{,}000$ psi is an economical choice in the area. Use Grade 60 bars for maximum economy in reinforcement. The outer face of edge columns will be flush with the edge of the slab.

SOLUTION. (Note that as with the direct design method, the equivalent frame method is a review. The designer must first select trial concrete dimensions and then make an equivalent frame analysis to check the assumed sizes for structural adequacy.)

Step 1. Plate thickness (h)

Same procedure as for direct design method ($h = 8'' \geqslant 7.47$).

Step 2. Trial column dimensions

For square columns $c_1 = c_2$. There are no limits on column size other than that the column be large enough to sustain the load with its eccentricity and provide a peripheral shear area that can transfer the load and a part of the unbalanced moment from the slab to the column.

If shear reinforcement is not used, a trial size exterior column can be obtained from Fig. 5-38 or by determining the area of the critical section shown in Fig. 5-4 that will limit the shear stress due to gravity load only to a conservative value, $\phi 2\sqrt{f'_c}$.

Determine trial size exterior column size.

From Fig. 5-38, for minimum interior *and* exterior columns, $c_1 = 17$ in. or required

$$A_c = \frac{(w_d + w_\ell)(\ell_2)(\ell_1/2)}{\phi(2)\sqrt{f'_c}} = \frac{[(1.4)(100) + 100](20)(20)/2}{(0.85)(2)(\sqrt{4000})}$$
$$= 446 \text{ in.}^2$$

For square exterior columns with exterior face flush with the exterior edge of the slab, the minimum size exterior square column can be calculated as follows:

$$d = h - 1.5 = 8 - 1.5 = 6.5 \text{ in.}$$
$$(2b_1 + b_2)d = 446 \text{ in.}^2$$
$$[(2)(c_1 + 3.5) + (c_1 + d)]d = 446 \text{ in.}^2$$
$$[(2)(c_1 + 3.5) + c_1 + 6.5]6.5 = 446 \text{ in.}^2$$
$$c_1 = 18.4 \text{ in.}$$

If both interior and exterior columns are made to the same size, the trial column size selected should lie between the two sizes shown for minimum exterior and interior columns shown in Figs. 5-35 through 5-40. Figure 5-38 shows minimum exterior square column, 17 in., minimum interior column. 12 in. Try 16 in. square columns for both.

A trial size for the interior column can be made as for the exterior one, except limiting the shear stress for gravity load only, V_u/A_c (neglecting unbalanced gravity load moment), to $\phi 4\sqrt{f'_c}$. A 16 in. square column should be conservative.

Step 3. Determine column stiffness, K_c

The stiffness of the exterior column or interior columns can be obtained by the simple approximations outlined in Step 2 of the direct design method. If greater accuracy is desired, see Fig. 5-13. The stiffness can be also calculated directly by use of column analogy as follows:

$$K_c = \frac{1}{A_{ac}} + \frac{Mc}{I_{ac}} \text{ (see Fig. 5-14)}$$

A_{ac} = Area of analogous column

I_{ac} = Moment of inertia of analogous column about neutral axis *x-x*

M = Moment due to a unit load at the extreme fiber of the analogous column which is located at the center of the slab = $\ell_c/2$.

$$A_{ac} = \frac{\ell_c - h}{E_c I_c}$$

$$I_{ac} = \frac{(\ell_c - h)^3}{12 E_c I_c}$$

$c = \ell_c/2$ or the distance from the neutral axis to the extreme fiber of the analogous column.

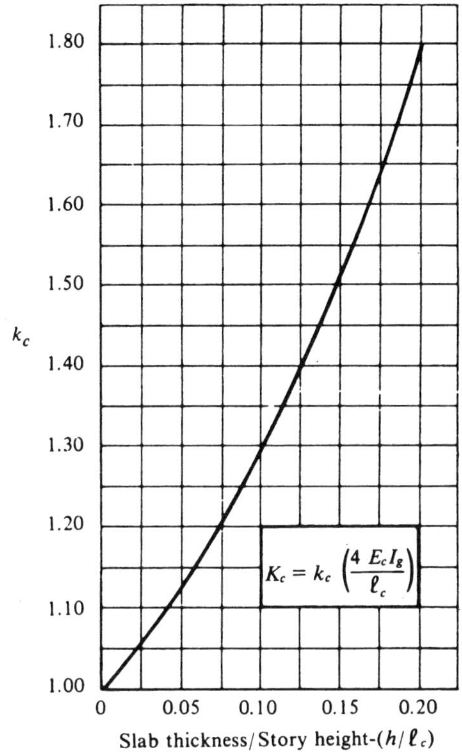

Figure 5-13 Column stiffness, K_c.

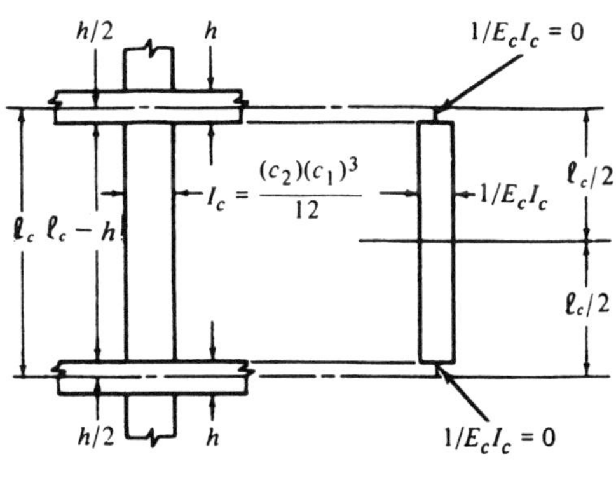

Figure 5-14 Column stiffness, K_c, by column analogy.

$$K_c = \frac{E_c I_c}{\ell_c - h} + \frac{3E_c I_c \ell_c^2}{(\ell_c - h)^3}$$

$$K_c/E_c = \frac{(I_c)}{(\ell_c - h)}\left[1 + \frac{3\ell_c^2}{(\ell_c - h)^2}\right]$$

For the trial column size of 16 in. from Step 2, the column stiffness will be determined by the column analogy procedure and compared with the approximate method outlined in Step 2 of the direct design method.

$$K_c/E_c = \frac{5461}{(144 - 8)}\left[1 + \frac{(3)(144)^2}{(144 - 8)^2}\right]$$

$$= 175 \text{ in.}^3$$

A value of 171 in.^{3} was obtained for K_c/E_c by the authors' approximation which is less than 3% in error.

Step 4. Slab stiffness, K_s

The flexural stiffness of the slab, like the flexural stiffnesses of columns, can be determined using the simple approximate methods outlined by using Fig. 5-15 or by the column analogy method.

Numerical values for the flexural stiffness of the slab will be determined by the column analogy method (see Fig. 5-15) and compared to the results obtained from Figs. 5-16 (*a*) and 5-16 (*b*).

$$K_s = \frac{1}{A_{ac}} + \frac{Mc}{I_{ac}}$$

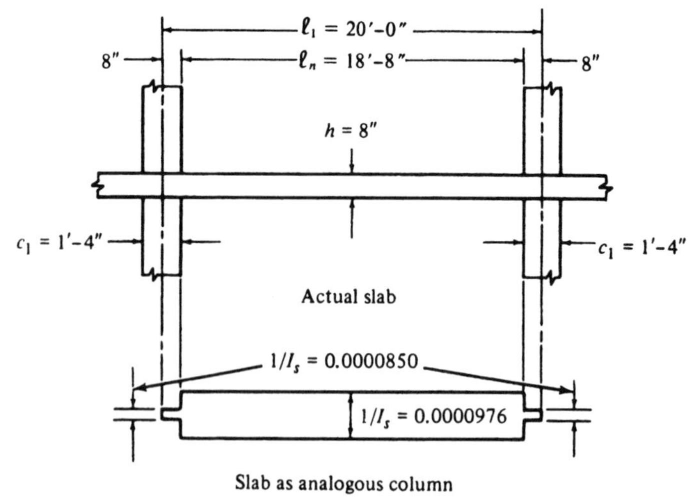

Figure 5-15 Slab stiffness, K_s, by column analogy.

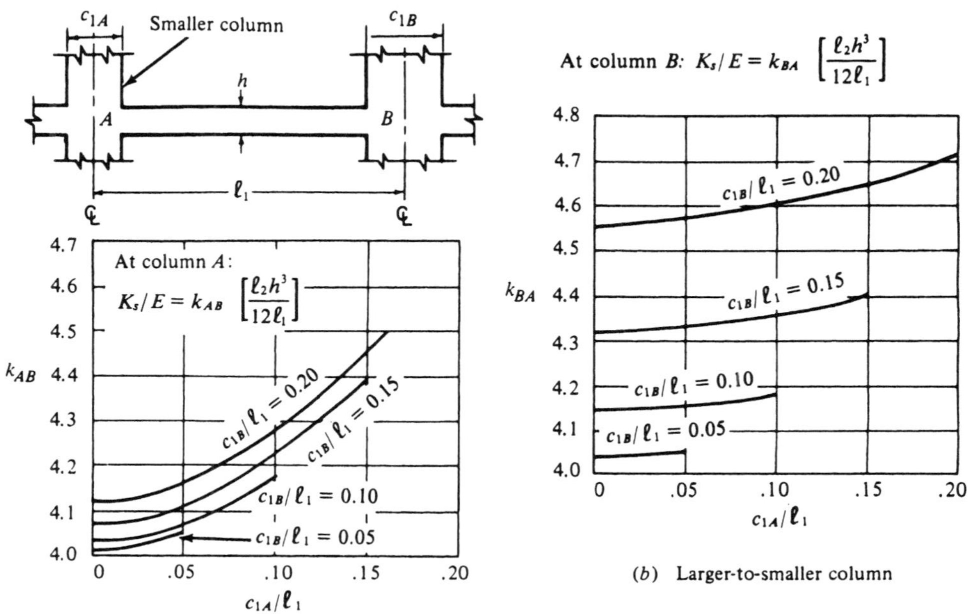

(a) Smaller-to-larger column

Figure 5-16 Flexural stiffness, K_s, for flat plates.

$$I_g \text{ (slab between columns)} = \frac{(240)(8)^3}{12} = 10{,}240 \text{ in.}^4;\ ^1\!/_{10{,}240} = 0.0000977$$

$$I_g \text{ (slab at column)} = \frac{I_g}{\left[1 - \dfrac{c_2}{\ell_2}\right]^2} = \frac{10{,}240}{\left[1 - \dfrac{16}{240}\right]^2}$$

$$= 11{,}755 \text{ in.}^4;\ 1/11{,}755 = 0.0000851$$

$$\begin{aligned} A_{ac} &= (0.0000977)(18.66)(12) &= 0.02186 \\ &\ (2)(0.0000851)(0.666)(12) &= \underline{0.00136} \\ & & 0.02322 \end{aligned}$$

$$\begin{aligned} I_{ac} &= (0.0000851)(240)^3/12 = 98.0 \\ &+ (0.0000126)(224)^3/12 = \underline{11.8} \\ & \qquad\qquad 109.8 \end{aligned}$$

$$K_s/E_c = \frac{1}{A_{ac}} + \frac{(1)(\ell_1/2)(\ell_1/2)}{I_{ac}}$$

$$= \frac{1}{0.02322} + \frac{(1)(120)(120)}{109.8}$$

$$= 43 + 131$$

$$= 174 \text{ in.}^3$$

K_s using Figs. 5-16(*a*) and 5-16(*b*)

$$c_{1A}/\ell_1 = c_{1B}/\ell_1 = \frac{16}{240} = 0.0667$$

$$k_{AB} = k_{BA} = 4.09 \text{ (See Figs. 5-16(a) or 5-16(b))}$$

$$I_g = \frac{\ell_2 h^3}{12} = \frac{(1)(20)(12)(8)^3}{12} = 10{,}240 \text{ in.}^4$$

$$\frac{K_s}{E_s} = \frac{k_{AB} I_s}{\ell_1} = \frac{k_{AB}(10{,}240)}{(20)(12)}$$

$$= (4.09)(42.7)$$

$$= 175 \text{ in.}^3 \text{ (174 in.}^3 \text{ obtained by column analogy)}$$

Step 5. Stiffness of exterior equivalent column, K_{ec}.

$$\frac{K_c}{E_c} = 175 \text{ in.}^3 \text{ (Refer to step 3 of equivalent frame method)}$$

An approximate expression for the stiffness of the torsional member at the exterior column shown in Fig. 5-10 is given below (Commentary R13.7.7.1).

$$K_t/E_s = \Sigma\, 9C/[\ell_2(1 - c_2/\ell_2)^3]$$

where C is the cross-sectional constant to define torsional properties of the slab plate section at each side of the exterior column (Section 13.0).

$$\begin{aligned} C &= \Sigma\,(1 - 0.63x/y)(x^3y/3) \\ &= 1[1 - 0.63(8/16)](8)^3(16)/3 = 1870 \end{aligned}$$

$$K_t/E_s = (2 \times 9)(1870)/(240)(1 - 16/240)^3 = 173 \text{ (See also Fig. 5-11).}$$

The authors recommend preparation of similar charts for the various plate thicknesses required in routine practice.

$$\frac{K_{ec}}{E_c} = \frac{\Sigma K_c}{1 + \dfrac{\Sigma K_c}{K_t}} = \frac{2 \times 175}{1 + \dfrac{(2)(175)}{173}}$$

$$= 115 \text{ in.}^3$$

Step 6. Moment distribution factors.

For distributing slab moments at exterior and interior columns. At the exterior column, the distribution factor is as follows:

$$\text{DF (Ext. Col.)} = \frac{K_s}{K_s + K_{ec}} = \frac{174}{174 + 115} = \frac{174}{289} = 0.602$$

$$\text{DF (Int. Col.)} = \frac{K_s}{K_s + K_{ec}} = \frac{174}{(2)(174) + 115} = \frac{174}{463} = 0.376$$

Step 7. Factored loads for moment distribution.

The specified live load is equal to 100/1.7 or 59 psf. The dead load is computed: $8/12 \times 150$ or 100 psf. The live load does not exceed three-quarters of the dead load ($59 < 0.75 \times 100$). The equivalent frame can be analyzed one floor at a time with the far ends of the columns assumed fixed for full factored gravity load on all spans and pattern loading neglected (Section 13.7.6.2).

Frame loads per foot of span (across 20 ft width ℓ_2)

$$w_d = \frac{(1.4)(100)\ell_2}{1000} = \frac{(1.4)(100)(20)}{1000}$$

$$= 2.80 \text{ klf}$$

$$w_\ell = \frac{(59)(1.7)\ell_2}{1000} = \frac{(100)(20)}{1000}$$

$$= 2.00 \text{ klf}$$

Step 8. Carryover factors.

COF_{AB} and COF_{BA} for the equivalent frame elastic analysis will be obtained by the column analogy method (see Fig. 5-15).

$$COF_{AB} = COF_{BA} \text{ (for column analogy method)}$$

$$= \frac{\dfrac{1}{A_{ac}} - \dfrac{(1)(\ell_1/2)(\ell_1/2)}{I_{ac}}}{\dfrac{1}{A_{ac}} + \dfrac{(1)(\ell_1/2)(\ell_1/2)}{I_{ac}}} = \frac{43 - 131}{43 + 131}$$

$$= \frac{-88}{174} = -0.506$$

Step 9. Fixed end moment factors.

M_{AB} and M_{BA} can be obtained by the column analogy method (see Fig. 5-17); area M/I diagram (P_{ac})

1. $\frac{(2)}{3}(18.67)(12)(0.245) = 36.59$

2. $(18.67)(12)(0.036) = 8.06$

3. $(2)\frac{(1)}{2}(0.667)(12)(0.0314) = 0.25$

4. $(2)\frac{(2)}{3}(0.667)(12)(0.0004) = \text{—}$

$\overline{\quad 44.90 \quad}$

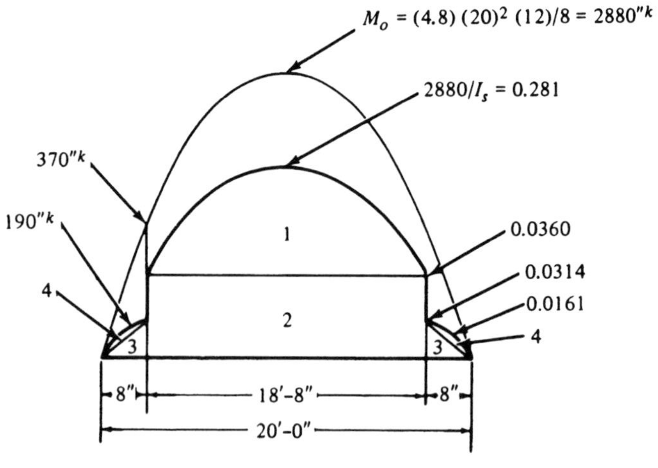

Figure 5-17 Slab fixed end moments by column analogy.

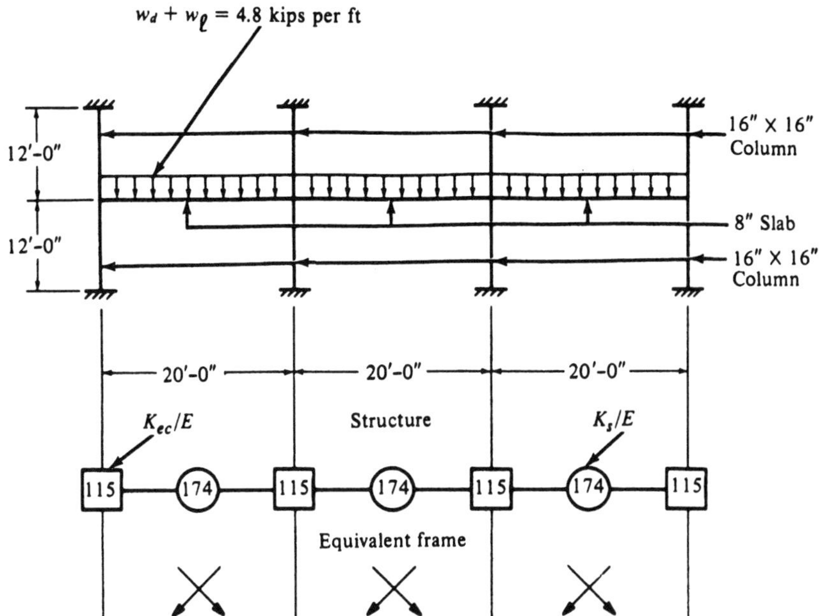

	Span 1 (left)	Span 1 (right)	Span 2 (left)	Span 2 (right)	Span 3 (left)	Span 3 (right)
COF	-0.506		-0.506		-0.506	
DF	0.602	0.376	0.376	0.376	0.376	0.602
FEM	-161	-161	-161	-161	-161	-161
D	+97	—	—	—	—	+97
CO	—	-49	—	—	-49	—
D	—	+18	-18	-18	+18	—
CO	-9	—	+9	+9	—	-9
D	+5	+3	-3	-3	+3	+5
CO	-2	-3	+2	+2	-3	-2
D	+1	—		—	—	+1
M_u	-69'k	-192	-171'k	-171	-192'k	-69'k
	+113'k		+69'k		+113'k	
	48.0	48.0	48.0	48.0	48.0	48.0
	-6.2	+6.2	—	—	+6.2	-6.2
V_u	41.8	54.2	48.0	48.0	54.2	41.8

Figure 5-18 Moment distribution—100% load on all spans.

$$FEM_{AB} = FEM_{BA} \text{ (symmetrical moment diagram)} = \frac{P_{ac}}{A_{ac}} = \frac{44.9}{0.0232}$$

$$= 1935 \text{ in.-kips} = 161 \text{ ft-kips}$$

Step 10. Moment distribution analyses.

1. Moment distribution for full load on all spans (see Fig. 5-18).
2. Compare results to those by the direct design method: see Fig. 5-5.

Step 11.

Compute slab shear at exterior trial column. The same procedure shown in Step 5 of the direct design method is used in the equivalent frame method (see Fig. 5-5). The shear (V_{cs}) is equal to the exterior column reaction which includes the slab load between the column centerline and the exterior face of the slab at the exterior face of the column. The moment at the centroid of the critical section is equal to the unbalanced moment at the column centerline minus (V_{cs}) (e) where e is equal to the distance between the column centerline and the centroid of the critical section. For shears and moments, see Fig. 5-19.

$$V_{cs} = 41.8 + (4.8)\left(\frac{8}{12}\right)$$

$$= 41.8 + 3.2 = 45.0 \text{ kips}$$

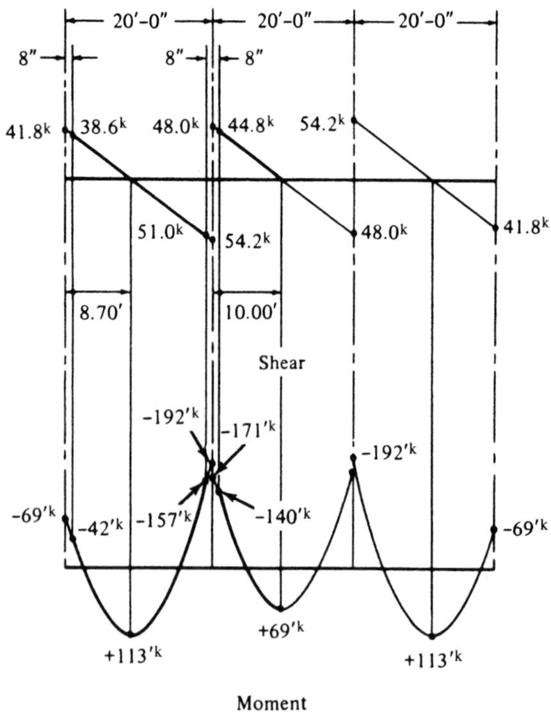

Figure 5-19 Moment distribution results—100% load on all spans.

$$M_{cs} = 69 - (4.8)\left(\frac{8}{12}\right)\left(\frac{4}{12}\right) - (V_{cs})(e)$$

$$= 69 - 1 - (45.0)\left(\frac{5.18}{12}\right) = 69 - 1 - 19$$

$$= 49 \text{ ft-kips}$$

Using the geometrical properties of the critical section as determined in Steps 5 and 6 of the direct design method (see Fig. 5-4), the shear stresses at the exterior column based on the equivalent frame analysis method are calculated as follows:

$$v_{u(AB)} = \frac{V_{cs}}{A_c} + \frac{\gamma_f c_{AB}}{J_c} \quad \text{(Section 13.5.3.2)}$$

$$\gamma_f = 1/[1 + \tfrac{2}{3}\sqrt{b_1/b_2}] = 1/[1 + \tfrac{2}{3}\sqrt{19.25/22.5}] = 0.619$$

$$\gamma_v = 1 - \gamma_f = 1 - 0.619 = 0.381 \text{ (Section 13.5.3.1)}$$

$$v_{u(AB)} = \frac{45{,}000}{(396)} + \frac{(0.381)(49{,}000)(12)(6.07)}{(17{,}160)}$$

$$= 114 + 79$$

$$= 193 \text{ psi} < \phi 4\sqrt{f_c'} = (0.85)(4)\sqrt{4000} = 215 \text{ allowed.}$$

$$v_{u(CD)} = \frac{V_{cs}}{A_c} - \frac{\gamma_v M_{cs} c_{cd}}{J_c}$$

$$= 114 - \frac{(0.381)(49{,}000)(12)(13.18)}{(17{,}160)}$$

$$= 114 - 172$$

$$= -58 \text{ psi} < \phi 4\sqrt{f_c'} = 0.85(4)\sqrt{4000} = 215 \text{ allowed.}$$

Step 12. Slab shears based on interior trial column.

Where the same size columns are used, the shear stress is not usually critical for the interior columns. The maximum shear stress at the critical section of the first interior column with total load on all spans and shears and moments as shown in Fig. 5-19 is as follows:

$$V_u = 48.0 + 54.2 = 102.2 \text{ kips}$$

$$M_v = 192 - 171 = 21 \text{ ft-kips}$$

$$v_u = \frac{V_{cs}}{A_c} + \frac{\gamma_v M_{cs} c}{J_c}$$

$$= \frac{102{,}200}{585} + \frac{(0.40)(21{,}000)(12)(11.25)}{(50{,}389)}$$

$$= 174 + 23 = 197 \text{ psi} < \phi 4\sqrt{f_c'} = 215 \text{ psi}$$

The maximum shear stress based on a pattern loading with three-fourths of w_ℓ on alternate spans as shown in Fig. 5-20 is calculated as follows:

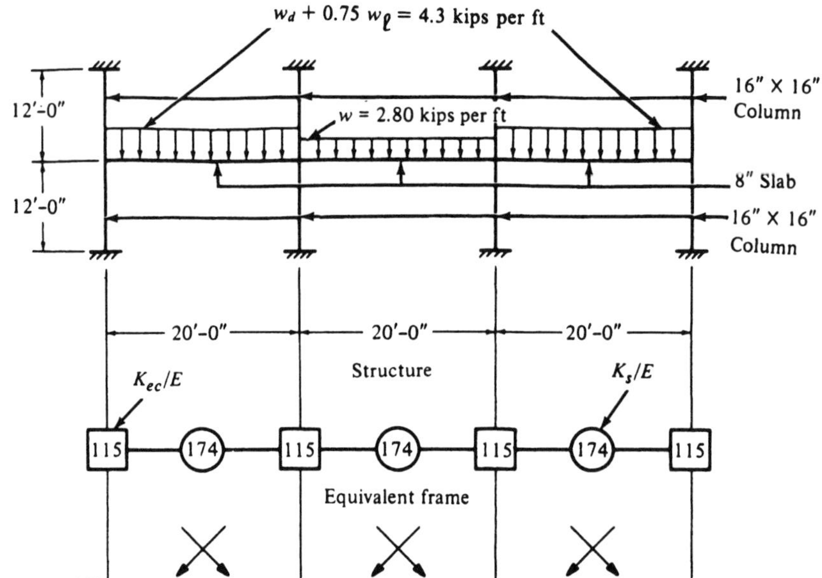

COF	-0.506		-0.506		-0.506	
DF	0.602	0.376	0.376	0.376	0.376	0.602
FEM	-144	-144	-94	-94	-144	-144
D	+87	+19	-19	-19	+19	+87
CO	-10	-44	+10	+10	-44	-10
D	+6	+20	-20	-20	+20	+6
CO	-4	-8	+8	+8	-8	-4
D	+2	+3	-3	-3	+3	+3
CO	-2	-1	+2	+2	-1	-2
D	+1	+1	-1	-1	+1	+1
M_u	-64'k	-154	-117'k	-117	-154'k	-64'k
	+108'k		+22'k		+108'k	
	43.0	43.0	28.0	28.0	43.0	43.0
	-4.5	+4.5	—	—	+4.5	-4.5
V_u	38.5	47.5	28.0	28.0	47.5	38.5

Figure 5-20 Moment distribution —75% w_ℓ on alternate spans.

$$V_{cs} = 47.5 + 28.0 = 75.5 \text{ kips}$$

$$M_{cs} = 154 - 117 = 37 \text{ ft-kips}$$

$$v_u = \frac{V_{cs}}{A_c} + \frac{\gamma_v M_{cs} c}{J_c}$$

$$= \frac{75{,}500}{585} + \frac{(0.40)(37{,}000)(12)(11.25)}{50{,}389}$$

$$= 129 + 40 = 169 \text{ psi} < \phi 4\sqrt{f_c'} = 215 \text{ psi O.K.}$$

Step 13. Beam shear.

The maximum beam shear for total factored load is as follows:

$$v_u = \frac{\left[51{,}000 - (4{,}800)\left(\frac{6.5}{12}\right)\right]}{(240)(6.5)}$$

$$= \frac{48{,}400}{(240)(6.5)}$$

$$= 31 \text{ psi} < \phi 2\sqrt{f_c'} = 108 \text{ psi O.K.}$$

Both exterior and interior trial size columns have now proved to be adequate in shear at the critical slab sections. Assuming the columns are adequate for axial load and moment, the reinforcing steel for column and middle strips of the flat plate can now be determined by the same method as demonstrated in Steps 7 and 8 of the direct design method.

Total end panel moments, obtained by both the direct design method and the equivalent frame analysis method are shown in Table 5-1.

TABLE 5-1 End Panel Moments

Method of Design	Direct Design Method	Equivalent Frame Method
$-M_u$ at face of exterior column	$-54'^k$	$-42'^k$
$+M_u$	$+108'^k$	$+113'^k$
$-M_u$ at face of 1st interior column	$-148'^k$	$-157'^k$

DESIGN MODIFICATIONS FOR SMALLER COLUMNS, LONGER SPANS AND/OR HEAVIER LOADS

Reason for Modification

The example used in the explanations for the "Direct Design Method" and the "Equivalent Frame Method" was a straightforward flat plate. Load, span, and column size were all in the range for which many flat plate designs are employed. As loads or spans become greater or smaller column sizes are required, modifications of the simple flat plate become necessary. In order to extend the range of the very economical flat plate concept, the transfer of vertical load and a part of the unbalanced moment by shear, particularly at exterior columns, becomes critical. Nearly as critical is the transfer of the remaining part of the unbalanced moment by flexure into the exterior column.

Types of Modification

Perhaps the simplest design modification to utilize flat plates for heavier loads is the cantilever edge. As the cantilever moment approaches that of an interior column (balanced layout) the problem of moment transfer to exterior column disappears and the cross-sectional area of the shear perimeter is almost evenly stressed by balanced vertical shear (see Fig. 5-32). Where the cantilevered edge layout cannot be employed, as for architectural reasons, and a flush edge is required, the 1995 Code permits at least three design modifications without resort to excessively large exterior columns: (1) structural steel or reinforcing steel shearheads (Section 11.12.4); (2) brackets (Sections 13.1.2, 13.6.2.5, 13.6.3.1); and (3) spandrel beams (Sections 13.0, 13.6.3.5, 13.6.4.5, 13.6.5, 13.6.8, 13.7.5).

Smaller size columns can sometimes be obtained by making adjustments to the amount of unbalanced moment, between slab and column, that is transferred by flexure and by shear (Section 13.5.3.3). When the fraction transferred by flexure, γ_f, is increased, the fraction transferred by shear, γ_v, will be decreased, provided that the ratio of flexural reinforcement $\rho = A_c/[(c_1 + 3h)d \leqslant \frac{3}{8}\rho_b$. For $f_c' = 4{,}000$ psi, $\rho_b = 0.0285$ and $\frac{3}{8}\rho_b = 0.0107$. For an exterior column, γ_f can be increased up to a maximum of 1.00 provided that $V_u \leqslant 0.75\ \phi V_c$ and for interior and corner columns γ_f can be increased up to 25% provided that $V_u \leqslant 0.4\,\phi V_c$.

Example of Column Size Reduction for Exterior, Interior and Corner Columns

Refer to the "Direct Design Method" example for an 8-in.-thick flat plate supporting a factored live load of 100 psf on 20-ft spans with 16-in.-square columns. It is desired to reduce the column sizes if possible. This is a trial-and-error process and as a first trial all columns will be kept the same size for uniformity and reduced from 16 in. square to 14 in. square and the slab-column connections checked for shear and flexural strength.

Exterior Span

$$M_o = w_u \ell_2 \ell_n^2/8$$

$$\ell_n = 20 \text{ ft} - 14/12 = 18.83 \text{ ft}$$

$$M_o = 0.24(20)(18.83)^2/8 = 213 \text{ ft-kips}$$

$$-M_u \text{ (exterior column face)} = -0.26(213) = -55 \text{ ft-kips}$$

$$+M_u \text{ (positive)} = +0.52(213) = +111 \text{ ft-kips}$$

$$-M_u \text{ (interior column face)} = -0.70(213) = -149 \text{ ft-kips}$$

$$-M_{un} \text{ (unbalanced)} = -0.30(213) = -64 \text{ ft-kips}$$

Exterior Column Critical Section

$$V_u = 4.80(18.83)/2 - (149 - 55)/18.83 + 4.8(14)/12$$

$$= 45.19 - 4.99 + 5.60 = 45.80 \text{ kips}$$

$$A_c = [(14 + 3.25)(2) + (14 + 6.5)](6.5) = 357.5 \text{ in.}^2$$

$$\gamma_f = 1.00 \text{ (Assuming 100\% } M_{un} \text{ transferred by flexure)}$$

$$\gamma_v = 0$$

$$0.75\phi V_c = (0.75)(0.85)(4)[(\sqrt{4000})/(1000)](357.5)$$

$$= 57.7 \text{ kips} > V_u = 45.8 \text{ kips OK}$$

$$A_s = M_u/(\phi f_y jd) = 64(12)/[0.9(60)(0.91)(6.5)]$$

$$= 2.40 \text{ in.}^2 \text{ Use 8-\#5} = 2.48 \text{ in.}^2 \text{ @4'' spacing in 38 in. width.}$$

$$\rho = 2.40/[(38)(6.5)] = 0.0097 < 0.375\rho_b = 0.0107 \text{ OK}$$

Interior Column Critical Section

$$V_u = 0.24(20)^2 + 4.99 = 101.0. \text{ kips (1st interior column)}$$

$$M_{un} = 0.07[(w_d + 0.5w_1)\ell_2\ell_n^2 - w_d'\ell_2'(\ell_n)^2] \text{ (Section 13.6.9.2)}$$

$$= 0.07\{[0.14 + (0.5)(0.10)](20)(18.83)^2 - 0.14(20)(18.83)^2\}$$

$$= 0.07[(0.19)(20)(18.83)^2 - 0.14(20)(18.33)^2]$$

$$= 0.07(1347 - 993)$$

$$= 24.78 \text{ ft-kips}$$

$$A_c = (14 + 6.5)(4)(6.5) = 533 \text{ in.}^2$$

$$J_c = d(b_1)^3/6 + (b_1)d^3/6 + d(b_2)(b_1)^2 \text{ (Commentary R11.12.6.2)}$$

$$= 6.5(20.5)^3/6 + 20.5(6.5)^3/6 + 6.5(20.5)(20.5)^2/2$$

$$= 9{,}333 + 938 + 27{,}999 = 38{,}270 \text{ in.}^4$$

$$\gamma_f = 1/[1 + (\tfrac{2}{3})(\sqrt{b_1/b_2})] \text{ (Assuming no increase)}$$
$$= 1/[1 + (\tfrac{2}{3}\sqrt{20.5/20.5})] = 0.60 \text{ (Section 13.5.3.2)}$$
$$\gamma_v = 1 - \gamma_f = 1 - 0.60 = 0.40$$
$$\phi v_n = \phi 4(\sqrt{4000}) = (0.85)(4)(\sqrt{4000})$$
$$= 215 \text{ psi}$$
$$v_u = V_u/A_c + \gamma_v M_{un} c/J_c \text{ (Commentary R11.12.6.2)}$$
$$= 101{,}000/533 + 0.40(24{,}780)(12)(20.5/2)/38{,}270$$
$$= 189 + 32$$
$$= 221 \text{ psi} \approx \phi v_n = 215 \text{ psi OK (within 3\%)}$$

Corner Column (See Fig. 5-21)

$$M_o = w_u \ell_2 \ell_n^2/8$$
$$\ell_n = 20 - 14/12 = 18.83 \text{ ft.}$$
$$w_u = (0.240)(10.58) \qquad = 2.54 \text{ (floor load)}$$
$$+ (0.040)(10)(1.4) \qquad = \underline{0.56} \text{ (exterior wall)}$$
$$3.10 \text{ kips/ft}$$

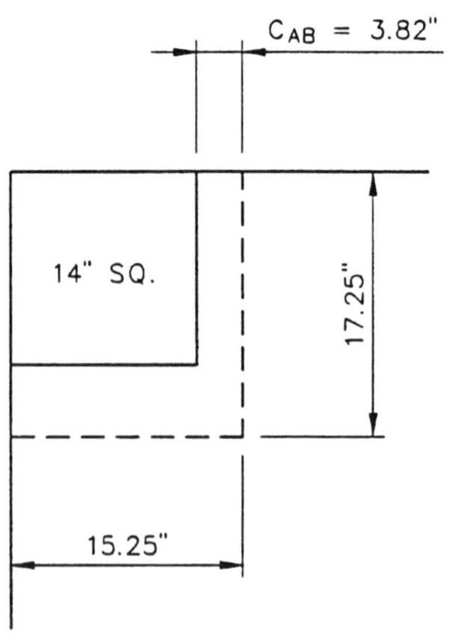

Figure 5-21 Corner Column.

$$M_o = (3.10)(18.83)^2/8 = 137 \text{ ft-kips}$$

$$-M_u \text{ (exterior column face)} = -(0.26)(137) = -36 \text{ ft-kips}$$

$$+M_u \text{ (positive)} = +(0.52)(137) = +71 \text{ ft-kips}$$

$$-M_u \text{ (interior column face)} = -(0.70)(137) = -96 \text{ ft-kips}$$

$$-M_{un} \text{ (unbalanced)} = -(0.30)(137) = -41 \text{ ft-kips}$$

$$V_u = (3.10)(18.32)/2 - (96 - 36)/18.83 + (3.10)(14/12)$$
$$+ (0.56)(18.83)/2 = 28.40 - 3.19 + 3.62 + 5.27$$
$$= 34.10 \text{ kips}$$

$$A_c = (b_1 + b_2)d = (17.25 + 17.25)(6.5) = 224.25 \text{ in.}^2$$

$$c = [(b_1 d)(b_1/2)]/A_c$$
$$= [(17.25)(6.5)(17.25/2)]/224.25 = 4.31 \text{ in.}$$

$$\gamma_f = 1.00 \text{ (Assuming 100\% } M_{un} \text{ transferred by flexure)}$$

$$\gamma_v = 0$$

$$0.75\phi V_c = (0.75)(0.85)(4)[(\sqrt{4000})/(1000)](224.25)$$
$$= 36.2 \text{ kips} > V_u = 34.10 \text{ kips OK}$$

$$A_s = M_{un}/(\phi f_y jd) = (41)(12)/[(0.90)(60)(0.92)(6.5)]$$
$$= 1.52 \text{ in.}^2 \text{ Use 5-\#5} = 1.55 \text{ in.}^2 \text{ @5.2'' spacing in 26 in.}$$
$$\text{width } (c_2 + 1.5h) = 14 + (1.5)(8) = 26 \text{ in.}$$

$$\rho = 1.52/[(26)(6.5)] = 0.0090 < 0.375\rho_b = 0.107 \text{ OK}$$

Effect of Load and Story Height on Column Size

As an indication of the size increase required for exterior columns without these modifications, consider the example used in "Direct Design." (Story height 12′-0″, 8 in. plate floor on 20 ft square panels with factored live load 100 psf). Minimum column sizes for shear and stiffness were computed for a range of factored live loads of 50 to 400 psf for floors with 8′-0″, 12′-0″, and 22′-8″ story heights using the modified stiffness method of ACI 318-89 Commentary R.13.6.3.3 utilizing α_{ec}. (The 22′-8″ story height column stiffness for a floor slab is about equivalent to a roof slab with 12′-0″ story height.) See Table 5-2.

Shearheads and Shear Reinforcement

The Code has extensive provisions for design of shear reinforcement in flat plates around interior columns (Section 11.12). The term "shearhead" is defined

TABLE 5-2 Effect of Load and Story Height on Flush Edge Square Column Sizes Required for Shear

w_ℓ Factored Live Load psf	$\ell_c = 8'\text{-}0''$			$\ell_c = 12'\text{-}0''$			$\ell_c = 22'\text{-}0''$		
	EXT. COLUMN		Int. Column Size $c_1 = c_2$ in.	EXT. COLUMN		Int. Column Size $c_1 = c_2$ in.	EXT. COLUMN		Int. Column Size $c_1 = c_2$ in.
	Size $c_1 = c_2$ in.	$-M_u$ ft-kips		Size $c_1 = c_2$ in.	$-M_u$ ft-kips		Size $c_1 = c_2$ in.	$-M_u$ ft-kips	
50	10	21	10	10	17	10	10	11	10
100	18	67	12	17	59	12	10	14	12
150	24	99	15	23	94	16	22	83	17
200	28	123	19	28	121	19	27	114	19
250	33	147	22	32	144	22	32	141	22
300	36	166	26	36	165	26	36	162	26

in the Code (Section 11.12.4.1) as a full penetration welded assembly of steel "shapes" at right angles and continuous through the column section. Provision for bar or wire shear reinforcement is also made (Section 11.12.3). Critical section sketches for common styles are illustrated (Commentary Fig. R11.12.4.7). If bars or wires are used as shear reinforcement, the maximum nominal shear strength, V_n, permitted on critical section $d/2$ from the face of the column can be increased from $4\sqrt{f'_c}\, b_o d$ to $6\sqrt{f'_c}\, b_o d$ (Section 11.12.3.2). The nominal shear strength of the concrete, V_c, shall not exceed $2\sqrt{f'_c}\, b_o d$ at the critical section or at any successive critical sections where the shear reinforcement is reduced or ended. If steel I- or channel shapes are used as shearheads, the nominal shear strength, V_n, can be further increased to $7\sqrt{f'_c}\, b_o d$ at the critical section $d/2$ from the column. Shear on an assumed critical section located inside the end of the shearhead reinforcement (Section 11.12.4.7) must not exceed $4\sqrt{f'_c}\, b_o d$ (Section 11.12.4.8). These increases, of course, are allowed only if the shearhead is designed as prescribed (Section 11.12.4), and if the shear on the minimum critical slab section on the perimeter at three-quarters the effective length of the shearhead does not exceed $4\sqrt{f'_c}\, b_o d$ (Section 11.12.4.7). Locations of critical section for various shapes of shearheads are shown (Commentary Fig. R11.12.4.7).

The design engineer enjoys great freedom to exercise ingenuity in shearhead design because of the variety of possible solutions, and the many proprietary devices with load test data, that are available. Specific Code provisions for shearhead design using four identical arms do not apply to exterior columns. Where shear is transferred to a column at an edge or a corner of a slab, special shearhead designs are required (Section 11.12.4).

EXAMPLE. STEEL SHEARHEADS. The sequence of references applicable to the design of an interior column structural steel shearhead will be illustrated by an example of a low-rise building with a superimposed factored live load w_ℓ of 300 psf to be supported by an 8 in. flat plate floor with square panels, spans of 20′-0″ between column centerlines and story heights above and below the floor of 12′-0″. This example is based on the desirability of a maximum interior column size of 12 in. Table 5-2 shows shearheads are required for square columns smaller than 26 in.

Shear at first critical section (d/2 from columns)

The total factored shear is

$$V_u = [(1.4)(0.100) + 0.300](20)^2$$

$$= 176 \text{ kips}$$

The design shear strength at a distance $d/2$ from the column face by Code (Section 11.12.4.8) when steel shearheads are used is

$$\phi V_n = \phi 7\sqrt{f'_c}\, b_o d = (0.85)(7)(\sqrt{4000})(4)(12 + 6.5)(6.5)/1000$$
$$= 181 \text{ kips OK} > 176 \text{ kips}$$

Size of shearhead

The required minimum perimeter of the critical section of the shearhead can be determined by limiting the maximum shear force to $V_u \leqslant \phi 4\sqrt{f'_c}\, b_o d$. The required b_o is

$$b_o = \frac{V_u}{\phi 4\sqrt{f'_c}\, d} = \frac{176{,}000}{(0.85)(4)(\sqrt{4000})(6.5)} = 126 \text{ in.}$$

For square columns and a shearhead with each arm made from a structural shape, the perimeter can be defined by the following expression in which W_s equals the width of the compression flange of the structural steel shearhead and c is the square column size. See Fig. 5-22 (Commentary Fig. 11.12.4.7) for definitions and location of the critical section,

$$b_o = 4\sqrt{2}\left[\left(\frac{3}{4}\right)\left(\ell_v - \frac{c}{2}\right) + \frac{c}{2} - \frac{W_s}{2}\right] + 4W_s$$

Solving for ℓ_v and substituting $c = 12''$ and $b_o = 126''$ and assuming $W_s = 4''$, the required length of each arm of the shearhead is:

$$\ell_v = \frac{[b_o - 4W_s]}{3\sqrt{2}} - \frac{c}{6} + \frac{2W_s}{3}$$
$$= \frac{[(126) - (4)(4)]}{3\sqrt{2}} - \frac{12}{6} + \frac{(2)(4)}{3}$$
$$= 21.3 \text{ in.; use 22 in.}$$

Figure 5-22 shows the location of the critical sections of the structural steel shearhead of our example.

In order to determine the required plastic resisting moment, M_p, of the shearhead (Section 11.12.4.6) it is necessary to assume a trial size structural shape and check if $\alpha_v \geqq 0.15$. α_v is defined as the ratio of the stiffness of one arm of the shearhead to the stiffness of the surrounding composite cracked slab section of width $(c_2 + d)$. To determine the stiffness of the composite cracked slab section, it is necessary to know the amount of reinforcing steel within the slab width $(c_2 + d)$. This amount of reinforcing steel can be determined as follows:

$$M_o = \frac{w\ell_n^2}{8} = \frac{(0.400)(20)(19)^2}{8} = 397 \text{ ft-kips}$$

$$-M_u \text{ (Int. span col. strip)} = 0.65\, M_o(0.75) = (0.65)(397)(0.75)$$
$$= 193.5 \text{ ft-kips}$$

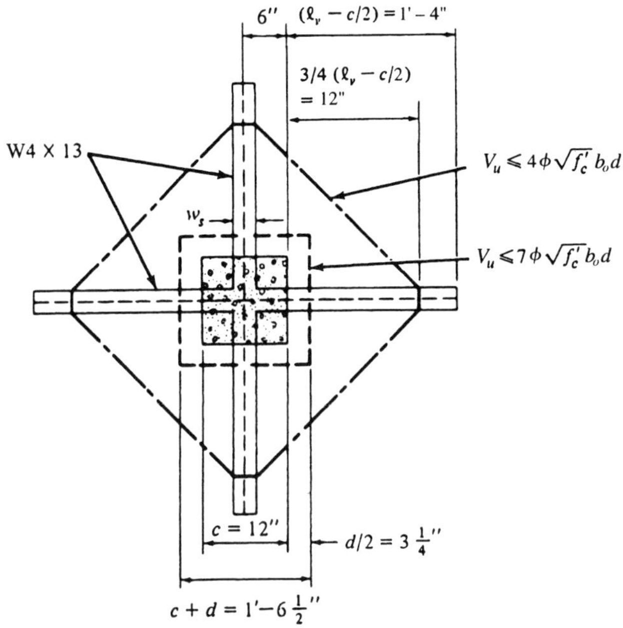

Figure 5-22 Location of critical sections of steel shearhead.

$$A_s = (\text{for assumed } j = 0.92) = \frac{M_u}{\phi f_y jd} = \frac{(193{,}500)(12)}{(0.90)(60{,}000)(0.92)(6.50)}$$

$$= 7.19 \text{ in.}^2 \quad 23 - \#5 = 7.13 \text{ in.}^2$$

If 50% of the column strip top bars are placed within the middle third of the column strip, five bars will fall within the width $c + d = 18.5$ in.

The stiffness of the composite cracked slab section with a trial size W4 × 13 structural steel shearhead is shown in Fig. 5-23. The moment of inertia and stiffness of this section can be easily obtained by transforming the concrete into

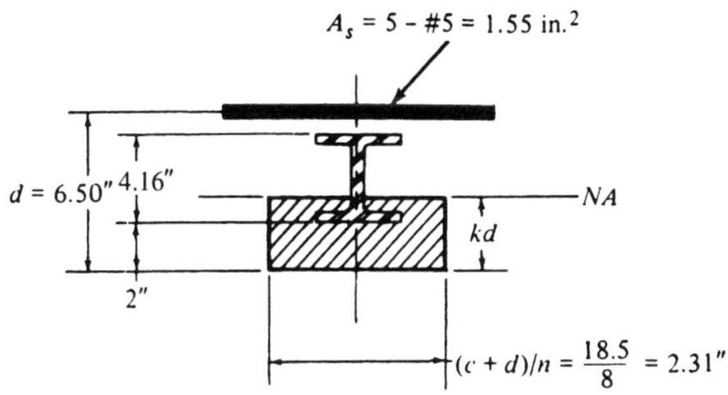

Figure 5-23 Composite cracked slab.

equivalent steel. $2.31\ kd^2/2 - (3.83)(4.08 - kd) - (1.55)(6.5 - kd) = 0$ and $kd = 2.93$ in. Knowing kd, the composite section moment of inertia can be calculated as follows:

$$\frac{(2.31)(2.93)^3}{3} = 19.37$$

$$(1.55)(3.57)^2 = 19.75$$

$$\text{W4} \times 13 = 11.30$$

$$(3.83)(1.15)^2 = \underline{\ 5.07}$$

$$I = 55.49 \text{ in.}^4$$

Because the concrete of the composite cracked slab has been transformed into equivalent steel, the ratio α_v of the stiffness of one steel shearhead arm to the stiffness of the composite cracked slab section is

$$\alpha_v = \frac{I(\text{W4} \times 13)}{I(\text{composite cracked section})} = \frac{11.30}{55.49}$$

$$= 0.204 \text{ OK} \geqslant 0.15 \text{ Code (Section 11.12.4.5)}$$

Moment Resistance for Selecting Shape

The plastic moment strength required for each arm can now be calculated (Code Eq. 11-39):

$$M_p = \frac{V_u}{\phi 2\eta}\left[h_v + \alpha_v\left(\ell_v - \frac{c_1}{2}\right)\right] = \frac{176}{(0.90)(2)(4)}\left[4.16 + 0.204\left(22 - \frac{12}{2}\right)\right]$$

$$= 182 \text{ in.-kips}$$

From the plastic moment, the required plastic section modulus can be determined for structural steel with a yield strength of 36 or 50 ksi.

$$Z = M_p/f_y$$

$$Z(f_y = 36 \text{ ksi}) = 182/36 = 5.06 \text{ in.}^3$$

$$Z(f_y = 50 \text{ ksi}) = 182/50 = 3.64 \text{ in.}^3$$

A W4 $\times$ 13 has a plastic section modulus of 6.28 in.^{3} and if made from a steel with a yield strength of 36 ksi is satisfactory.

Adjustment of slab reinforcement

If sufficient flexural capacity is provided in the shearhead to assure that the shear capacity of the slab is reached before the flexural capacity of the shearhead is exceeded, and if the shear capacity of the slab at the critical section of the shearhead is sufficient, then the shearhead can be used to reduce the column strip negative reinforcing steel (Section 11.11.4.9).

The shearhead is allowed to contribute a negative resisting moment, M_v, to each column strip:

$$M_v = \frac{\phi \alpha_v V_u}{2\eta}\left(\ell_v - \frac{c_1}{2}\right) \quad \text{(Section 11.12.4.9, Eq. 11-40)}$$

$$= \frac{(0.90)(0.204)(176)(22 - 12/2)}{(2)(4)}$$

$$= 65 \text{ in.-kips}$$

$$= 5.4 \text{ ft-kips}$$

M_v however, cannot exceed 30% of the column strip factored moment, the change in moment over the length, ℓ_v, of the shearhead, or M_p (Section 11.12.4.9).

$$30\% \text{ of column strip moment} = (0.30)(193.5) = 58.0 \text{ ft-kips}$$

$$\text{Change in moment in length, } \ell_v, \text{ is } \left[-\frac{V_u}{2}(\ell_v) + \frac{w\ell_2\ell_v^2}{2}\right](0.75)$$

$$\Delta M_u = \left[-\frac{176}{2}(1.83) + \frac{(0.440)(20)(1.83)^2}{2}\right](0.75)$$

$$\Delta M_u = (-161 + 42)(0.75)$$

$$\Delta M_u = 146 \text{ ft-kips}$$

$$\text{Plastic moment strength, } \phi M_p = \frac{182}{12} = 15.2 \text{ ft-kips}$$

The full reduction of 9.9 ft-kips can be used.

BRACKETS OR CAPS. Small brackets or caps at the exterior columns may be used to provide the required shear perimeter, equal to minimum column size, $c_1 = c_2$ in Table 5-2. The Code provides limiting ratios of cap geometry.

If the cap is dimensioned as in Fig. 5-24 to maintain the same minimum shear section perimeter $2(c_1 + d/2) + (c_2 + d)$, the thickness of cap, h_c, must be at least equal to the projection of cap, p_c. The Code permits a maximum angle of 45° within which the section is considered effective for shear (Section 13.1.2). Note that the critical section for negative moment in the direction c_1 is now located $1/2\ p_c$ from the face of the reduced column (Section 13.7.7.2).*

---

* As the Code is written, this provision applies only to design by the "Equivalent Frame Method". It is conservatively recommended by the authors, however, for use in both methods of design.

For exterior columns only, the critical sections for negative moment do not coincide with ends of the clear span when brackets or capitals are used (Section 13.6.2.5, clear span, and 13.7.7.2, critical section). The effective clear span, ℓ_n, has been increased by $1/2\ p_c$ which will increase moments. Offsetting this increase, however, is the reduced stiffness of the column itself, K_c, so that the original shear perimeter dimensions for the plate will probably be adequate without calculation. It cannot be assumed, however, that the original panel moments, other than the negative moment at the exterior column, are conservative. Figure 5-24 shows shear caps proportioned for half-size columns. Negative moment at the edge, most sensitive to column stiffness, will be reduced, but both negative moment at first interior column and positive moment will increase. It cannot be assumed that reduction in these moments are within the 10% adjustment permitted (Section 13.6.7) (see Fig. 5-3).

If a cap is employed with a reduced column size because the column required without a shear cap would be too large, it will be necessary to repeat Direct Design Steps 2 through 5 and to adjust the bar selections for the exterior panel accordingly. Computation of K_c will become more complicated by the cap shown and still more complicated if a sloping bracket of variable depth is used.

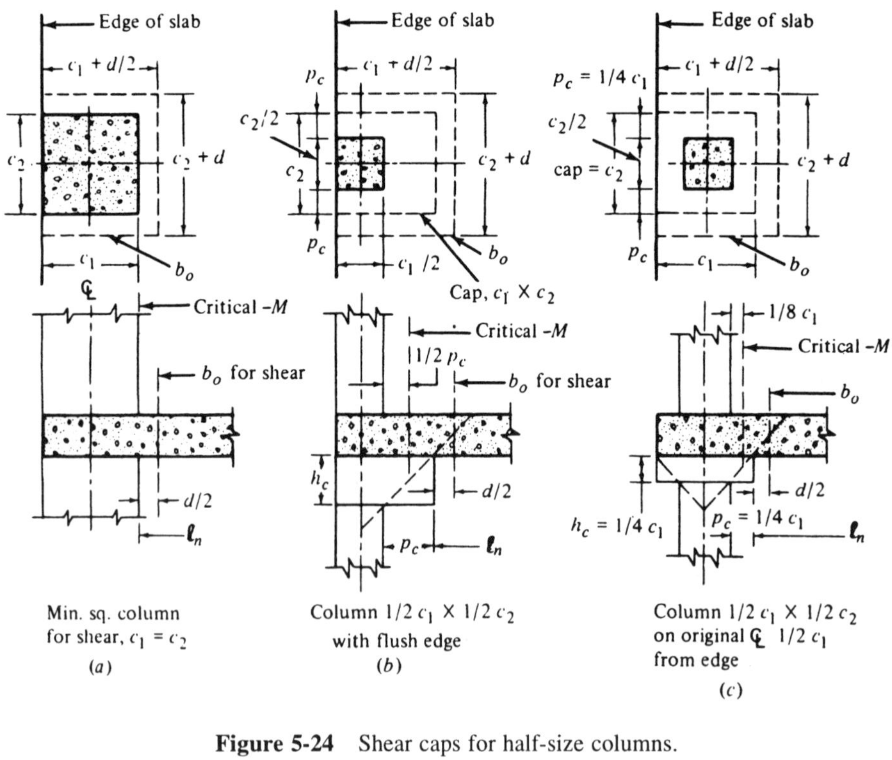

Figure 5-24 Shear caps for half-size columns.

The Code is ambiguous in prescribing this operation. The moment of inertia "outside of the capital" is based on column cross section (Section 13.7.4.1). The moment of inertia is assumed infinite within the joint depth of the slab-beam (Section 13.7.4.3). It is recommended that the contribution to stiffness, K_c, provided by a capital be taken into account, which is undoubtedly the intent of the Code. A crude approximation which will permit the user to take account of the cap contribution to column stiffness by a simple calculation is to add one-fourth the cap depth to the plate thickness, h, and enter Fig. 5-13 with the large value of h/ℓ_c, computing a new K_c for the reduced column section. The slab stiffness, K_s, need not be changed from that computed for the original large column. The same correction for width of support is provided from the face of column, bracket, or capital (Section 13.7.3.3). The remaining design Steps 3 through 5 can be followed using Figs. 5-11 through 5-13 with the revised value of K_c.

Spandrel Beams

The use of spandrel beams for the reduction of shear stress in heavily loaded flat plates will permit reduction of the exterior column size. The stiffening effect at the edges is also beneficial in controlling edge deflections and providing moment transfer by torsion from the plate to the edge column. Spandrel beams can also be used with columns as frames to develop resistance to lateral forces and thus may eliminate need of shear walls.

With the many possible permutations of depth, width, L- or T-shapes, location within the depth of the column, proportion of beam width to column width, etc., only a few general suggestions applicable to all can be given here. The 1995 Code states that *torsion* reinforcement *where required* consist of closed stirrups and longitudinal bars (Section 11.6.4.1). Torsion effects may be neglected if the factored torsional moment, T_u is $\phi\sqrt{f'_c}\,(A_{cp}^2/p_{cp})$ or less for non-prestressed members (Section 11.6.1). Thus, the Code will eliminate many unnecessary closed stirrups but it does require additional calculations to determine the torsional moment. Except for torsion reinforcement, the Code specifically permits the use of closed stirrups formed in two pieces, a practice recommended for simplified assembly in the field (Section 12.13.5). See Chapter 9, Fig. 9-17, for various types of two-piece closed stirrups, both recommended and unsuitable styles.

The nominal torsional moment strength, T_n, is based on a thin-walled tube space truss analogy. A beam resisting torsion is treated as a thin-walled tube with the core concrete part of the cross section neglected. The portions of plate considered as participating in the resistance to torsion are equal to the largest projection of the spandrel beam, above or below the plate, but not more than $4h_f$ (Section 13.2.4). Conservatively, assume that the *sum* of the plate portions if spandrel is T-shaped, is limited to $4h_f$.

For spandrel beams at the edge of a flat plate, "their share of the exterior negative factored moment" must be resisted by torsion (Section 13.6.3.5). Logically, that portion of the unbalanced moment at the exterior column not transferred directly by flexure to the column must be transferred by torsion through the spandrel beams. To apply the expression for the percentage of moment transferred directly (Section 13.5.3), the plan dimensions of the shear section, i.e., the values of b_1 and b_2, are best determined by a sketch of the proposed spandrel and column (see Figs. 5-25, 5-26 and 5-27). Consider the spandrel depth on either side of the column within a 45° angle as a part of $b_2 = c_2 + 2h$; and $b_1 = c_1 + d$. Note that a spandrel flush with the inside face of the column will change the shear behavior of the plate from two-way to one-way (beam) shear (Section 11.12.1.1) (see Figs. 5-26 and 5-27).

The vertical load for the design of the spandrel includes all loads on the tributary areas between the centerline of the span parallel to the spandrel and lines at 45° angles from the corners of the column (see Fig. 5-26). Spandrels for which $\alpha_1 (\ell_2/\ell_1) \geqq 1.0$ carry this full tributary load; where $\alpha_1 = 0$, zero load. Interpolation is required for intermediate values of α_1. Also, the spandrel must carry all directly applied loads (Section 13.6.8.3).

EXAMPLE. For the simplest case, consider an interior span spandrel flush with the outer face of the column and equal in width of the column. Let this spandrel be employed to reduce the size required for the exterior column (Table 5-2) in a flat plate floor 8 in. thick, carrying 300 psf (w_ℓ) factored load on square panels $\ell_1 = \ell_2 = 20$ ft with story height 12 ft. The sprandel beam supports an exterior wall (12 in. wide) load of 660 plf. The required column size for a direct connection is tabulated as 36 in. square. Use the spandrel beam to accommodate a maximum column size of 20 in. square (see Fig. 5-27).

$$\alpha \approx \frac{(20)(20)^3}{(120)(8)^3} \geqq 1.0 \qquad \text{(Section 13.0)}$$

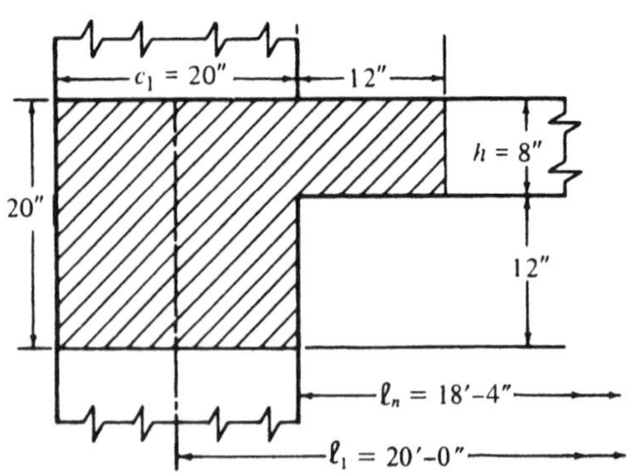

Figure 5-25 Effective section for torsion—spandrel example.

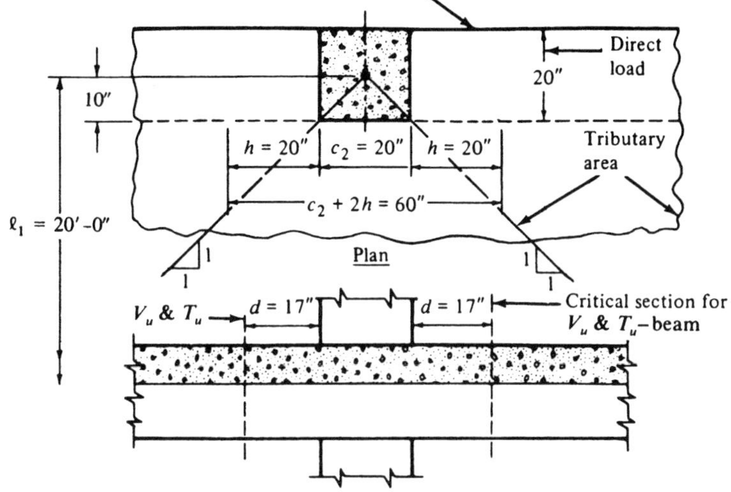

Figure 5-26 Factored loads and critical sections—Beam.

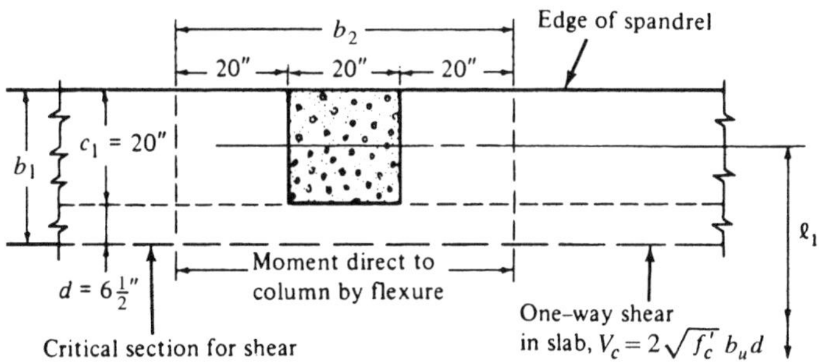

Figure 5-27 Critical shear in slab with spandrel.

and so full tributary area loads are assigned to the spandrel (Section 13.6.8.1). Let the spandrel total load $= W$ (see Fig. 5-26).

$$W = (0.440)(18.33)(9.17/2) + \left(\frac{20 \times 20}{144}\right)(0.150 \times 1.4)(18.33)$$

$$+ \left(0.300 \times \frac{20}{12}\right)(18.33) + 0.66(18.33)(1.4)$$

$$= 36.98 + 10.69 + 9.16 + 16.94$$

$$= 73.77 \text{ kips in a triangular and uniform load pattern}$$

$$M_0 = \frac{(0.440)(20)(18.33)^2}{8} = 370 \text{ ft-kips (Span } \perp \text{ to spandrel)}$$

$$\begin{aligned} -M_u \text{ (exterior column face)} &= (0.30)(370) = 111 \text{ ft-kips} \\ +M_u \text{ (positive)} &= (0.50)(370) = 185 \text{ ft-kips} \\ -M_u \text{ (interior column face)} &= (0.70)(370) = 259 \text{ ft-kips} \end{aligned}$$

The fraction of the exterior moment transferred by flexure directly to the column (Section 13.3.3.2) can be assumed =

$$\frac{1}{\left(1 + 2/3\sqrt{\dfrac{b_1}{b_2}}\right)} = \frac{1}{\left(1 + 2/3\sqrt{\dfrac{c_1 + d}{c_2 + 2h}}\right)} = \frac{1}{\left(1 + 2/3\sqrt{\dfrac{26.50}{60}}\right)} = 0.69$$

(See Fig. 5-26) moment within the 60 in. plate width centered on the column centerline = 0.69 × 111 = 77 ft-kips.

One-way (beam) shear on a 60 in. width at critical section of plate

$$V_u = 0.69(\tfrac{1}{2})(0.440)(20)\left(9.17 - \frac{6.50}{12}\right) - (259 - 111)/18.33 = 18.13 \text{ kips}$$

Area of shear section, A_c

$$\begin{aligned} \phi V_c &= \phi 2\sqrt{f_c'}\, b_w d \\ &= (0.85)(2)\sqrt{4000}\,(60)(6.50)/1000 \\ &= 41.9 \text{ kips} \end{aligned}$$

$$41.9 > 18.13 \text{ O.K.}$$

(Sections 11.1 and 11.3.1)

Torsional moment per spandrel,

$$T_u = (0.31)(111) = 34.41 \text{ ft-kips}$$

Beam cross-sectional torsion properties (Section 11.0)

$$\begin{aligned} A_{cp} &= \text{area enclosed by perimeter} \\ &= (20)^2 + (12)(8) = 400 + 96 = 496 \text{ in.}^2 \\ p_{cp} &= \text{outside perimeter of concrete cross section} \\ &= 20 + 20 + 20 + 12 + 8 + 12 + 12 = 104 \text{ in.} \end{aligned}$$

Maximum torsional moment that can be neglected

$$\begin{aligned} \phi(\sqrt{4000}/1000)(A_{cp})^2/(p_{cp}) &= 0.85(\sqrt{4000}/1000)(496)^2/104 \\ &= 127.17 \text{ in.-kips} \\ &= 10.6 \text{ ft-kips} \end{aligned}$$

Factored torsional moment = 34.41 ft-kips > 10.6 ft-kips so torsion must be considered.

The spandrel selected for this example illustrates the advantages of selecting a nearly square cross section when Code requirements for torsion must be considered. A square shape is, of course, not most efficient for purely vertical loads, but the spandrel is a special case. Where the width is not restricted by architectural considerations, the authors suggest that increased beam width be employed to avoid the design, detailing, inspection, and field placing problems, and ensuing time and expense, involved in the use of torsion reinforcement.
Spandrel shear at *d* from column face

$$V_u = [(1.70 + 5.73)/2](18.33/2) - [(1.70 + 2.34)/2](1.46)$$

$$= 34.05 - 2.95 = 31.10 \text{ kips}$$

$$\phi V_c = 0.85(2)(\sqrt{4000})(20)(17.5)/1000 = 37.63 \text{ kips}$$

$$V_u > \phi V_c/2;\ 31.10 > 37.63/2 = 18.82 \text{ kips}$$

For factored shear force, V_u, greater than one-half the shear strength ϕV_c, minimum stirrups are required (Section 11.5.5.1), except where the beam depth does not exceed $2\frac{1}{2}$ times flange thickness or $\frac{1}{2}$ times the web thickness (Section 11.5.5.1-c). The example exactly meets both limits on depth and so stirrups are required for torsion only. See Figs. 5-28 and 5-29.

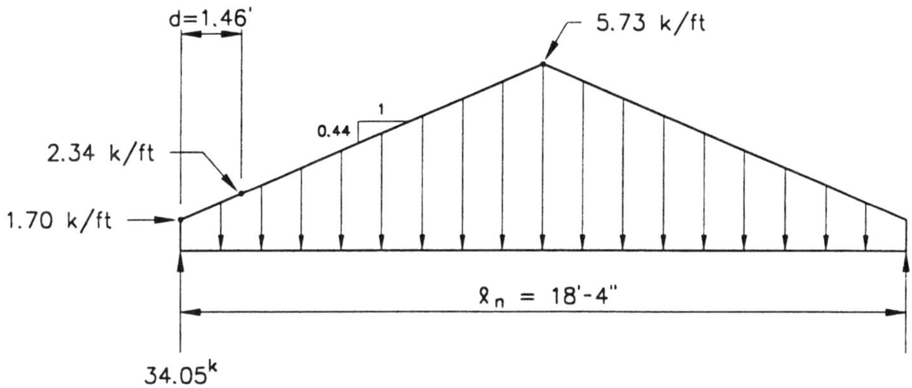

Uniform load	Wall (0.66 × 1.4)	= 0.92 k/ft
	Beam $[(20)^2/144](0.150)(1.4)$	= 0.58 k/ft
	Live (0.30)(8)/12	= 0.20
		1.70 k/ft
Triangular	(0.440)(9.17)	= 4.03
		5.73 k/ft

Figure 5-28 Spandrel load diagram.

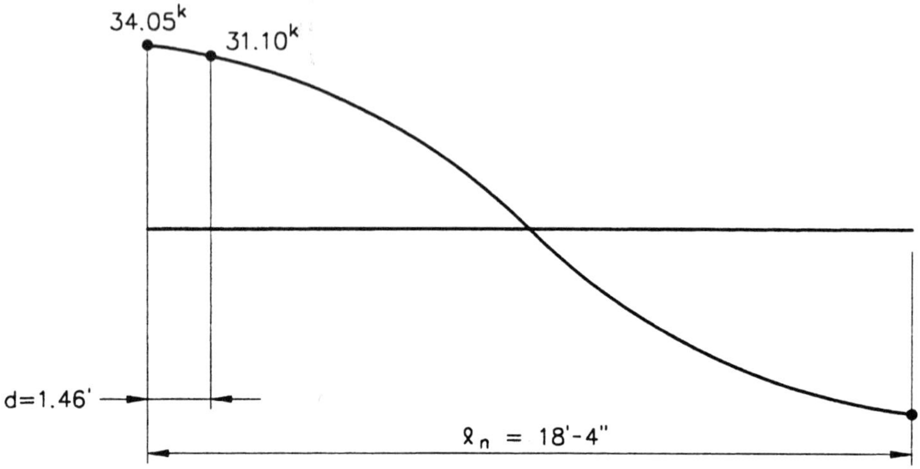

Figure 5-29 Spandrel shear diagram.

The factored torsional moment, T_u, in an indeterminate structure can be reduced to

$$T_u = \phi 4\sqrt{f'_c}\,(A_{cp})^2/p_{cp} \text{ (Section 11.6.2.2a)}$$
$$= (0.85)(4)(\sqrt{4000})[(496)^2/(104)]/12{,}000$$
$$= 42.39 \text{ ft-kips} > 34.41 \text{ ft-kips use } 34.41 \text{ ft-kips}$$

Cross-sectional dimensions of spandrel:

$$p_h = \text{perimeter of centerline of torsional reinforcement}$$
$$= 16.5(4) = 66 \text{ in.}$$
$$A_{oh} = \text{area enclosed by centerline of torsional reinforcement}$$
$$= (16.5)^2 = 272.25 \text{ in.}^2$$

Factored shear stress ⩽ shear cracking stress (Section 11.6.3.1a)

$$V_c = 2\sqrt{f'_c}\,b_w d = 2\sqrt{4000}\,(20)(17.5)/1000 = 44.27 \text{ kips}$$
$$\{[(V_u)/(b_w d)]^2 + [(T_u p_h)/(1.7A^2_{oh})]^2\}^{0.5} \leqslant \phi(V_c/b_w d + 8\sqrt{f'_c})$$
$$\{[29{,}100/(20 \times 17.5)]^2 + [(34{,}470)(66)/(1.7 \times 272.25^2)]^2\}^{0.5}$$
$$\leqslant 0.85[(44{,}270)/(20 \times 17.5) + 8\sqrt{4000}]$$
$$(6{,}913 + 325)^{0.5} \leqslant 0.85(126 + 506)$$
$$85 \text{ psi} \leqslant 537 \text{ psi OK}$$

Reinforcement for torsion (Section 11.6.3.5).

$$A_o = 0.85A_{oh} = 0.85(227.25) = 193.16 \text{ in.}^2$$

$$\phi T_n = (\phi 2A_o A_t f_{yv}/s)(\cot\theta) \geqslant T_u$$

$$\therefore\ A_t/s = 34.41 \times 12/[(0.85)(2)(193.16)(60)(1)] = 0.0210 \text{ in.}^2/\text{in./leg}$$

$$s_{\max} = p_h/8 = 66/8 = 8.25 \text{ say } 8 \text{ in.}$$

$$2A_{t\min} = 50b_w s/f_{yv} = 50(20)(8)/60{,}000$$

$$= 0.13 \text{ in.}^2$$

$A_t = 0.0210(8) = 0.17$ in.^{2} Use 28-#4 stirrups spaced at 8 in. centers.

Longitudinal torsional reinforcement (Section 11.6.3.7).

$$A_l = (A_t/s)(p_h)(f_{yv}/f_{yt})\cot^2\theta$$

$$= [(0.20/8)(66)](60/60)(1) = 1.65 \text{ in.}^2$$

Check the minimum area of longitudinal torsion reinforcement (Section 11.6.5.3). Since $(25\ b_w/f_{yv}) =$

$25(20)/60{,}000 = 0.0083$ in.^{2}/in., use $A_t/s = 0.0210$ in.^{2}/in.

$$A_{\ell,\min} = \frac{5\sqrt{f_c'}\,A_{cp}}{f_{yt}} - \left(\frac{A_t}{S}\right) p_h \frac{f_{yv}}{f_{yt}} \qquad \text{(Eq. 11-24)}$$

$$= \frac{5\sqrt{4000}\,(496)}{60{,}000} - (0.0210)(66)\frac{60{,}000}{60{,}000}$$

$$= 1.35 \text{ in.}^2 < 1.65 \text{ in.}^2$$

For $A_\ell = 1.65$ in.^{2}, use 4-#6 bars (provide 1.76 in.^{2}) with one bar centered on each side and the top and bottom. See Fig. 5-30.

Maximum spacing of longitudinal torsion reinforcement = 12 in. (Section 11.6.6.2). Minimum bar size is $s/24 = 8/24 = 0.33$ in. For #6 bars, $d_b = 0.75$ in. > 0.33 in., Okay.

Flexural reinforcement for spandrel beam (Sections 13.6.4.1 and 13.6.4.4).

$$M_o = 5.73(18.33)^2/8 = 241 \text{ ft-kips (Span parallel to spandrel)}$$

$$\alpha_1 = [(20)^4/12]/[(110)(8)^3/12] = 2.84$$

$$-M_u \text{ and } +M_u = 0.75M_o(0.85) \text{ (Section 13.6.5.1)}$$

$$= (0.75)(241)(0.85) = 154 \text{ ft-kips}$$

$$-A_s \text{ and } +A_s = M_u/(\phi f_y jd) = (154)(12)/[(0.9)(60)(0.95)(17.5)]$$

$= 2.06$ in.^{2} use 2-#9 top bars over column and 2-#9 bottom bars. See Fig. 5-30.

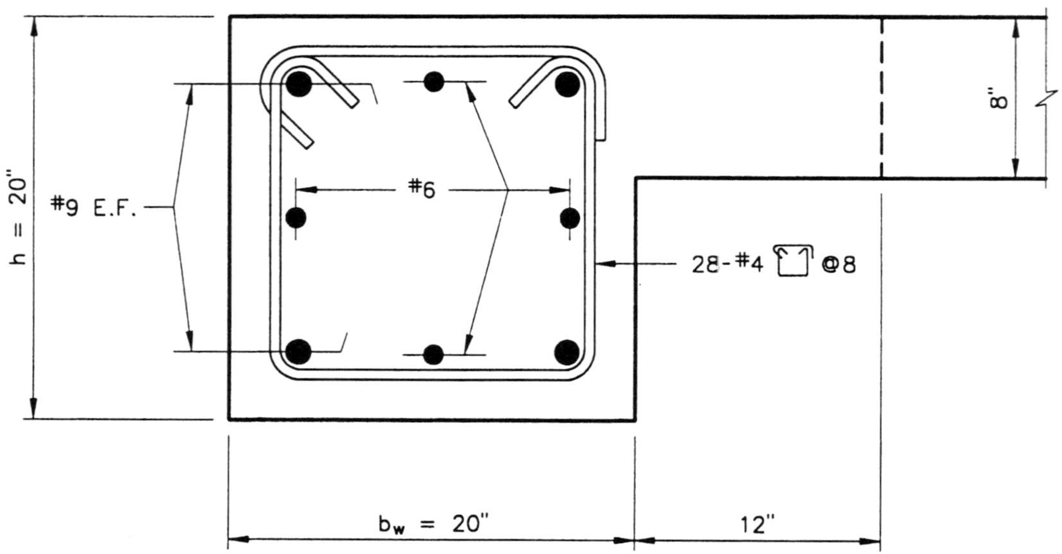

Figure 5-30 Spandrel beam with torsion.

DESIGN NOTES ON SHEAR IN TWO-WAY SLABS

Shear Due to Vertical Loads and Moment Transfer Combined

Figure 5-31 shows a trace of the computer operations to select a minimum size square exterior column located flush with the edge of the plate. For a specified concrete strength = 4,000 psi, allowable shear on the concrete equals 253 psi, which is the critical limit governing minimum size of the column. The shear stresses shown are the sum of the stresses due to gravity loads and transfer of part of the unbalanced slab moment. For each trial size of column, shear is computed for two load conditions: (1) gravity shear maximum and (2) moment transfer shear maximum on interior columns.

This insight into the interior working of a computer operation provides the explanation for the difficulty encountered in a direct manual solution for selection of minimum column size. If the preliminary design selected is too small, successively larger trial sizes do not necessarily reduce shear—the trials do not converge directly to a solution. Note that shear stresses become larger for all sizes above 10 in. square to 12.5 in. square columns. The solutions for successively larger sizes do not converge toward the allowable shear until a size above about 12.5 in. is used. This effect results from increasing column stiffness in the ratio $(c + \Delta c)^3/c^3$, with proportionate increase in moment to be transferred by shear to the column while the increase in the section resisting shear is a linear function.

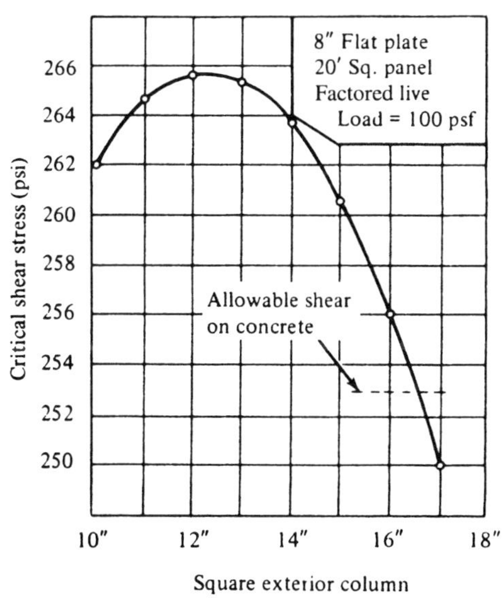

Figure 5-31 Shear vs. column size.

Effect of Column Shape

This insight also provides the designer a quick guide to reduce further trials required for a solution. The pattern for square columns is shown in Fig. 5-31. If architectural requirements permit increase in the width dimensions, c_2, only, the increase in column stiffness and shear resistance are both linear so that convergent solutions may be expected from the first trial. On the other hand, if architectural requirements prescribe further increases in width, much larger increments in depth, c_1, will be required, since divergence of results for small increases will be more pronounced than for square columns.

See Fig. 5-32, cases (*a*), (*b*), (*c*), and (*g*), for a comparison of shear perimeters for square versus rectangular edge columns and corner columns flush with slab edges. The Code is very specific in defining the critical shear section for these cases (Sections 11.12.1.1 and 11.12.1.2 and corresponding Commentary sections).

Two-way slab concrete shear strength, V_c, is controlled by the smaller value of three Code equations (Section 11.12.2.1):

$$V_c = (2 + 4/\beta_c)\sqrt{f'_c}\, b_o d \quad \text{(Eq. 11-35)}$$

where β_c = ratio of long side to short side of column

and b_o = perimeter of critical shear section

$$V_c = (\alpha_s d/b_o)\sqrt{f'_c}\, b_o d \ \text{(Eq. 11-36)}$$

where α_s = 40 for interior columns

= 30 for edge columns

= 20 for corner columns

$$V_c = 4\sqrt{f'_c}\, b_o d \ \text{(Eq. 11-37)}$$

See Fig. 5-33. For β_c ratios greater than 2, V_c varies from $4\sqrt{f'_c}\, b_o d$ at $\beta_c = 2$ to $2.4\sqrt{f'_c}\, b_o d$ at $\beta_c = 10$. For d/b_o ratios greater than 0.02, V_c varies from $4\sqrt{f'_c}\, b_o d$ at $d/b_o = 0.02$ to $V_c = 2.4\sqrt{f'_c}\, b_o d$ at $d/b_o = 0.10$.

To utilize the maximum shear strength of the concrete at $V_c = 4\sqrt{f'_c}\, b_o d$ in design it is necessary to limit β_c values to a maximum of 2 and use d/b_o ratios equal to or greater than the following:

$d/b_o \geqslant 0.050$ for interior columns

$d/b_o \geqslant 0.066$ for edge columns

$d/b_o \geqslant 0.010$ for corner columns.

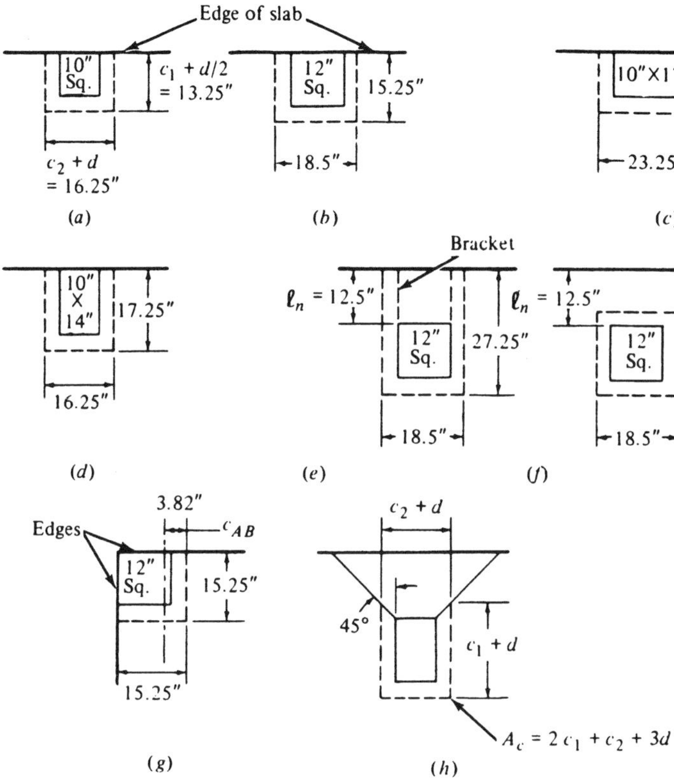

Numerical Values with d = 6.5 in.		
Case	Shear Section Area A_c (in.^{2})	Shear Coefficient for Moment Transfer J_c (in.^{4})
(*a*)	43.0 d = 279	6,000
((*b*)	49.0 d = 319	8,800
(*c*)	50.0 d = 325	6,650
(*d*)	51.5 d = 335	11,550
(*e*)	74.0 d = 481	41,900
(*f*)	74.0 d = 481	28,300
(*g*)	30.5 d = 198	5,144

Figure 5-32 Column shape and location—effect on shear coefficients.

Location of Exterior Columns—Cantilevered Edges

For exterior columns at cantilevered edges the Code is somewhat ambiguous in defining the critical shear section perimeter. Openings in the slab wider than the critical section and parallel to it within a distance less than ten times slab

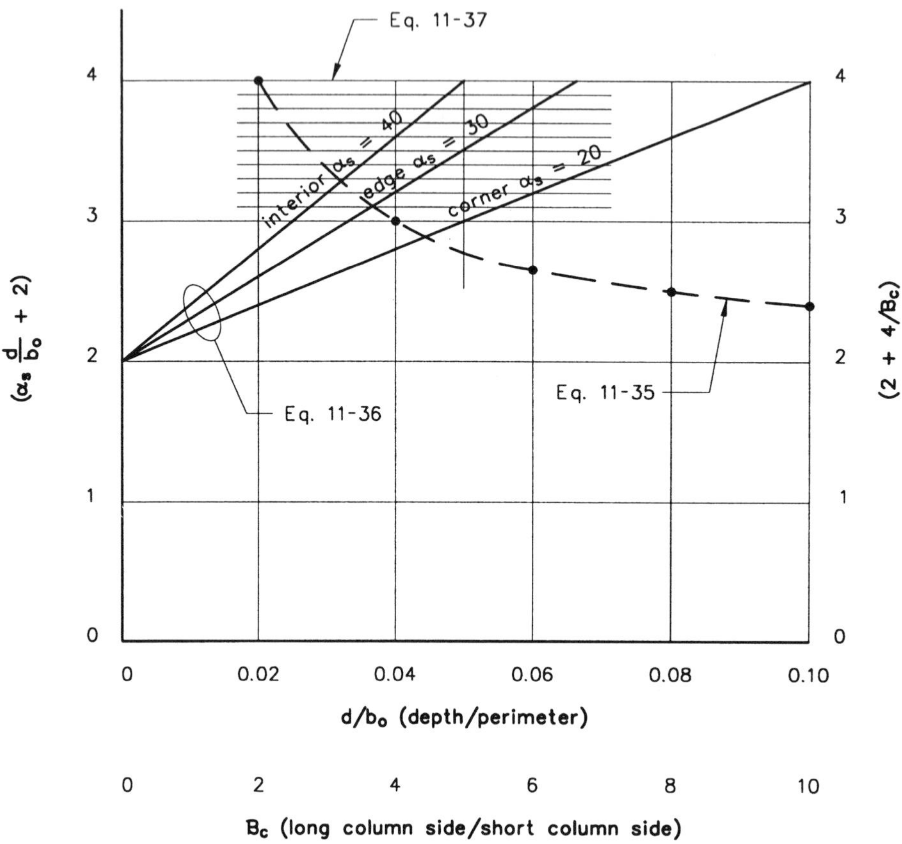

Figure 5-33 Two-way slab shear.

thickness ($10h$) (Section 11.12.5) are to be regarded as a free edge, but the resulting shear perimeter must be a "minimum" (Section 11.12.1.2). The reader is also advised that the criterion for using an open 3-sided figure, Fig. 5-32, case (e), or a closed 4-sided figure, case (*f*), is that the perimeter be a minimum in Commentary reference report of ACI Committee 426. This criterion leads to using the closed shape, case (*f*), when the cantilever span $\ell_n = \frac{1}{2}(c_2 + d)$. Note that this limit is usually far less than $10h$.) For the numerical example, cases (*e*) and (*f*), note that, although the shear perimeter section areas, A_c, are equal, the moment transfer shear coefficient, J_c, is much larger for the 3-sided figure.

The authors consider both solutions, cases (*e*) and (*f*), unrealistic and likely to overestimate the shear strength since tests have shown that the outer face of a column does not participate equally in resisting shear due to loads on the other side of the column. For a prudent design, it is recommended that the critical section perimeter for computation of both area, A_c, and shear coefficient for moment transfer, J_c, be taken conservatively 3-sided as shown in case (*h*). For small cantilever loads and spans the cantilever dead load moment will, of course, reduce the moment to be transferred to the edge column and resulting shear stress. When the cantilever dead load moment is equal to the dead load moment at an interior column from a span not less than two-thirds the exterior span, it would seem reasonable to use a 4-sided critical shear perimeter for both A_c and J_c, as in case (*f*). The section shown in case (*e*) is, of course, proper and conservative when a bracket as wide as the column is used.

Derivations of Formulas for Shear Constants

The shear coefficient for moment transfer, J_c, is not defined in Code or Commentary. A formula for computing J_c is given (Commentary R11.12.6.2) for the case of an interior rectangular column only. J_c is the sum of the moments of inertia about the *X*-axis and *Z*-axis for faces of the critical section resisting varying torsional shear and the moments of inertia about the *X*-axis for the faces resisting uniform vertical shear. See Fig. 5-34 for derivation of formulas for J_c at rectangular interior, edge, and corner columns. (For circular columns, the authors suggest computation of J_c as in Figs. 6-24, 6-25, and 6-26. As a quick approximation, take A_c and J_c for an equivalent square column.)

CASE 1: **Interior Column**

Critical shear section perimeter = A-B-C-D-A

$$\text{Area of critical shear section} = (AB + BC + CD + DA)(\text{depth})$$

$$A_c = (2c_1 + 2c_2 + 4d)(d)$$

By symmetry, c.g. lies at center each way: $C_{AB} = \frac{1}{2}(c_1 + d)$

J = polar moment of inertia of critical shear section

For moment about the *X*-axis,

$$J_c = I_{xx} + I_{zz} \text{ for two faces, } 2(c_1 + d) + I_{xx} \text{ for two faces } 2(c_2 + d)$$

For faces $(c_1 + d)$:

$$I_{xx} = \frac{2d(c_1 + d)^3}{12} = \frac{d(c_1 + d)^3}{6}$$

$$I_{zz} = \frac{2(c_1 + d)d^3}{12} = \frac{(c_1 + d)d^3}{6}$$

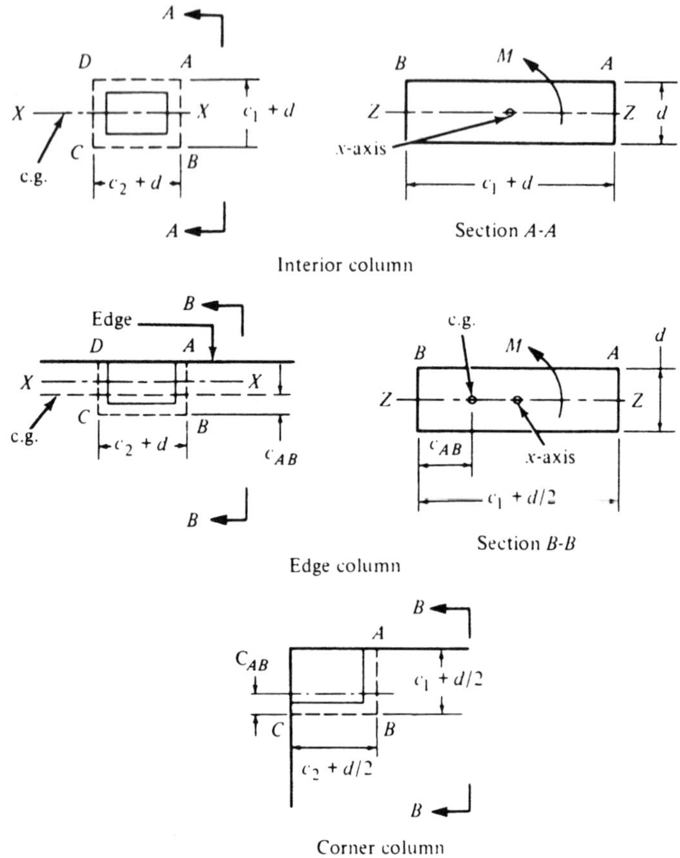

Figure 5-34 Derivation of shear constants.

For faces $(c_2 + d)$:

$$I_{xx} = 2(c_2 + d)(d)\left[\frac{c_1 + d}{2}\right]^2$$

$$J_c = \frac{d(c_1 + d)^3}{6} + \frac{(c_1 + d)d^3}{6} + \frac{(c_2 + d)(d)(c_1 + d)^2}{2}$$

(This expression appears as the definition of J_c in the Commentary R.11.12.6.2)

CASE 2: **Edge Column**

Critical shear section perimeter (3-sided) $= AB + BC + CD$

$$A_c = (2c_1 + c_2 + 2d)(d)$$

$$C_{AB} = \frac{(2)(c_1 + \frac{1}{2}d)(d)(\frac{1}{2})(c_1 + \frac{1}{2}d)}{A_c}$$

$$C_{AB} = \frac{d(c_1 + \frac{1}{2}d)^2}{A_c}$$

$$J_c = I_{xx} + I_{zz} \text{ for two faces, } 2(c_1 + \tfrac{1}{2}d) + I_{xx} \text{ for face } (c_2 + d)$$

For (2) faces $(c_1 + \frac{1}{2}d)$:

$$I_{xx} = \frac{(2)(d)(c_1 + \frac{1}{2}d)^3}{12} + (2)(d)(c_1 + \tfrac{1}{2}d)\left[\frac{(c_1 + \frac{1}{2}d)}{2} - C_{AB}\right]^2$$

$$= \frac{d(c_1 + \frac{1}{2}d)^3}{6} + 2d(c_1 + \tfrac{1}{2}d)\left[\frac{(c_1 + \frac{1}{2}d)}{2} - C_{AB}\right]^2$$

$$I_{zz} = \frac{2(c_1 + \frac{1}{2}d)d^3}{12} = \frac{(c_1 + \frac{1}{2}d)d^3}{6}$$

For (1) face $(c_2 + d)$:

$$I_{xx} = (c_2 + d)(d)(C_{AB})^2$$

Combining terms,

$$\boxed{J_c = \frac{(c_1 + \frac{1}{2}d)d^3}{6} + \frac{d(c_1 + \frac{1}{2}d)^3}{6} + 2d(c_1 + \tfrac{1}{2}d)\left[\frac{(c_1 + \frac{1}{2}d)}{2} - C_{AB}\right]^2 + d(c_2 + d)(C_{AB})^2}$$

CASE 3: **Corner Column**

Critical shear section perimeter (2-sided) $= AB + BC$

$$A_c = (c_1 + c_2 + d)(d)$$

$$C_{AB} = \frac{(c_1 + \frac{1}{2}d)(d)(\frac{1}{2})(c_1 + \frac{1}{2}d)}{A_c} = \frac{d(c_1 + \frac{1}{2}d)^2}{2A_c}$$

$$J_c = I_{xx} \text{ for face } (c_2 + \tfrac{1}{2}d) + I_{xx} + I_{zz} \text{ for face } (c_1 + \tfrac{1}{2}d)$$

For face $(c_2 + \frac{1}{2}d)$:

$$I_{xx} = d(c_2 + \tfrac{1}{2}d)(C_{AB})^2$$

For face $(c_1 + \frac{1}{2}d)$:

$$I_{xx} = \frac{d(c_1 + \frac{1}{2}d)^3}{12} + (c_1 + \tfrac{1}{2}d)(d)\left[\frac{c_1 + \frac{1}{2}d}{2} - C_{AB}\right]^2$$

$$I_{zz} = \frac{(c_1 + \frac{1}{2}d)d^3}{12}$$

$$J_c = \frac{(c_1 + \frac{1}{2}d)d^3}{12} + \frac{d(c_1 + \frac{1}{2}d)^3}{12} + d(c_1 + \tfrac{1}{2}d)\left[\frac{(c_1 + \frac{1}{2}d)}{2} - (C_{AB})\right]^2 + d(c_2 + \tfrac{1}{2}d)(C_{AB})^2$$

First Trial Column Sizes Versus Load, Span, and Plate Thickness

See Figs. 5-35 through 5-40 for a guide to the selection of first trial square column sizes for columns supporting flat plates without shearheads. See also Fig. 5-32, cases (*a*) through (*h*), for a guide to the effect of varying the column shape and location upon resulting shear coefficients.

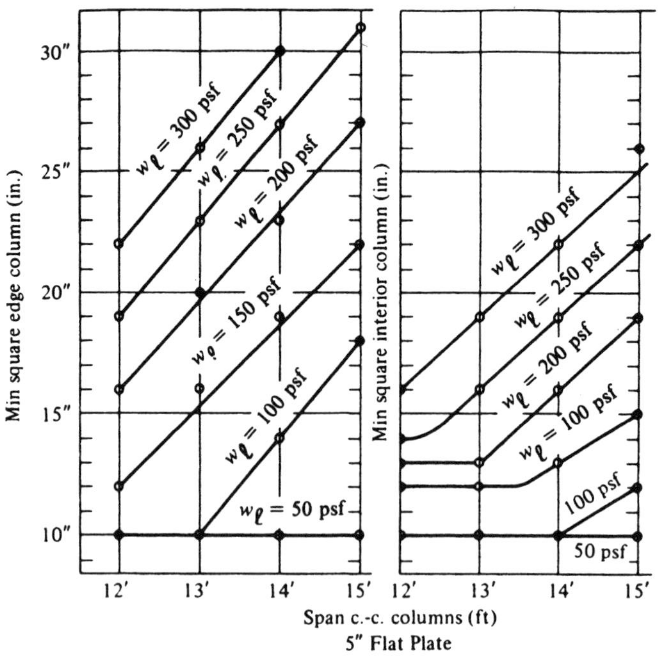

Figure 5-35 Minimum square columns, 5″ flat plate.

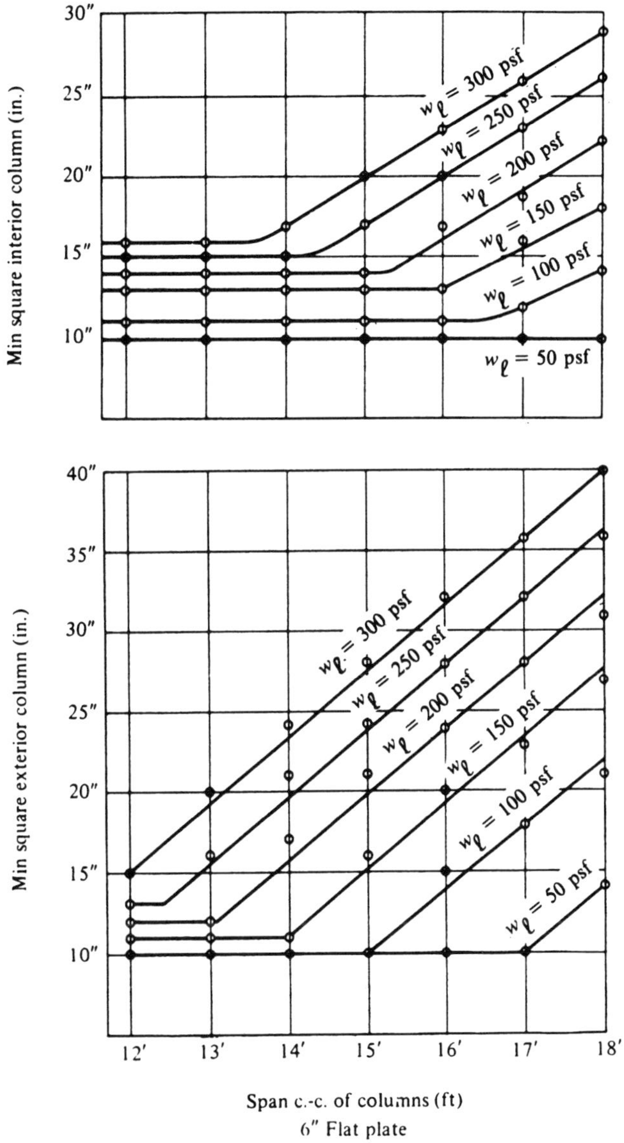

Figure 5-36 Minimum square columns, 6″ flat plate.

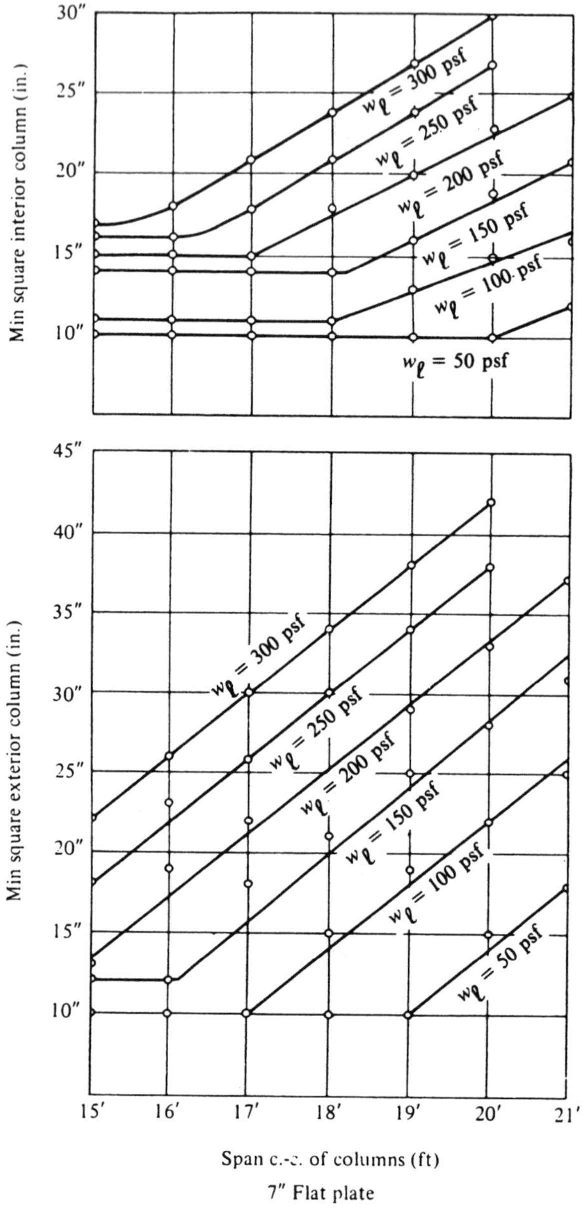

Figure 5-37 Minimum square columns, 7″ flat plate.

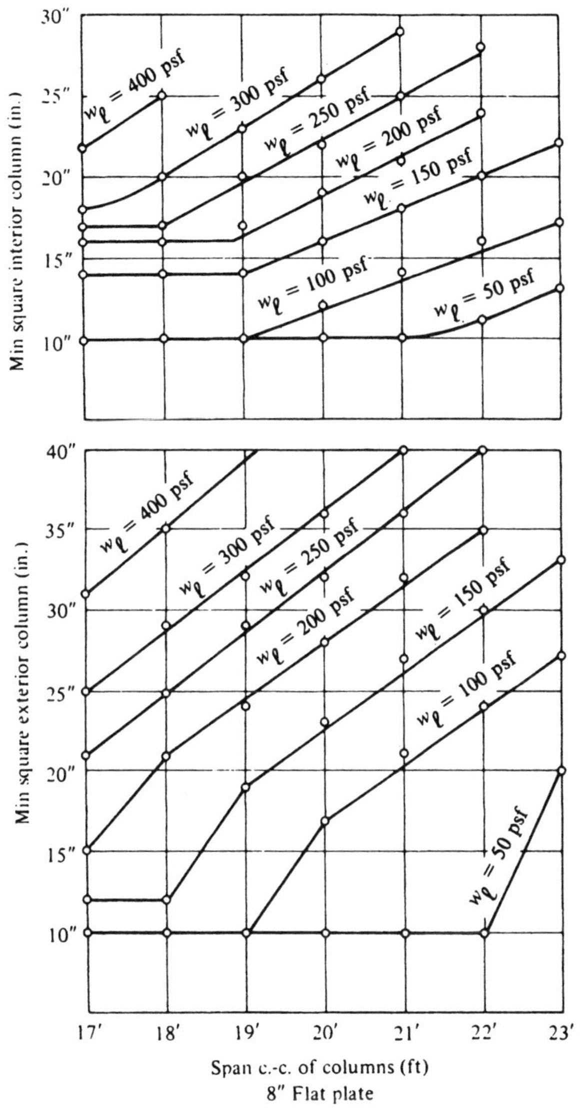

Figure 5-38 Minimum square columns, 8" flat plate.

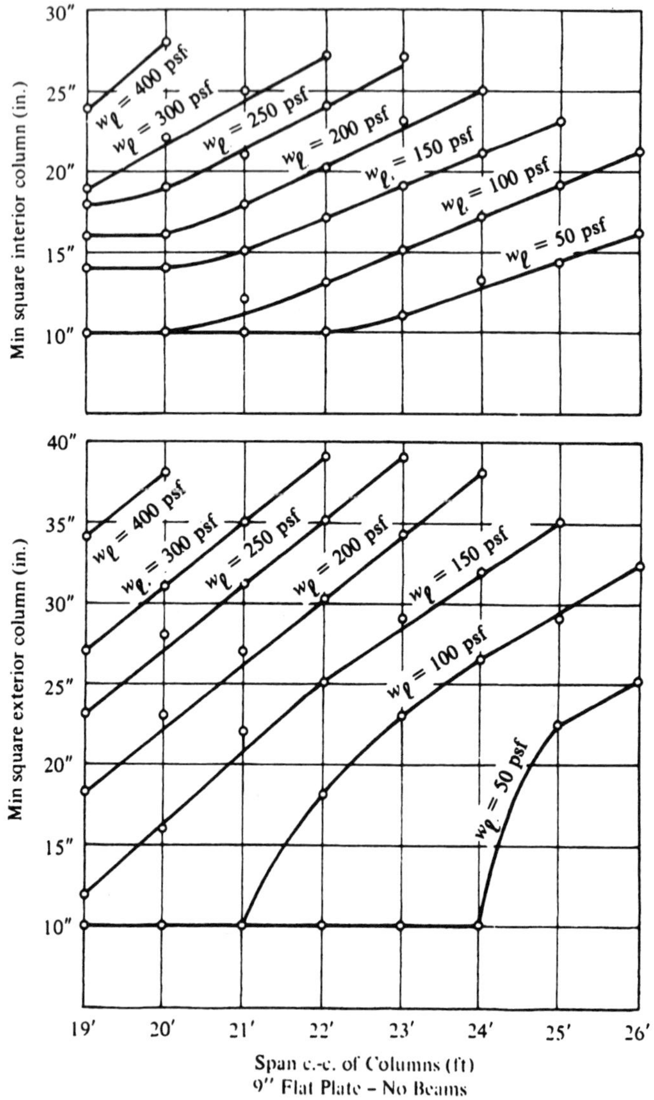

Figure 5-39 Minimum square columns, 9″ flat plate.

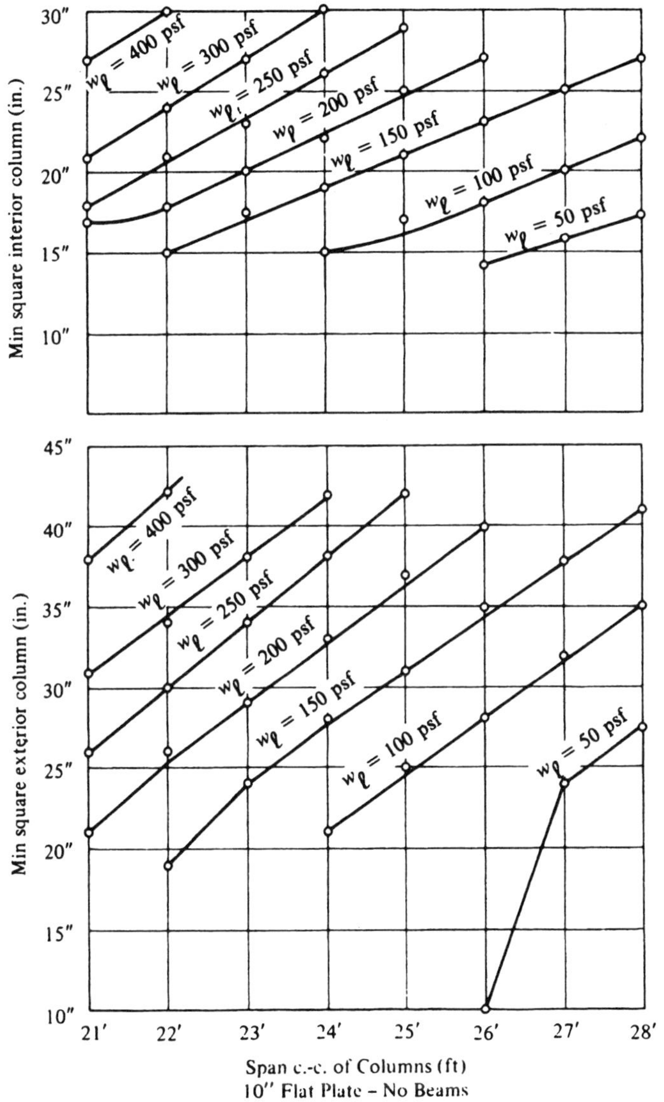

Figure 5-40 Minimum square columns, 10″ flat plate.

6 Two-Way Solid Flat Slab Design

GENERAL

Traditionally, the term "flat slab" has represented a solid uniform depth slab with thicker drop panels supported upon "mushroom head" conical capitals with round columns or various prismatic brackets with square columns. Early codes distinguished between four-way reinforcement and two-way reinforcement systems. The four-way system, intricate by today's standards, was soon abandoned because of the excessive field labor required, particularly with the truss bar systems popular then. Also, effective depth was sacrificed when four layers of the reinforcement intersected at the columns. As forming costs increased, most flat slab design eliminated either the drop panels or the capitals and brackets. Today, the flat plate form is preferred for light loads and short spans; the flat slab with capitals and brackets only, in heavy duty designs where few partitions are used; and the flat slab with drop panels only, for heavy duty design with or without partitions. Edge beams can be used with all three forms of the solid flat slab, but they are usually avoided where not structurally required to avoid the added forming expense. The waffle flat slab (Chapter 7) extends both the span and load limits without excessive dead weight.

LIMITING SIZE OF COLUMNS

Either the drop panel or the capital can be used to reduce the column size required by the limits on design shear strength (two-way) around the column ("punching shear"). See Table 5-2 for comparison of column sizes required for shear in flat plate designs. See also Figs. 5-35 through 5-40. If capitals are used, approximately the same total shear periphery as that of the column at the flat plate must be furnished by the capital within a 45° angle to the vertical axis of the column. As the column size below the capital is reduced, the total stiffness of the equivalent column will be reduced and less unbalanced moment will be required to be transferred by shear at the edge of the column. Thus the size of the capitals required will usually be somewhat smaller than the minimum size exterior columns tabulated.

DROP PANELS

The use of a drop panel will increase the shear capacity of the slab system. It will also increase the relative stiffness of the slab which reduces the unbalanced moments transferred to the columns. The Code requires that the drop panel extend at least one-sixth of the span length in each direction and project below the slab at least one-quarter of the slab thickness in order for the full effective depth of the drop to be used for negative moment reinforcement (Sections 13.3.7.1 and 13.3.7.2). The depth of the drop should not be assumed more than one-fourth of the distance between the edge of the drop and the face of the column or edge of the capital (Section 13.3.7.3).

In the slab with drop panel examples presented herein, it will be assumed that the designer, faced with a live load and span requiring an excessively large column in a flat plate design, will proceed with a flat slab design as follows: (1) establish a maximum size of column tolerable, say, 0.075 ℓ, (2) assume a drop panel $\ell/3$ in length, (3) calculate the depth through the drop panel required for vertical two-way shear strength (Sections 11.12.1.2 and 11.12.2.1), and (4) calculate the minimum slab thickness beyond the drop panel to avoid the deflection calculations otherwise required (Section 9.5.3) and to satisfy the one-way (beam) shear at the edge of the drop panel (Section 11.3.1).

As a simplification of the forming, if possible, the slab dimension between the drop panels should be an even multiple of 2 ft and preferably of 4 ft. If the span is not an even multiple in feet, it will usually be preferable to add the few odd inches to the drop panel dimension. For edge columns it will usually be necessary to proceed through Steps 3 and 4 twice, since the alternate loading condition requires calculation of the combined gravity load shear and shear developed by transfer of part of the unbalanced moment from the slab to the column.

DIRECT DESIGN METHOD

The best way to explain the application of the "direct design method" to flat slabs is by numerical examples. It is intended that the user of this guide first proceed through the simpler direct design of flat plates (Chapter 5) before using this section. The following numerical examples are intended to illustrate the differences between the Direct Design Method of past codes which were based on the stiffness ratio of an equivalent column (α_{ec}), and the Direct Design Method of ACI 318-95. The use of the stiffness ratio, α_{ec}, is permitted by Commentary R13.6.3.3. The examples are also intended to show, for the stiffness ratio, the complications added by the drop panels or capitals with the least possible repetition of calculation routines explained in Chapter 5. The chart and short-cut approximate calculations for the slab stiffness of flat plates (Chapter 5) do not apply to flat slabs with drop panels.

For general use where no available design aids are applicable, as with longer drop panels, diamond shaped drops (oriented at 45° to the column grid), tapered depth drops, waffle slab solid heads, or other variations for architectural effect, design examples are illustrated in this chapter. A short cut approximation for trial sections with standard drops is compared to the 1995 method. All design examples selected to illustrate applications of the 1995 Code can be compared directly to tabulated designs.[†]

EXAMPLE 1. *Direct design method—flat slab with drop panel—square exterior panel*

$\ell_1 = \ell_2 = 25'$-0″; $w_\ell = 300$ psf; maximum column acceptable: square $c_1 = c_2 = 0.067\ \ell_1$; use $f'_c = 4{,}000$ psi; Grade 60 bars; concrete weight 150 pcf. (See similar design in *CRSI Handbook, 1996*); minimum column 20-in. square; with slab thickness $= 8\frac{1}{2}$ in.; drop depth $= 6\frac{1}{2}$ in. Story height, ℓ_c, is 12′-0″; columns above and below slab.

In Fig. 6-1 columns are taken as interior columns past the first line. There are no beams and so $I_b = 0$; $\alpha = 0$ (Section 13.0). When the provisions of Table 9.5(c) are met, no deflection calculations are required by the Code (Section 9.5.3.2).

Step 1. Thickness to avoid deflection calculations

$$h \text{ need not be greater than } = \ell_n/33$$

The exterior panel clear span $= 300 - 20 = 280$ in.; the minimum thickness is $h = 280/33 = 8.48$ say $8\frac{1}{2}$ in.

Step 2. Shear for gravity load only

$$w_\ell = 300 \text{ psf } (L = 300/1.7 = 176);$$

$$\text{slab weight} = (8.5/12)(0.150) = 0.106 \text{ ksf}$$

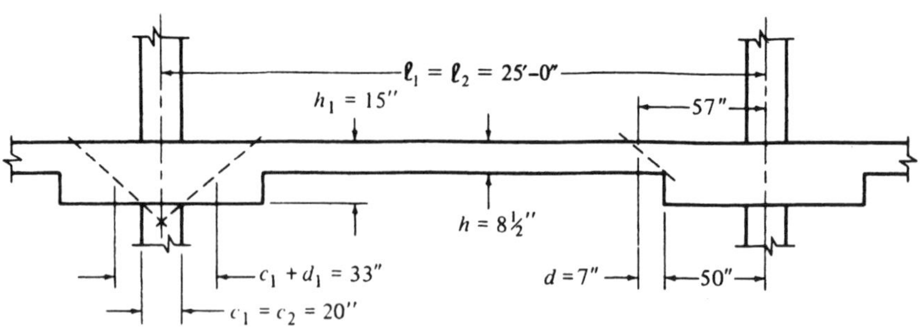

Figure 6-1 Dimensions—trial section.

[†] *CRSI Handbook*, 1996.

$$w_d = 0.106 \times 1.4 = 0.148 \text{ ksf}; \; w_u = 0.300 + 0.148 = 0.448 \text{ ksf}$$

Try drop panel, $h_1 = 15$ in. For 1 in. cover and #8 bars, $d_1 = 13$ in.

$$\ell/3 = (1/3)(25'\text{-}0'') = 8.33 \text{ ft}; \; c_1 = c_2 = 20 \text{ in.}; \; (c + d_1) = 33 \text{ in.}$$

Total interior panel factored shear, $V_u = (0.448)[(25)^2 - (33/12)^2]$

$$+ (8.33)^2(6.5/12)(0.150)(1.4)$$

$$= 277 + 8 \text{ kips} = 285 \text{ kips}$$

Design shear strength at interior column (Section 11.2.2.1-c, Eq. 11-37). ϕV_c based on Section 11.12.2.1-c,Eq. 11-37 governs; Section 11.12.2.1-b is not applicable because $(\alpha_s d/b_o + 2) = (40)(13)/(4 \times 33) + 2 = 5.94 > 4$

$$\phi V_c = \phi b_o d 4\sqrt{f_c'} = (0.85)(4)(33)(13)(4)\sqrt{4000}/1000$$

$$= 369 \text{ kips} > V_u = 285 \text{ kips} = \text{O.K.}$$

Factored shear at the edge of the drop: $V_u = 285$ kips − weight of drop − loads on drop − loads on strip of width d (slab) around drop. Width = 8′-4″ + (2 × 7)″ = 9.50 ft.

$$V_u = 285 - 8 - (0.448)(9.50)^2 = 237 \text{ kips}$$

$$\phi V_c = (0.85)(4)(114)(6)(2)\sqrt{4000}/1000$$

$$= 294 \text{ kips} > V_u = 237 \text{ kips O.K. (Section 11.1.1).}$$

Note that the 15 in. total drop panel thickness is about right with some reserve capacity for combined shear to be calculated in Step 4. Slab thickness is satisfactory provided that the assumed cover and one and a half bar diameters do not exceed 2 in.* Use: Slab $h = 8\frac{1}{2}$ in.; drop panel, $h_1 = 15$ in., 8′-4″ square; columns 20 in.

Step 3. Negative and positive moments (Section 13.6.3)

Slab load between drops = 0.448(25) = 11.20 k/ft

Slab load at drops = 11.20 + (6.5/12)(0.150)(8.33)(1.4) = 12.15 k/ft

M_o (Section 13.6.2.2) See Fig. 6-2 = 768 ft-kips

---

* The slab thickness need not be increased even if the one-way factored shear at the edge of the drop, V_u, were to exceed the design shear strength of the concrete ($\phi V_c = 294$ kips). It will usually be more economical to increase the minimum drop panel width dimension, $\ell/3$, slightly. In this case, using the minimum dimension, $\ell/3$ = 8′-4″, the distance between the drop panels is (25′-0″) − (8′-4″) = 16′-8″. As previously noted, it would be desirable to use a drop panel 9′-0″ square so that the distance between drop panels for slab forms becomes a multiple of 4′-0″, the standard width of plywood sheets, especially if many repetitions of this panel will occur.

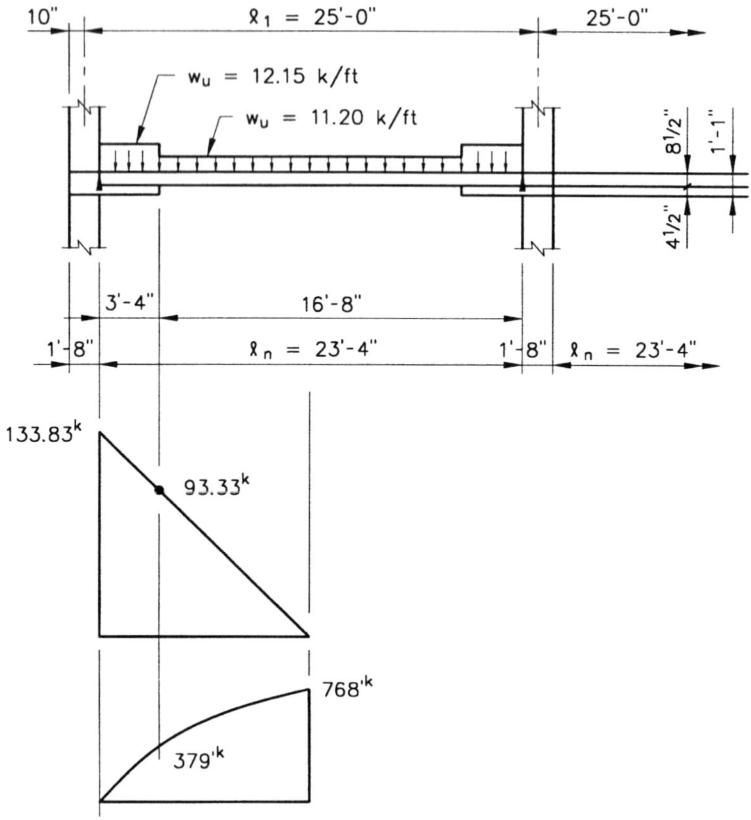

Figure 6-2 Simple span beam for M_o.

End span moments by coefficients (Section 13.6.3.3):

$-M_u$ (exterior support) $= 0.26M_o = 0.26(768) = 200$ ft-kips
$+M_u$ (exterior support) $= 0.52M_o = 0.52(768) = 399$ ft-kips
$-M_u$ (1st interior support) $= 0.70M_o = 0.70(768) = 538$ ft-kips
$-M_{un}$ (Section 13.6.3.6) $= 0.30M_o = 0.30(768) = 230$ ft-kips

Interior span moments by coefficients (Section 13.6.3.2):

$-M_u = 0.65M_o = 0.65(768) = 499$ ft-kips
$+M_u = 0.35M_o = 0.35(768) = 269$ ft-kips

Step 4. Combined shear due to gravity load and unbalanced moment at interior column

Up to this point in the example, all calculations have been based on the tentative trial section dimensions. These dimensions for the slab depth and the column size have been checked to satisfy all requirements except that of two-way shear at a distance $d/2$ from the columns due to combined gravity loads and transfer of unbalanced moments by shear. Note that the drop panel depth selected was ample to allow about 29% excess design shear strength for gravity loads only at the section $d/2$ from the column. Since unbalanced moment to interior columns with equal spans on each side is not large, this design shear strength will probably suffice for the combined condition, at interior columns beyond the first line, and perhaps for the first interior columns also.

For square panels, the prescribed formula for unbalanced moment to the interior columns reduces to

$$\text{Code } M_{un} = (0.07)\{[(w_d) + 0.5(w_\ell)](\ell_2)(\ell_n)^2 - (w_d')(\ell_2')(\ell_n')^2\}$$

(Section 13.6.9.2)

$$= (0.07)\{[(0.148) + (0.5)(0.300)](25)(23.33)^2 - (0.148)(25)(23.33)^2\}$$

$$= 143 \text{ ft-kips}$$

Combined shear

The fraction of the unbalanced moment to be transferred by flexure is prescribed by the following formula (Section 13.5.3.2):

$$\gamma_f = 1/[1 + (\tfrac{2}{3})\sqrt{(b_1/b_2)}] \qquad \text{(Eq. 13-1)}$$

$$= 1/[1 + (\tfrac{2}{3})\sqrt{31/31}] = 0.60$$

The fraction of the unbalanced moment to be transferred by shear is

$$\gamma_v = 1 - \gamma_f = 1 - 0.60 = 0.40$$

The moment of inertia of the shear section (Fig. 6-3), J_c is

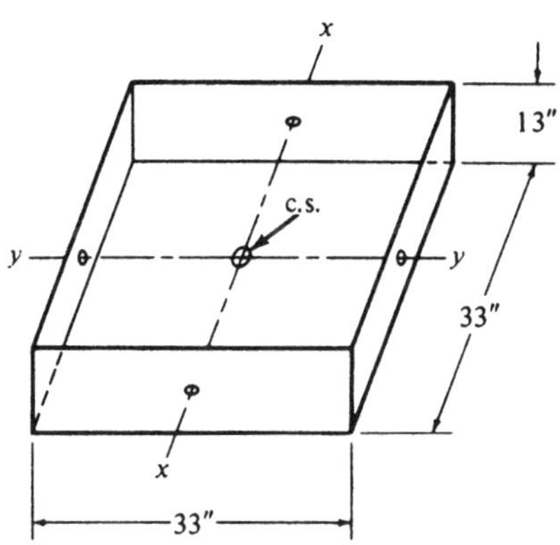

Figure 6-3 Shear section dimensions for interior column.

$$J_c = \frac{(2)(33)(13)^3}{12} + \frac{(2)(13)(33)^3}{12} + (2)(33)(13)\left(\frac{33}{2}\right)^2$$

$$= 323{,}537 \text{ in.}^4$$

The area of the shear section is

$$A_c = (4)(33)(13) = 1716 \text{ in.}^2$$

The unbalanced moment transferred by shear is

$$M_{un} = (0.40)(143) = 57 \text{ ft-kips}$$

Maximum factored unit shear stress caused by gravity load and transfer of unbalanced moment.

$$v_u = V_u/A_c + \gamma_v M_{un} c/J_c \leqslant \phi 4\sqrt{f_c'}$$

$$\text{(See Fig. 6-2) } V_u = 133.83(2) + 12.15(20/12) - 0.448(33/12)^2$$

$$= 267.66 + 20.25 - 3.39 = 284.52 \text{ kips}$$

$$M_{un} = 143 \text{ ft-kips}$$

$$v_u = 284{,}520/1716 + 0.40(143{,}000)(12)(16.5)/323{,}537$$

$$= 166 + 35 = 201 \text{ psi} < 0.85(4)\sqrt{4000} = 215 \text{ psi OK}$$

Step 5. Combined shear due to gravity load only and unbalanced moment at exterior column.

Area of critical shear section, $A_c = b_o d = [(26.5)(2) + 33](13) = 1{,}118 \text{ in.}^2$ Axis Z-Z passes through the center of gravity of the shear area; point c. s. on the Z-axis is the centroid of shear areas.
See Figs. 6-4 and 6-5.

$$c_{AB} = (2)(13)(26.5)(\tfrac{1}{2})(26.5)/1{,}118 = 8.32 \text{ in.}$$

$$c_{CD} = 26.5 - 8.32 = 18.18 \text{ in.}$$

$$J_c = (26.5)(13)^3/6 + (13)(26.5)^3/6$$

$$+ (2)(13)(26.5)(26.5/2 - 8.32)^2 + (13)(33)(8.32)^2$$

$$= 9{,}703 + 40{,}321 + 16{,}746 + 29{,}666 = 96{,}436 \text{ in.}^4$$

The fraction of the unbalanced moment to be transferred by flexure is prescribed by the following formula (Section 13.5.3.2):

$$\gamma_f = 1/[1 + (\tfrac{2}{3})\sqrt{(b_1/b_2)}] \qquad \text{(See Figs. 6-4 and 6-5).}$$

$$= 1/[1 + (\tfrac{2}{3})\sqrt{(26.5/33)}] = 0.626$$

The fraction to be transferred by shear;

$$\gamma_v = 1 - \gamma_f = 1 - 0.626 = 0.374$$

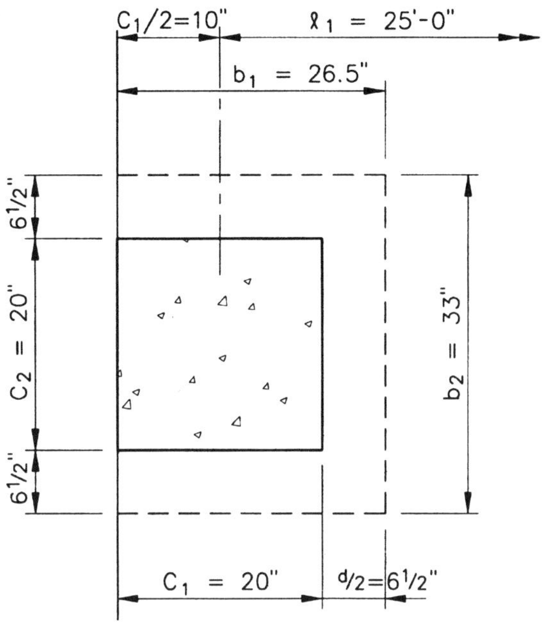

Figure 6-4 Plan at exterior column.

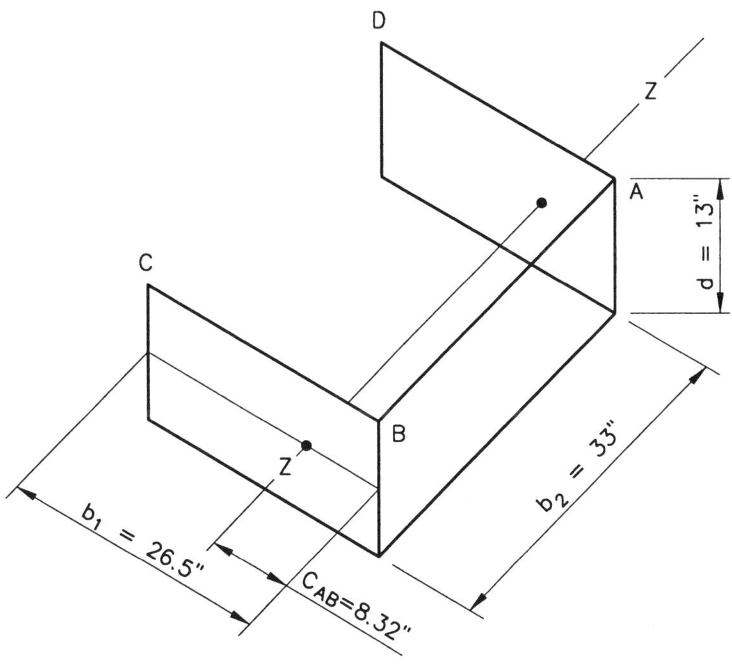

Figure 6-5 Shear section dimensions at exterior column.

Factored floor load at the exterior column;

From Fig. 6-2	=	133.83
From column interior face to slab edge		
(12.15)(1.67)	=	13.82
Exterior curtain wall (0.22)(25)	=	5.50
Shear decrease (moment difference)	=	−13.52
V_u	=	139.63 kips

Maximum combined factored shear stress due to gravity loads and to moment transfer;

$$v_u = V_u/A_c + \gamma_v M_{un} c_{AB}/J_c$$

$$= 139{,}630/1{,}118 + (0.374)(230{,}000)(12)(8.32)/96{,}436$$

$$= 125 + 89 = 214 \text{ psi} < \phi 4\sqrt{f_c'} = (0.85)(4)(\sqrt{4000}) = 215 \text{ psi OK}$$

(Note that these unit shear calculations are slightly more critical at the first interior column and much more critical at the exterior column.)

The completion of the design for this flat slab example is very similar to that for flat plate designs (see examples in Chapter 5), and will not be repeated here. Bar length detail requirements for the flat slab with drop panels are similar to those for flat plates also, and will not be repeated.

Recommendations for Selection of Flat Slab Thickness

To avoid complex calculations, the authors suggest that selection of slab thickness conform to the minimums, except for elements such as cantilevers (see Chapter 2). The application of concepts such as M_{cr}/M_a (Section 9.0) ratio, cracked transformed moment of inertia, etc. (Section 9.5.2.3) to deflection calculations for two-way slabs complicated by drop panels, plus lack of specific Code requirements for treatment of same creates special problems outside the scope of this guide.

If a thickness less than the minimum is desired, however, deflections, both immediate and long-term, must be calculated (Section 9.5.3.4). Such computed immediate deflections for service live load must not exceed $\ell_n/180$ for roofs nor $\ell_n/360$ for floors; long-term deflections for roofs or floors supporting or attached to nonstructural elements likely to be damaged, $\ell_n/480$, and if not attached to such elements, $\ell_n/240$ (Table 9.5-b). Where approximately equal spans are employed without edge beams, the minimum thickness for corner panels will control. For large areas with a high number of interior panels, the panel minimum thickness may be employed throughout if edge beams are sufficiently stiff, $\alpha \geqslant 0.80$ (Section 9.5.3.3).

For a three-bay wide structure with unequal spans, and the exterior spans larger, the minimum thickness will usually be established by the corner panel.

Special Conditions

OPENINGS. Openings of any size in a two-way slab are permitted (Section 13.4) if all applicable general requirements for strength and serviceability are satisfied (Section 13.4.1). The Code specifically permits openings within the following limits without a special analysis (Section 13.4.2):

1. Any size within the area common to both middle strips
2. One-eighth of the width of each column strip in the area common to both column strips
3. One-quarter of the width of either strip in the area common to one column and one middle strip

Significant variations in the wording of the provisions for cases (1), (2), (3) will be noted. For case (1) the "total amount of reinforcement required for the panel *without* the opening" must be maintained. Thus, if the entire center quarter panel were omitted, the center half span of both middle strips, reinforcement for the factored loads removed as well as the loads remaining must be supplied in the column strips. In case (2), "the equivalent of the reinforcement interrupted shall be added on all sides of the openings." In case (3), the actual wording is that "not more than one-quarter of the reinforcement in either strip shall be interrupted by the opening. The equivalent of reinforcement interrupted shall be added on all sides of the openings." The single specific requirement for analysis of the effect of openings meeting limitations of cases (1), (2), and (3) is that design shear strength requirements must be met. The special analysis for openings not within the three limitations, of course, includes shear (Section 11.12) and deflection limits (Section 9.5.3.4).

Effect of openings on design shear strength

The shear requirements related to openings in flat slabs are quite specific (Section 11.12). When the openings are located within a column strip, cases (2) and (3), "that part of the periphery of the critical shear section (around the column head) which is enclosed by radial projections of the opening of the centroid of the loaded area (column) shall be considered ineffective" (Section 11.12.5.1). "For slabs with shearheads, the ineffective portion of the perimeter shall be one-half of that defined in (a)." Note that *all* shear capacity constants (A_c, J_c) are affected (see Fig. 6-6; see also Fig. 7-6 for the effect of waffle slab voids near the column).

Openings in waffle flat slabs

For the special case of waffle flat slabs, any opening entirely in the top slab, that is between joists, can be regarded as satisfying all requirements without special analysis. Any top bars which would otherwise be interrupted can simply be spread on both sides of the hole. Where larger openings are required in waffle slabs, cases (2) and (3), and the joist ribs must be interrupted, the design

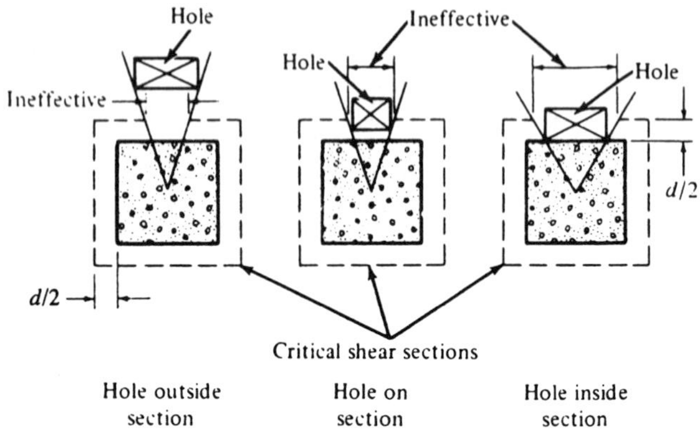

Figure 6-6 Critical shear sections reduced by openings.

moment strength requirements can be satisfied by adding bottom steel equivalent to that interrupted in the adjacent ribs.

Openings in two-way slabs on beams

For two-way slabs on beams, openings in the slabs larger than those around which the reinforcement can be spread without interruption, special analysis may be avoided by framing the opening with small auxiliary beams to carry tributary loads to the main two-way beams (see Fig. 6-7).*

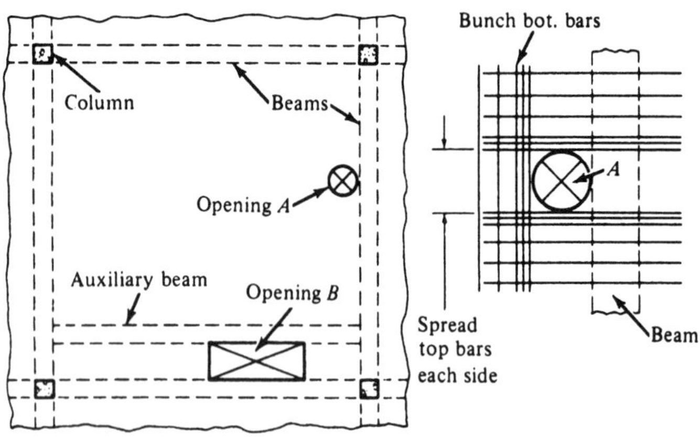

Figure 6-7 Openings in two-way slabs with beams.

* "Determining Safe Locations for Holes in Concrete Floors," Paul F. Rice; *Plant Engineering*, April 14, 1977.

TRANSFER OF MOMENT AND SHEAR TO CORNER COLUMNS. A special problem exists in the design for transfer of unbalanced moment and shear to corner columns where no edge beam or cantilever is present. The prescribed critical shear section for two-way shear is two-sided, so that the centroid is off the column center in both directions, and the column center itself may be off center in the half-column strip. Both conditions are inefficient. The Code provides the same critical sections as for all columns: for moment, at the face of the column or capital; for shear, a distance $d/2$ from the column face for two-way shear, and a section across the half-column strip and half the middle strip for one-way (beam) shear. For two-way slabs with beams, $\alpha > 1.0$, in each direction additional steel is required top and bottom (Section 13.3.6).

For the special case of a flat slab exterior corner with or without drop panel, and no edge beams, the authors suggest that the user investigate a critical section for moment on a diagonal in addition to the orthogonal critical sections specified in the Code (see Fig. 6-8).

The additional analysis is intended to ensure that sufficient reinforcement is provided and effectively developed to avoid negative moment (top) diagonal cracks at the diagonal section. The sections prescribed for one-way shear parallel to the edges extending across the entire column half-strip and half the middle strip seem unrealistic.

The exterior moments assigned to the two column half-strips can be combined and oriented at right angles to a critical section (at 45° for a square panel). Similarly, the components of force carried in the two sets of orthogonal top bars can be combined. Only those bars fully developed beyond the critical section can be considered fully effective. Usually, some additional top steel will be

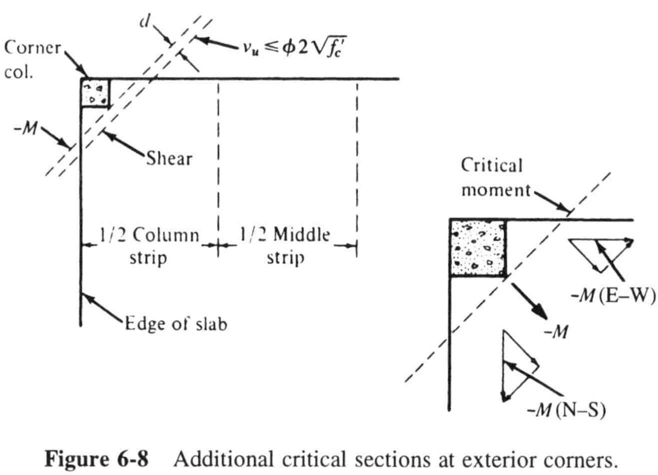

Figure 6-8 Additional critical sections at exterior corners.

found desirable. To avoid building up additional layers of top steel and the consequent decrease in the effective depth, such additional bars can be placed parallel to the edges within a width equal to column side, c, plus 1.5 h on one side of the column (Section 13.5.3.2).

In addition to the beam (one-way) shear investigation on the suggested diagonal section, some adjustment in the fraction of the unbalanced moment transferred by shear may be required. See (Section 13.5.3.3) for corners with beams.

ROUND COLUMNS AND CONICAL CAPITALS. Round columns with conical capitals using standard reusable steel forms were very commonly used with the original flat slab and drop panel system. Single use standard round column forms are now available, and the round columns are frequently used with capitals *or* drop panels.

Two special problems with the use of round columns and round capitals arise under the Code: (1) determination of the torsional shear constant, J_c, of a circular section for moment transfer by shear (Section 11.12.6), and (2) determination of the column stiffnesses, K_c, when a conical capital is furnished at the upper end (Section 13.7.4).

Shear transfer constant, J_c, for round interior columns

For round interior columns, $J_c = 4dr^3 + \dfrac{rd^3}{3}$, where d = the effective depth of the slab or drop panel above, r = the radius of the round shear section = $1/2\ (t + d)$, and t = diameter of the round column (or capital) at the bottom of the slab or drop panel above (for derivation, see Figs. 6-19, 6-20 and 6-21 at the end of this chapter).

Round and square columns, without capitals or brackets, may be considered equivalent when the diameter of the round column equals 1.14 times the side dimension of the square column. Cross-sectional area, flexural stiffness, and the shear constant, J_c, will be closely equivalent. See Figs. 6-22 through 6-26 for minimum size square columns with flat slabs.

SHEAR TRANSFER CONSTANT, J_C, FOR ROUND EXTERIOR COLUMNS. For round columns (or capitals) at or near the edge of the slab, the exact solution for J_c involves a semicircular shear section plus two straight tangents perpendicular to and extending to the edge (see Fig. 6-9). A conservative value for this open section should be used. The authors suggest $J_c = 2.2\ dr^3 + rd^3/3$ for all round edge columns whether tangent to the edge or set back slightly, until the cantilever setback develops sufficient moment so that the column can be considered as an interior column. See Fig. 5-32 for the similar case of square edge columns.

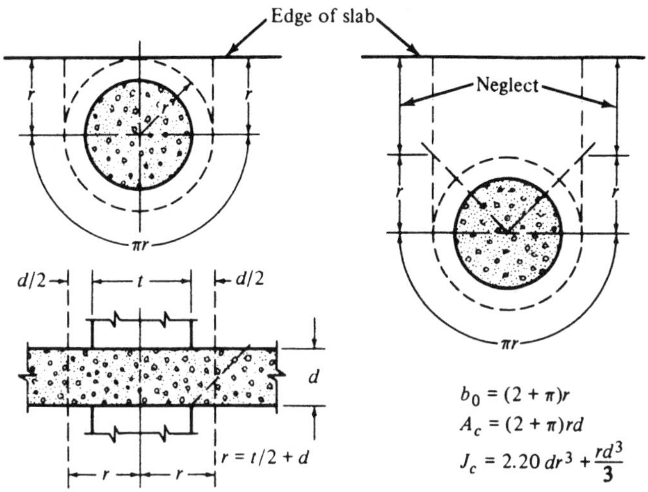

Figure 6-9 Torisional constant, J_c, for round edge exterior columns.

Effect of Capitals or Brackets upon Column Flexural Stiffness, K_c

An exact solution can be obtained using either the moment-area or column-analogy method. In the exact solution, an infinite moment of inertia within the depth of the slab or drop panel, a variable moment of inertia for the depth of the capital, and a constant moment of inertia for the gross cross section of the column for the height from the floor line to the bottom of the capital must be considered (Section 13.7.4). The exact solution will be complicated by the variable moment of inertia portion and its eccentric location from the midpoint of the column length. As a result of this eccentricity, separate carry-over factors will result at the top and bottom of the column for moment distribution in the equivalent frame method. Separate stiffness factors, K_c, will result for use with the equivalent frame method (Section 13.7).

An approximate average value for K_c, satisfactory for most practical design applications, can be quickly determined (see Fig. 6-10). Replace the actual col-

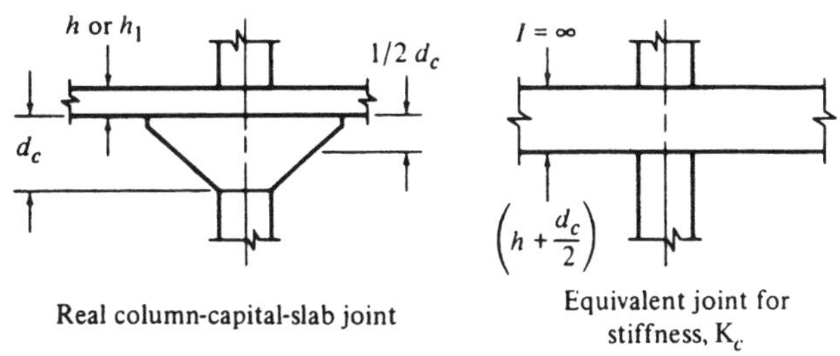

Figure 6-10 Column flexural stiffness, K_c, for round columns with capitals.

umn capital with an "equivalent drop panel" equal to the depth of slab (or actual drop panel if used) plus *one-half* the depth of the capital. Use the "equivalent drop panel" depth for h and solve for an average K_c which may be used at the joints above and below the column. It will be necessary to use the column-analogy formula since the ratio of "equivalent drop panel depth" to story height will often exceed the range of the chart (Fig. 5-13).

This approximation assumes an infinite stiffness within the joint to the midpoint of the capital depth and neglects the smaller bottom half of the capital. The approximation would seem justifiable since the Code does not specifically prescribe the exact solution: "Determination of the moment of inertia of columns at any cross section outside of joints or capitals using the gross section area of concrete shall be permitted" (Section 13.7.4.1). "Variation in moment of inertia along axis of columns shall be taken into account" (Section 13.7.4.2). This approximation does "take into account" the addition to column stiffness contributed by a capital as well as the proportions of the capital itself.

EQUIVALENT FRAME METHOD

The equivalent frame method of analysis, for flat slabs with drop panels is complicated by the presence of the drop panel, which causes a variation in the flexural and torsional stiffness of the slab. An example of three spans with cantilevers and similar concrete dimensions as the end span designed by the direct design method will be shown to demonstrate the equivalent frame method. With the equivalent frame method there is no limitation on the ratio of live load to dead load (unfactored) as with the direct design method of $L/D \leq 2$ (Section 13.6.1.5). The factored live load is increased to 400 psf resulting in a ratio of live load to dead load (unfactored) of $(400/1.7) \div 100 = 2.35$.

EXAMPLE. Figure 6-11 shows the dimensions of the structure with square drop panels and square bays.

Step 1. Determine the slab thickness, h

The Code makes no recommendations for minimum slab thickness for slabs with cantilevers when deflections are not calculated. If the cantilever factored moment approaches 0.65 M_o, the minimum slab thickness can be based on that for an interior span (see Code Table 9.5(c)).

$$h(\text{minimum}) = \ell_n/36 = \frac{300 - 22}{36} = 7.72 \leq 8''$$

Step 2. Trial column dimensions

Use square columns with $c_1 = c_2 = 22$ in.

Step 3. Determine column flexural stiffness, K_c

Column flexural stiffness can be determined from Fig. 5-13 used with the direct design method.

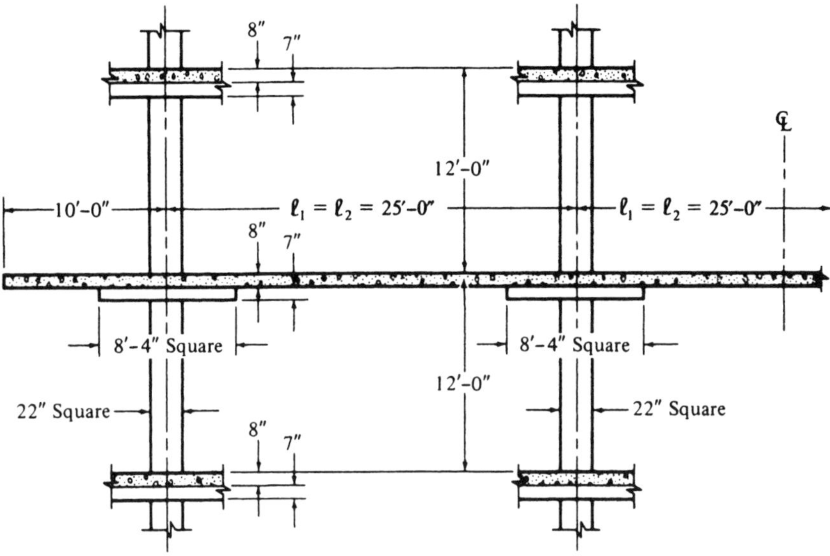

Figure 6-11 Flat slab equivalent frame.

$$K_c/E_c = \frac{I_c}{(\ell_c - h)}\left[1 + \frac{3\ell_c^2}{(\ell_c - h)^2}\right] = \left[\frac{(0.0833)(22)^4}{(144 - 15)}\right]\left[1 + \frac{(3)(144)^2}{(144 - 15)^2}\right]$$

$= 717 \text{ in.}^3$ (710 in.^3 by Fig. 5-13 and direct design method)

This method of determining the column stiffness assumes that the column has an infinite moment of inertia for the depth of the drop panel (Section 13.7.4.3).

Step 4. Determine slab flexural stiffness, K_s

The slab stiffness will be determined by the column analogy method. Moment of inertia of slab cross section: See Fig. 6-12

I_g (slab between drops) $= (300/12)(8)^3 = 12{,}800 \text{ in.}^4$

I_g (slab at drop panel)

(300)(8) =	2,400 × 4	=	9,600
(100)(7) =	700 × 11.5	=	8,050
	3,100 (5.7)		17,650

$$I_g = (300/3)(5.7)^3 + [(2)(100)/(3)](2.3)^3 + (100/(3)(9.3)^3$$

$$= 18{,}519 + 811 + 26{,}812 = 46{,}142 \text{ in.}^4$$

I_g (slab at column)

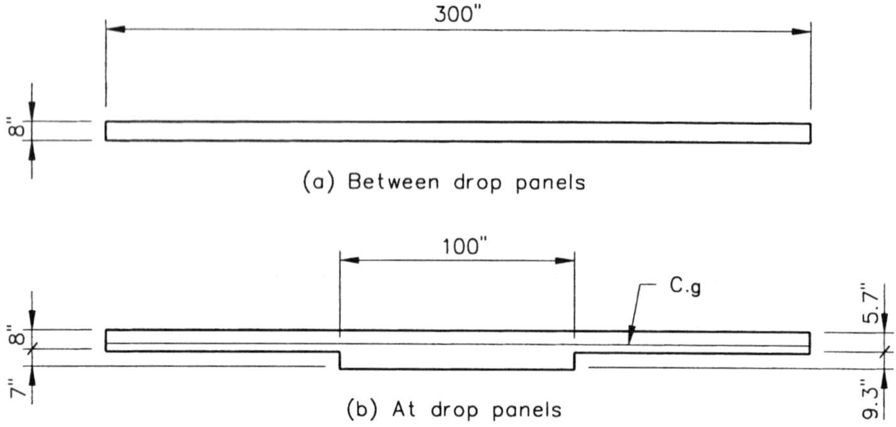

Figure 6-12 Slab sections.

$$I_g = I_g/[1 - c_2/\ell_2]^2 \quad \text{(Section 13.7.3.3)}$$
$$= 46{,}142/[1 - 22/300]^2 = 53{,}734 \text{ in.}^4$$

$1/I_g$ (slab between drops) $= 1/12{,}800 = 0.0000781$

$1/I_g$ (slab at drop panel) $= 1/46{,}142 = 0.0000217$

$1/I_g$ (slab at column) $= 1/53{,}734 = 0.0000186$

A_{ac} = area of analogous column. See Fig. 6-13.

$$\begin{aligned} &= (0.0000781)(16.67)(12) &&= 0.0156 \\ &+ (0.0000217)(3.25)(12)(2) &&= 0.0017 \\ &+ (0.0000186)(0.917)(12)(2) &&= \underline{0.0004} \\ & &&\ 0.0177 \end{aligned}$$

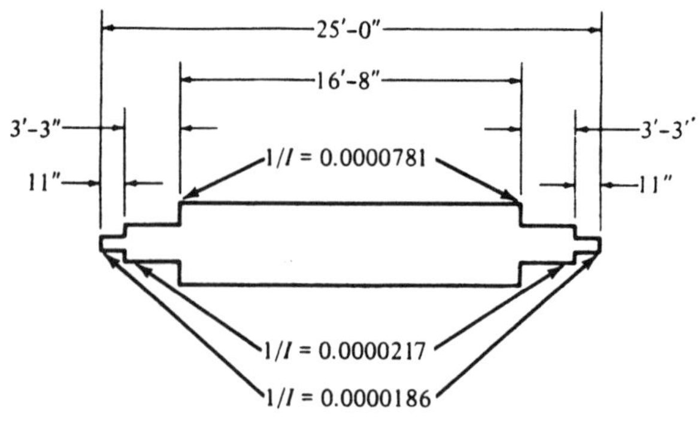

Figure 6-13 Flat slab flexural stiffness, K_s, by column analogy.

$$I_{ac} = \frac{(0.0000186)(300)^3}{12} + \frac{(0.0000031)(278)^3}{12} + \frac{(0.0000564)(200)^3}{12}$$

$$= 41.8 + 5.6 + 37.6 = 85.0$$

$$K_s/E_c = \frac{1}{A_{ac}} + \frac{(1)(\ell_1 2)(\ell_1/2)}{I_{ac}}$$

$$= \frac{1}{0.0177} + \frac{(1)(150)(150)}{85.0} = 56.5 + 264.7$$

$$= 321 \text{ in.}^3$$

Step 5. Torsional stiffness of the slab, K_t

The member considered for torsional stiffness is one that consists of a part of the slab perpendicular to the direction in which moments are being considered. This portion of slab is considered to have a width c_1 (width of column in direction moments are being considered) and a constant depth equal to that of the slab (Section 13.7.5.1). K_t can vary considerably depending on the depth used in calculating the torsional constant, C. Calculated values for, K_t/E_c, can be based on a C-value, using the depth of slab equal to that between drop panels, a varying depth of slab, an average depth of slab, or the depth of the slab at the drop panel. The "exact" solution, K_t, of a varying depth torsional member was based on the average effective angle of rotation and was calculated as shown by Eberhardt and Hoffman.*

$$\frac{1}{K_t} = \theta_t = \frac{\ell_2(1 - c_2/\ell_2)^5}{40GC_D} + \ell_2(1 - 2x_2/\ell_2)^5\left[\frac{1}{C_s} - \frac{1}{C_D}\right] \qquad \text{Eq. (6-1)}$$

where

x_2 = distance from centerline of span to edge of drop panel

C_s = torsional constant for slab cross section of width c_1 and depth equal to slab thickness between drop panels

C_D = torsional constant for slab cross section of width c_1 and depth equal to total slab thickness at drop panel

If E_c is assumed equal to 2 G, then Eq. (6-1) becomes

$$\frac{E_c}{K_t} = \frac{\ell_2(1 - c_2/\ell_2)^5}{20C_D} + \frac{\ell_2(1 - 2x_2/\ell_2)^5}{20}\left[\frac{1}{C_s} - \frac{1}{C_D}\right] \qquad \text{Eq. (6-2)}$$

$$C_s = \left[1 - 0.63\frac{x}{y}\right]\frac{x^3(y)}{3} = \left[1 - 0.63\frac{8}{22}\right]\frac{(8)^3(22)}{3} = 2890 \text{ in.}^4$$

* Arthur C. Eberhardt and Edward S. Hoffman, "Equivalent Frame Analysis for Slab Design," *ACI Journal*, May 1971.

$$C_D = \left[1 - 0.63\,\frac{15}{22}\right]\frac{(15)^3(22)}{3} = 14{,}120 \text{ in.}^4$$

$$\frac{E_c}{K_t} = \frac{300(1 - 22/300)^5}{(20)(14120)} + \frac{300(1 - 100/300)^5}{20}\left[\frac{1}{2890} - \frac{1}{14120}\right]$$

$$= 0.00127$$

$$K_t/E_c = 787 \text{ in.}^3$$

Step 6. Stiffiness of equivalent columns

$$K_{ec}/E_c = \frac{\Sigma K_c}{1 + \Sigma K_c/K_t} = \frac{(2)(717)}{1 + \dfrac{(2)(717)}{787}}$$

$$= 508 \text{ in.}^3$$

Step 7. Factored loads

The service live load is 400/1.7 or 235 psf. The dead load of the slab beyond the drop panel is 8/12 × 150 or 100 psf. The service live load exceeds three-quarters of the dead load [235 > 3/4 (100)]. When live load is greater than three-quarters of the dead load, the equivalent frame method requires that moments be determined for two patterns of loading, using three-quarters of factored live load (Section 13.7.6.3). Alternate panels are loaded for maximum and minimum positive moments. Two adjacent panels only are loaded for maximum negative moments. In no case should factored moments be less than would occur with factored live load, w_ℓ, on all spans (Section 13.7.6.4).

$$w_d(\text{slab beyond drop panel}) = \frac{8}{12}\left[\frac{(150)(1.4)(25)}{(1000)}\right] = 3.50 \text{ klf}$$

$$w_d(\text{drop panel}) = \frac{7}{12}\left[\frac{(150)(1.4)(8.33)}{(1000)}\right] = 1.02 \text{ klf}$$

$$100\%\ w_\ell = \frac{(400)(25)}{1000} = 10.00 \text{ klf}$$

$$0.75\ w_\ell = \frac{(0.75)(400)(25)}{1000} = 7.50 \text{ klf}$$

Step 8. Carry-over factors for moment distribution (see Fig. 6-13)

$$COF = \frac{1/A_{ac} - M_C/I_{ac}}{1/A_{ac} + M_C/I_{ac}} = \frac{\dfrac{1}{0.0177} - \dfrac{(150)(150)}{85.0}}{\dfrac{1}{0.0177} + \dfrac{(150)(150)}{85.0}} = \frac{56.5 - 264.7}{56.5 + 264.7} = \frac{-208.2}{321.2}$$

$$= -0.648$$

Step 9. Fixed end moments

The fixed end moments will be determined by the column-analogy method for the equivalent frame (see Fig. 6-13).

Area of M/I diagram for slab uniform load of 1.0 klf (See Fig. 6-14)

1. $(\frac{2}{3})(200)(0.0325) = 4.333$
2. $(200)(0.0407) = 8.140$
3. $(2)(39)(0.0029) = 0.226$
4. $(2)(39)(0.0084)(\frac{1}{2}) = 0.328$
5. $(2)(39)(0.0005)(\frac{2}{3}) = 0.026$
6. $(2)(11)(0.0025)(\frac{1}{2}) = 0.028$
7. $(2)(11)(0.0001)(\frac{2}{3}) = 0.001$

Slab $P_{ac} = 13.082$

Area of M/I diagram for drop panel factored dead load of 1.02 klf (See Fig. 6-15).

1. $(200)(0.00830) = 1.660$
2. $(2)(39)(0.00090) = 0.070$
3. $(2)(39)(0.0014)(\frac{1}{2}) = 0.055$
4. $(2)(39)(0.00035)(\frac{2}{3}) = 0.018$
5. $(2)(11)(0.00077)(\frac{1}{2}) = 0.008$
6. $(2)(11)(0.00003)(\frac{2}{3}) = 0.001$

Drop panel $P_{ac} = 1.812$

The fixed end moments will be equal to the stress at each end of the analogous column. (See Figs. 6-14 and 6-15).

$$\text{FEM} = P_{ac}/A_{ac}$$

$$\text{FEM}\ (w_d) = [(13.082)(3.5) + 1.812]/0.0177$$

$$= 2{,}689 \text{ in.-kips} = 224 \text{ ft-kips}$$

$$\text{FEM}\ (w_\ell) = (13.082)(10)/0.0177$$

$$= 7{,}391 \text{ in.-kips} = 616 \text{ ft-kips}$$

$$\text{FEM}\ (w_d + w_\ell) = 224 + 616 = 840 \text{ ft-kips}$$

$$\text{FEM}\ (w_d + 0.75\, w_\ell) = 224 + (0.75)(616)$$

$$= 686 \text{ ft-kips}$$

$$\text{FEM (cantilever } w_d) = (3.5)(10)^2/2 + (1.02)(4.17)^2/2$$

$$= 175 + 9 = 184 \text{ ft-kips}$$

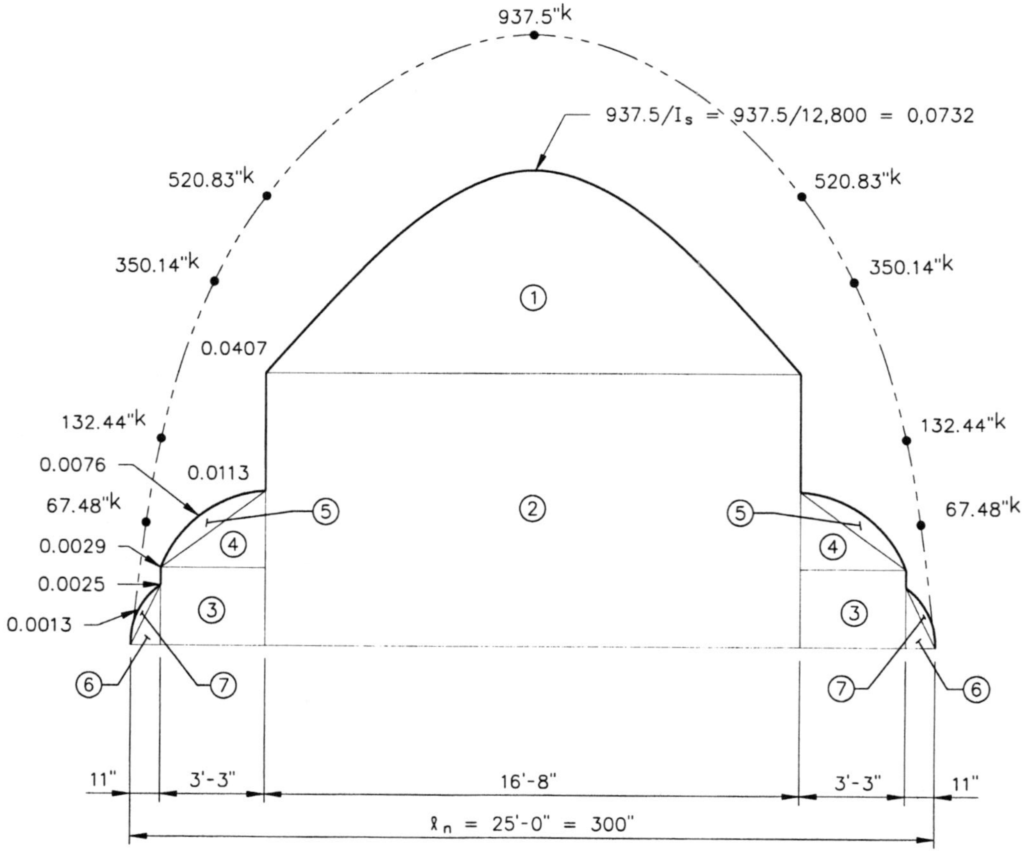

Figure 6-14 M/I diagram for uniform load of 1.0 klf on flat slab.

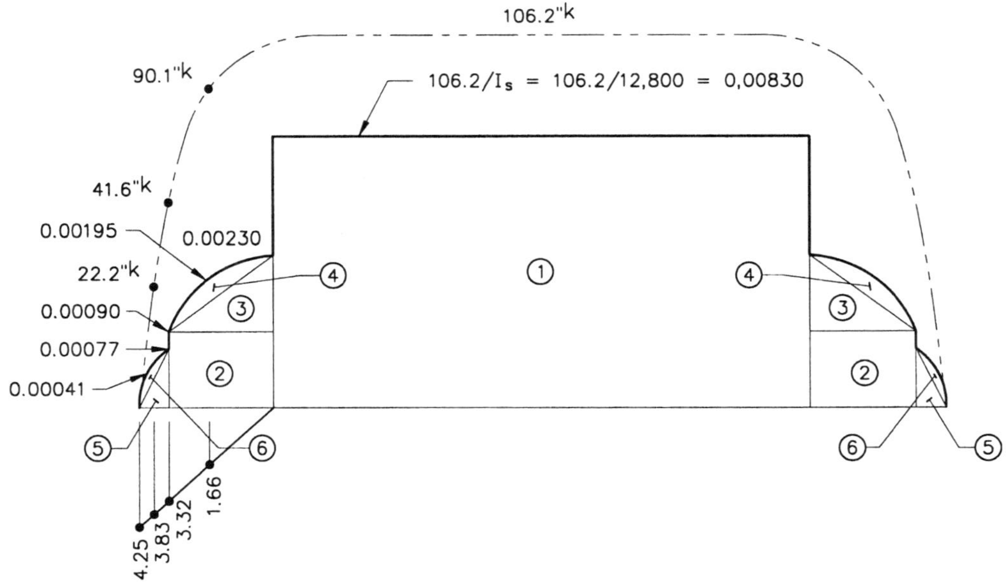

Figure 6-15 M/I diagram for additional dead load of drop panel = 1.02 klf.

$$\text{FEM (cantilever } w_d + w_\ell) = (13.50)(10)^2/2 + (1.02)(4.17)^2/2$$
$$= 675 + 9 = 684 \text{ ft-kips}$$
$$\text{FEM (cantilever } w_d + 0.75\, w_\ell) = (11.0)(10)^2/2 + 9$$
$$= 550 + 9 = 559 \text{ ft-kips}$$

Step 10. Moment distribution factors

For distributing slab moments at exterior and interior columns.

$$DF \text{ (Ext. Col.)} = \frac{K_s}{K_s + K_{ec}} = \frac{321}{321 + 508} = 0.387$$

$$DF \text{ (Int. Col.)} = \frac{K_s}{\Sigma K_s + K_{ec}} = \frac{321}{(2)(321) + 508} = 0.279$$

Elastic analyses of the equivalent frame for the structure shown in Figs. 6-16 and 6-17 can be made by moment distribution using the carry-over factor of Step 8, the fixed end moments of Step 9, and the distribution factors of Step 10. To determine the factored moments in accordance with the Code, it is necessary to analyze the structure with $w_d + w_\ell$ on all spans, with $w_d + 0.75\, w_\ell$ on alternate spans for maximum (and minimum) positive factored moments, and with $w_d + 0.75\, w_\ell$ on adjacent spans for maximum negative factored moment (Section 13.7.6.3).

The results of these four analyses are shown in Table 6-1, along with factored moments for an interior span obtained by the direct design method. Note that $w_d + w_\ell$ on all spans produces the maximum negative factored moment at the face of the column and that $w_d + 0.75\, w_\ell$ on alternate spans produces the maximum positive factored moment. Note also that the direct design method gives a 45 percent larger positive factored moment and a 17% smaller negative factored moment than the equivalent frame method for an interior span. This comparison also shows that negative factored moments develop at the midspan of unloaded panels. See "Approximate Method versus Frame Analysis," Chapter 2, for a similar comparison applied to one-way construction with beam-column or slab and beam frames. The authors' recommendations in Chapter 2 are equally applicable to the use of the approximate direct design method versus the "exact" equivalent frame method and will not be repeated here.

SUPPLEMENTAL FLAT SLAB DESIGN DATA

Shear Constants

Brief derivations of shear constants for the area of the critical shear section, A_c, and the flexure-torsion moment of inertia, J_c, are given in Figs. 6-18, -19, -20, and -21, for rectangular and round columns. Since the constants are reduced near openings or edges, values for various locations of the columns near edges and corners are given. The authors' recommendation (page 133) for a conservative

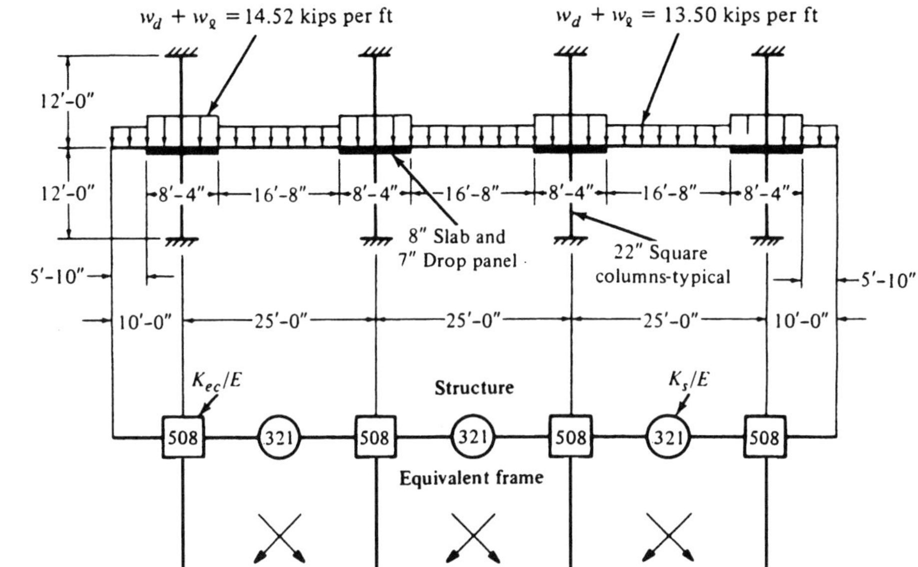

COF		−0.647		−0.647		−0.647		
DF	0	0.387	0.279	0.279	0.279	0.279	0.387	0
FEM	−684	−840	−840	−840	−840	−840	−840	−684
D	—	+60	—	—	—	—	+60	—
CO			−39	—	—	−39		
D			+15	−15	−15	+15		
CO		−10		+10	+10		−10	
D		+4	+3	−2	−2	+3	+4	
CO		−2	−3	+1	+1	−3	−2	
D		+1	+1	−1	−1	+1	+1	
M_u @ ℄	−684	−787	−863	−847	−847	−863	−781	−684
	139.2	173.0	173.0	173.0	173.0	173.0	173.0	139.2
		−3.0	+3.0			+3.0	−3.0	
V_u @ ℄	139.2	170.0	176.0	173.0	173.0	176.0	170.0	139.2
M_u @ face	−563	−637	−707	−694	−694	−707	−637	−563
M_u @ midspan		+239		+217		+239		

Figure 6-16 Elastic analysis—full factored load all panels.

upper limit on the shear constants for an open critical shear section with square columns should be extended to round columns near edges. Unless the cantilever edge moment is sufficient to create an interior column condition (closed shear section), the portions of the critical shear section on the cantilever side of the column cannot be considered fully effective. For these small cantilevers, the authors advise use of no larger value than that shown for a circular shear section tangent to the edge (labeled "Maximum Recommended" in Fig. 6-20.

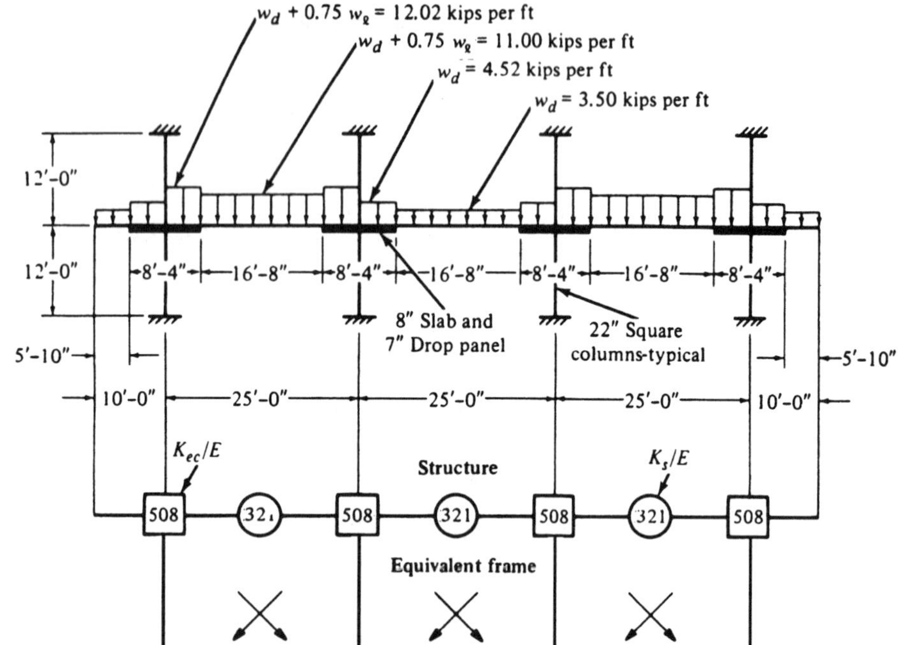

COF		−0.647		−0.647		−0.647		
DF	0	0.387	0.279	0.279	0.279	0.387	0.387	0
FEM	−184	−686	−686	−224	−224	−686	−686	−184
D		+194	+129	−129	−129	+129	+194	
CO		−83	−126	+83	+83	−126	−83	
D		+32	+58	−58	−58	+58	+32	
CO		−38	−21	+38	+38	−21	−38	
D		+15	+16	−16	−16	+16	+15	
CO		−10	−10	+10	+10	−10	−10	
D		+4	+6	−6	−6	+6	+4	
CO		−4	−3	+4	+4	−3	−4	
D		+2	+2	−2	−2	+2	+2	
M_u @ ℄	−184	−574	−737	−300	−300	−737	−574	−184
		+243		−28		+243		
V_u fixed	39.2	141.8	141.8	48.0	48.0	141.8	141.8	39.2
		−6.5	+6.5			+6.5	−6.5	
V_u @ ℄	39.2	135.3	148.3	48.0	48.0	148.3	135.3	39.2
M_u @ face	−150	−471	−606	−258	−258	−606	−471	−150
M_u @ midspan		+213		−18		+213		

(*a*)

Figure 6-17 (*a*) Elastic analysis load pattern for maximum positive factored moment; (*b*) Elastic analysis load pattern for maximum positive factored moment; (*c*) Elastic analysis load pattern for maximum negative factored moment; (*d*) Elastic analysis load pattern for maximum negative factored moment.

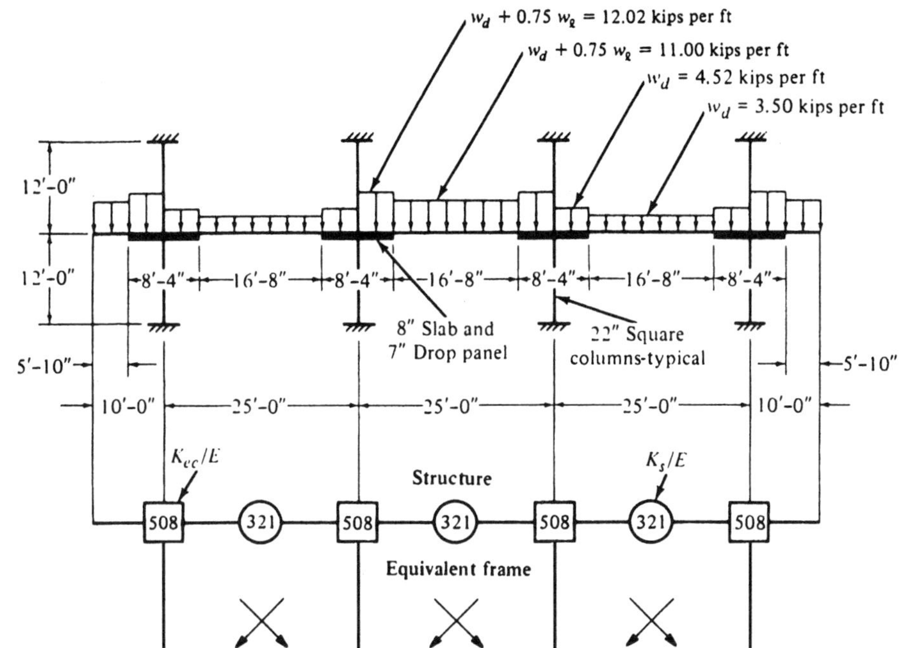

COF		−0.647		−0.647		−0.647		
DF	0	0.387	0.279	0.279	0.279	0.279	0.387	0
FEM	−559	−224	−224	−686	−686	−224	−224	−559
D		−130	−129	+129	+129	−129	−130	
CO		+83	+84	−83	−83	+84	+83	
D		−32	−47	+47	+47	−47	−32	
CO		+30	+21	−21	−21	+21	+30	
D		−12	−12	+12	+12	−12	−12	
CO		+8	+8	−8	−8	+8	+8	
D		−3	−4	+4	+4	−4	−3	
CO		+3	+2	−3	−3	+2	+3	
D		−1	−1	+1	+1	−1	−1	
M_u @ ℄	−559	−278	−302	−608	−608	−302	−278	−559
V_u fixed	114.2	48.0	48.0	141.8	141.8	48.0	48.0	114.2
		−1.0	+1.0			+1.0	−1.0	
V_u @ ℄	114.2	47.0	49.4	141.8	141.8	49.0	47.0	114.2
M_u @ face	−459	−237	−259	−483	−483	−259	−237	−459
M_u @ midspan		−8		+260		−8		

(*b*)

Figure 6-17 (*continued*)

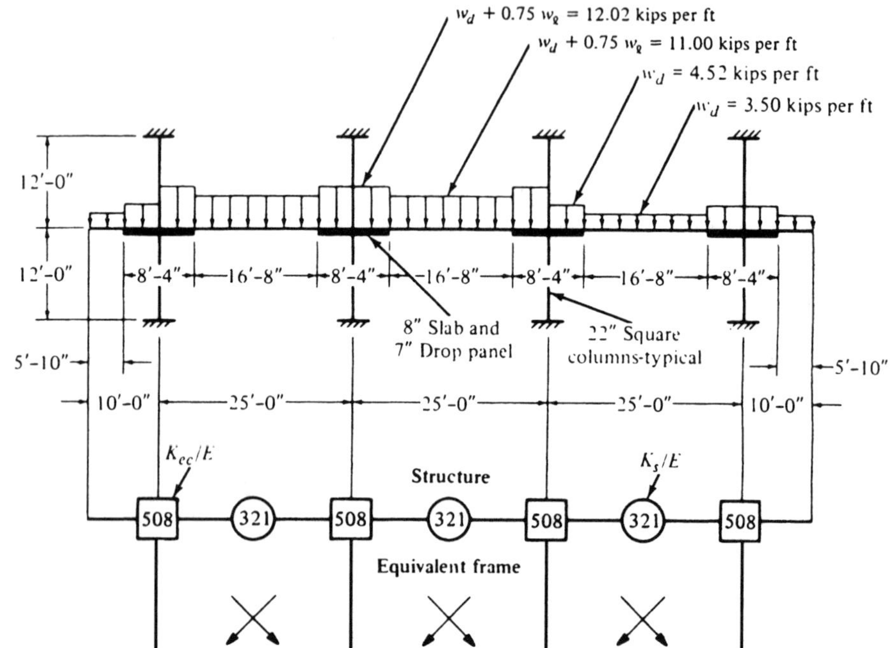

COF		−0.647		−0.647		−0.647		
DF	0	0.387	0.279	0.279	0.279	0.279	0.387	0
FEM	−184	−686	−686	−686	−686	−224	−224	−184
D		+194	—	—	+129	−129	+15	
CO		—	−126	−83	—	−10	+83	
D		—	+12	−12	−3	+3	−32	
CO		−8	—	+2	+8	+21	−2	
D		+3	+1	−1	+4	−4	+1	
CO		−1	−2	−3	+1	−1	+3	
D		—	—	—	—	—	−1	
M_u @ ℄	−184	−498	−801	−783	−574	−344	−157	−184
V_u fixed	39.2	141.8	141.8	141.8	141.8	48.0	48.0	39.2
		−12.1	+12.1	+8.4	−8.4	+7.5	−7.5	
V_u @ ℄	39.2	129.7	153.9	150.2	133.4	55.5	40.5	39.2
M_u @ face	−150	−384	−694	−650	−457	−295	−122	−150
M_u @ midspan		+219		+190		+32		

(*c*)

Figure 6-17 (*continued*)

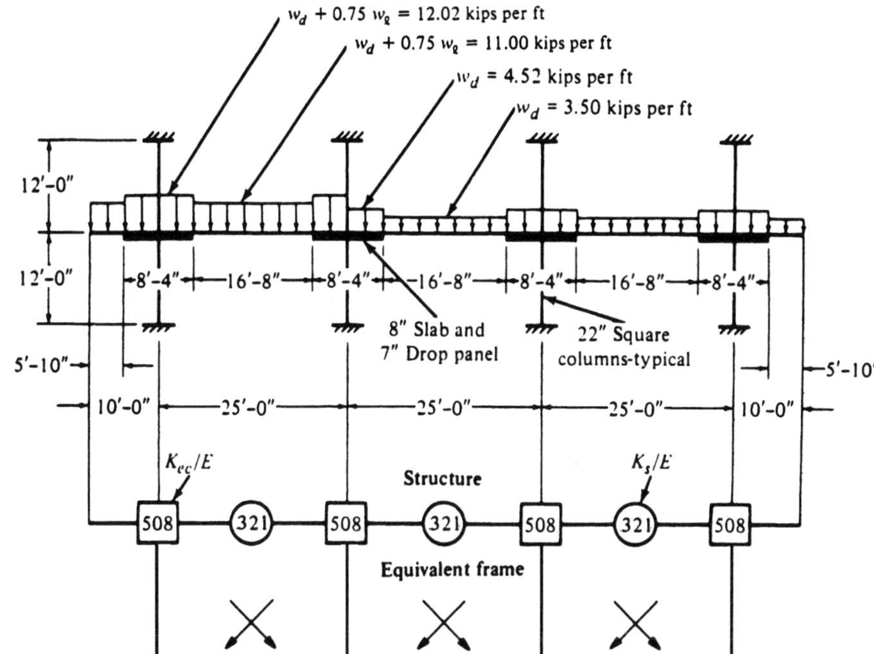

COF		−0.647		−0.647		−0.647		
DF	0	0.387	0.279	0.279	0.279	0.279	0.387	0
FEM	−559	−686	−686	−224	−224	−224	−224	−184
D		+49	+129	−129	—	—	+15	
CO		−83	−32	—	+83	−10	—	
D		+32	+9	−9	−26	+26	—	
CO		−6	−21	+17	+6	—	−17	
D		+2	+11	−11	−2	+2	+7	
CO		−7	−1	+1	+7	−5	−1	
D		+3	+1	−1	−3	+3	—	
CO		−1	−2	+2	+1	—	−2	
D		—	+1	−1	—	—	+1	
M_u @ ℄	−559	−697	−591	−355	−158	−208	−221	−184
V_u fixed	114.2	141.8	141.8	48.0	48.0	48.0	48.0	39.2
		+4.2	−4.2	+7.9	−7.9	−0.5	0.5	
V_u @ ℄	114.2	146.0	137.6	55.9	40.1	47.5	48.5	39.2
M_u @ face	−459	−568						
M_u @ midspan		+224		+26		+68		

(*d*)

Figure 6-17 (*continued*)

TABLE 6-1 Comparison of Factored Moments by the Equivalent Frame and Direct Design Methods

Design Method and Loading	TOTAL PANEL MOMENTS					
	MAXIMUM NEGATIVE FACTORED MOMENT AT FACE OF COLUMN (ft-kips)		MAXIMUM POSITIVE FACTORED MOMENT (ft-kips)		MINIMUM POSITIVE FACTORED MOMENT (ft-kips)	
	Exterior column	Interior column	Exterior span	Interior span	Exterior span	Interior span
Equivalent frame $w_d + w_\ell$ on all spans	−637	−707	+239	+217	+239	+217
Equivalent frame $w_d + 0.75\ w_\ell$ pattern loading (Section 13.7.6.3)	−568	−694	+224	+260	−8	−18
Direct design interior Span with $-M_u = -0.65\ M_o$ and $+M_u = +0.35\ M_o$	—	−592	—	+319	—	—

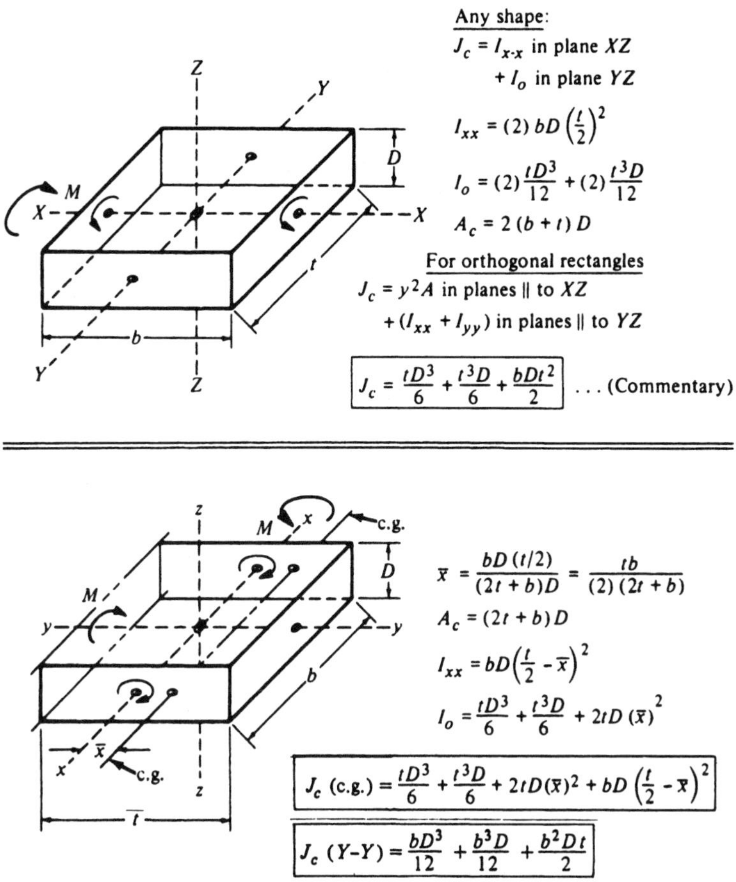

Figure 6-18 Shear constants—derivation for rectangular columns.

Similarly, no larger value than that shown in Fig. 6-21 is recommended for round corner columns.

For irregular column shapes, such as octagonal, hexagonal, etc., values for the shear constants can be interpolated between those for round and square shapes. Quick estimates, suitable for preliminary design, can be similarly made for L- or T-shaped columns.

Preliminary Trial Proportions

For any practical application of flat slabs to be designed in conformance with the Code, the selection of preliminary sizes as near to final sizes as possible is vital. Since the Code design procedures are essentially reviews, not necessarily convergent, a poor initial selection may require several adjustments each of which requires a complete review—only to prove inadequate or overdesigned.

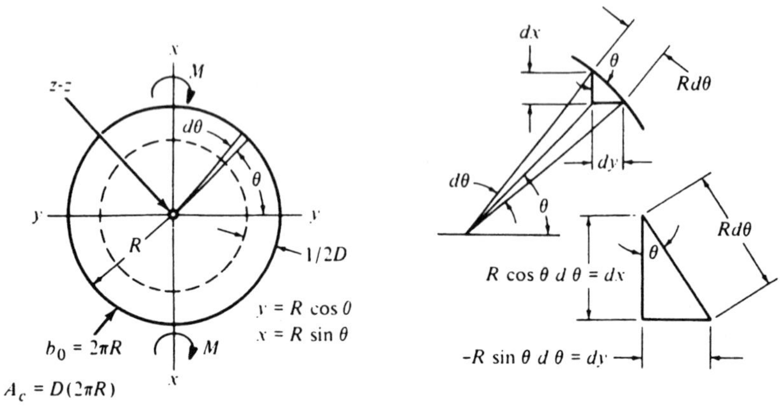

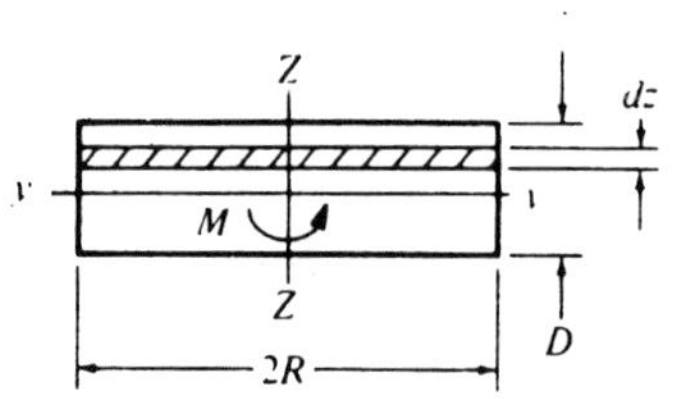

$J_c = I_{x\text{-}x}$ for projection on X-Z plane
$+ I_{z\text{-}z}$ for projection on Y-Z plane

$$I_{x\text{-}x} = D\int_0^{2\pi} y^2\,ds = D\int_0^{2\pi} (R\cos\theta)^2\,R\,d\theta$$

$$I_{x\text{-}x} = R^3 D\int_0^{2\pi} (\cos\theta)^2\,d\theta = R^3 D\left(\frac{\theta}{2} + \frac{\sin 2\theta}{4}\right)\Bigg|_0^{2\pi}$$

$$I_{x\text{-}x} = \pi R^3 D$$

$$I_{z\text{-}z} = (2)(2RD)\int_{-D/2}^{D/2} Z^2\,dz = 4RD\left(\frac{Z^3}{3}\right)\Bigg|_{-D/2}^{D/2} = \frac{RD^3}{3}$$

$$\boxed{J_c = \pi R^3 D + \frac{RD^3}{3}}$$

Figure 6-19 Shear constants—derivation for round interior columns.

Minimum values for the drop panel thickness (slab depth plus the projection of 25 to 100 percent of the slab depth) from a number of computer studies are shown in Figs. 6-22, -23, -24, -25, and -26 for *factored* live loads of 100, 200, 300, and 400 psf. Column sizes were established as a minimum of 12 in. square, considered by the authors to be a minimum for loads sufficiently heavy and spans sufficiently long to require the flat slab with drop panel instead of a flat plate. As load and span increase, the column size was increased up to a maximum of 1/12 span, still using the minimum depth of the drop panel, 1.25 slab thickness. For heavier loading-longer span combinations, the drop panel depth was increased to a maximum of twice the slab thickness. When shear in the slab at the edge of the drop panel became critical, the drop panel size was increased from the minimum of 0.33 span to 0.40 span. The preliminary trial proportion

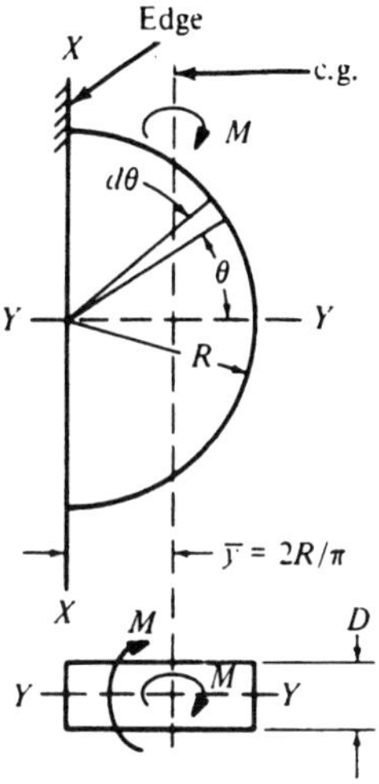

$$x = R\sin\theta \qquad y = R\cos\theta$$
$$dx = R\cos\theta\, d\theta \qquad dy = -R\sin\theta\, d\theta$$

$$\bar{y} = \frac{\int_0^{\pi/2} y R\, d\theta}{\int_0^{\pi/2} R\, d\theta} = \frac{2R}{\pi} = 0.636R$$

$$J_c = \frac{1}{2}\left(\pi R^3 D + \frac{RD^3}{3}\right) - (\pi RD)\left(\frac{2R}{\pi}\right)^2$$

$$\boxed{J_c\,(\text{c.g.}) = 0.296\, R^3 D + \frac{RD^3}{6}} \quad \text{.... Minimum practicable}$$

$$\boxed{J_c\,(Y\text{-}Y) = R^3 D\,\frac{\pi}{2} + \frac{RD^3}{12}}$$

$$A_c = \pi RD$$

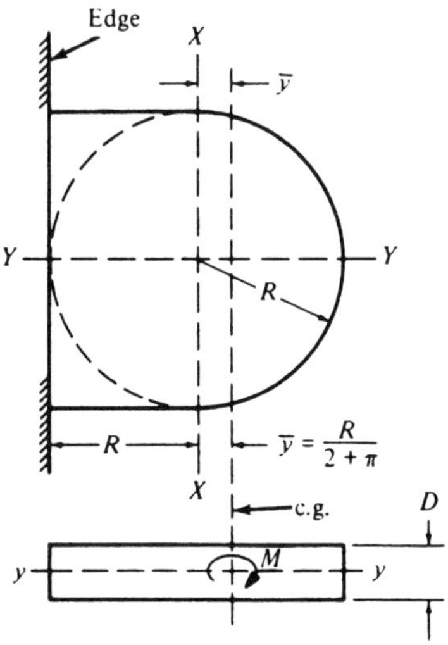

$$\bar{y} = \frac{2R^2 - (2R)(R/2)}{\pi R + 2R} = \frac{R^2}{R(2+\pi)} = \frac{R}{5.14}$$

$$J_c = \left(\frac{\pi}{2} R^3 D + \frac{RD^3}{6}\right) - (\pi RD)\left(\frac{R}{2+\pi}\right)^2 + \frac{2RD^3}{12} + \frac{2R^3 D}{3} + 2RD\left(\frac{R}{2+\pi}\right)^2$$

$$\boxed{J_c\,(\text{c.g.}) = 2.043\, R^3 D + \frac{RD^3}{3}}$$

$$\boxed{J_c\,(Y\text{-}Y) = R^3 D\left(\frac{\pi}{2} + 2\right) + \frac{RD^3}{6}}$$

(Maximum recommended)

$$A_c = (2 + \pi)RD$$

Figure 6-20 Shear constants—derivation for round edge columns.

charts were prepared for square exterior panels, since the first interior column is more heavily loaded than other interior columns with all panels equal in size. The minimum column sizes shown are for flush edge exterior columns with no edge beams and accompanying minimum first interior columns. Story height was 12 ft with the same size columns assumed above and below the slab. All charts are for 3,000 psi concrete strength of concrete weighing 150 pcf and Grade 60 rebars. Spans shown are center-to-center of columns. The symbols used are h_D = drop panel thickness; h = slab thickness, c_1 and c_2 = square column cross-section dimensions.

Maximum recommended

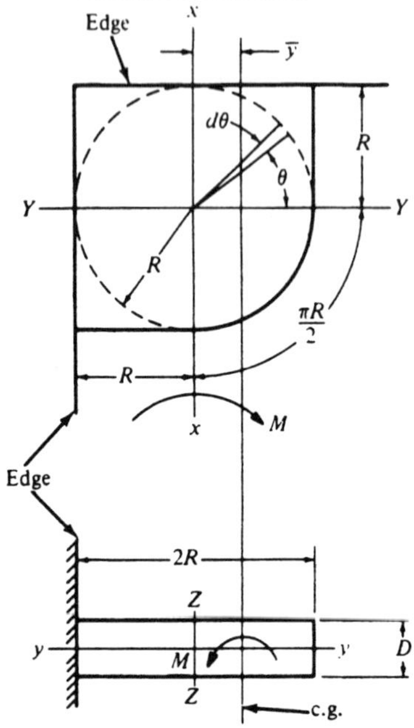

$$ds = R\,d\theta$$
$$x = R\sin\theta$$
$$dx = R\cos\theta\,d\theta$$
$$y = R\cos\theta$$
$$dy = -R\sin\theta\,d\theta$$

$$\bar{y} = \frac{\int y\,ds}{\int ds}$$

$$= \frac{(R - \frac{1}{2}R)R + \int_{3/2\pi}^{2\pi} R\cos\theta\,R\,d\theta}{2R + \frac{\pi R}{2}}$$

$$= \frac{\frac{1}{2}R^2 + R^2\sin\theta\Big]_{3/2\pi}^{2\pi}}{R\left(2 + \frac{\pi}{2}\right)}$$

$$\bar{y} = \frac{1.5R}{2 + 0.5\pi} = 0.4205R$$

$$J_c = I_{x\text{-}x} \text{ in } X\text{-}Y + I_{y\text{-}y} \text{ in } Y\text{-}Z$$

$$J_c = (RD)(0.58R)^2 + \frac{R^3D}{12} + (RD)(0.92R)^2 + \int_{3/2\pi}^{2\pi} (R\cos\theta - 0.42R)^2\,R\,d\theta + \frac{(2R)D^3}{12}$$

$$J_c = R^3D\,(0.336 + 0.0833 + 0.846 + \pi/3 - 0.84) + \frac{RD^3}{6}$$

$$\boxed{J_c = 1.472\,R^3D + \frac{1}{6}\,RD^3} \qquad A_c = (RD)\left(2 + \frac{\pi}{2}\right)$$

Minimum recommended

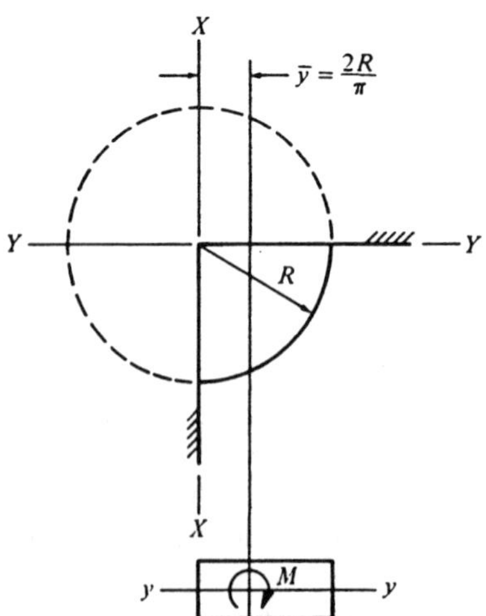

$$A_c = \frac{\pi RD}{2}$$

$$\bar{y} = \frac{2R}{\pi} \quad \text{(same as ½ circle)}$$

$$J_c = \frac{\pi R^3D}{4} + \frac{RD^3}{12} - \frac{\pi RD}{2}\left(\frac{2R}{\pi}\right)^2$$

$$= R^3D\left[\frac{\pi}{4} - \left(\frac{\pi}{2}\right)\frac{4}{\pi^2}\right] + \frac{RD^3}{12}$$

$$\boxed{\begin{aligned} J_c &= R^3D\left[\frac{\pi}{4} - \frac{2}{\pi}\right] + \frac{RD^3}{12} \\ J_c &= 0.149\,R^3D + \frac{RD^3}{12} \end{aligned}}$$

Figure 6-21 Shear constants—derivation for round corner columns.

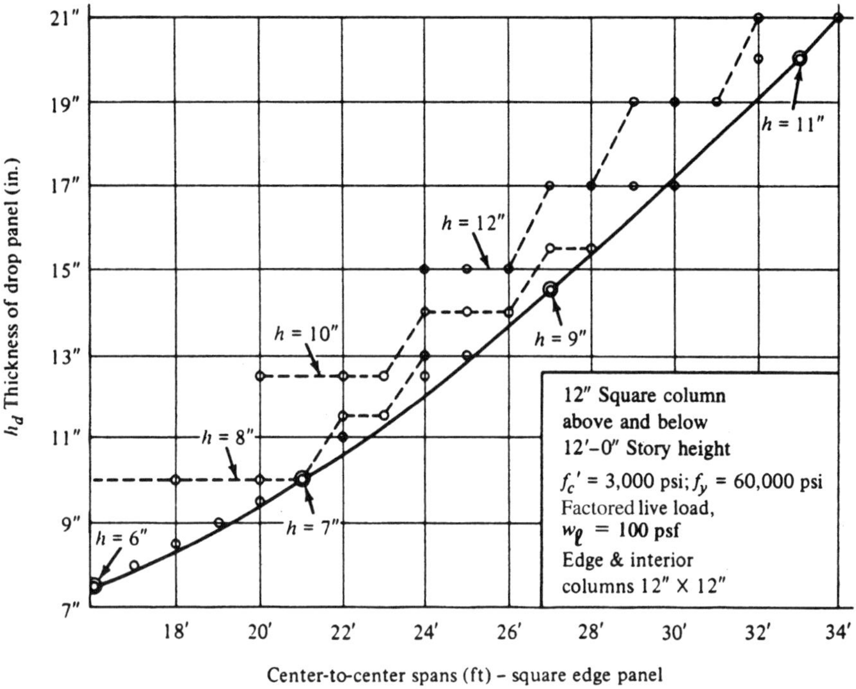

Figure 6-22 Required drop panel thickness for minimum size square columns, exterior.

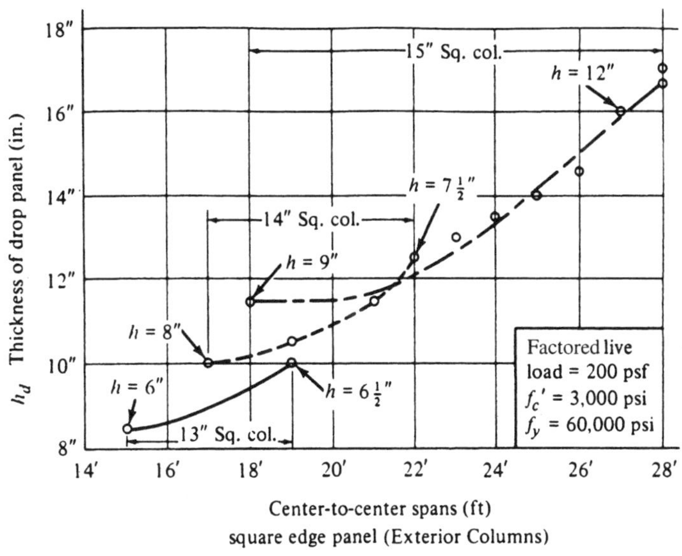

Figure 6-23 Required drop panel thickness for 12 in. square columns.

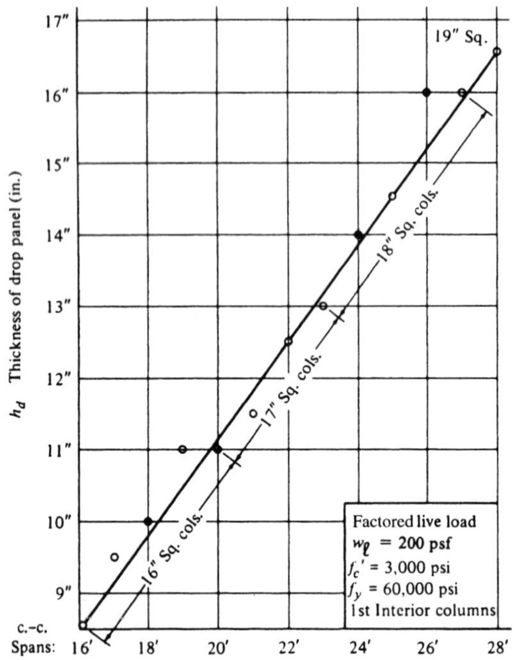

Figure 6-24 $w_\ell = 200$ psf drop panel thickness for minimum size square columns.

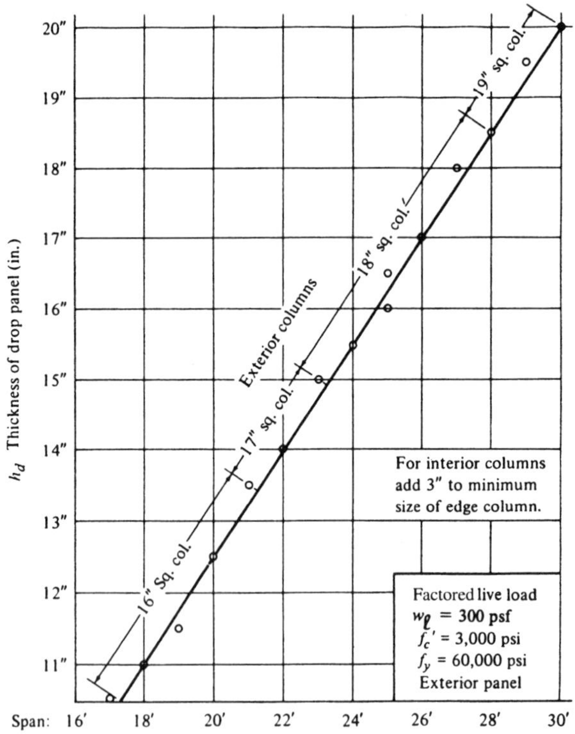

Figure 6-25 $w_\ell = 300$ psf drop panel thickness for minimum size square columns.

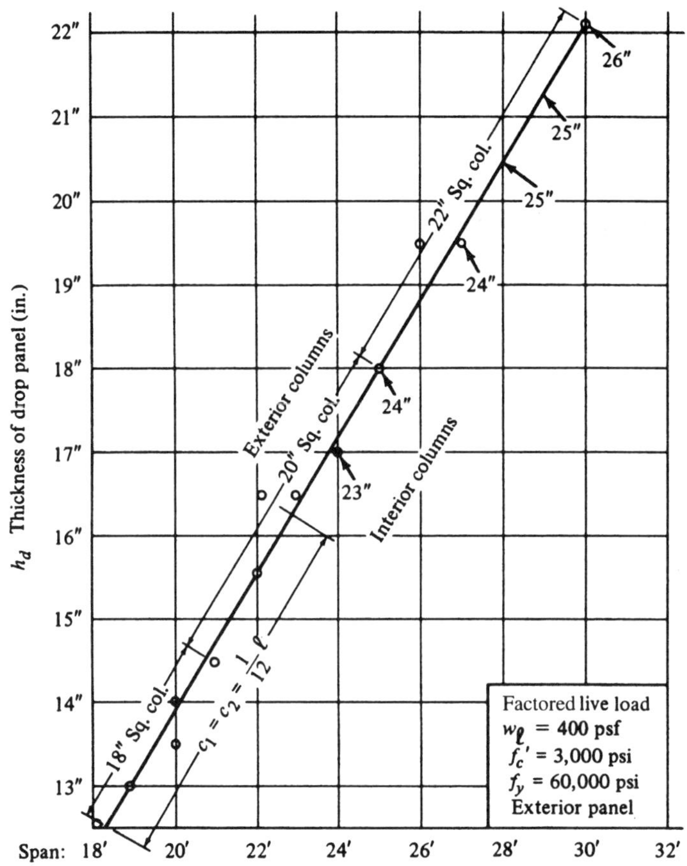

Figure 6-26 $w_\ell = 400$ psf drop panel thickness for minimum size square columns.

7 Two-Way (Waffle) Flat Slab Design

GENERAL

A "waffle" flat slab is a two-way joist system. Under the general provisions of the 1995 Code, Chapter 13, the two-way joist portion may be combined with a solid column head or with solid wide beam sections on the column centerlines for uniform depth construction. These are the usual applications, favored for economy of formwork and the unobstructed level ceiling with a regular pattern of void spaces utilizable for lighting or ventilating fixtures. The arrangement of two-way joists with a solid head results in a slab stiffness distribution similar to that for a solid slab with a drop panel. The use of solid "beam" sections on the column centerlines also provides a solid full depth for shear transfer to the columns and permits concentration of moment reinforcement at the column. Other common variations are cantilevered edges or use of spandrel beams deeper than the uniform depth elsewhere.

A large amount of waffle flat slab construction employs standard size reusable forms for economy. Standard form sizes, conforming to the Code limitations defining joist construction (Section 8.11), provide for ribs at least 4 in. wide, spaced not more than 30 in. clear, and for a depth not more than $3\frac{1}{2}$ times minimum width. Within these limitations it has been feasible to establish industry-wide standard sizes for reusable square void forms,* 19×19 and 30×30 in., in 6, 8, 10, and 12 in. depths and 8, 10, 12, 14, 16, and 20 in. depths, respectively. For nonstandard sizes, within the limitations of the Code or exceeding these limits, custom-made forms are required. The Code permits use of 10 percent higher shear strength of concrete for joists within the established dimensional limits (Section 8.11.8). Waffle slab construction outside these limits is designed as multiple two-way slabs and beams. Waffle slab forms with depths up to 24 in. for spans up to 60 ft are available in 40-in. square voids for 8-in. wide beams and a 4-ft modular spacing, and 52-in. wide square voids for 8-in. wide beams and a 5-ft modular spacing.

* "Recommended Industry Practice for Concrete Joist Construction," CRSI, *Manual of Standard Pracrice*, 26th Edition, 1997. Also ANSI Standard ANSI/CRSI A48.2.

The principal complications to two-way slab design introduced by the waffle slabs are the additional computations required for stiffness of the solid and ribbed areas, and the additional critical shear sections in the ribs. These complications will be illustrated in the design examples following.

DIRECT DESIGN METHOD

EXAMPLE 1. *Waffle slab with solid head—square interior panel*

$$\text{Spans } \ell_1 = \ell_2 = 39'0''$$

Factored live load 180 psf

Use standard 30″ × 30″ × 20″ forms for 6″ ribs @ 36″ with a void depth of 20″ plus a 3″ top slab

Solid head (or drop panel) $D = 5 \times 36''$ module + 6″ outside rib = 15′-6′ square

For this example, an interior panel, the distribution of the panel moments is made using the direct design method: $\alpha_1 = \alpha_2 = 0$ (Section 13.0); $\ell_1 = \ell_2$; $\alpha_1 \ell_1/\ell_2 = 0$; $\alpha_2 \ell_1/\ell_2 = 0$.

Panel moments (Sections 13.6.3.2, 13.6.4.1 and 13.6.4.4):

$$\text{Column strip } -M_u = -(0.65)(0.75)\,M_o = -0.488\,M_o$$

$$\text{(6 ribs) } +M_u = +(0.35)(0.60)\,M_o = +0.210\,M_o$$

$$\text{Middle strip } -M_u = -(0.65)(0.25)\,M_o = -0.162\,M_o$$

$$\text{(7 ribs) } +M_u = +(0.35)(0.40)\,M_o = +0.140\,M_o$$

Critical sections for shear; see Fig. 7-1:

1. Two-way shear at columns ("punching shear") (Sections 11.12.1.2 and 11.12.2). Design shear strength of concrete, $\phi V_c = \phi 4\sqrt{f_c'}\, b_o d$ on the square section $(c + d)$ in the solid head, where d = average effective depth in the solid head.
2. One-way shear in the 24 joists (ribs) at a distance d from the edge of the solid head (Sections 11.12.1.1 and 8.11.8) $\phi V_c = \phi 2.2\sqrt{f_c'}\, b_w d$. Note that the Code permits an increase in the one-way design shear strength of the concrete (Section 11.3.2.1) up to $\phi 3.5\sqrt{f_c'}\, b_w d$ upon a "detailed analysis" with design shear strength computed as

$$\phi V_c = \phi 1.1(1.9\sqrt{f_c'} + 2500\rho_w V_u d/M_u) b_w d$$

The edge of the solid head will be at approximately one-sixth of the span, the factored moment, M_u, at this point will be negative but quite low, and the steel

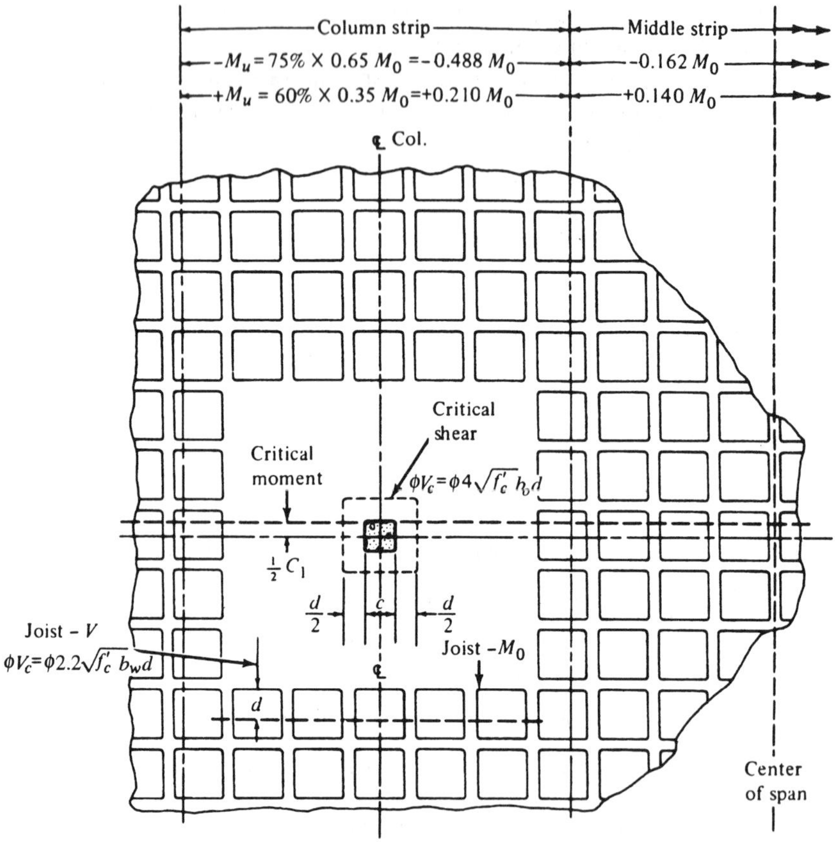

Figure 7-1 Layout, panel moments, and critical sections—waffle slab with solid head.

percentage, ρ_w, will be relatively high, so that the design shear strength on this critical section by the detailed analysis will usually be higher than the value $\phi 2.2\sqrt{f'_c}\ b_w d$. For economy, stirrups in any type of joist ribs are to be avoided wherever possible, and so the authors advise use of the "long" formula here to determine the maximum design shear strength of the concrete without stirrups.

Critical sections for moment

1. The critical section for negative factored moment of the panel is at the face of the column as for flat plates (Section 13.6.3.1). See Fig. 7.1. If the required ratio of negative reinforcement in the joist ribs of the middle strip is excessive, the bottom bars can be extended to serve as compressive reinforcement (Section 10.3.4). Practice recommended by the industry (CRSI) is to extend at least one of the two bottom bars in the ribs to overlap on the column centerlines in all full-length joists. This practice avoids an abrupt stress con-

centration resulting from cutting all bars at the column centerline and possible concrete shrinkage cracking at this point.

2. Column strip joists develop a critical section for negative moment at the face of the solid head. Again the bottom bars may be considered effective for compression since the Code requires all straight bottom bars to be continuous with Class A tension lap splices permitted in the region of negative moment (Section 13.3.8.5).

EXAMPLE 2. *Waffle slab with solid beam sections on column centerlines—square interior panel*
Use the same dimensions as in previous example for comparison. See Fig. 7-2.

$\ell_1 = \ell_2 = 39'\text{-}0''$ 36″ module; 6″ ribs @36″; 20″ + 3″ depth

$w_\ell = 180$ psf Beams and joist-slab are monolithic, and so $E_c = E_{cs}$

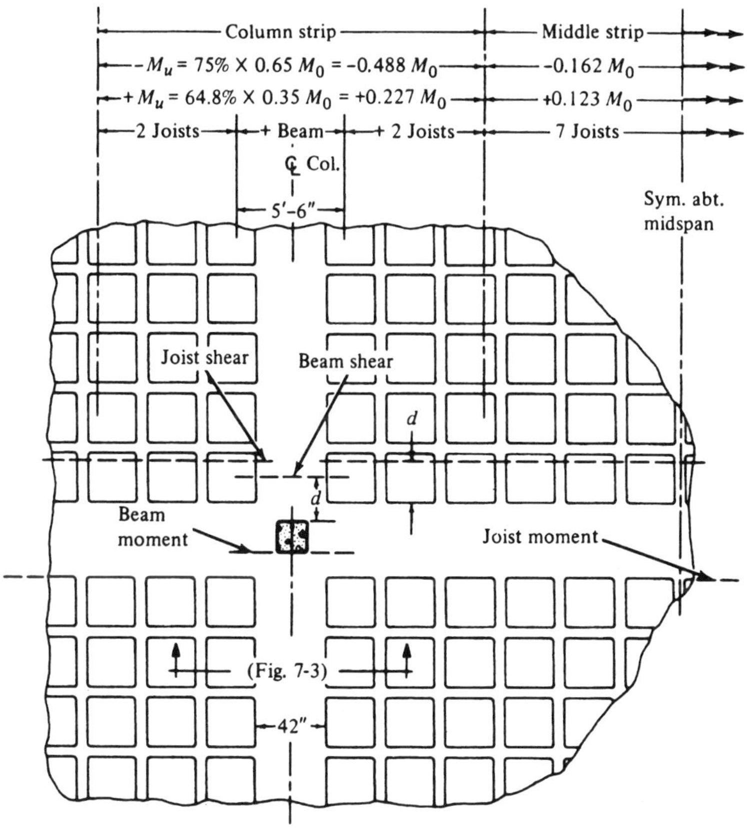

Figure 7-2 Layout, panel moments, and critical sections—waffle slab with beams.

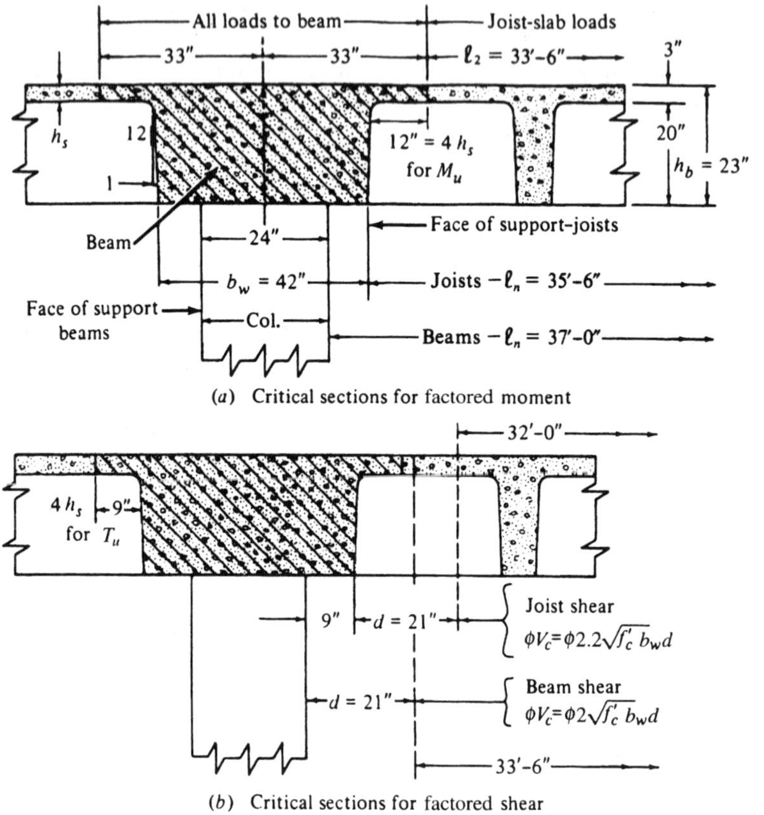

Figure 7-3 Cross-section dimensions and critical sections.

Omit one line of the dome forms on the column centerlines to form a solid beam section 42 in. wide, plus overhanging flanges of $4h_s = 12$ in. on each side (Section 13.2.4) (see Fig. 7-3).

Code Interpretations

The problems of determining the factored moments for this particular application are aggravated by incomplete, interdependent provisions in the Code, some ambiguous and some apparently conflicting. The location of critical moment sections and the determination of clear spans for the calculation of factored moments at these critical sections can be interpreted differently. The authors' interpretations are shown in Fig. 7-3.

"Supports" may be walls or columns, but not beams (Section 13.1.2)! Clear span, ℓ_n, is measured between faces of supports (Section 13.0), or clear span, ℓ_n, extends from the faces of columns, capitals, brackets, or walls (*note:* no beams) (Section 13.6.2.5). For the purpose of strength and relative stiffness, the width of a T-beam may include a portion of the slab up to 4 h in width on each side of the beam (Section 13.2.4). However, only the loads located within the width of the beam stem should be considered as directly assigned to the beams.

The authors' interpretations of the above Code provisions are as follows:

1. The columns support the beams; clear span, ℓ_n, for beams is the distance between faces of the columns (= 37′-0″).
2. The beams support the "slab system," which in this case is the two-way joists and top slab. Clear span, ℓ_n, for the joists is the distance between the solid faces of the beams (= 35′-6″).
3. The "slab system" loads to the beams (M_o) are transmitted by the joist-slab portion having a transverse width between beam flanges (= 33′-6″), and a clear span, ℓ_n = 35′-6″, which is loaded in the center 33′-6″) portion.

Location of the critical shear sections and the distribution of factored loads causing the critical factored shear thereon are equally complex problems, particularly since they do not coincide with the critical sections for factored moment nor, therefore, the factored loads for factored moment (see Fig. 7-3). The authors' interpretations are as follows:

1. Joists behave as beams in shear and so the one-way shear provisions apply. Critical shear sections are at a distance, d, from the supports of the joists (joist d from the face of the beam webs). (Even though the joists are part of a two-way system and one Code provision (Section 13.5.2), taken literally, would indicate that two-way joist systems are designed for factored shear and factored moment at the same critical sections as a solid two-way slab, but it does not seem realistic to accept this interpretation because the resulting "critical section" for joist shear could then lie inside a wide beam supported by slender columns.)
2. Beams behave in one-way shear since the solid area about the column is less than $d/2$ from its face (Section 13.6.8). Critical shear sections in the beams, therefore, are at a distance, d, from the face of the beam supports (face of the columns).
3. Beams are to be designed for the factored loads applied directly upon them (Section 13.6.5.3). For this purpose the authors chose to use the full width of the beams, including the flanges ($4h$), consistent with a transverse width of 33′-6″ for loads upon the joist-slab system. For convenience in the calculations, the authors recommend applying *all* uniform loads on the beam surface to the beam and using the same division into positive and negative moments as prescribed for the loads considered in the slab design—0.35 and 0.65, respectively.

PANEL MOMENTS. In order to determine the effect of the relative stiffness of the beams on the distribution of the panel moments, numerical values must be used. See Fig. 7-3 for cross-sectional dimensions.

Cross-sectional properties:

$I_s = (12{,}470)(13) = 162{,}110$ in.^{4} (for 13 joists @36 in., gross section) (See Fig. 7-4)

$I_b = 51{,}660$ in.^{4} (for 66 in. × in. 23 in. beam section) (See Fig. 7-5)

$$\alpha_1 = \alpha_2 = \frac{E_{cb}I_b}{E_{cs}I_s} = \frac{I_b}{I_s} = \frac{51{,}660}{162{,}110} = 0.319 \quad \text{(Section 13.0)}$$

$$\alpha_1 \ell_2/\ell_1 = \alpha_2 \ell_1/\ell_2 = (1.0)(0.319) = 0.319$$

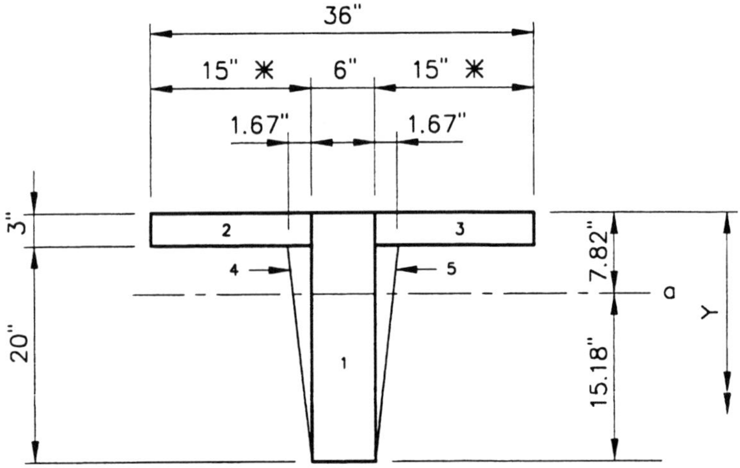

Centroid of cross section

			Area				y
①	(6)(23)	=	138	×	11.5	=	1,587
② & ③	(2)(15)(3)	=	90	×	1.5	=	135
④ & ⑤	$(2)(\frac{1}{2})(20)(1.67)$	=	33.4	×	9.67	=	323
			261.4				2,045

$y_{c.g.} = 2{,}045/261.4 = 7.82$ in.

Moment of inertia, I_g:

①	$(6)(23)^3/12$	=	6,083
	$+\ (138)(3.68)^2$	=	1,869
② & ③	$(2)(15)(3)^3/12$	=	67
	$+\ (2)(45)(6.32)^2$	=	3,595
④ & ⑤	$(2)(1.67)(20)^3/36$	=	742
	$+\ (2)(16.70)(1.85)^2$	=	114
		I_g =	12,470 in.^{4}

Figure 7-4 Joist cross-sectional properties.

* Although Section 13.2.4 limits the overhanging slab dimension to $4h = 12$ in., the minor difference between using one-half of a dome void of 15 in. or $4 \times 3 + 1.67 = 13.67$ in. is neglected in this example to simplify calculations.

TABLE 7-1 Percentages of the Total Panel Moments to the Column Strip (Sections 13.6.4.1, 13. 6.4.4)

RATIO	NEGATIVE FACTORED MOMENT (%) (0.65 M_o)				POSITIVE FACTORED MOMENT (%) (0.35 M_o)			
ℓ_2/ℓ_1	0.5	1.0	1.5	2.0	0.5	1.0	1.5	2.0
$\alpha_1 \ell_2/\ell_1 = 0$	75	75	75	75	60	60	60	60
$\alpha_1 \ell_2/\ell_1 \geqslant 0$	90	75	60	45	90	75	60	45

Using the above constants, the moments are to be distributed among column strips, middle strips, and beams as prescribed by the Code (Sections 13.6.4, 13.6.5, and 13.6.6).

The Code prescribes interior panel negative and positive factored moments as 0.65 M_o and 0.35 M_o respectively. The distribution to column and middle strips is prescribed for ℓ_2/ℓ_1 ratios of 0.5 and 2.0 and for $\alpha_1 \ell_2/\ell_1$ ratios of 0 and 1.0 or more. Linear interpolation between each of the two sets of parameters is required. See Table 7-1.

Column strip and beam: $-M_u = -(0.65)(0.75)M_o = -0.488M_o$

Middle strip: $-M_u = -(0.65 - 0.488)M_o = -0.162M_o$

Column strip and beam:
$$+M_u = +(0.35)[0.60 + (0.75 - 0.60)(0.319)]M_o$$
$$= +(0.35)(0.648)M_o$$
$$= +0.227M_o$$

Middle strip: $+M_u = +(0.35 - 0.227)M_o = +0.123M_o$

Since the ratios $\alpha_1 \ell_2/\ell_1 < 1.0$, the proportion of the column strip moment assigned to the beams must be determined by interpolation between 0% (for ratio = 0) to 85% (for ratio = 1.0) (Section 13.6.5.1).

Factored moments to beam:
$$= (0.319)(0.85) \times \text{column strip moments}$$
$$= 0.271 \times \text{column strip moments}$$
$$-M_u = -0.488M_o(0.271)$$
$$= -0.132M_o$$
$$+M_u = -0.227M_o(0.271)$$
$$= +0.061M_o$$

Column strip outside beam:
$$-M_u = (-0.488 + 0.132)M_o$$
$$= -0.356M_o$$
$$+M_u = (0.227 - 0.061)M_o$$
$$= +0.166M_o$$

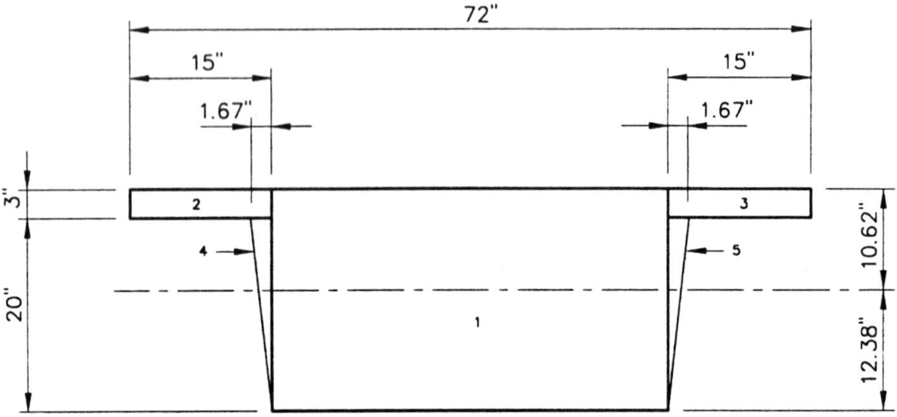

Centroid of cross section

			Area		y		
①	(42)(23)	=	966	×	11.5	=	11,109
② & ③	(2)(15)(3)	=	90	×	1.5	=	135
④ & ⑤	$(2)(\frac{1}{2})(1.67)(20)$	=	33.4	×	9.67	=	323
			1,089.4				11,567

$$y_{c.g.} = 11{,}567/1{,}089.4 = 10.62 \text{ in.}$$

Moment of inertia, I_g:

①	$(42)(23)^3/12$	=	42,585
	$+ (966)(0.88)^2$	=	748
② & ③	$(2)(15)(3)^3/12$	=	67
	$+ (2)(45)(9.12)^2$	=	7,486
④ & ⑤	$(2)(1.67)(20)^3/36$	=	742
	$+ (33.4)(0.95)^2$	=	30
		I_g =	51,660 in.^{4}

Figure 7-5 Beam cross-sectional properties.

BEAM DESIGN

Shear critical sections

Shear at a concentrated load or reaction is governed by the more severe of two conditions: (1) one-way (beam) shear at a critical section, d, from the face of the support, or (2) two-way (punching) shear at a critical section, $d/2$, from support (Sections 11.12.1.2 and 11.12.2.1-b). Two-way design shear strength, $\phi V_c = \phi 4\sqrt{f_c'}\, b_o d$, is twice the one-way design shear strength by the "short" formula (Section 11.12.1.1). For the dimensions of beam, column, and slab depth used in this example (Fig. 7-3), the two-way shear critical section happens

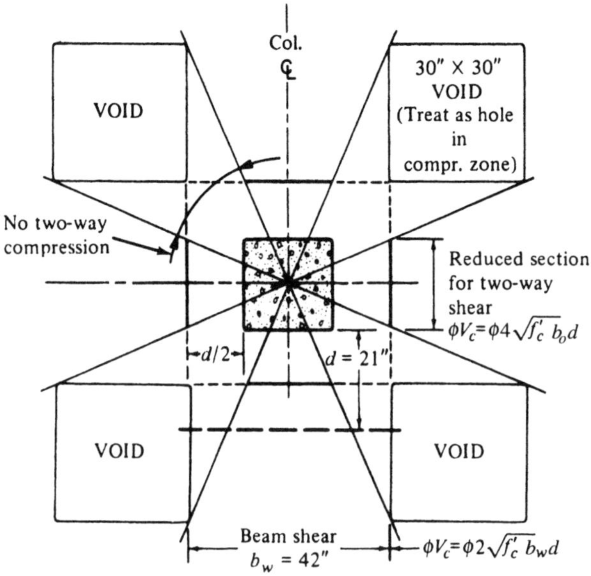

Figure 7-6 Interior column shear sections—waffle slab with beams.

to be located practically at the face of the beam sections: $c_1 + (2)(d/2) = 45$ in. (see Fig. 7-6).

For two-way shear, the void areas must be regarded as partial depth holes which reduce the effective section falling between the radial lines to the center of the reaction (Section 11.12.5). In this case, the reduced section effective for two-way shear strength is approximately half the beam widths effective for one-way shear strength. If wider beam sections had been used, or smaller columns, two sets of shear calculations would have been required. Here, however, due to the essentially one-way behavior of these beams, one-way shear strength seems more appropriate and sufficiently conservative if the short formula, $\phi V_c = \phi 2\sqrt{f'_c}\, b_w d$, is used (Section 11.3.1.1).

SHEAR STRENGTH—REINFORCEMENT:

Use $f'_c = 4{,}000$ psi; $f_y = 60{,}000$ psi; concrete weight = 150 pcf.

Total factored load, $w_u = (0.135)(1.4) + 0.180 = 0.369$ ksf on the waffle portion.

$$w_u = (1089.4)\frac{0.150}{144}(1.4) + (0.180)\frac{(72)}{(12)}$$

$$= 1.59 + 1.08 = 2.67 \text{ klf}$$

directly to the beam (Section 13.6.8.3) (see Fig. 7-7).

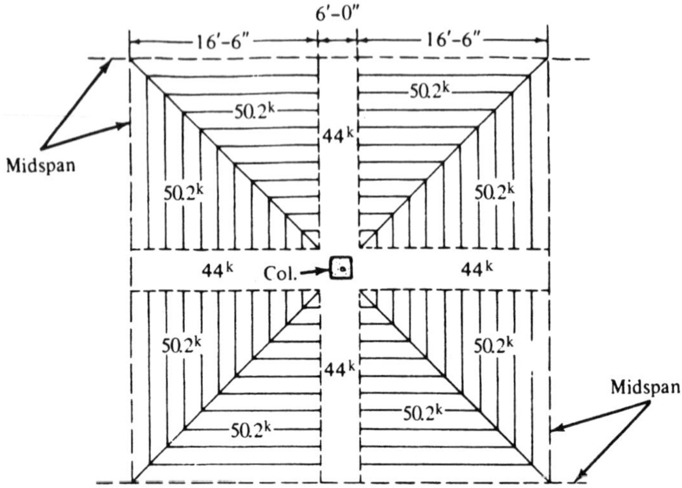

Figure 7-7 Tributary areas and panel shears—waffle slab with beams.

$$\text{Factored beam shear, } V_u = 1/4\ [(0.369)(33.0)^2 + (2.67)(33.0)(1/2)$$
$$= 100.5 + 44.0$$
$$= 144.5 \text{ kips} \quad \text{(Section 13.6.8.2)}$$

In locating the critical sections an average value for the effective depth, $d = 21$ in., was estimated and used for convenience. In calculations for the shear strength and for proportioning stirrups more conservatively, the variable actual value of d is used. Here $d = 20$ in. Average width, b_w (at mid-depth) is used.

$$\phi V_c = \phi 2 \sqrt{f'_c}\, b_w d$$
$$= (0.85)(2)\sqrt{4{,}000}\,(43.7)(20)/1000$$
$$= 94.0 \text{ kips}$$

Minimum shear reinforcement is required where $V_u > \phi V_c/2$ in beams with a total depth greater than 10 in., $2\frac{1}{2}$ times the flange thickness (3 in.), or half the web width (43.7 in. average) (Section 11.5.5.1). In this case the beam depth is greater than the third limit which would govern. Shear reinforcement is, therefore, required for $\phi V_s = V_u - \phi V_c = 143.7 - 94.0 = 49.7$ kips at the critical section, and minimum stirrups must be continued to the point where $V_u = \phi V_c/2 = 47.0$ kips.

$$\text{For vertical stirrups: } A_v = \frac{\phi V_s s}{\phi f_y d} = \frac{(49{,}700)(8)}{(0.85)(60{,}000)(20)} = 0.39 \text{ in.}^2$$

$$\text{Minimum stirrups: } A_v = (50)(b_w)(s)/f_y = (50)(43.7)(10)/60{,}000$$
$$= 0.364 \text{ in.}^2 \quad \text{(Section 11.5.5.3, Eq. 11.13)}$$

For minimum A_v, a spacing, $s = 10$ in. is used, as the maximum spacing allowed is $1/2d$ (Section 11.5.4.1). Minimum stirrups are sufficient for $\phi V_s = \phi(50)b_w d = (0.85)(50)(43.7)(20)/1000 = 37.2$ kips. The point where $\phi V_c + \phi V_s = 94.0 + 37.2 = 131.2$ kips is approximately $\dfrac{143.7 - 131.2}{143.7} \times \dfrac{33.5 \times 12}{2} = 18$ in. from the critical section. Beyond this point minimum stirrups will be required to approximately $\dfrac{143.7 - 47.0}{143.7} \times \dfrac{33.5 \times 12}{2}$ $= 135$ in. from the critical section. The critical section is 21 in. from the face of the column, approximately 12 in. from the face of the intersecting beam (see Fig. 7-8. Stirrups for $\phi V_s = 49.7$ kips will be used for the distance 12 in. + 21 in. beginning 1 in. from the face of the intersecting beam, with minimum stirrups for the remaining distance.

Use: 5-#3 @ 8″

12-#3 @10″

at each end of each beam.

TORSION. To complete this example, an investigation for the effect of torsion in the beams is required. Under balanced loadings or equal loadings in adjacent panels of the same size, the negative moments from any panel under consideration are balanced by identical negative moments from the adjacent panels. Columns then receive no moment from the floor system. When negative moments are unequal due to unequal loads or spans or both, the unbalanced difference between slab moments in adjacent panels is resisted by the columns. This unbalanced moment is transferred from the slab system to the columns principally by direct flexure from the slab system (in this case, from the beams) parallel to the direction of the unbalanced moments. The remainder of this

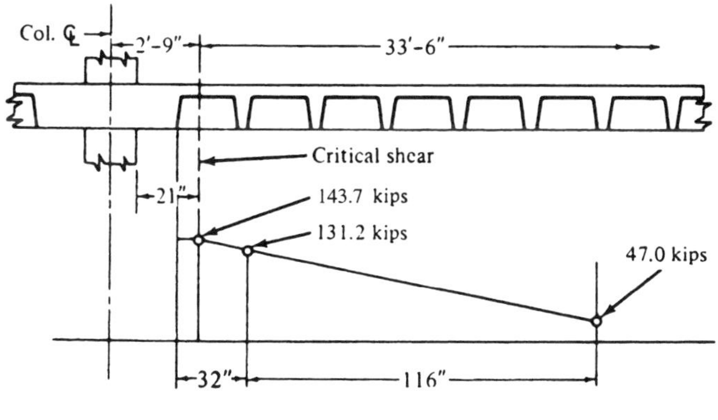

Figure 7-8 Factored shear diagram—beams.

unbalanced moment is transferred to the column by torsion between the slab system (in this example, the beams at right angles to the direction of the unbalanced moment) and the sides of the column parallel to the moment. With equal spans and loads as in this example, half of the remaining moment is transferred by torsion on each side of the column.

The Code prescribes a factored unbalanced moment, M_{un}, to columns for use in the direct design method (Section 13.6.9.2, Eq. 13-4). For flat plates, flat slabs, and waffle flat slabs without beams, the Code also prescribes the percentage of M_{un} which is assumed to be transferred by flexure (Section 13.5.3.2, Eq. 13-1) and shear (Section 13.5.3.1). For two-way systems with beams, as in this example, the Code is less explicit and it becomes necessary to compute this percentage. An approximate procedure illustrated here consists of two steps:

1. Compute M_{un} (Section 13.6.9.2, Eq. 13-4)
2. Distribute the unbalanced moment, M_{un}, to the beam and panel joists on each side of the column, in accordance with Section 13.6.4.5.

The unbalanced moment resisted by the joists on each side of the column creates torsion within the beam supporting the joists, which is equal to the factored torsional moment, T_u. The torsional moment is assumed to be zero at the midspan and a maximum at the critical section located at a distance, d, from the face of the support (Section 11.6.2.4).

TORSION. Calculate the factored unbalanced moment transferred to an interior column by Code Equation 13-4 (Section 13.6.9.2).

The unbalanced moment which must be transferred to the interior columns (Section 13.6.9.2) is

$$\begin{aligned} M_{un} &= 0.07(w_d + 0.5w_\ell)\ell_2\ell_n^2 - w_d'\ell_2'(\ell_n')^2] \\ &= (0.07)(0.5)(0.180)(39)(37)^2 \\ &= 336 \text{ ft-kips} \end{aligned}$$

The total unbalanced moment is transferred between the column and the floor by the beam in flexure and the joists on each side of the column to beams at right angles.

The portion of the total unbalanced moment to be transferred by torsion by the beams at right angles to the direction in which moments are being considered is in the same proportion.

$$T_u = (1 - 0.319)(\tfrac{1}{2})M_{un} = (0.340)(336) = 114.2 \text{ ft-kips}$$

Beam cross-section properties for torsion:

A_{cp} = area enclosed by perimeter (See Fig. 7-5).
= 1,089 in.^{2}

p_{cp} = perimeter = 72 + 42 + 2(3 + 13.33 + 20.04) = 186.74 in.

Neglect torsion when $T_u < \phi\sqrt{f'_c}\,(A_{cp})^2/p_{cp}$ (Section 11.6.1)

$< (0.85)(\sqrt{4000})(1089)^2/[(186.74)(1000)]$

< 341.40 in.-kips

T_u = 114.2 ft -kips = 1,370 in.-kips > 341.40 in.-kips, therefore torsion must be considered and torsion reinforcement consisting of closed stirrups provided for vertical shear plus torsional shear instead of the #3 open double stirrups previously required for vertical shear only.

In statically indeterminate structures, the torsional moment, T_u = 1,370 in.-kips can be reduced due to a redistribution of internal forces upon cracking to four times the moment below which torsion can be neglected, or (4)(341.40) = 1,366 in.-kips (Section 11.6.2.2). T_u = 1,370 in.-kips > 1,366 in.-kips ∴ use 1,366 in.-kips for design of torsional reinforcement.

Beam cross-section properties of torsional reinforcement.

p_h = perimeter of centerline of torsional reinforcement

= 38.5 + 38.5 + 19.5 + 19.5 = 116 in.

A_{oh} = Area inside of the centerline of the torsional reinforcement

= (38.5)(19.5) = 750.75 in.^{2}.

Beam cross-section dimensions for torsion reinforcement:

$V_c = 2\sqrt{f'_c}\,b_w d = 2\sqrt{4000}\,)(42)(20)/1000 = 106.25$ kips

Maximum shear stress ⩽ cracking shear stress (Section 11.6.3.1)

$\{[V_u/(b_w d)]^2 + [T_u p_h/(1.7A_{oh}^2)]^2\}^{(0.5)} \leqslant \phi[V_c/(b_w d) + 8\sqrt{f'_c}\,]$

$\{[144{,}500)/(45 \times 20)]^2 + (1{,}366{,}000)(116)/(1.7)(750.75)^2\}^{(0.5)}$

$\leqslant (0.85)[(106{,}250)/(42 \times 20) + 8\sqrt{4000}]$

$(25{,}780 + 27{,}350)^{(0.5)} \leqslant (0.85)(126.5 + 506.0)$

230 psi < 538 psi ∴ cross section dimensions are OK.

Torsion reinforcement:

A_o = (area enclosed by centerline of the outermost closed stirrups can be taken as $0.85A_{oh}$ Commentary R11.6.3.6)

$= 0.85A_{oh} = 0.85(750.75) = 638.14$ in.^{2}

A_t = (area of one leg of a closed hoop)

$= T_n s/(2A_o f_{yv} \cot\theta)$ (Section 11.6.3.6)

$= (1{,}366s)/[(2)(638.14)(60)(1)]$

$= 0.0178s$

$$s_{max} \leqslant 12 \text{ in. (Section 11.6.6.1)}$$
$$\leqslant p_h/8 = 116/8 = 14.5 \text{ in.}$$
$$\leqslant d/2 = 20/2 = 10 \text{ in.}$$
$$V_n = V_c + V_s$$
$$V_c = 2\sqrt{4000}\,(b_w d) = 2\sqrt{4000}\,(42)(20) = 106.25 \text{ kips}$$
$$V_s = V_u/\phi - V_c = 144.5/0.85 - 106.25 = 63.8 \text{ kips}$$
$$A_v = V_s s/(f_{yv} d) = [(63.8)(s)]/[(60)(20)] = 0.0532s$$

Shear reinforcement required for one closed stirrup = $A_t + A_v/2$.

s, in.	A_t = 0.0178s	A_v = 0.0532s	Req'd $A_v/2 + A_t$, in.^{2}
4	0.071	0.213	0.178 #4 @4 = 0.20
6	0.107	0.319	0.266 #5 @6 = 0.31
8	0.142	0.425	0.354 #6 @8 = 0.44
9	0.160	0.478	0.399 #6 @9 = 0.44

Use #6 closed stirrups @ 9 in. centers.
Check minimum area of closed stirrups (Section 11.6.5.2).

$$A_v + 2A_t = 50b_w s/f_{yv} = 50(42)(9)/60{,}000 = 0.315 \text{ in.}^2 < 2(0.44) = 0.88 \text{ in.}^2, \text{ OK.}$$

Longitudinal torsion reinforcement (Section 11.6.3.7).

$$A_l = \frac{A_t}{s} p_h (f_{yv}/f_{yl}) \cot^2\theta \quad \text{(Section 11.6.3.7)}$$
$$= (0.0178)(116)(60/60)(1)^2 = 2.06 \text{ in.}^2 \quad \text{(See Fig. 7-9).}$$

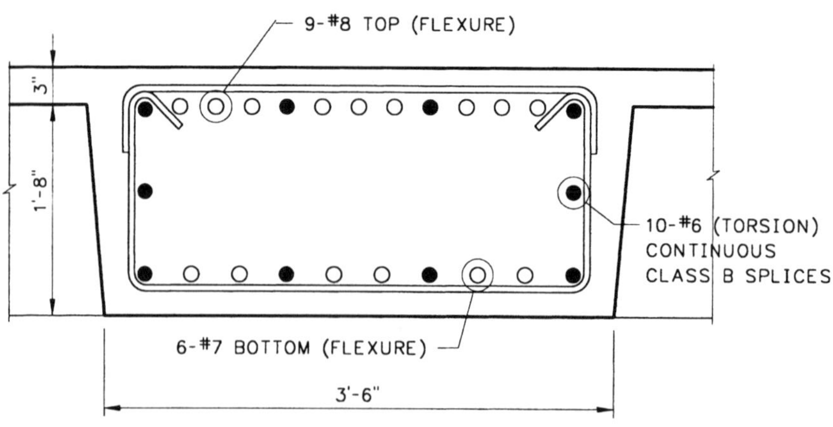

Figure 7-9 Beam reinforcement details.

Check minimum area of longitudinal torsion reinforcement (Section 11.6.5.3). Since $25b_w/f_y = 25(42)/60{,}000 = 0.0175 < 0.0178$, use $A_t/s = 0.0178$.

$$A_{\ell,\min} = \frac{5\sqrt{f_c'}\,A_{cp}}{f_{y\ell}} - \left(\frac{A_t}{s}\right)p_h\frac{f_{yv}}{f_{y\ell}}$$

$$= \frac{5\sqrt{4{,}000}\,(1{,}089)}{60{,}000} - 0.178(116)\frac{60}{60}$$

$$= 5.74 - 2.06 = 3.68 \text{ in.}^2 > 2.06, \text{ use } 3.68 \text{ in.}^2$$

MOMENT REINFORCEMENT FOR BEAMS. Moment reinforcement for the beams can be selected as illustrated in Chapter 9 of this guide. It should be noted, however, that the moment due to the factored loads applied directly to the beams must be added to the percentage of the two-way waffle slab moment resisted by the beams (Section 13.6.5.3). The total panel moment is distributed as follows:

Beams (Directly applied loads, 100%):	$-0.132M_o$; $+\ 0.061\ M_o$
Middle strip joists (7 joists):	$-0.162M_o$; $+\ 0.123\ M_o$
Column strip joists (4 joists):	$-0.356M_o$; $+\ 0.166\ M_o$

BEAM MOMENTS AND REINFORCEMENT

Minimum flexural reinforcement (Section 10.5)

For $f_c' \leqslant 4{,}440$ psi, min $\rho = 200/f_y$

For $f_y = 60{,}000$ psi, min $\rho = 200/60{,}000 = 0.0033$

Factored positive and negative moments

Min $A_s = 0.0033\ bd = 0.0033(42)(21) = 2.91$ in.^{2} unless the amount provided is one-third greater than that required by analysis.

Total factored static moment, M_o

M_o = (M_o from joists) + (M_o from factored loads applied directly to beams).

M_o (for joists) $= (0.369)(33)(37)^2/8 = 2{,}084$ ft-kips

M_o (for beams) $= (2.67)(37)^2/8 = 457$ ft-kips

Positive moment reinforcement

$$+M_u = 0.061(2084) + (0.35)(457)$$

$$= 127.1 + 160.0 = 287.1 \text{ ft-kips}$$

$+A_s = M_u/\phi f_y jd$ where $j = 1 - a/2$; assume $j = 0.98$

$+A_s = (287.1 \times 12)/[(0.90)(60)(0.98)(21)] = 3.08$ in.^{2}

Or calculate $+A_s$ using the expression from page 258

$$+A_s = 0.85\left(1 - \sqrt{1 - \frac{(2.35 \times 287.1 \times 12)}{0.90(4)(42)(21)^2}}\right)\frac{(42)(21)(4)}{60} = 3.14 \text{ in.}^2$$

$$+A_s = 3.14 \text{ in.}^2 > \min A_s = 2.91 \text{ in.}^2\text{; use 6-\#7 bars} = 3.60 \text{ in.}^2$$

Negative moment reinforcement

$$-M_u = 0.132(2084) + (0.65)(457)$$

$$= 275.1 + 297.1 = 572.2 \text{ ft-kips}$$

$$-A_s = (572.2 \times 12)/[(0.90)(60)(0.94)(21)] = 6.44 \text{ in.}^2$$

or

$$-A_s = 0.85\left(1 - \sqrt{1 - \frac{(2.35 \times 572.2 \times 12)}{0.90(4)(42)(21)^2}}\right)\frac{(42)(21)(4)}{60} = 6.47 \text{ in.}^2$$

$$-A_s = 6.47 \text{ in.}^2 \gg \min A_s = 2.91 \text{ in.}^2\text{; use 9-\#8 bars} - 7.11 \text{ in.}^2$$

See Fig. 7-9 for beam reinforcement details.

SHEAR IN BEAMS. The design of the two-way beams in this example follows the detailed procedures in Chapter 9 of this guide. The location of the critical section for one-way shear at $d = 21$ in. from the face of the column is shown in Fig. 7-2. The calculation of the total external shear force at this point is easily visualized as one-fourth of the sum of the total panel load minus the loads inside the square formed by the four critical shear sections.

One Code provision peculiar to two-way beams with two-way slabs is that the total external shear must be apportioned to these beams and slabs in the ratio of their relative stiffnesses when the quantity $(\alpha_1 \ell_2/\ell_1)$ is less than 1.0 (Section 13.6.8.2). In our example, with square panels $\ell_2/\ell_1 = 1.0$, and with the beams used, $\alpha_1 = \alpha_2 = 0.319$. The Code provides that 100% of the total shear due to loads in the tributary areas plus that directly applied to the beam (see Fig. 7-7) be assigned to the beams for values of $\alpha_1 \ell_2/\ell_1 \geqslant 1.0$ and that for values such that $0 < \alpha_1 < 1.0$, the percentage be determined by linear interpolation between zero and 100% (Section 13.6.8.2).

SHEAR IN JOISTS. Under the conditions of this example, the beams are wider than the columns and have an equal stiffness in both directions so that they obviously carry equal loads. The voids in the waffle system area become *holes* (in the compression flange) and both the beams and the joists behave as one-way members in shear. No two-way shear investigation is required. Beam shear is one-fourth of the loads outside the critical shear section. The proportion of the one-way shear assigned to the *joists*, however, can be reduced to (1.000 − 0.319) of the total shear on the waffle area (Section 13.6.8.2).

The Code specifically requires that the two-way beams be proportioned to resist the factored *shear* caused by 100% of the directly applied factored loads (Section 13.6.8.3) whereas, for factored moment, 100% of only those loads applied on the beam and *not considered in the slab design* need be added to the beam's share of the column strip moment (Sections 13.6.5.1 and 13.6.5.3). The practical result of these requirements is to *reduce* the load for moment in the beam which would be resisted by fewer larger bars at less expense, and to *increase* the load for shear in the beam (consider stirrups to cost 50¢/lb in place). In our example, all shear is obviously carried to the column by the beams (see Fig. 7-10). As shown by the following calculations, the joists selected here did not require stirrups. Stirrups in joists are not generally economical.

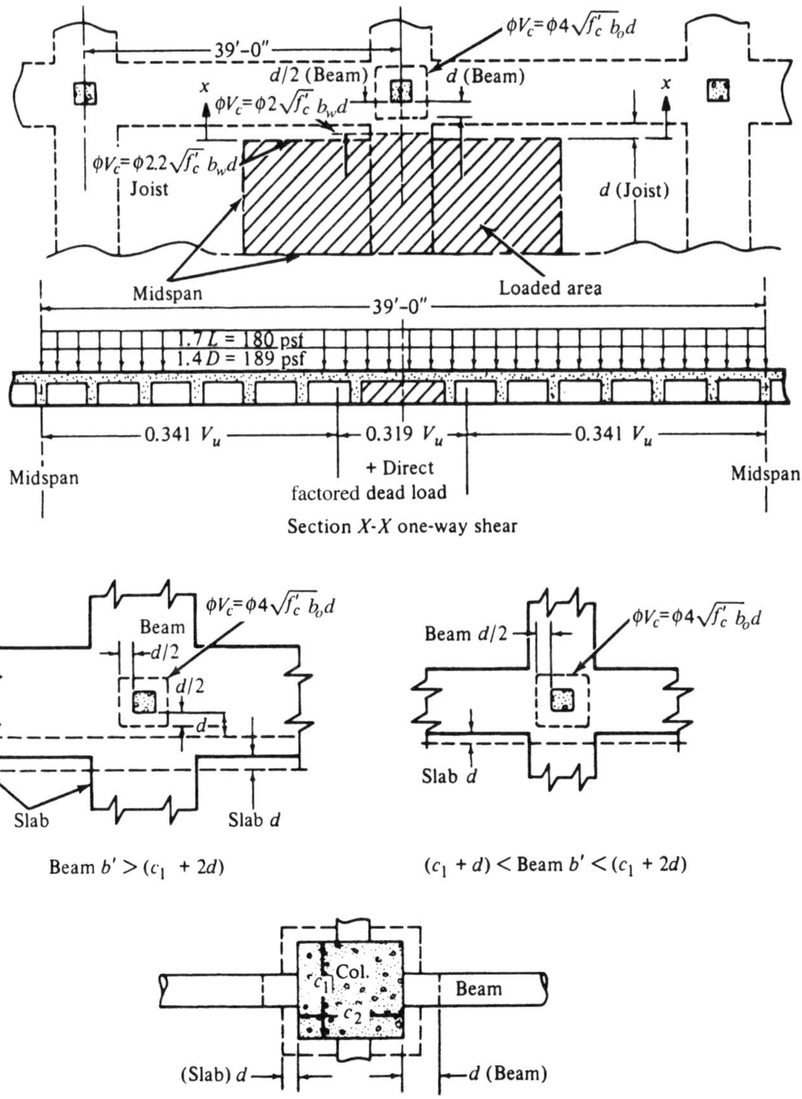

Figure 7-10 Factored loads and critical sections for beams.

Total one-way (joist) shear at a constant distance, $d = 21$ in., from the face of the beams can be distributed as follows (see Fig. 7-11):

$$V_u = 198 \text{ kips}$$

Using this maximum value, the average factored shear per joist is

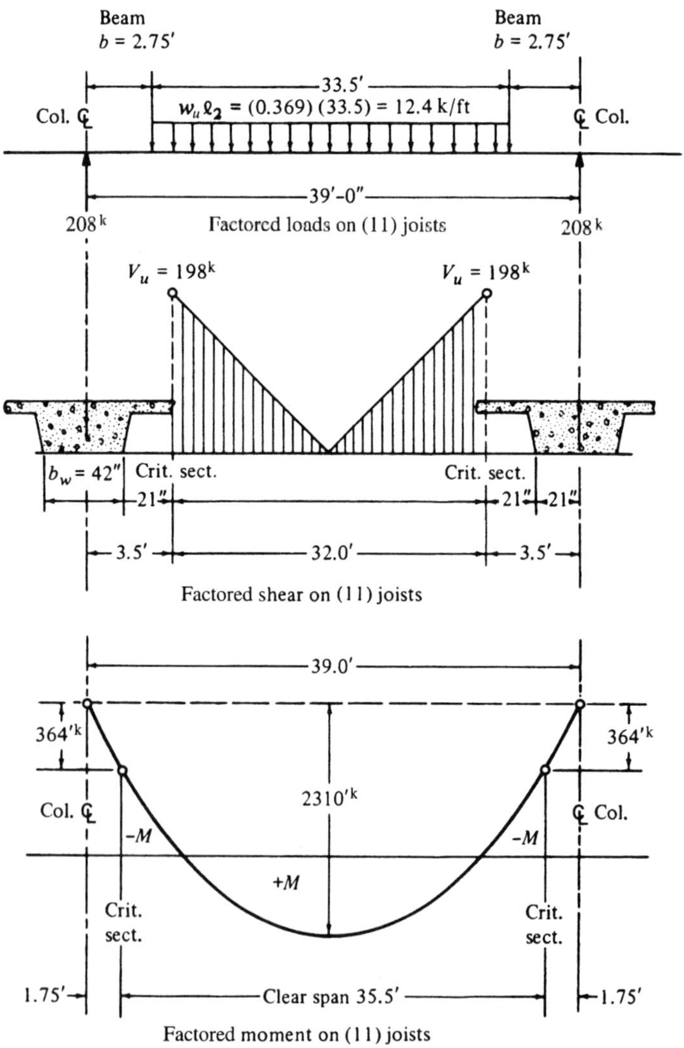

Figure 7-11 Factored loads and critical sections for design of joists.

$$V_u = \frac{1}{11}(198) = 18.0 \text{ kips per joist}$$

The average width of the joist web for design shear strength,

$$b_w = 6 + \frac{20}{(12)(2)} = 6.83 \text{ in.}$$

$$\phi V_c = \phi 2.2\sqrt{f_c'}\,b_w d \qquad \text{(Sections 8.11.8 and 11.3.1.1)}$$

$$= (0.85)(2.2)\sqrt{4000}\,(6.83)(21)/1000$$

$$= 17.0 \text{ kips} < V_u = 18.0 \text{ kips}$$

Stirrups are required in the joists.

To take advantage of the reduction allowed for joists, the total factored shear could be calculated for uniform loads at a section 21 in. from the face of the beam stem as 198 kips plus the added dead load of the beam.

$$(1.59)(32)/2 = 25.4 \text{ kips}$$

Factored shear on the beam: $V_u = 25.4 + (0.319)(198) = 88.6$ kips

Factored shear on (11) joists: $V_u = (0.681)(198) = 134.8$ kips

Factored shear per joist $= 134.8/11 = 12.3$ kips < 17.0 kips.

No stirrups are required in the joists.

To avoid these confusing double calculations for beam loads, one for moment and one for shear, the authors' recommendation is to consider all loads on the entire width of the beam, including the flanges, to be carried by the beam in both shear and moment and to add the different percentages prescribed for shear and moment from the remainder of the panel. Where beams are wider than the critical section for one-way shear ($c_l + 2d$), investigate the two-way shear in the beam section, $(\frac{1}{2})d$ from the column face and one-way shear across the entire width of the panel at a distance equal to the *slab* depth from the face of the beams at right angles. Where beam width is less than ($c_1 + 2d$) but equal or greater than ($c_1 + d$), investigate one-way shear only and consider the beam section only. Where beams are less than the column width, an exact solution requires an investigation of both one-way and two-way shear at prescribed distances on the combined slab and beam section. Usually, it will be more practical to assign all shear to the beams. Regard void spaces (such as waffles) in the compression flange (bottom) as holes (see Fig. 7-10).

These recommendations will safely account for all shear to the column (Section 13.6.8.4), conform to the shear distribution provisions (Section 13.6.8.2), and be conservative for the moment in the beams and shear in the joists (Section 13.6.5.3). In addition to the practical reduction of design time (and possible error), these procedures tend to locate more steel in the beam in fewer, larger sizes.

JOIST MOMENTS AND REINFORCEMENT

Column strip joists (4)

Negative moment

$$-M_u = 0.356M_o = (0.356)(2084) = 742 \text{ ft-kips}$$

$$-A_s = M_u/\phi f_y jd \qquad \text{assume } j = 0.86$$

$$= (742 \times 12)/[(0.90)(60)(0.86)(21.75)] = 8.82 \text{ in.}^2$$

or

$$-A_s = 0.85\left(1 - \sqrt{1 - \frac{(2.35 \times 742 \times 12)}{(0.90)(4)(24)(21.75)^2}}\right)\frac{(24)(21.75)(4)}{60} = 8.91 \text{ in.}^2$$

$$\text{Min } A_s = 0.0033(24)(21.75) = 1.72 \text{ in.}^2$$

Use 15-#7 bars $= 9.0$ in.^{2}

Positive moment

$$+M_u = 0.166M_o = (0.166)(2084) = 346 \text{ ft-kips}$$

$$+A_s = M_u/\phi f_y jd \qquad \text{assume } j = 0.98$$

$$+A_s = (346 \times 12)/[(0.90)(60)(0.98)(21.75)] = 3.61 \text{ in.}^2$$

or

$$+A_s = 0.85\left(1 - \sqrt{1 - \frac{(2.35 \times 346 \times 12)}{(0.90)(4)(144)(21.75)^2}}\right)\frac{(144)(21.75)(4)}{60} = 3.57 \text{ in.}^2$$

$$\text{Min } A_s = 0.0033(24)(21.75) = 1.72 \text{ in.}^2$$

Use 2-#6 bars in each of the 4 joists providing a total $A_s = 2(0.44)(4) = 3.52$ in.^{2} which is $\approx A_s$ required, OK.

Middle strip joists (7)

Negative moment

$$-M_u = 0.162M_o = (0.162)(2084) = 338 \text{ ft-kips}$$

$$-A_s = (338)(12)/[(0.90)(60)(0.92)(21.75)] = 3.75 \text{ in.}^2$$

or

$$-A_s = 0.85\left(1 - \sqrt{1 - \frac{(2.35 \times 338 \times 12)}{(0.90)(4)(42)(21.75)^2}}\right)\frac{(42)(21.75)(4)}{60} = 3.57 \text{ in.}^2$$

$$\text{Min}\, A_s = 0.0033(42)(21.75) = 3.01 \text{ in.}^2$$

Use 12-#5 bars = 3.72 in.^{2}

Positive moment

$$+M_u = 0.123M_o = (0.123)(2084) = 256 \text{ ft-kips}$$

$$+A_s = (256)(12)/[(0.90)(60)(0.98)(21.75)] = 2.67 \text{ in.}^2$$

or

$$+A_s = 0.85\left(1 - \sqrt{1 - \frac{(2.35 \times 256 \times 12)}{(0.90)(4)(252)(21.75)^2}}\right)\frac{(252)(21.75)(4)}{60} = 2.62 \text{ in.}^2$$

$$\text{Min}\, A_s = 0.0033(42)(21.75) = 3.01 \text{ in.}^2$$

Use 1-#4 bar and 1-#5 bar in each of the 7 joists providing a total $A_s = 7(0.51) = 3.57$ in.^{2} which is 36% greater than the A_s required by analysis (Section 10.5.3), OK.

See Fig. 7-12 for joist reinforcement details.

BAR LENGTHS. Design details for reinforcement of flat slabs are described in (Section 13.3 and Code Fig. 13.3.8). In order to satisfy maximum spacing limits for control of crack width (Section 13.3.2 and Commentary R10.6) the authors selected straight top bars uniformly spaced across the strip in which they are required without reference to the location of the joist ribs. See Fig. 7-12 for minimum bar lengths (Code Fig. 13.3.8).

EXTERIOR PANELS

The design of the exterior panels for waffle slabs involves the same considerations illustrated for solid two-way flat plates and flat slabs with drop paneis (Chapters 5 and 6). The additional considerations necessary with the waffle system make the use of small sketches helpful for design, as well as for this explanation (see Fig. 7-13). The cantilever arrangements with edge columns set back from the building line as in Fig. 7-13 (*a*) and (*c*) assist in developing the top

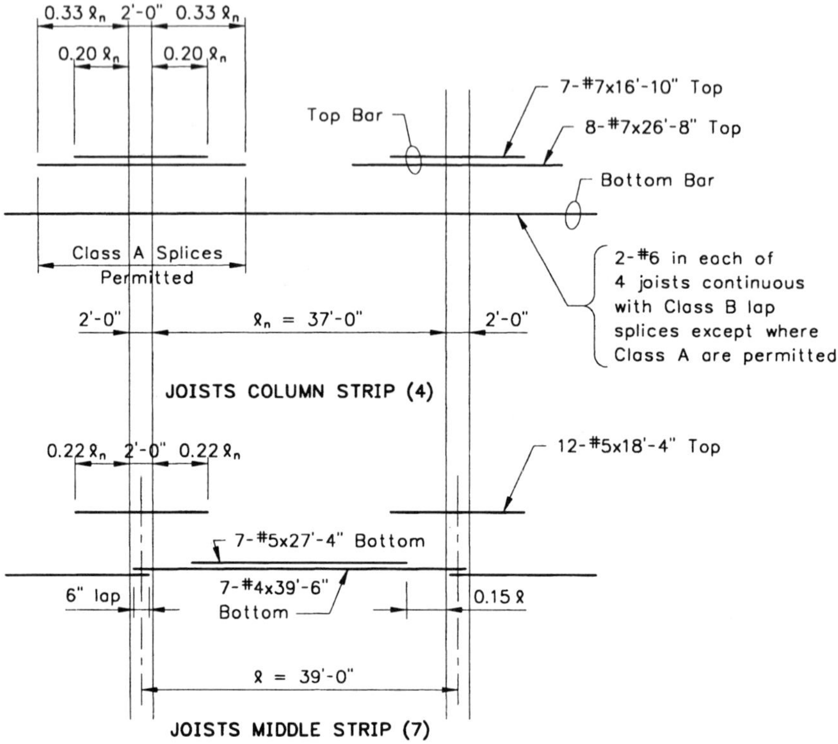

Figure 7-12 Joist reinforcement details.

bars for negative factored moment, reduce the unbalanced negative moments at the columns, and provide a symmetrical four-sided shear section which simplifies computations. The Code has no specific requirements for the consideration of the variable cross sections of spandrel beams (or slabs) illustrated. It specifically provides that the torsional member shall be assumed to have a constant cross section equal to the width of the column, c_1, and depth of the slab (Section 13.7.5.1-a). The next provision permits it to consist of a beam plus portions of the slab, again as if constant (Section 13.7.5.1-b) .

In view of the variations possible within these definitions, some of which are shown in Fig. 7-13, it would seem more realistic to evaluate the torsional strength by the procedure used for flat slabs with drop panels (see Fig. 6-8). This permissible application (Section 13.7.5.1-a) permits consideration of portions of the flange in the waffle slab up to 4 h on either side of a rib or solid beam section (Section 13.2.4).

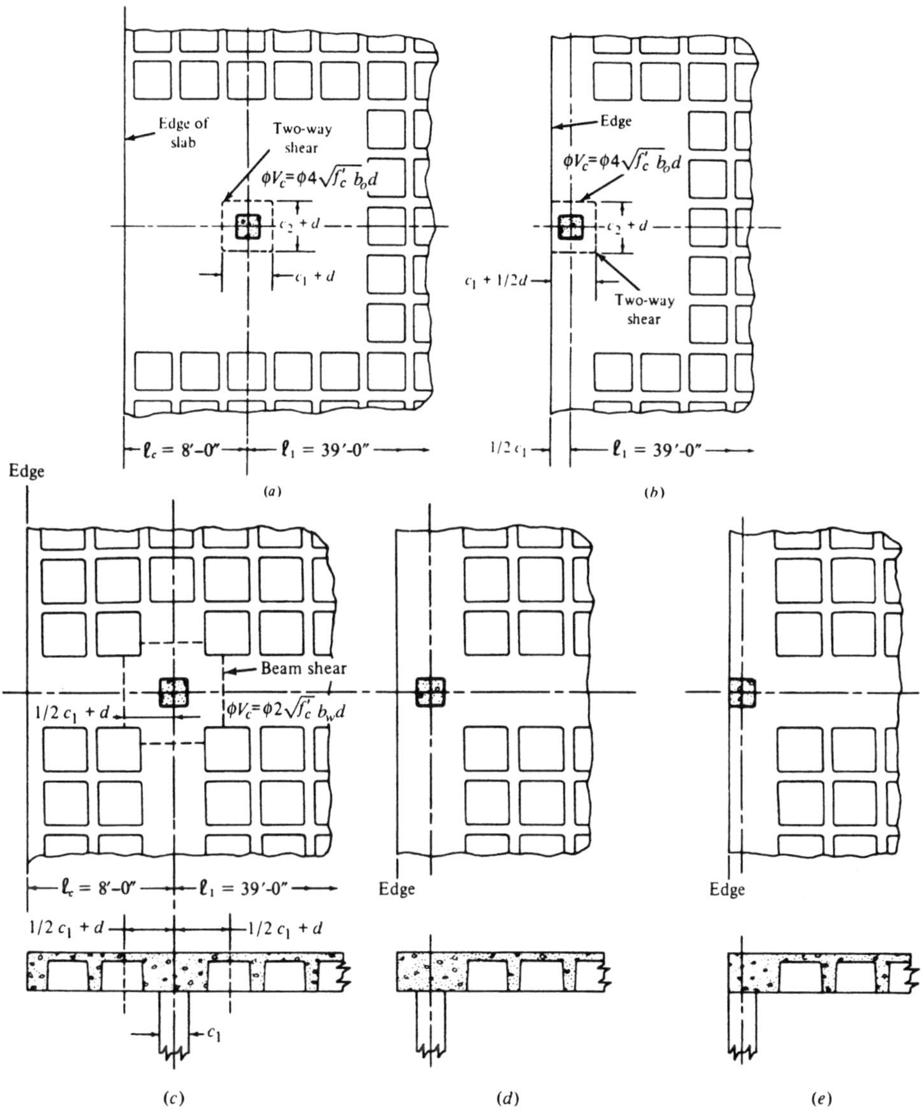

Figure 7-13 Exterior column locations with waffle slabs; (*a*) waffle flat plate with solid head cantilever edge; (*b*) waffle flat plate with solid head flush edge; (*c*) waffle flat plate with beams and with cantilever edge; (*d*) edge column center of beam; (*e*) beam edge flush with column.

For conditions (*b*), and (*d*) it should be sufficiently accurate to consider the torsional member similar to interior beams with a constant section with dimensions $x = 23$ in., $y = 33$ in. for (*b*) and $y = 42$ in. for (*d*) as for the interior beams.

Torsional Stiffness

Computations for torsional stiffness, K_t, for use with the Equivalent Frame Method (Section 13.7) are shown in a Chapter 6 example for a two-way solid flat slab with drop panels and will not be repeated here.

8 Two-Way Slab-Beam Design

GENERAL

Dissatisfaction with results of previously required procedures of analysis and design for two-way slab-beam systems was the original reason for the massive research upon two-way slab systems begun about 1955. All available load test results indicated that the flat plate or flat slab was a most efficient design with an adequate safety factor. The two-way slab on beams, analyzed as separate elements, appeared overdesigned with unnecessarily high safety factors. Since 1971, all two-way reinforced systems, with or without beams, have been subject to the same integrated analysis and design procedures (Chapter 13). The requirements have been adjusted to provide nearly uniform strength design capacity for all two-way designs (Chapter 13).

DESIGN AIDS

Most of the charts presented in Chapter 5 for the determination of relative stiffnesses apply also to the two-way system with beams. In this chapter, only the additional procedures required for utilization of the beams will be presented. To avoid duplication, applications of Chapter 5 charts will be indicated by references.

ADVANTAGES

Practical economics indicate consideration of beams in a two-way system when the spans and/or loads are large enough so that required slab thickness and minimum column sizes for a slab without beams become large enough to offset the forming cost and added story height with the beams. The examples presented apply the two-way slab-beam to span and load combinations suitable to illustrate this effect. See Table 5-2 for comparative flat plate minimum size columns.

DIRECT DESIGN

All of the limitations upon the application of the direct design method (Section 13.6.1) previously discussed apply when beams are included. The one additional limitation applicable *only* for a slab panel supported by beams *on all sides* is that the relative stiffness factors of the beams in the two perpendicular directions, $\alpha_1 \ell_2^2/\alpha_2 \ell_1^2$, shall not be less than 0.2 nor more than 5.0 (Section 13.6.1.6). The requirement does not preclude the use of the direct design method for beams in one direction only or on one side only, such as spandrel beams.

INTERIOR PANELS

For the design of interior panels, the important effect is to assume more of the factored moment is resisted by the beams and less by the slabs. In square panels, the total negative panel moment assigned to the column strip (including the beam) is unchanged and total positive panel moment assigned is increased from 60% to 75%. For rectangular panels, however, as the span ratio, ℓ_2/ℓ_1, approaches the limits, 0.5 and 2.0, the required distribution of the resisting moment per strip changes significantly due to the presence of beams (Sections 13.6.4.1 and 13.6.4.4) (see Fig. 8-1).

The following examples illustrate only the required distribution of the total panel moments for interior panels with beams. Design of the slabs and beams

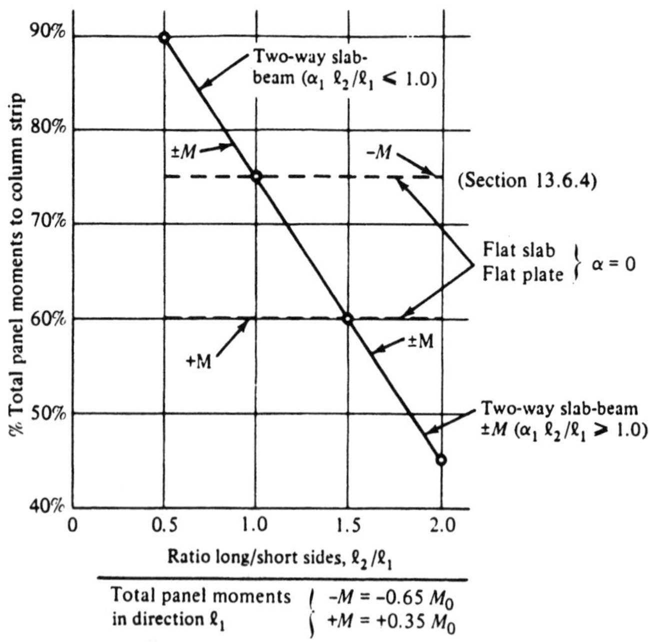

Figure 8-1 Interior panel—column strip moments.

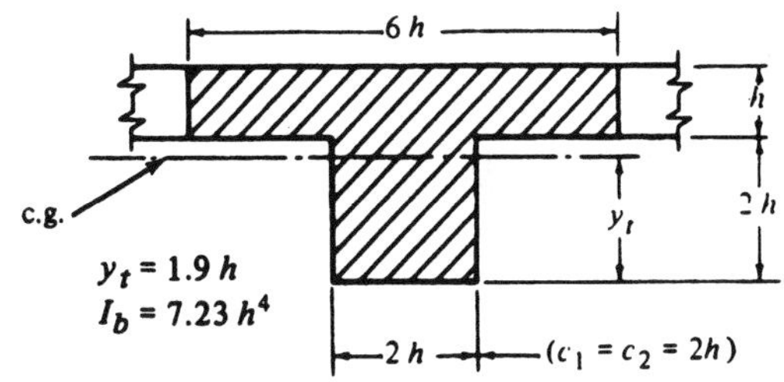

Figure 8-2 Beam dimensions and properties.

after the factored moments are determined follows the same procedures shown in Chapters 5 and 9.

EXAMPLE 1. *Square interior panel.* Gross section properties are to be used (Section 13.0). See Fig. 8-2. Assume the same beam section on all sides of the panel; same spans in adjacent interior panels. The limiting total width of top flange for this beam is $6h$ (Section 13.2.4).

Determination of minimum slab thickness (Section 9.5.3.1)

Conformance to the limitations of the following three equations relieves the designer of deflection computations otherwise required by the Code. A span must be assumed for the example in order to determine moment of inertia, I_s, for the slab and the ratio of beam/slab stiffnesses, α, since the first two equations are in review form. Try clear span $\ell_n = 40\,h$.

$$\ell_1 = \ell_2;\ I_s = 3.33\,h^4;\ \text{(Section 13.0)}\ \alpha_1 = \alpha_2 = \alpha_m = E_{cb}I_b/E_{cs}I_s = 2.18$$

$$\text{(Section 9.0)}\ \beta = \ell_1/\ell_2 = 1.0 \text{ for square panel}$$

$$\text{(Section 9.0)}\ \beta = 1.0 \text{ for interior panel, beams on four sides}$$

$$\text{Min } h = \ell_n(0.8 + f_y/200{,}000)/[36 + 5\beta(\alpha_m - 0.2)]$$

$$= \ell_n(0.8 + 60{,}000/200{,}000)/[36 + (5)(1)(2.18 - 0.2)]$$

$$= \ell_n/41.7 \text{ Eq. (9-11)}$$

The controlling requirement (Eq. 9-11) is $\ell_n \leqslant 41.7\,h$. $\ell_n = 40\,h \leqslant 41.7\,h$. $\alpha_1\ell_2/\ell_1 = \alpha_2\ell_1/\ell_2 \geqslant 1.0$ and so beam takes 85% of the column strip moments (Section 13.6.5.1).

Distribution of positive and negative panel moments (Sections 13.6.4.1 and 13.6.4.4).

$$-M_u = 0.65\,M_o;\ +M_u = +0.35\,M_o \text{ (Section 13.6.3.2)}$$

Total of beam and column strip: $+75\%\ (0.35\,M_o)$; $-75\%\ (0.65\,M_o)$

Middle strip: $+25\%\ (0.35\,M_o)$; $-25\%\ (0.65\,M_o)$

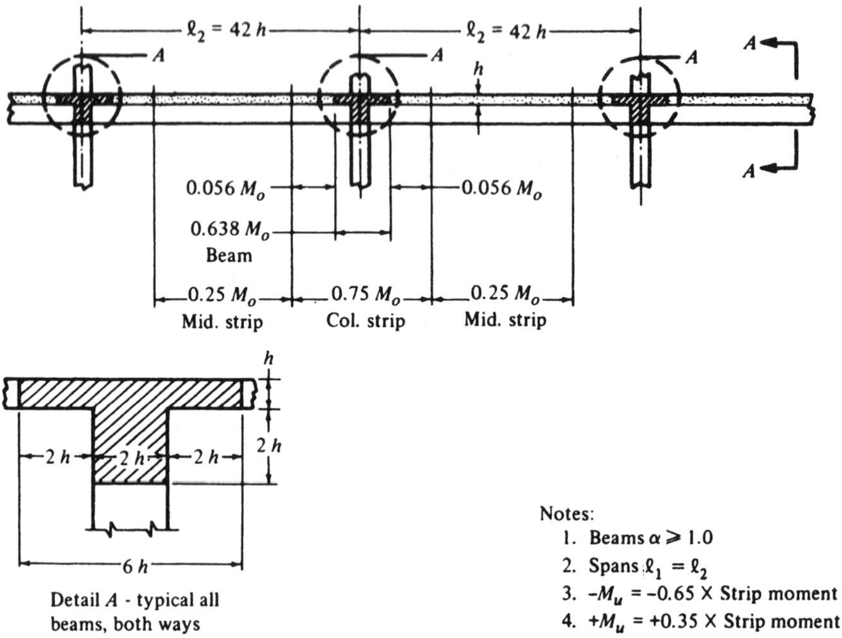

Figure 8-3 Square interior panels—total moments per strip.

Beam: $+63.8\%$ $(0.35\ M_o)$; -63.8% $(0.65\ M_o)$ (Section 13.6.5.1)

Each side of the column strip outside of the beam:

$$+5.6\%\ (0.35\ M_o);\ -5.6\%\ (0.65\ M_o)$$

See Fig. 8-3 for summary layout.

EXAMPLE 2. *Rectangular interior panel*

Maximum $\ell_1/\ell_2 = 2.0$ (Section 13.6.1.2). Same beam proportions as in the previous example. Beams on all sides. See Fig. 8-4.

$$\beta = \ell_1/\ell_2 = 2.0 \text{ (Section 9.0)}$$

Try $h = 6''$; $\ell_1 = 25'\text{-}0''$ $\ell_2 = 12'\text{-}6''$ with beam web $= 2\ h$

$I_{s1} = 2.08\ h^4$; $I_{s2} = 4.07\ h^4$; $\alpha_1 = 3.48$; $\alpha_2 = 1.73$; $\alpha_m = 2.60$ (Section 9.0)

$\ell_2 = 25\ h$; $\ell_1 = 50\ h$; $\ell_n = 48\ h$ (Section 9.0).

Review for minimum thickness per Section 9.5.3.3-c:

$$\text{Min } h = \ell_n(0.8 + f_y/200{,}000)/[36 + 9\beta] \qquad \text{Eq. (9-12)}$$

$$= \ell_n(0.8 + 60{,}000/200{,}000)/[36 + (9)(2)]$$

$$= \ell_n/49.1$$

Check ratio of relative beam stiffnesses in the two directions:

$$\alpha_1\ell_2^2/\alpha_2\ell_1^2 = 0.5; \quad \alpha_2\ell_1^2/\alpha_1\ell_2^2 = 2.0 \text{ (Section 13.6.1.6)}$$

$$0.2 < \text{Both} < 5.0$$

For use in assigning moments to each strip (Section 13.6.4):

$$\alpha_1\ell_2/\ell_1 = 1.74 \quad \alpha_2\ell_1/\ell_2 = 3.46$$

Moments in the short direction (ℓ_2): Column Strip ± 45% total

Middle Strip ± 55%

Beam ± 38.2% (Section 13.6.5.1)

Column strip each side of beam ± 3.4%

Moments in the long direction (ℓ_1): Column Strip ± 90% total

Middle Strip ± 10%

Beam ± 76.5% (Section 13.6.5.1)

Note that width of column strips equals $\ell_1/2$ in both ℓ_1- and ℓ_2-directions (Section 13.2.1).

Column strip each side of beam ± 6.75%

See Fig. 8-4 for a summary layout.

SUCCESSIVE TRIALS

If the moment or shear is excessive in the beams for the dimensions assumed as a first trial, and any increase of the dimensions is desired, the distribution of the panel moments shown in Fig. 8-4 will be unchanged provided that the relative stiffnesses are not less than 0.2 nor greater than 5.0 and $\alpha_1\ell_2/\ell_1 \geqq 1.0$ (Section 13.6.1.6). Note that the expression, $\alpha_1\ell_2/\ell_1 \geqslant 1.0$ determines the distribution to the column and middle strips as well as to the beams (Sections 13.6.4.1, 13.6.4.4, 13.6.5). Reduction of the beam sizes will not change the distribution unless the ratio, $\alpha_2\ell_1/\ell_2$, or $\alpha_1\ell_2/\ell_1$, falls below 1.0. It should also be noted that the tabulated percentages of moment for various values of α, β, ℓ_2/ℓ_1, and $\alpha_1\ell_2/\ell_1$ are end points and that straight-line interpolation across and vertically in these tables is intended for intermediate values of these quantities.

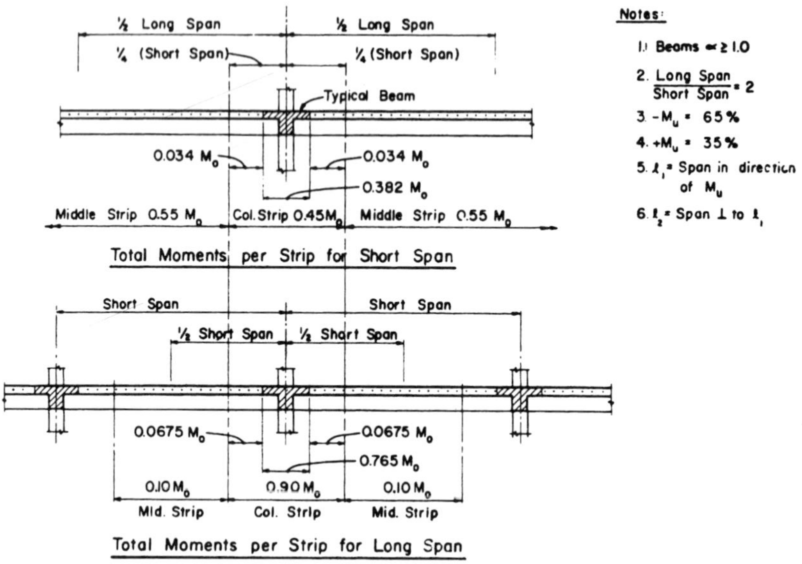

Figure 8-4 Rectangular interior panel (2:1)—strip moments.

EXAMPLE 3. *Increase depth of beam in span ℓ_1 (see Fig. 8-5).*

$$I_{b1} = 18.88\,h^4 \qquad I_{b2} = 7.23\,h^4$$

$$\alpha_1 = \frac{18.88}{2.08} = 9.14 \qquad \alpha_2 = \frac{7.23}{4.17} = 1.73$$

$$\alpha_1\ell_2/\ell_1 = 4.57 \qquad \alpha_2\ell_1/\ell_2 = 3.46$$

The distribution of the moments is unchanged since both ratios are greater than 1.0 Check relative beam stiffness factors,

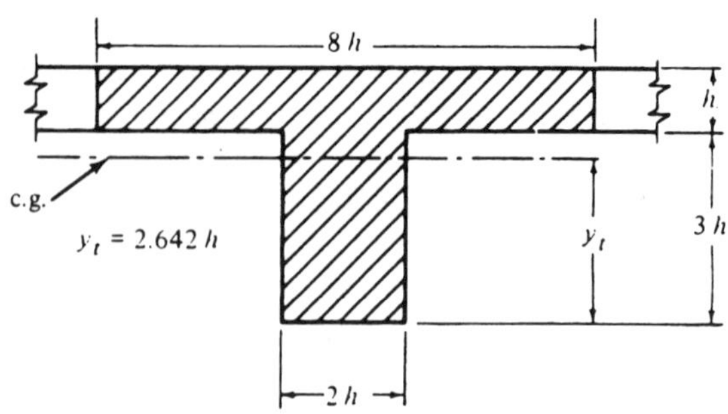

Figure 8-5 Beam dimensions.

$$\frac{\alpha_1 \ell_2^2}{\alpha_2 \ell_1^2} = 1.32; \quad 0.2 < 1.32 < 5.0 \text{ (Section 13.6.1.6)}$$

Check minimum thickness; $\ell_1/\ell_2 = \beta = 2.0$; $\beta_s = 1.0$ interior

$$\alpha_m = \frac{9.14 + 1.73}{2} = 5.44$$

Eq. 9-12: $\quad \text{Min } h = \ell_n(0.8 + 60{,}000/200{,}000)/[36 + (9)(2)]$

$$= \ell_n/49.1$$

$$\text{Min } h \geqslant \ell_n/49.1 = 5.87 \text{ in. (Section 9.5.3.3)}$$

Design for Factored Shear and Moment

Completion of any design after the determination of the applied bending moments for thicknesses within the above limits and actual loads follows the same procedures illustrated in Chapter 9. A number of specific minor questions of interpretation for the provisions of Chapter 13, however, will be encountered. For convenience, wide beams on column centerlines should be assumed to transmit all shear to the column (see pages 194–198). A number of common variations of slab, beam, and column proportions are shown in Fig. 8-6. In all cases, it will be necessary to check the beam (one-way) shear with design shear strength of concrete, $\phi V_c = \phi 2\sqrt{f_c'}\, b_w d$ (or the "long" formula) at a critical section a distance equal to the beam depth, d, from the column face (Sections 11.3.1 and 11.3.2). For cases where all beams are much wider or much narrower than the column, it may be necessary to check the two-way shear (Section 11.11.1) as for a slab, with $\phi V_c = \phi 4\sqrt{f_c'}\, b_o d$, at a critical section 0.5 d from the column face (Sections 11.12.1.2 and 11.12.2). For the small narrow beam, the shear section will lie partly in the beams and partly in the slab (see Fig. 7-10). Loads within tributary areas inside lines at 45° between the column centers and the midspan comprise the maximum external shear load assumed per beam (Section 13.6.8.1). Note that the wording used in many general codes prescribing live load reductions for beams would preclude use of such reductions for beams which are designed as merely deeper but integral parts of a slab. Except as noted, the slab supported by beams on four sides will be subjected to one-way shear only, on all sides, at a critical section one slab depth from the face of the beam *web*. For this purpose, the overhanging flanges are not to be considered as "faces of supports."

Torsion

For torsional stiffness, K_t, and strength, the beam section is assumed to consist of the sum of the rectangles, $(b_w)(h_b) + (h_s)$ (projection $\leqslant 4\ h_s$). The projec-

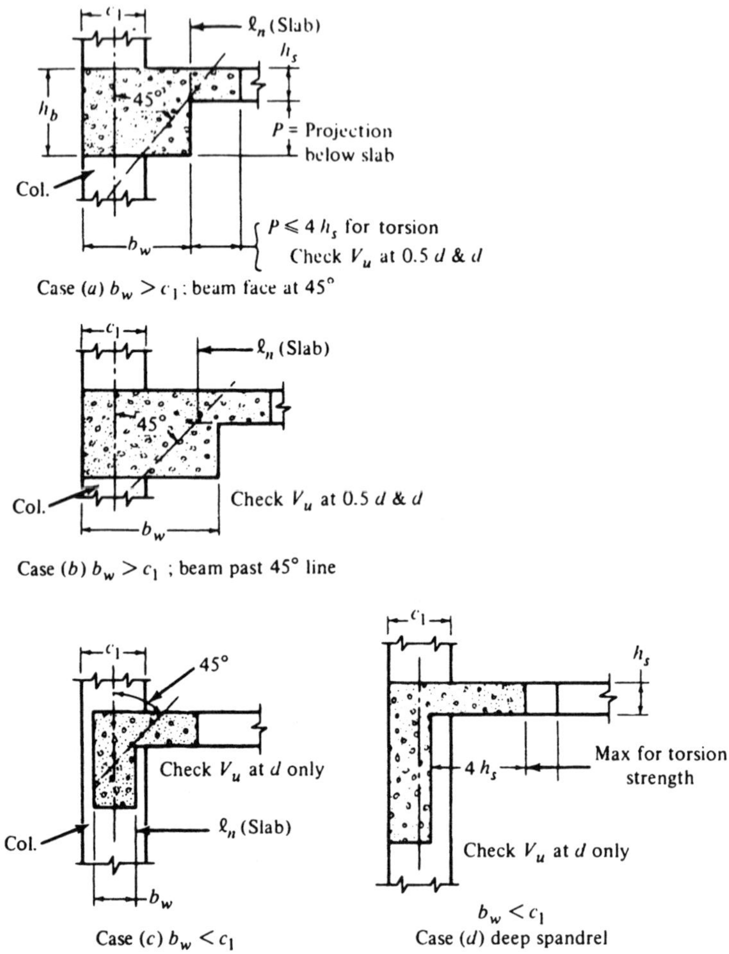

Figure 8-6 Edge beam variations.

tion is taken as the larger above and below the slab where a spandrel beam projects thus (Sections 11.6.1 and 13.2.4). Unless closed stirrup torsion reinforcement can be provided in the overhanging flanges, the torsional strength capacity required should be based on the reinforced (stem) portion only.

Moment and Shear

The clear span of the slab is measured from the face of the column, capital, bracket, or wall (Section 13.6.2.5). When the slab is supported upon beams, the

Code is ambiguous. The authors recommend the most conservative interpretation for each case (see Fig. 8-6). Where the beam is wider than the column, the clear span, ℓ_n, can be taken from the face of the beam or the intersection of the extension of the slab soffit line with a 45° angle in a vertical plane from the center of the column, whichever is longer. This interpretation is justified because the portion of the beam within this angle satisfies the definition of a capital (Section 13.1.2) and presumably, a bracket, although no definition of a bracket is provided. For beams narrower than the column, the Code seems to permit the clear span for the slab to be measured from the face of the column (Section 13.6.2.5), but the more conservative interpretation is recommended, taking ℓ_n from the face of the beam (Fig. 8-6).

EXTERIOR PANELS

Coefficients for distributing M_0 are given in Table 2-1 for the following conditions:

1. Slabs with their exterior edge simply supported on masonry or concrete walls are considered unrestrained.
2. Two-way slab systems with beams supported on all four sides.
3. Flat plate slab systems without edge beams.
4. Flat plate slab systems with edge beams.
5. Slabs with their exterior edge built integrally into a very stiff wall and which have their exterior edge fully restrained.

Intermediate restraint can be evaluated by linear interpolation between the simply supported condition and the fully restrained condition. The interpolation factor is the ratio of edge rotation of the partially restrained slab to the edge rotation of the simply supported slab.

All slabs must conform to the restrictions imposed for the direct design method (Section 13.6.1). An alternative to the coefficients in Section 13.6.3.3 is to use a modified stiffness method presented in the Commentary and which were the provisions in the 1977 Code (Commentary R13.6.3.3).

In design of beams for shear for exterior panels supported by beams on all sides, or for a flat plate with an edge beam, the torsional moments which must be combined with the vertical shear for the edge beam become significant. See Chapter 9 for an example and design aids for stirrup reinforcement to resist the combined torsion and shear.

OTHER TWO-WAY SLABS

The 1995 Code has left the design requirements for two-way slabs supported by beams not on the column centerlines (see Fig. 8-7), as well as two-way slab design for walls with supports on all sides somewhat in limbo. The two-way slab

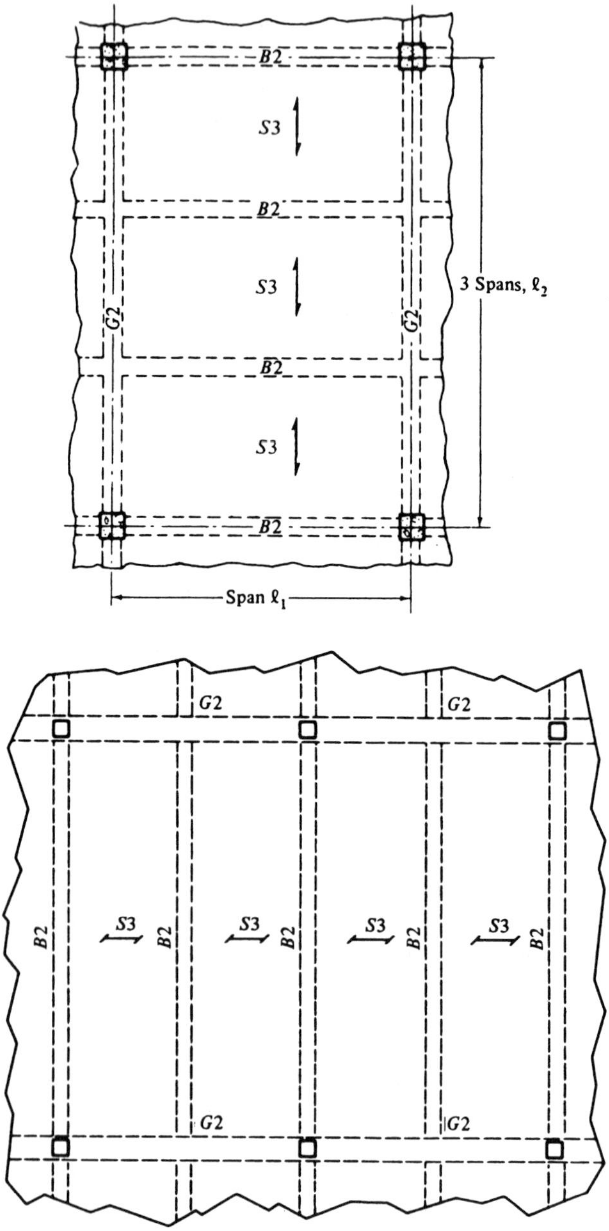

Figure 8-7 Slab-beam-girder system.

provisions (Chapter 13) apply specifically to two-way floor slabs, uniformly loaded, with beams if any on column centerlines and not offset more than 10 percent of the span as part of the "column strip" (Sections 13.6.5.1 and 13.6.1.4). The Code states that two-way slab systems may be supported by walls or columns (Section 13.1.2), and most of the specific provisions, such as taking clear span between faces of the columns or walls, obviously intend that beams, if any, be on column centerlines.

Walls

Walls resisting earth or other lateral pressure and spanning vertically between floor and footing and horizontally between cross-walls or pilasters are seldom continuous over three or more spans vertically, and the direct design method for two-way behavior is not applicable. Such walls can be designed as two-way slabs to comply with the Code provisions for walls (Section 14.2.1), but for practical purposes this problem is outside the scope of the specific provisions for two-way slabs and must be designed under the permissive procedure clause (Section 13.3.1). For practical design of such walls, the authors recommend the use of the 1963 ACI Code two-way slab provisions with appropriate adjustments for yielding or unyielding supports and nonuniform distributed loads. This procedure can qualify for approval under 1995 Code provisions (Sections 1.4, 13.1.1, 13.3.1, and 14.2.1).

Two-way Slab-Beam-Girder Systems

For floor slabs supported on beams and girders as in Fig. 8-7, the authors recommend design as one-way slabs for practical reasons. When the ratio of sides is greater than 1:2, direct design as two-way slabs is not permitted. Even the ratio is expressed in terms of "spans" which could be interpreted as clear spans or as center-to-center of supports. In Fig. 8-7(a), the ratio of center-to-center spans is 2.0; if beams $B2$ were wider or if clear span were taken from column faces, the ratio of clear spans would be greater than 2.0. Equivalent frame analysis is permitted, but it is not easily extended to the conditions in Fig. 8-7. The use of one-way slab design will eliminate these and other ambiguities in the interpretation of the Code provisions for two-way slabs and thereby greatly speed up the design.

More importantly, it will eliminate ambiguities in the interpretation and application of the live load reductions permitted by the general codes, which will then become increasingly important for slab, beam, and girder as intended (see Chapter 9). The minimum shrinkage and temperature reinforcement (Section 7.12.1) in long span direction will suffice for actual two-way bending (not considered in the one-way design) if it is placed properly in the top and bottom of the slab at the supporting girders and midspan, respectively.

This simple design aproach for satisfying equilibrium is limited, however, by the need to satisfy also compatibility of deflections. Any analysis is acceptable which satisfies both (Section 13.3.1). The layout in Fig. 8-7(a) with subpanels, $\ell_1 \times \ell_2$, has a rectangular ratio of long to short sides approximately two, which may be regarded as an upper limit. When this ratio is between one (square) and two, two-way slab action will prevail in the subpanels. Relative stiffnesses of $S3$, $B2$, and $G2$ must be carefully considered in relation to the entire panel (ℓ_1, 3 ℓ_2) to determine whether the slabs, $S3$, can behave independently as either one- or two-way members. Two other patterns of over-all behavior for the entire panel, ℓ_1, 3 ℓ_2, are possible: (1) as a two-way system with spans ℓ_1 and 3 ℓ_2 and (2) as a pure one-way system with span ℓ_1. If beams, $B2$, have a stiffness ratio, $\alpha_B = I_b/I_s$ greater than 1.0, the subpanels, $S3$, will develop two-way action over spans ℓ_1, ℓ_2. Similarly, if girders $G2$ have minimum stiffnesses relative to the slab $\alpha_{G2} \geqslant 1.0$ and relative to beams $B2$, $\alpha_{G2}\ell_1^2/\alpha_{B2}9\,\ell_2^2 \geqslant 1.0$, two-way action as a panel, ℓ_1, 3 ℓ_2, is unlikely. In design of the individual end slabs, $S3$, some further consideration of "warping" stresses due to relative deflections, at columns, $\Delta = 0$, at third points $\Delta > 0$. Under full load, the center subpanel will be free of these stresses. Where such stiffness requirements become impracticable and both girders and beams are relatively flexible, the equivalent frame two-way slab design method for panels, ℓ_1, 3 ℓ_2, should be applied: girders, $G2$, part of the column strip in span 3 ℓ_2; beams $B2$ on column lines similarly in span ℓ_1; and intermediate beams, $B2$, as part of the middle strip in span ℓ_1.

The layout in Fig. 8-7(b) presents much less difficulty. The plan geometry guarantees one-way behavior in the short slabs, $S3$, and effectively precludes two-way behavior, except in the vicinity of the girders, $G2$.

SUMMARY EVALUATION—TWO-WAY CONSTRUCTION

From numerical evaluations of the four principal variations of the two-way construction prescribed in the 1995 Code (Chapter 13), the authors offer the following general observations as to the practicable range for economical applications:

1. Flat plate with no beams—thicknesses 5 in. to 10 in.—spans 15′-0″ to 30′-0″ maximum—factored live loads up to 300 psf maximum. All maximum limits subject to suitability of required minimum exterior column size.
2. Flat plate with shear heads to reduce column size. Same range as flat plate, except column size can be reduced. Higher load/span combinations within above maximum limits can be utilized with reasonable size columns.

3. Flat plate with edge beam or cantilever. Same range as flat plate with shear heads at exterior columns; limited by interior column size.
4. Flat slab with drop panels and/or capitals. Slab thicknesses, 4 in. minimum to 8 in. minimum, spans 20′-0″ to 36′-0″ maximum. Factored loads to 1000 psf.
5. Flat slab with drop panels and/or capitals and spandrel beams. Thicknesses 4 in. to 8 in. maximum.
6. Two-way slabs with beams on four sides on column centerlines. Thicknesses 3-1/2 in. to 12 in. maximum.

9 Beams and Girders

GENERAL

Although the terms "beam" and "girder" are frequently used in all codes, and, specifically, in the 1995 ACI Code, it was considered unnecessary to provide definitions for these terms in the Code. The common usage of these terms was evidently considered sufficiently definitive. For the purposes of this book, beams and girders are considered to consist of flexural members, generally horizontal, which act as a primary part of the structural framing system. A beam is herein considered to carry distributed loads such as parallel walls or tributary areas of the floor or roof systems; a girder, although it too may support these distributed loads, supports major concentrated loads such as columns or beam reactions. Most of the special provisions of the ACI Code will be applicable to both beams and girders under the authors' definitions of common usage, even where the Code refers merely to beams. Most of the requirements of the general codes in effect, such as fire protection and reduction of live loads, will also apply equally to both. For convenience in this chapter, all references to beams will apply equally to girders; provisions peculiar to girders will be identified.

CODE REQUIREMENTS

Code requirements applicable only to beams and girders may be grouped roughly into four categories, subdivided as follows:

1. *Analysis*—live load reductions (general code), Fig. 2-1; flange effects, "T-beams" (Section 8.10); approximate frame analysis (Section 8.3.3); elastic frame analysis (Section 8.3.1); limit analysis (Section 8.4); depth effect for depth/span ratios greater than 0.4 for continuous spans or 0.8 for simple spans (Section 10.7); lateral support effect (Section 10.4). Beams with two-way slabs, see Chapter 8 in this book.
2. *Design*—limiting maximum reinforcement ratio, ρ, (Section 10.3.3); limiting minimum ρ (Section 10.5); torsion reinforcement (Section 11.6); minimum stirrups (Section 11.5.5); crack control for "beam" cover (Section 10.6); deflection limits (Section 9.5); development of end anchorage (Sections 12.11.1, 12.11.2, and 12.12.1).

3. *Details*—ties for compression reinforcement (Section 7.11.); confinement within joints (Sections 7.9, 7.10.4.7, and 7.10.5.5); minimum cover (Section 7.7.1); bundled bars (Section 7.6.6); continuous bars and closed stirrups (or 135° hooks) for structural integrity (Section 7.13.2); fireproofing cover (general codes).
4. *Construction*—casting concrete (Section 5.10); joints (Section 6.4).

Some general requirements will be discussed first, but interpretation of most of these Code requirements where several apparent interpretations are open is best undertaken by examples.

ANALYSIS

Loads

Service loads are required to be increased by load factors (Section 9.2) for the following three combinations of load effects:

1. $U = 1.4D + 1.7L$ (Code Eq. 9-1)
2. $U = 0.75\,(1.4D + 1.7L + 1.7W)$ (Code Eq. 9-2)
3. $U = 0.9D + 1.3W$ (Code Eq. 9-3)

where U = required strength; D = dead load effect; L = specified live load effect; and W = specified wind load effect. Equation (9-1) must always be investigated (Section 9.2.1). Equations (9-2) and (9-3) must be investigated when the beam is part of a structural frame expected to resist lateral loads and not fully braced against movement due to these (Section 9.2.2). Equation (9-3) will control design only in unusual cases where the magnitude of the wind force is relatively very large. Its use can usually be avoided by inspection. It can be shown that Equation (9-3) will control design with live load present only if: $0.15D + 1.275L < 0.025W$ (unlikely); or when live load is entirely absent if: $0.15D < 0.025W$ (true when D, L, and W have additive effects and $W > 6.0D$, or when D and L have effects opposite those of W, and $-W < 0.692D$). Only the last possibility need be considered in practice, and it is not likely to be encountered often.

Note that the three basic load combinations are to be employed to compute the required strength for design to resist a variety of load effects with various special load factors for each (Sections 9.2.3, 9.2.4, 9.2.5, and 9.2.6). The Code also requires consideration of differential settlement, creep, shrinkage, expansion of shrinkage-compensating concrete, and temperature change effects where they are significant at an effective load factor of $0.75 \times 1.4 = 1.05$ (Section 9.2.7). These effects will be significant when 1.05 times the most adverse combination of these effects realistically expected to occur and realistically estimated exceeds $0.25\,(1.4D + 1.7L)$ or $1.275W$.

Frame Analysis Methods

Analysis for moments and shears in beams may utilize the assumptions and procedures for an (exact) elastic frame analysis (Section 8.3.1) or may consist of the approximate frame analysis coefficients (Section 8.3.3). The reader is referred to Chapter 2 for a discussion of the limitations on the approximate method and an explanation of the "exact" elastic method. General-purpose computer programs usually use the "exact" elastic method. The approximate method is useful for preliminary design, checking results of computer analyses, and for hand calculations of small projects. The provisions for redistribution of negative moments (Section 8.4) may be applied to the results of any elastic frame analysis. Redistribution of negative moment is necessary to obtain economy of finished designs by the exact elastic analysis method; it is the only concession to principles of limit analysis specifically permitted.

Unbraced Frames With Slender Columns

In addition to the elastic frame analysis moments, including effects of both vertical and lateral loading conditions, the Code requires that the magnified end moments required for slender columns be used in the design of the beams only in frames not otherwise braced against sidesway (Section 10.13.7).

Slender Beams

A special provision must be observed for slender beams (Section 10.4). Slender beams are defined as those in which the spacing of lateral supports to the compression flange or face is more than 50 times the least width of same. Since lateral support equal to a minimum of 2 percent of the total (internal) compression in the compression flange is usually considered adequate, and may often be satisfied by simple friction of separate floor materials transmitting the load to the beam, the beam slenderness problem will be encountered only rarely in routine design. It may arise in design of long span upturned roof beams or in the design of thin deep panels used as spandrel beams. Such spandrels are often precast and added to the frame as simply supported beams with a span equal to the column spacing and width of the compression flange as little as 4 in.

When a slender beam condition arises, the simple solution is to provide lateral bracing within the spacing limits, thus eliminating the condition. Where lateral bracing is impracticable, the load capacity must be reduced.*

* Hansell, William, and Winter, George, "Lateral Stability of Reinforced Concrete beams," *ACI Journal*, **56** (Sept. 1959), 193-214. Also, Sant, J., and Bletzacker, R. W. "Experimental Study of Lateral Stability of Reinforced Concrete Beams," *ACI Journal*, **58** (Dec. 1961), 713-736.

Deep Beams

The ordinary beam formulas which assume a linear variation of strain are inadequate for the design of deep beams. Nonlinear stress distribution at service loads and the effect of vertical stresses must be taken into consideration for the analysis of beams with depths greater than 0.4 of the span for continuous spans and 0.8 for simple spans (Section 10.7). Special requirements for shear reinforcement are prescribed for deep beams loaded only at the top. Loads applied at the sides or the bottom of a deep beam do not have the larger capacity attributed to deep beams and should be designed for shear as for ordinary beams (Commentary R11.8.1). Deep beams are required to have minimum reinforcement as prescribed for shear (Sections 11.8.9 and 11.8.10) or walls (Section 14.3) in both directions, whichever is larger (Section 10.7.4), in addition to the minimum flexural reinforcement (Section 10.5).

Deep beams present a complex problem in analysis. As the depth-to-span ratio increases, the effect of the load and stress in the vertical direction cannot be neglected as in shallow beams. The assumption of straight-line stress distribution at service loads does not apply to deep beams and the curve of stress distribution takes on radically different shapes.

For flexural design purposes, the Code defines deep beams as those with total depth greater than 0.4 clear span for continuous beams and 0.8 for simple beams (Section 10.7). No separate limit is provided for end spans. The Code wording is "depth-span" ratios; the authors' recommendation is that "depth" be taken as total depth and "span" be taken as clear span. It is difficult to determine the effective depth in deep girders and the width of the support can make the center-to-center span meaningless.

For shear design purposes, a deep beam is defined in the Code as a beam in which the ratio of effective depth to clear span is greater than 0.2 (Section 11.8.1). Effective depth, d, is defined as the distance from the extreme compression fiber to the centroid of the tension reinforcement (Section 11.0). Due to the nonlinear distribution of stress under elastic analysis, there is often no extreme compression fiber; the maximum compression stress may lie inside the beam depth with tension areas above and below. For practical purposes, the authors recommend application of the special provisions for shear in deep beams (Section 11.8) to beams in which the overall depth is greater than 0.2 the clear span for simple or continuous beams (Section 11.8.1). The special provisions for shear apply only to beams loaded at the top. Beams loaded otherwise are designed for shear as ordinary beams (Commentary R11.8.1).

Where very important deep beams are encountered or where the design involves sufficient repetition to become important, the use of models to determine the elastic stress distribution or strength will obviously be helpful. Such use of models is specifically recognized for use as a supplement to the calculations which may be required by a building official (Section 1.2.2). Such calculations must include consideration of the end support width, position, and extent—top, bottom, or full depth; support continuity (if any), position, and extent; and type of loads—uniform, concentrated, applied at the top, bottom, or between; and hole and opening locations, for practical applications. Finite element solutions are often required.

The design of deep beams is a complex procedure involving use of basic theory for analysis of the particular conditions encountered in each practical application. The Code itself contains only a few general provisions (Sections 10.7 and 11.8). The subject is generally outside the scope of this Guide, except for the single application of two-pile cap footings (Chapter 12), but the user will find the references following Chapter 11, Code Commentary, useful.*

Structural Integrity

Detailing for structural integrity requires that bottom bars of joists terminate with standard end hooks, or at least one bottom bar must be lap-spliced (Section 7.13.2.1). Structural integrity for perimeter beams require continuous bars for one-sixth of the top reinforcement and one-quarter of the bottom reinforcement, and that closed stirrups or stirrups with 135° hooks be used (Section 7.13.2.2). Structural integrity for interior beams require one-quarter of the bottom reinforcement to be continuous (Section 7.13.2.3).

EXAMPLE 1: INTERIOR BEAM B1. Select beams suitable for the support of a joist-slab floor system of alternate spans 20′–0″ and 24′–0″ center-to-center of the supporting beams. Use normal weight concrete. Factored loads for the joist system are w_d = 100 psf and w_ℓ = 250 psf. The beams span 21′–4″ between 16 in. square columns. Assume the lateral load resistance is provided by walls. The beams are to be 15 in. deep to maintain a level ceiling with the 12 + 3 joist-slab system (see Fig. 9-1).

Beam Loads

Since the joist-slab spans alternate 20 ft and 24 ft, the beams on interior column lines (*B*-1) receive the following loads: Factored live load $w_\ell = \frac{1}{2}(20 + 24)(0.250) = 5.5$ klf. Factored dead load, for a beam 24 in. wide as a first trial,

$$w_d = [\tfrac{1}{2}(20 + 24) - 2](0.100) + (2)\frac{(15)}{12}(0.150)(1.4) = 2.5 \text{ klf}$$

* Additional references are: Design of Deep Girders, No. ST-66, Portland Cement Association. Shear Behavior of Deep Reinforced Concrete Beams, Vols. I-V (A continuing series), Dec. 1970, by Robert A. Crist, Eric C. Wang Civil Engineering Research Facility, University of New Mexico, Albuquerque, New Mexico. "Suggested Revisions to ACI Building Code Clauses Dealing with Shear Friction and Shear in Deep Beams and Corbels," J. G. McGregor and N. M. Hawkins, *ACI Journal*, Nov. 1977, p. 537.

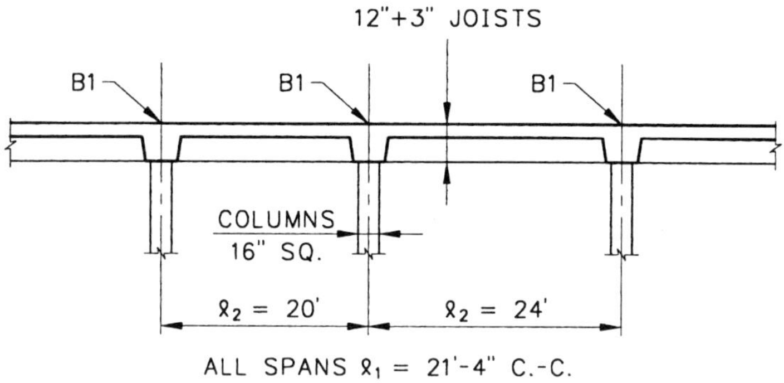

Figure 9-1 Interior span beam *B*-1.

The foregoing factors satisfy the load factor combination prescribed as: $U = 1.4D + 1.7L$ (Section 9.2.1, Eq. 9-1). $w_u = 2.5 + 5.5 = 8.0$ klf. Live load reductions are usually followed in the general codes only for live loads less than 100 psf, and so such reductions will not be considered in this example (see Fig. 2-1). The approximate analysis method will be used, since this example is within the limitations thereon (Section 8.3.3).

Moments

Negative factored moment at faces of interior supports other than the first line, $-M_u = 1/11\ w_u(\ell_n)^2$; positive factored moment in interior spans, $+M_u = 1/16\ w_u(\ell_n)^2$ (Section 8.3.3).

$$-M_u = \frac{(8.0)(20)^2}{11} = -291 \text{ ft-kips}$$

$$+M_u = \frac{(8.0)(20)^2}{16} = +200 \text{ ft-kips}$$

Shear

Maximum factored shear at critical section, $d = 12$ in., from face of support,

$$V_u = \frac{(8.0)(18)}{2} = 72 \text{ kips at 12 in. from support}$$

Minimum factored shear at midspan for live load on half the span

$$V_u = \frac{(5.5)(10)}{4} = 13.75 = \text{kips at midspan}$$

Required cover $= 1\frac{1}{2}$ in. min (Section 7.7.1); minimum stirrups are required since $h > \frac{1}{2}b_w$ (Section 11.5.5.1-c); effective depth to the top bars, $d = 12.5$ in.; to the bottom bars, $d = 12.6$ in. (see Fig. 9-2).

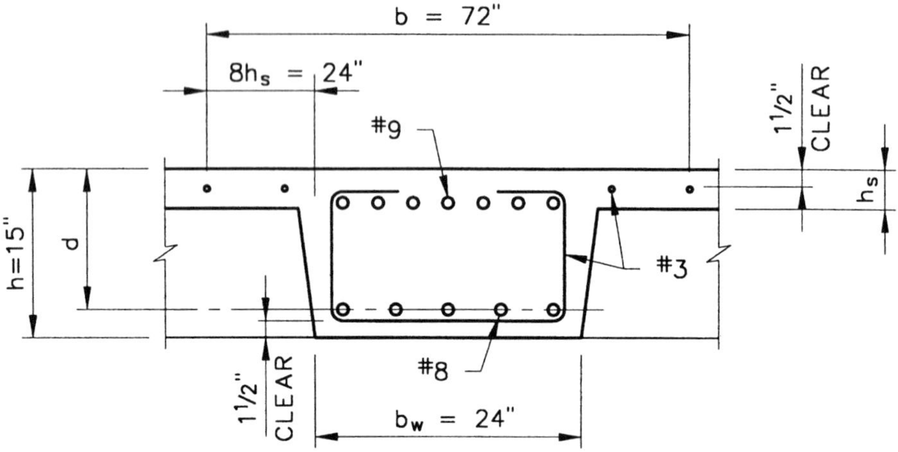

Figure 9-2 Section dimensions—1st trial—*B*-1.

STRENGTH DESIGN. Use $f_c' = 4{,}000$ psi; $f_y = 60{,}000$ psi. Moment reinforcement

$$A_s = \frac{M_u}{\phi f_y(d - a/2)}; \quad a = \frac{A_s f_y}{0.85 f_c' b_w}$$

See Fig. 9-28 (b).

$$0.75\rho_b = 0.0214$$

For top bars,

$$R = \frac{M_u}{b_w d^2} = \frac{(291)(12{,}000)}{(24)(12.5)^2} = 931 > 930 \text{ for } 0.75\rho_b$$

Required $\rho = 0.0215$; top bars, $A_s = (0.0215)(24)(12.5) = 6.4$ in.^{2} Try 7-#9 bars. Unless the web width, $b_w = 24$ in., is increased about 2 percent, the bottom bars must be extended for compression so that the net tensile reinforcement ratio, $\rho - \rho' \leqslant 0.75\rho_b$ (Section 10.3.3) (see Fig. 2-8).

For bottom bars, the width of slab taken as a T-beam flange is 24 in., as determined by Code Section 8.10.2-a, and

$$R = \frac{(200)(12{,}000)}{(72)(12.6)^2} = 210 < 930; \quad \rho = 0.0040$$

Bottom bars,

$$A_s = (0.0040)(72)(12.6) = 3.65 \text{ in.}^2$$

Try 5-#8 bars.

Shear

$$\phi V_n \geqslant V_u \qquad \text{(Section 11.1.1, Eq. 11-1)}$$

$$V_n = V_c + V_s \qquad \text{(Section 11.1.1, Eq. 11-2)}$$

$$V_c = 2\sqrt{f'_c}\, b_w d \qquad \text{(Section 11.3.1.1, Eq. 11-3)}$$

$$V_s = A_v f_y d/s \qquad \text{(Section 11.5.6.2, Eq. 11-15)}$$

These are the four basic equations for shear investigations in ACI 318-95 for shear in routine applications for reinforced concrete using vertical stirrups for A_v. The Code provisions for shear are all written in terms of total forces: shear from applied factored loads, V_u; "nominal" shear strength V_n; and "nominal" shear strengths of the concrete, V_c, and the stirrups V_s.

Experienced practical designers will immediately note that actually employing these equations in daily use involves extra steps in calculation and masks the level of stress implied. It also complicates the use of the design aids to save calculations, since such aids are, almost necessarily, expressed in terms of unit stresses for reinforcement and concrete. The design aids in this chapter are naturally based on stresses, but their application within the framework of the formulas for force, as given in the Code, requires only the following common definitions for stresses:

$$\frac{V_u}{\phi b_w d} = \text{total shear stress due to factored loads}$$

$$\frac{V_c}{b_w d} = \text{nominal shear stress allowed on concrete}$$

$$\frac{V_s}{b_w d} = \text{nominal shear stress carried by reinforcement}$$

The design equation becomes

$$\frac{V_c}{b_w d} + \frac{V_s}{b_w d} \geqslant \frac{V_u}{\phi b_w d}$$

Hereafter in this chapter, the stress notation will be employed with appropriate references to the Code sections from which allowable "nominal" stresses are derived.

For Example 1: $V_u = 71.7\text{k}$ at 12.5 in. from the face of the column (Section 11.1.3.1), and $\dfrac{V_c}{b_w d} = 2\sqrt{f'_c}$ (Section 11.3.1.1).

$$\frac{V_u}{\phi b_w d} = \frac{(71.7)(1{,}000)}{(0.85)(24)(12.5)} = 281 \text{ psi}$$

Allowable shear on the concrete, $\dfrac{V_c}{b_w d} = 2\sqrt{f'_c} = 126$ psi. The required shear stress to be carried by the reinforcement is

$$\frac{V_s}{b_w d} = \left(\frac{V_u}{\phi b_w d} - \frac{V_c}{b_w d}\right) = 281 - 126 = 155 \text{ psi}$$

With stirrups for shear reinforcement at a maximum spacing 0.50d for vertical legs (Section 11.5.4.1), the maximum allowable shear stress in the reinforcement is

$$\frac{V_s}{b_w d} \leqslant 4\sqrt{f'_c} = 253 \text{ psi} \quad \text{(Section 11.5.4.3)}$$

for the spacing of vertical legs at 0.25d maximum (Section 11.5.4.3) and the maximum allowable shear stress in the reinforcement is

$$\frac{V_s}{b_w d} \leqslant 8\sqrt{f'_c} = 506 \text{ psi} \quad \text{(Section 11.5.6.8)}$$

Spacing of stirrups at the face of the support would be calculated as

$$s = \frac{A_v f_y}{b_w}\left(\frac{b_w d}{V_s}\right) = 2 \times 0.11 \frac{60{,}000}{24 \times 155} = 3.5 \text{ in.}$$

At locations of the beam closer to midspan, the magnitude of the nominal shear stress carried by the reinforcement is less than the magnitude at the support. Thus, the spacing of the stirrups may be increased.

At midspan with live load on half the span, the total shear stress

$$\frac{V_u}{\phi b_w d} = \frac{13.75 \times 1000}{24 \times 12.6 \times 0.85} = 54 \text{ psi}$$

Minimum stirrups must be provided at all locations of the beam where total shear stress, $V_u/(\phi b_w d)$, exceeds half the nominal shear strength provided by the concrete, that is, where $V_u/(\phi b_w d) \geqslant \frac{1}{2}(126) = 63$ psi. Although the total shear stress at midspan, 54 psi is less than 63 psi, the shear stress within 1 ft of the midspan exceeds 63 psi. Thus, this beam would require minimum reinforcement over the full length with a spacing calculated as (Section 11.5.5.3)

$$s = (2 \times 0.11)\frac{60{,}000}{50 \times 24} = 11 \text{ in.}$$

A maximum spacing of $d/2 = 6$ in. must be used (Section 11.5.4.1).

Several design options are available to reduce the number of stirrups:

1. Increase b_w (Section 11.5.5.1-c) so that $b_w \geqslant 2h = 30$ in. This design change will eliminate the minimum stirrups where

$$\frac{1}{2}\frac{V_c}{b_w d} \leqslant \frac{V_u}{\phi b_w d} \leqslant \frac{V_c}{b_w d}$$

Stirrups will still be required where $V_u/\phi \geqslant V_c$ near the ends of the span.

2. Increase b_w so that at the end of the span,

$$\frac{V_u}{\phi b_w d} < \frac{1}{2}\frac{V_c}{b_w d}$$

for the entire length of the beam. The required width is:

$$b_w = 24\left(\frac{281}{126}\right) = 54 \text{ in.}$$

3. Increase the depth. In order to totally eliminate stirrups, the required depth would be

$$d = 12.5\left(\frac{281}{63}\right) = 56 \text{ in.}$$

This depth is obviously too large to be a reasonable design. A moderate depth increase from 24 in. to 36 in., however, would greatly reduce the amount of stirrups required.

4. Recalculate width for the first two options based upon allowable shear,

$$\frac{V_c}{b_w d} = 1.9\sqrt{f_c'} + (2500)(\rho_w)\frac{V_u d}{M_u}$$

(This last option involves tedius calculations for little gain, except at ends of simple spans where full bottom reinforcement is extended into the support. WARNING: in cases such as tapered end joists, the calculation will require use of a computer as it is difficult to predict critical sections.)

Open Stirrups Required for Vertical Shear, $V_u/\phi b_w d$

In this example, the width, b_w, will be increased from 24 in. to 28 in., in order to eliminate the need for compression reinforcement indicated in the analysis for flexure. This increase in web width reduces $V_u/\phi b_w d$ to 241 psi $< 4\sqrt{f_c'}$. The vertical stirrups may be spaced at 0.50d maximum. At the critical section, a distance d from the support:

$$A_v = \frac{V_u/\phi b_w d - V_c/b_w d}{f_y} b_w s = \text{area of stirrups required} \quad \text{(Section 11.5.6.1)},$$

where $V_c/b_w d$ = 126 psi; b_w = 28 in.; s = stirrup spacing; and f_y = 60,000 psi. For #3 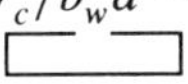, $A_v = 2 \times 0.11 = 0.22$ in. Solving the equation for spacing, s:

$$s = \frac{(0.22)(60{,}000)}{(241 - 126)(28)} = 4 \text{ in.}$$

At the maximum spacing, $s = 6$ in., the minimum area of stirrups required (Section 11.5.5.3, Eq. 11-14), A_v is

$$A_v = \frac{(50)(b_w)(s)}{f_y} = \frac{(50)(28)(6)}{60{,}000} = 0.14 \text{ in.}^2$$

Minimum factored shear, at midspan for half-panel loads, as simple span:

$$V_u = \frac{(5.5)(10)}{4} = 13.75 \text{ kips}$$

$$\frac{V_u}{\phi b_w d} = \frac{(13.75)(1{,}000)}{(28)(12.6)(0.85)} = 46 \text{ psi}$$

(See Fig. 9-3.) If this value is taken as midspan shear, and a shear envelope is made, $(V_u/\phi b_w d)$ decreases at 21 psi/ft. Space #3 ⊔ @4 until

$$\left(\frac{V_u}{\phi b_w d} - \frac{V_c}{b_w d}\right) = \frac{(2)(0.11)(60{,}000)}{(28)(6)} = 79 \text{ psi}$$

to permit use of the maximum spacing, 6 in.

$$\frac{V_u}{\phi b_w d} = 79 + 126 = 205 \text{ psi} \qquad \frac{(241 - 205)}{21} = 1.71 \text{ ft} = 21 \text{ in.}$$

Distance from face of support = 12.5 in. (d) + 21 = 33.5 in. Use 10 − #3 @4, 13 − #3 @6 ⊔. These stirrups would suffice if torsion is negligible.

Torsion Requirements for Design

The design for shear must include an investigation of torsion (Section 11.6.1) due to unbalanced loading on any beam. A factored torsional moment T_u produces a shear stress due to torsion of $T_u/(A_{cp}^2/p_{cp})$. The relative magnitudes of stresses due to shear forces and torsional moments are most easily understood

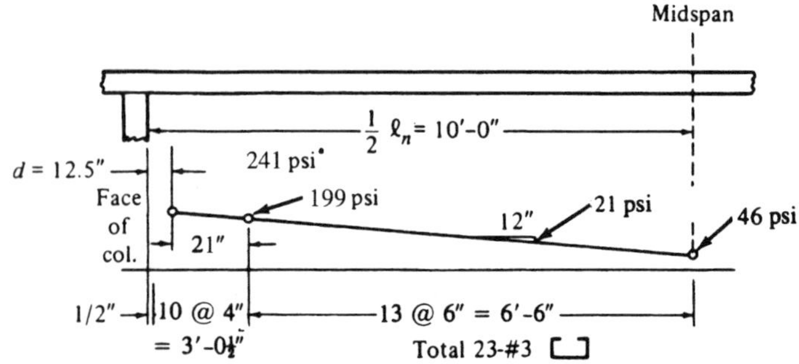

Figure 9-3 Shear reinforcement—beam *B*-1.

if the calculations are performed in terms of stresses rather than forces. Thus, the stress concept will be used for torsion as it was for shear. The effect of torsion may be neglected in design if the torsional stress $T_u/(\phi A_{cp}^2/p_{cp})$ does not exceed 1.0 $\sqrt{f_c'}$ (Section 11.6.1), and so this calculation will seldom be required for interior beams supporting slabs with approximately equal spans and loads on either side. If torsion reinforcement is required

$$\frac{T_u}{(\phi A_{cp}^2/p_{cp})} > 1.0\sqrt{f_c'},$$

it must be provided in addition to that required for shear and flexure (Section 11.6.3.8), but may be combined with other reinforcement. In this example, if torsion reinforcement is required, closed stirrups only will be provided at a closer spacing or in larger sizes than required for shear alone.

ANALYSIS. Only the unbalanced joist-slab moment which will be transferred to the columns by the beam will create torsion in the beam. The total unbalanced moment at the centerline of the beam can be obtained from an elastic analysis or from the difference in the fixed-end moments from the short span with factored dead load only and from the long span with factored dead plus factored live load (see Fig. 9-4).

Assume relative flexural stiffnesses as follows: $K_{ec} = 61$; $K_s = 100$ for the short span and $K_s' = 83$ for the long span. The ratio of equivalent column stiffnesses to total stiffnesses at the joint

$$\frac{K_{ec}}{K_{ec} + \Sigma K_s} = \frac{61}{61 + 100 + 83} = 0.25$$

The total unbalanced fixed end moment, FEM_{un}, is:

$$\text{FEM}_{\text{un}} = (21.33)\left[\frac{(0.100 + 0.250)(24)^2}{12} - \frac{(0.100)(20)^2}{12}\right]$$

$$\text{FEM}_{\text{un}} = 287 \text{ ft-kips}$$

From Fig. 9-4, read total unbalanced moment to be transferred to an interior column equals 0.4 $\text{FEM}_{\text{un}} = (0.4)(287) = 115$ ft-kips. Note that the equivalent column stiffness, K_{ec} was used to estimate the total unbalanced moment at the joint to avoid overestimating it by neglecting the torsional flexibility of the beam.

The next design task is to divide the unbalanced moment into portions which must be transferred to the column by direct flexure through the joists and by torsion through the beam. In this example, the direct flexural connection consists of one 10 in. wide joist on the column centerline since the span $\ell_1 = 21'$-$4''$ is 4 in. more than an odd multiple of the joist module.

For two-way slab connections, the Code specifically prescribes a formula for separating the flexure and torsion (Section 13.5.3.2). Application of this formula using $b_1 = 10$ in. for the flexural connection and $b_2 = 16$ in. for the torsional

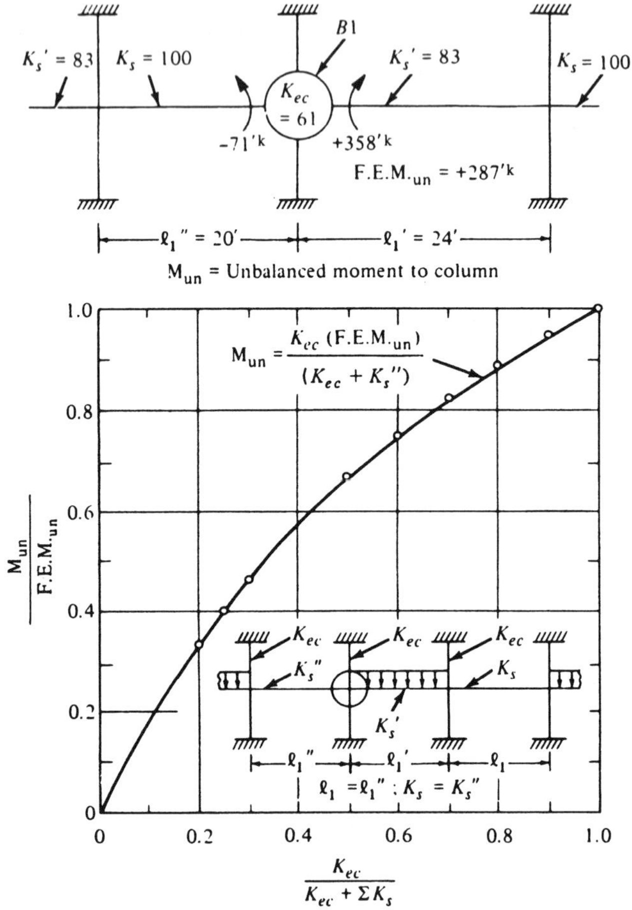

Figure 9-4 Relative joist stiffnesses and unbalanced moment.

connection results in assigning 65 percent of the unbalanced moment to flexure. The remainder of the unbalanced moment, 35 percent, is assigned by Code Equation 11-41, to be transferred by torsional shear stresses. The authors believe this Code provision is applicable only for two-way slabs or beams that are wide enough in both directions to include a critical perimeter section for two-way shear, $(c_1 + d)$ by $(c_2 + d)$. Since there is no specific Code requirement for the condition in this example, the authors believe that a conservative estimate of the unbalanced moment transferred by torsion must be made.

The critical sections for torsion are located in the beams on each side of the column at a distance $(c_2/2 + d) = 1.71$ ft from the center of the column (Section 11.6.2.4). Assume all unbalanced moment from the joist-slab system outside the length between critical sections is transferred by the beams in torsion, half on each side.

$$T_u = \tfrac{1}{2}(\mathrm{M}_{un})\left[\frac{(\tfrac{1}{2}\ell_2 - 1.71)}{\tfrac{1}{2}\ell_2}\right] = (\tfrac{1}{2})(115)\left[\frac{(10.67 - 1.71)}{10.67}\right] = 48.3 \text{ ft-kips}$$

Torsional Stress

Torsion (Section 11.6)

$$\phi T_n \geqslant T_u \qquad \text{(Eq. 11-20)}$$

A new provision of the 1995 Code is that no torsion is resisted by concrete, $T_c = 0$ (Commentary Section R11.6.3.5). As a result, the entire nominal torsion strength is provided by the reinforcement.

$$T_n = \frac{2A_o A_t f_{yv}}{s} \cot\theta \qquad \text{(Eq. 11-21)}$$

Torsion is negligible when the total torsional stress

$$\frac{T_u}{\phi A_{cp}^2/p_{cp}} \leqslant 1.0\sqrt{f_c'} \qquad \text{(Section 11.6.1)}$$

Note that the actual expression in Section 11.6.1 obscures the negligible level of torsion stress. For a T-beam, A_{cp} and p_{cp} can include portions of the adjacent slabs conforming to Section 13.2.4.
(Neglect the overhanging flanges when stirrup reinforcement may be required as it is impracticable to reinforce the 3 in. flanges.)

$$A_{cp} = 15 \times 28 = 420 \text{ in.}^2$$

$$p_{cp} = 2(15 + 28) = 86 \text{ in.}$$

$$\frac{T_u}{(\phi A_{cp}^2/p_{cp})} = \frac{(48.3)(12{,}000)(86)}{(0.85)(420^2)} = 332 \text{ psi} = 5.26\sqrt{f_c'}$$

At this point the designer must consider several Code requirements:

1. If

$$\frac{T_u}{(\phi A_{cp}^2/p_{cp})} \leqslant 1.0\sqrt{f_c'}\,,$$

torsion may be neglected (Section 11.6.1)

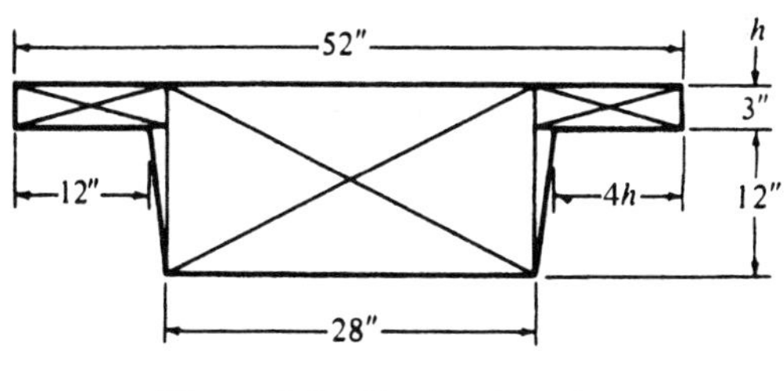

Figure 9-5 Torsional section.

2. If torsional moment, T_u, is not required to maintain equilibrium, design may be based upon a reduced torsional moment equal to the cracking moment at which

$$\frac{T_u}{(\phi A_{cp}^2/p_{cp})} = 4\sqrt{f_c'} \quad \text{(Section 11.6.2.2)}$$

3. If torsion stress can not be reduced for design, the upper limit for design is reached when

$$\sqrt{\left(\frac{V_u}{b_w d}\right)^2 + \left(\frac{T_u p_h}{1.7A_{oh}^2}\right)^2} \leqslant \phi\left(\frac{V_c}{b_w d} + 8\sqrt{f_c'}\right) \quad \text{(Section 11.6.3)}$$

where $V_c = 2\sqrt{f_c'}$.

Taking these points in order:

1.
$$\frac{T_u}{\phi A_{cp}^2/p_{cp}} - 5.26\sqrt{f_c'} > 1.0\sqrt{f_c'}$$

Shear reinforcement consisting of closed stirrups rather than open stirrups is required.

2. The example is an indeterminate structure where reduction of the torsional moment can occur due to redistribution of internal forces upon cracking. Reduce T_u for

$$\frac{T_u}{\phi A_{cp}^2/p_{cp}} = 4\sqrt{f_c'} \qquad \text{(Section 11.6.2.2)}$$

$$T_u = \phi(4\sqrt{f_c'})\frac{A_{cp}^2}{p_{cp}} = \frac{0.85(253)(420^2)}{86 \times 12{,}000} = 36.8 \text{ ft-kips}$$

NOTE: When taking advantage of redistribution of torsional moments, the end moments of the joist system in this example will be reduced likewise, and the positive moments in the joist system will be increased (Section 11.6.2.2.).

$$\frac{V_u}{(\phi b_w d)} = 241 \text{ psi}$$

$$A_{oh} = (28 - 3.375)(15 - 3.375) = 286 \text{ in.}^2$$

$$p_h = 2(24.625 + 11.625) = 72.5 \text{ in.}$$

$$\frac{T_u p_h}{\phi 1.7A_{oh}^2} = \frac{(36.8)(72.5)(12{,}000)}{(0.85)(1.7)(286^2)} = 271 \text{ psi}$$

Maximum allowable stress by Code Equation 11-18 is

$$\phi\left(\frac{V_c}{b_w d} + 8\sqrt{f_c'}\right) = 0.85(2\sqrt{f_c'} + 8\sqrt{f_c'}) = 8.5\sqrt{f_c'}$$

$$\sqrt{(241 \times 0.85)^2 + (271 \times 0.85)^2} = 308 \text{ psi} = 4.87\sqrt{f_c'} < 8.5\sqrt{f_c'} \text{ O.K.}$$

Selection of Stirrup Reinforcement

The next design decision is reduced to option (1) increasing width of beam, or option (2) providing shear reinforcement, using b_w = 28 in., for the full torsional shear

$$\frac{T_u}{\phi A_{cp}^2/p_{cp}} = 5.26\sqrt{f_c'}$$

or to take advantage of cracking and redistribution of internal forces so that the maximum torsional shear is 4 $\sqrt{f_c'}$. Option (1) would require an excessive width of 28 × 5.26/1.0 = 147 in. to reduce torsional shear stresses to negligible levels. Similarly, the torsional reinforcement required to maintain torsional stiffness assumed for end moment of the joist system

$$\frac{T_u}{(\phi A_{cp}^2/p_{cp})} = 5.26\sqrt{f_c'}$$

will usually be less cost-effective than adjusting the flexural reinforcement in the joist for the end moment of 36.8 ft-kips.

In most practical designs, it will be unnecessary to go through all of these calculations. Elastic frame analysis may indicate that beams in indeterminate structures with dimensions selected within architectural limits to provide adequate stiffness to control servicability deflections may be subjected to torsion stresses larger than $4\sqrt{f_c'}$. The designer may, however, take advantage of a reduction in torsional moment due to redistribution of internal forces (Section 11.6.3), and simply design for

$$\frac{T_u}{(\phi A_{cp}^2/p_{cp})} = 4\sqrt{f_c'}$$

Design aids are already available for usual beam sizes to give quick solutions for size and spacing of closed stirrups required for combined shear and torsion.* With such aids, for

$$1.0\sqrt{f_c'} < T_u/(\phi A_{cp}^2/p_{cp}) < 4.0\sqrt{f_c'}$$

the stirrup patterns for

$$\frac{T_u}{\phi A_{cp}^2/p_{cp}} = 4\sqrt{f_c'}$$

can be scaled down with satisfactory accuracy for practical use.

The calculations for this example will be completed merely to illustrate the technique of design for combined shear-torsion reinforcement conforming to the 1995 Code.

* CRSI Handbook, 1996.

Data required are $V_c/(b_w d)$, and $V_u/(\phi b_w d)$, and $T_u/(\phi A_{oh}^2/p_h)$. From previous calculations:

$$\frac{V_u}{\phi b_w d} = 241 \text{ psi} \quad \text{and} \quad \frac{T_u}{\phi 1.7A_{oh}^2/p_h} = 271 \text{ psi}$$

To enter Fig. 9-25, compute k_1, and k_2

$$k_1 = \frac{V_u/(b_w d)}{\phi\sqrt{f_c'}} = \frac{241}{63.25} = 3.81 \qquad k_2 = \frac{T_u p_h/1.7A_{oh}^2}{\phi\sqrt{f_c'}} = \frac{271}{63.25} = 4.28$$

Read in Fig. 9-25: (1) that combined shear and torsion does not exceed the maximum allowed, (2) that closed stirrups are required, and (3) that the maximum spacing of closed stirrups is 0.50 d = 6.25 in.

See Fig. 9-23 for $d = 2b_w$.* For our example, d = 15 in. and b_w = 28 in., y_1 = 24.5 in., x_1 = 11.5 in., and $y_1 \approx 2x_1$.

Read for $T_u/(\phi 1.7A_{oh}^2/p_h)$ = 271 psi, #4 stirrups @11.1 in. spacing for torsion. Maximum spacing of stirrups for torsion = 8.5 in. Note that for this example, the abscissa in Fig. 9-23 is x = 15 in. The required shear stress capacity of the stirrups is

$$\frac{V_s}{b_w d} = \frac{V_u}{\phi b_w d} - \frac{V_c}{b_w d} = 241 - 126 = 115 \text{ psi}$$

Fig. 9-19 shows that #4 @6.25 in., (0.5d), b_w = 28 in., and d = 12.5 in. provide capacity

$$\frac{V_s}{(b_w d)} = 145 \text{ psi}$$

Maximum spacing for shear, s = 6.25 in.

Combining, required spacing for strength for both shear and torsion.

$$s = \frac{1}{\dfrac{11.5}{6.25(145)} + \dfrac{1}{11.1}} = 4.6 \text{ in.}$$

Stirrups spaced at 4 in. satisfy the maximum spacings for both shear and torsion. Since maximum shear stirrup spacing is 0.5d = 6.25 in. (Fig. 9-25), use of multiple (4 leg) stirrups or #5 instead of #4 could save little and would merely complicate design.

Check the required spacing of stirrups for strength at a distance of 32 in. from the face of the column, $V_u/(\phi b_w d) - V_c/(b_w d) = 206 - 126 = 80$ psi, and

* For manual calculations to select stirrup size-spacing combinations, it is convenient to use the dimensions b_w and d as constants so that stirrup capacity is directly proportional to the size of the stirrup bars. The stirrup spacing design charts herein have been prepared upon the basis of the centerline stirrup dimensions using #4 stirrups (Section 11.0). These dimensions are conservative when used for #3 stirrups.

$T_u/(\phi 1.7A_{oh}^2/p_h)$ = 222 psi. Spacing of shear stirrups = 6.25(145)/80 = 11.3 in. and the spacing of torsion stirrups from Fig. 9-23 is 13.5 in. Combining, required spacing for strength for both shear and torsion,

$$s = \frac{1}{\dfrac{1}{11.3} + \dfrac{1}{13.5}} = 6.2 \text{ in.}$$

For s = 6 in. the minimum area of closed stirrups is calculated by Code Eq. 11-23 (Section 11.6.5.2) as

$$A_v + 2A_t = \frac{50b_w s}{f_{yv}} = \frac{50(28)6}{60{,}000}$$
$$= 0.14 \text{ in.}^2 < 0.20 \text{ in.}^2, \text{ OK}$$

Complete Fig. 9-3 by adding torsion stresses and their distribution. See Fig. 9-6. The controlling points are 75 in. from the column face for shear and 92 in. for torsion. At each end of the beam use closed stirrups 9 − #4 @4″; 10 − #4@6″.

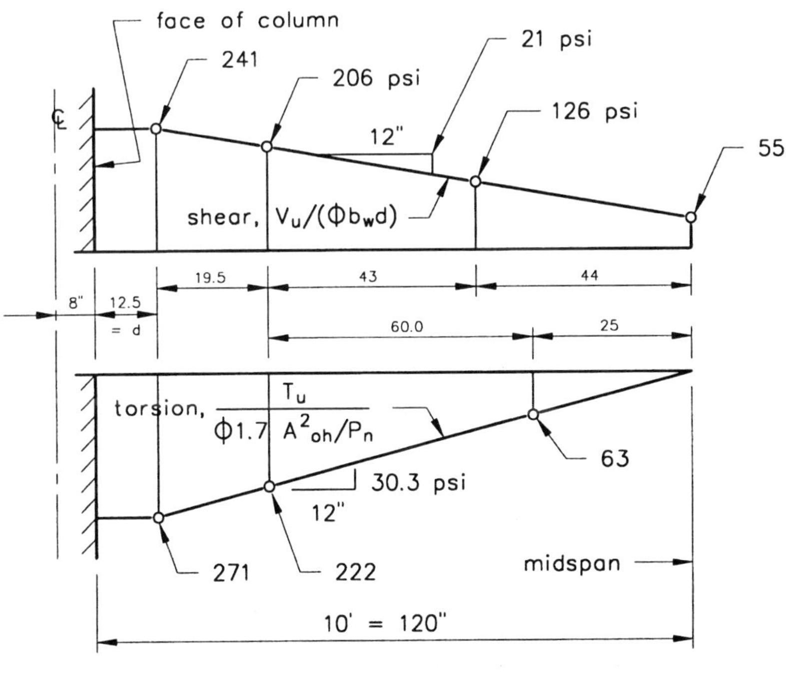

Figure 9-6 Combined shear and torsion.

Longitudinal Bars for Torsion

The Code requires longitudinal bars distributed around the inside perimeter of closed stirrups at a spacing not to exceed 12 in. with a size not less than #3 and a diameter at least 1/24 of the stirrup spacing, in each corner of the stirrups (Section 11.6.6). In this example, the perimeter is $2(15 - 3.5 + 28 - 3.5) = 72$ in., and so six bars will be required, three in the top and three in the bottom.

The minimum total area of the six bars, A_ℓ, is the larger of

$$A_\ell = \frac{A_t}{s} p_h \left(\frac{f_{yv}}{f_{y\ell}}\right) \cot^2 \theta \qquad \text{(Section 11.6.3.7, Eq. 11-22)}$$

$$A_\ell = \frac{5\sqrt{f'_c}\, A_{cp}}{f_{y\ell}} - \left(\frac{A_t}{s}\right) p_h \frac{f_{yt}}{f_{y\ell}} \qquad \text{(Eq. 11-24)}$$

where A_t/s shall not be taken less than $25\, b_w/f_{yv}$.

At the critical section for torsion, d from the face of the support, the top and bottom bars required for flexure will suffice. In this example, typically for most interior beams, loading for maximum flexure causes little torsion and loading for maximum torsion reduces both shear and flexure, and so the same bars will suffice for both loadings. This condition does *not* exist for edge or spandrel beams in torsion, and in such cases the torsional requirements for longitudinal bar areas (computed) are additive to flexural requirements (Section 11.6.3.8). In this example, the top bars are required for torsion beyond the cutoff point permitted by flexural requirements. The area of the longitudinal bars needs to be computed.

Note that A_ℓ is the value for six (equal) bars and that the spacing, s, at cutoff point is at the arbitrary maximum. In Eq. 11-22 for A_ℓ, use the *computed* spacing, s, at which the actual area of stirrups, A_t, would satisfy torsional stresses. (Otherwise, the designer providing stirrups larger than required, for simplicity of placing and detailing, may be computing heavier top than bottom bars at midspan where actual shear and torsion approach zero.) In Eq. 11-23, use the largest spacing actually provided to calculate $A_t/s = 0.2/6 = 0.033$. In this case, $25b_w/f_{yv} = 25(28)/60{,}000 = 0.012 < 0.033$. Use 0.033 for A_t/s in Eq. 11-24.

At the cutoff point of the top bars for flexure, usually $0.30\, \ell_n, v_u/\phi = 198$ psi and $\dfrac{T_u}{\phi 1.7 A_{oh}^2/p_h} = 212$ psi.

The *computed* spacing, s, required for torsion at this point

$$s = \frac{2A_o A_t f_{yv} p_{cp}}{212 A_{oh}^2} = \frac{2(0.85)(24.5 \times 11.5)(0.2)(60{,}000)2(24.5 + 11.5)}{212(24.5^2)(11.5^2)}$$

$$= 31.2 \text{ in.}$$

$$A_\ell = \frac{0.2}{31.2} 2(24.5 + 11.5) = 0.46 \text{ in.}^2$$

$$A_\ell = \frac{5(63.2)(28)(15)}{60{,}000} - (0.033)2(24.5 + 11.5) = 2.21 - 2.37 < 0 \text{ in.}^2$$

The required area for each of the six bars is 0.46/6 = 0.08 in.^{2} Use three #4 top bars. Extend top bars past the cutoff point of the top flexural bars 12 in. minimum for a Class B tension lap splice of top bars. (3-#4 × 10-0).

NOTE: The preceding example has been deliberately constructed to illustrate Code provisions regarding shear and torsion. The maximum ratio of adjoining spans for use of approximate frame analysis and a relatively heavy live load to create torsion are used. Torsion effects must be included with shear and bending where $T_u/(\phi A_{cp}^2/p_{cp})$ exceeds 1.0 $\sqrt{f_c'}$ (Section 11.6.1) and torsional stiffness, K_t, must be included in the equivalent column stiffness, K_{ec}, to determine K_{ec} for calculation of unbalanced moments to interior columns with two-way systems (Section 13.6.9.2).

In the preceding example, the approximate frame analysis method was applied (Section 8.3.3). It will be noted that torsional stiffness per se is not involved in the analysis for moment or shear, but that unbalanced moment to be transferred to the column will be excessive unless the equivalent column stiffness, K_{ec}, is considered. Part of this unbalanced moment is transferred to the column by flexure and part by torsion. For two-way slabs only, the Code prescribes the fraction transferred by torsion (Section 13.5.3.2). In this example, a very conservative interpretation has been employed to estimate the fraction transferred by flexure, leaving the remainder for transfer by shear.* It is further assumed that the entire live load is movable so that full live load and no live load can be present on adjoining long and short spans, respectively.

Under these conservative assumptions, it will be observed that torsion becomes very significant. With the beam dimensions which would usually be chosen for shear and moment only, the torsion stress developed requires closed stirrups at close spacings. To avoid cost of closed stirrups and labor to place them, a considerably wider beam would be required. The authors recommend that users of the Code assess the probability of pattern loading closely to avoid overestimating unbalanced moments transferred through torsion to columns; and, where full pattern loading is realistically expected on unequal adjoining spans, that use of wider interior beams be considered to avoid the need of closed stirrups. Where closed stirrups must be used, provision for use of two-piece closed stirrups (Section 7.11.3) will facilitate bar placing.

SELECTION AND DETAILS OF FLEXURAL BARS

Development (Bond)

No design is complete until the proper selection of bar sizes and lengths is made. Determination of the division of total factored moment into negative and positive bending moments and the areas of reinforcement for each is unique for each

* K. Saether and N. M. Prachand, "Torsion in Spandrel Beams," *ACI Journal*, **66** (Jan. 1969), 24-30.

assumed pattern of loads. As noted under "Redistribution of Negative Moments" (Section 8.4), reinforced concrete frames will adjust to various loading patterns provided the total area of steel top and bottom is sufficient and is extended through all possible critical tension sections for full development. Structures designed for seismic conditions that require energy dissipation in the nonlinear range of response have special provisions in Code Chapter 21. Details of reinforcement discussed in this section are not adequate for seismic design.

The Code provides an intricate set of overlapping requirements for minimum bar length (Sections 12. 11, 12.12, and 12.13). The "Explanations of Revisions" first published with the proposed 1971 Code gave an elaborate explanation of "shifted" maximum moment diagrams. Only for an element repeated identically many times can the expense of design time to compute absolute minimum bar lengths for minimum reinforcing steel weight be justified. A practical approach for routine design is to establish arbitrary cutoff points as ratios of span in a "Typical Bar Lengths Detail." These "standard" bar lengths can be made applicable for all uniformly loaded beams in a structure; special details are then necessary only for unusually short or long beams, or girders with odd locations of concentrated loads.

With this approach, the provision of proper development lengths is reduced to a simple comparison of actual length available, computed as a ratio of the span to the cutoff point, to the length required for full development, ℓ_d, for different bar sizes and selection of the largest size of bar usable. See Tables 13-4 and 5 for straight embedment, ℓ_d, and hooked end embedment, ℓ_{dh}, respectively.

Cutoff locations available in various manuals* should be adjusted for unequal spans and/or pattern loadings that occur for each individual project. Some engineers prefer to reference these typical details to clear span and some c.-c. span. Clear span is more nearly correct for a variety of support widths, but the c.-c. span is more convenient for the engineer, detailer, estimator, and placer. Whichever is used, the ratio employed should be adjusted to give approximately the same resulting length.

For this example, typical bar lengths are shown in Fig. 9-7 for an interior span using straight bars only and vertical stirrups. The arrangement in Fig. 9-7

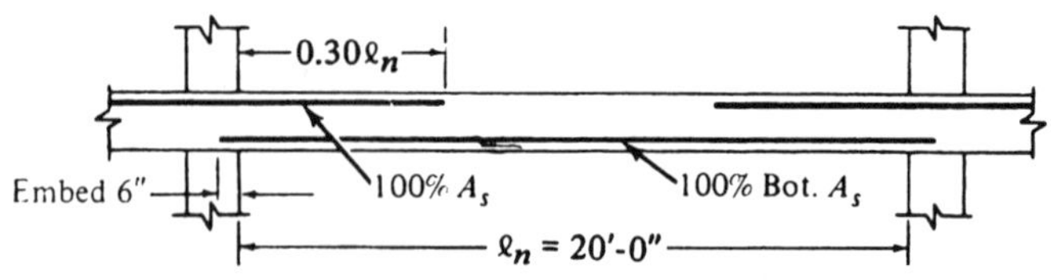

Figure 9-7 Typical bar lengths—*A*.

* *ACI Detailing Manual—1994* (SP-66).

is perhaps the least sophisticated system and therefore the most foolproof from design through construction. With it design time is reduced to a minimum. In this case, the ratio of cutoff 0.30 ℓ_n barely sufficed; an optimal equivalent more widely applicable would be 0.33 ℓ_n. The available top bar development length is 0.30 × 20′-0″ × 12 = 72 in. See Table 13-4, top bars, f_c' = 4,000 psi. The maximum size for these top bars is #9 (ℓ_d = 70 in.). This bar selection now satisfies the general requirements of development on each side of the critical section, face of support (Section 12.12.1).

To complete the investigation for the top bar selection, the point of inflection for maximum negative moment must be located (see Fig. 9-8). The distance from the face of the support to the point of inflection (P.I.) is 4′-9″. The bar length provided extends past the point of inflection so that it is *not* cut off in a tension zone, avoiding elaborate alternative computations required to justify same (Section 12.10.5). Since none of the top bars are cut off short, the requirements of the Code regarding "continuing reinforcement" are satisfied (Section 12.10.4). The top bars must extend beyond the point of inflection a distance equal to the largest value of: (1) d, (2) 12 bar diameters, or for $A_s/3$, (3) $\ell_n/16$ (Sections 12.10.3 and 12.12.3).

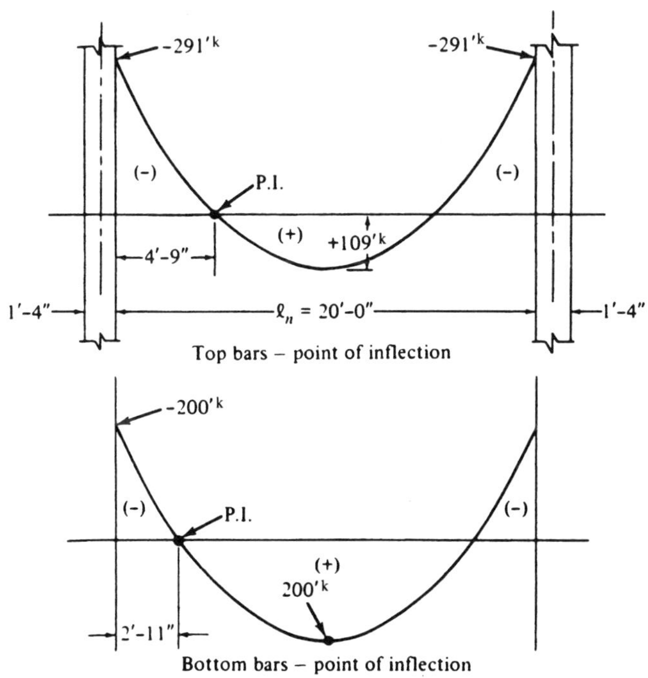

Figure 9-8 Factored moment diagrams.

Available length for development	= 72 in.
Distance to point of inflection	= 57 in.
Extension beyond point no longer required	= 15 in.

1. d = 12.5 in.	< 15 in.	Cutoff is sufficient length for #9 bars
2. 12 d_b (#9): 12 × 1.128 in.	= 13.5 in.	
3. ℓ_n /16 = 240/16	= 15 in.	

The area of top bars required = 6.4 in.^{2} Use seven #9 bars in the beam for A_s = 7.00 in.^{2} Crack control is found to be satisfied by calculating d_c = 2.44 in., A = 16.72 in.^{2}, and z = 124 kips/in.

The area of bottom bars required = 3.65 in.^{2} Five #8 bottom bars will be used. Since 100 percent of the bottom A_s is extended 6 in. into the support, assuming our example beam is part of a braced frame, no further development computations are required. For a beam which forms part of a primary lateral load resisting frame, 25 percent of the bottom steel must be extended into the support a distance ℓ_d or an equivalent embedment for a larger percentage. (Section 12.11.2). Structural integrity reguirements for this beam are satisfied by use of closed stirrups. If closed stirrups had not been used, then 25 percent of the bottom A_s would need to have been continuous or to have Class A tension lap splices.

Note that different sizes were selected for top and bottom bars as an aid to inspectors and placers to help avoid error. Preferably, these two sizes would be continued through all beams varying number but not size to suit spans and loads encountered. Since stirrups are supplied for the full length of the beam, some positive support for the stirrup cage and joist-slab top bars should be provided in the 8′-0″ gap between ends of the main top bars. The 3-#4 × 10′-0″ additional bars required for torsion with a lap of 12 in. at each end will provide sufficient support. See Table 13-6 for tension lap splice lengths.

Cutoff Points for Reduced Reinforcement

For a structure with many repetitions of the same beam design, two commonly used refinements of the simple typical bar lengths detail in Fig. 9-7 are shown in Fig. 9-9. The engineer must judge whether the indicated reductions in reinforcing steel weight are repeated sufficiently to justify the cost of added design, inspection, and placing time. The additional computations required are shown for illustration below.

The short bottom bars are cut off in a compression zone, inside the point of inflection for positive moment (Fig. 9-8) thereby avoiding computations for cutoffs in tension zones (Section 12.10.5). The Code requires that the following equation be satisfied at the point of inflection, for bars cut off in this region:

$$\ell_d \leqslant \frac{M_n}{V_u} + \ell_a$$

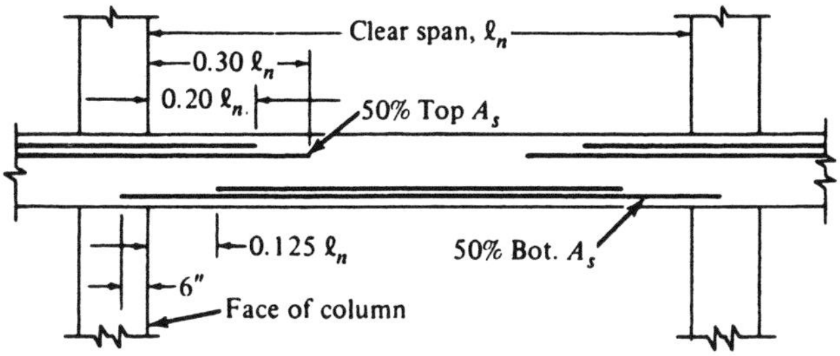

Figure 9-9 Typical bar lengths—*B*.

where ℓ_d is based on f_y: M_n is the positive nominal moment strength at the section (P. I.): V_u is the factored shear force at the P.I.; and ℓ_a is the embedment past the P.I., but not to exceed d or 12 bar diameters, whichever is greater for this calculation (Section 12.11.3). To save calculations, take M_n for 60 percent A_s as

$$\left(0.6\frac{M_u}{\phi}\right) = 0.6\frac{(200)}{0.90} = 133 \text{ ft-kips}$$

At the P.I.

$$V_u = (8)(10 - 2.92) = 57 \text{ kips}$$

for the continuing #8 bottom bars, ℓ_d = 48 in. in tension (Table 13-4); $\ell_a = d$ = 12.5 in.

$$\frac{(133)(12)}{(57)} + 12.5 = 28.0 + 12.5 = 40.5 > 38 \text{ in.}$$

Development length from bottom bar cutoff to point of maximum positive moment, at midspan, = 7.50 × 12 = 90 in. > 38 in. The proposed cutoff point and selection of #8 bottom bars are satisfactory.

The cutoff point for the top bars involves a number of more rigorous conditions to satisfy (Sections 12.10.3, 12.10.4, 12.10.5, 12.12.2, and 12.12.3). A complete redesign will be required to use smaller than #9 size bars with less than 100% of the bars extending 0.30 ℓ_n (see Fig. 9-10).

1. ℓ_d for the cutoff bars cannot exceed 48 in. (Sections 12.10.4 and 12.12.2). From Table 13-4, ℓ_d = 47 in. for #8, and 54 in. for #9, both "other" bars. #8 is the maximum size for this criterion.
2. Note that d = 125 in.; 12 bar diameters #8 = 12 in. < 24 in. (Section 12.10.3). This criterion is satisfied.
3. The long bars must be fully developed in 48 in. (Section 12.10.4). ℓ_d for #8 = 47 in. < 48 in.

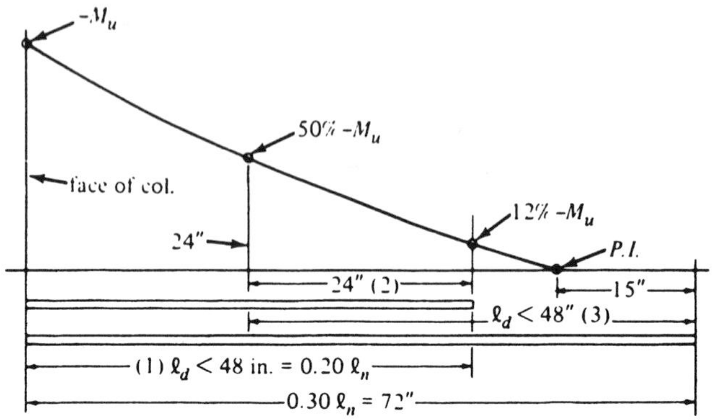

Figure 9-10 Critical lengths for top bar cutoff.

4. A final provision that must be met is that *one* of the following three conditions is satisfied at the cutoff point (Section 12.10.5):
   (a) Calculated $V_u \leqslant 2/3\phi V_n$ allowed
   (b) Excess stirrups are present for 60 psi capacity
   (c) that the continuing bars have an area at least twice the required area and that $V_u \leqslant 3/4\phi V_n$ allowed (see Fig. 9-3).

   Bar area in the continuing bars is approximately $50\%/12\% = 4.2$ times the required area.

Condition (c) is satisfied.

8-#8 top bars will satisfy the area required and meet development requirements for the typical bar lengths detail of Fig. 9-9. In the example beam, these bars would be spaced approximately 3 in. center-to-center. This closer spacing than the design for the bar lengths in Fig. 9-7 will complicate placing to fit between column verticals and may preclude use of this more sophisticated typical detail at some supports in a practical situation. This consideration will also be involved in the decision for typical details on a specific project.

At this point, all the requirements for the first stage of design, design for strength, are complete. The Code requirements for serviceability constitute a second stage of design, usually performed simultaneously with the strength design. By use of design aids limiting the maximum spacing of bars for crack control and tabulations of maximum load for the usual range of span and depth combinations to control deflection in selecting depth, the second stage of design can be satisfied without any additional routine calculations. For illustrative purposes, it will be performed here as a review of the design just completed to determine whether any adjustments are required.

SERVICEABILITY DESIGN

Performance Under Service Loads

CONTROL OF DEFLECTION. The immediate deflection of a beam must be computed if the beam thickness is less than the minimum thicknesses prescribed (Section 9.5.2.1, Table 9-5a).

$$\text{For interior spans, } \min h = \ell_n/21 = 240/21 = 11.4 \text{ in.} < 15 \text{ in.}$$

$$\text{For end spans, } \min h = \ell_n/18.5 = 240/18.5 = 13 \text{ in.} < 15 \text{ in.}$$

Deflection need not be computed for this example. For a complete example of deflection calculations, see Chapter 4.

CRACK CONTROL. Determine the maximum spacing of the top reinforcement, #9 bars, with #4 stirrups and $1\frac{1}{2}$ in. clear cover, to satisfy crack control requirements for interior exposure. Part of the top reinforcement will be placed in the 3 in. thick flanges (Section 10.6.3). The quantity, z, must not exceed 175 k/in. for interior exposure (Section 10.6.4).

$$z = f_s \sqrt[3]{d_c A}$$

where A = the tributary area of concrete in tension per bar; f_s = tensile stress in the bar at service load. If not calculated, f_s may be taken as $0.6 f_y$; d_c = the distance from the tensile face to the center of the bar.

Let S_b = the spacing of the bars. c.-c., then $A = (2)(d_c)(S_b)$. Service dead load = 2.5/1.4 = 1.78 k/ft; service live load = 5.5/1.7 = 3.23 k/ft; total service load = 1.78 + 3.23 = 5.01 k/ft.

$$f_s = \frac{\phi(60)}{(8.0/5.01)} = \frac{(0.90)(60)}{1.596} = 33.8 \text{ ksi}$$

$$d_c = 1.500 + 0.500 + (\tfrac{1}{2})(1.128) = 2.56 \text{ in.}$$

Solving for the maximum spacing of the #9 bars, S_b:

$$z = (33.8)\sqrt[3]{(2)(d_c^2)(S_b)} = 175 \text{ k/in.}$$

$$2(d_c^2)(S_b) = \left(\frac{175}{33.8}\right)^3 = 138.7$$

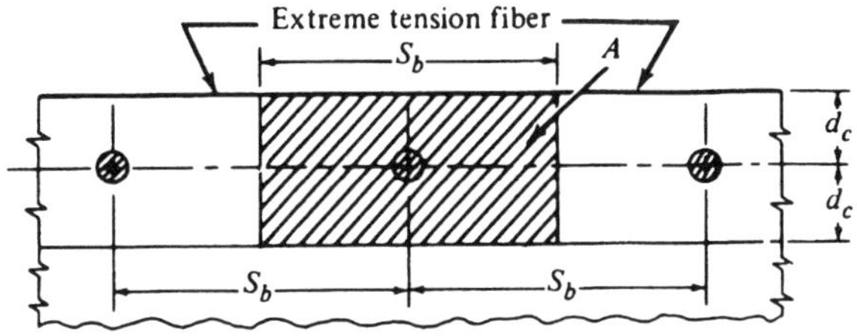

Figure 9-11 Crack control.

Maximum spacing,

$$S_b = \frac{138.7}{(2)(2.56)^2} = 10.6 \text{ in.}$$

Spacing for the #9 bars in the beam section (Fig. 9-2) is satisfactory, approximately 3 in. < 10 in. maximum. Note that it will not be possible to provide bars in the 3 in. thick flanges with the full required cover, $1\frac{1}{2}$ in., top and bottom (Section 7.7.1). In this case, #3 bars will be placed with $1\frac{1}{2}$ in. top cover so that they will fit under the top joist bars which require only $\frac{3}{4}$ in. cover (Section 7.7.1). Spacing of the #8 bottom bars is approximately 4.75 in., which is satisfactory without further check. These calculations complete the second stage of design, for serviceability, in Example 1.

EXAMPLE 2. *Spandrel Beam—B-3*. Design of beam *B*-3 for torsion (see Fig. 9-12): Continue Example 1 floor system loads, spans, and column size. Consider beam *B*-3 as the spandrel beam between two interior columns in the direction of its span. Successive spans, ℓ_1, are equal to 21′-4″ c.-c. of 16 in. × 16 in. edge columns. The floor system, span ℓ_2 = 20′-0″ c.-c. between beams *B*-2 and *B*-3, is a one-way joist-slab, 12 + 3; with loads, w_d = 100 psf and w_ℓ = 250 psf. Beam *B*-3 is to be 15 in. in total depth to maintain the level ceiling. Design for flexural and serviceability requirements will not be repeated because they will be similar to the design for *B*-1 in Example 1. Structural integrity for perimeter beams requires that one-sixth of the top bars and 25 percent of the bottom bars must be made continuous (Section 7.13.2.2.).

In Example 1, the design charts of this chapter were employed to illustrate the short-cut approach to design for combined shear and torsion. In Example 2, each applicable Code formula will be used for purposes of comparison and to present the basis of the design charts. For users wishing to develop their own design charts for other stresses, concrete cover, stirrup size, or lightweight aggregate for an extended range of beam sizes, or simply to a larger scale for greater accuracy, a study of Example 2 should make such conversions simple.

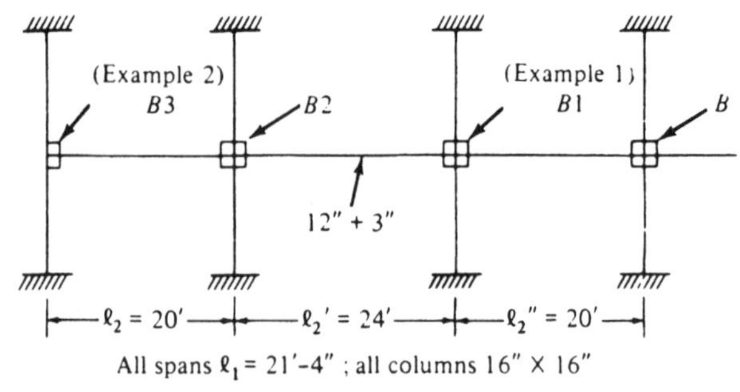

Figure 9-12 Beam-column frame with beam *B*-3.

Design for Strength—Shear and Torsion

Total factored torsional moment through $B3 \leqslant -M_{\text{ext}}$ of the floor system. Use the approximate frame analysis (Section 8.3.2) to determine the end moment where the support is a spandrel beam. First trial width,

$$b = 20 \text{ in.}$$

$$w_u = 0.100 + 0.250 = 0.350 \text{ ksf on the floor area}$$

$$-M_{u(\text{ext})} = \frac{1}{24}(w_u)(\ell_n)^2$$

$$-M_{u(\text{ext})} = \frac{1}{24}(0.35)(20 - 16/12)^2 = 5.08 \text{ ft-kips/ft}$$

Torsion on one side of the column from midspan to the critical section at d = 12.5 in. from the face of the column (Section 11.6.2.4) = T_u.

$$T_u = (5.08)(10 - 1.04) = 45 \text{ ft-kips}$$

The resisting section for torsion (Section 11.6.1) = A_{cp}^2/p_{cp} (see Fig. 9-13). NOTE: Neglect the small rectangle in the flange, since it will not be reinforced.

$$\frac{A_{cp}^2}{p_{cp}} = \frac{(15 \times 20)^2}{2(15 + 20)} = 1286 \text{ in.}^3$$

$$\frac{T_u}{\phi A_{cp}^2/p_{cp}} = \frac{(45{,}000)(12)}{(0.85)(1286)} = 494 \text{ psi} > 1.0\sqrt{f_c'} = 63 \text{ psi}$$

allowable without torsion reinforcement, and so torsion reinforcement must be provided (Section 11.6.1).

Reduce $T_u/(\phi A_{cp}^2/p_{cp})$ to $4\sqrt{f_c'} = 253$ psi (Section 11.6.2.2)

Design joists for reduced end moment (Section 11.6.2.2.)

$$-M_{(\text{ext})} = 5.08\frac{253}{494} = 2.60 \text{ ft-kips/ft}$$

$$T_u = 45\frac{253}{494} = 23 \text{ ft-kips}$$

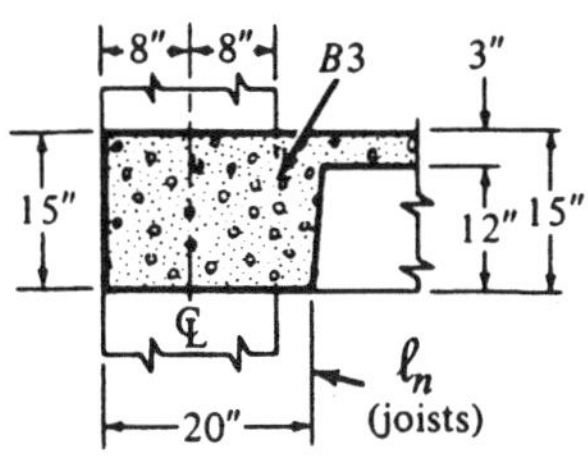

Figure 9-13 1st trial size, *B*-3.

The location where torsion reinforcement can be eliminated has factored torsion of

$$(63)(1286)/12{,}000 = 6.75 \text{ ft-kips}$$

It is located $12 \times 6.75/2.60 = 31$ in. from midspan.

Shear

If the torsion-shear reinforcement is to be combined, the shear stresses must now be determined (Sections 11.3.1, and 11.6.6). For the beam widths *B*3, 20 in.; and *B*2, 28 in., $\ell_n = [20 - (10 + 14)/12] = 18.0$ ft.

$$\text{Total shear, } V_u = \left[\frac{(0.35)(18.0)}{2} + \frac{(0.250)(20)}{2} + \frac{(20 \times 15)(0.150)(1.4)}{144}\right](8.96)$$

$$= (3.15 + 0.417 + 0.437)(8.96)$$

$$= 35.9 \text{ kips}$$

Take $d = 12.5$ in.

$$\frac{V_u}{\phi b_w d} = \frac{(35.9)(1000)}{(0.85)(20)(12.5)} = 169 \text{ psi}$$

Shear at midspan for partial panel loads (simple span basis):

$$V_u = \tfrac{1}{4}(0.250)(10)(10.67) = 6.7 \text{ kips}$$

Midspan

$$\frac{V_u}{\phi b_w d} = \frac{6.7(168.9)}{(35.9)} = 32 \text{ psi}$$

(See Figs. 9-15 and 9-16 for stirrup dimensions and torsion-shear diagrams.)

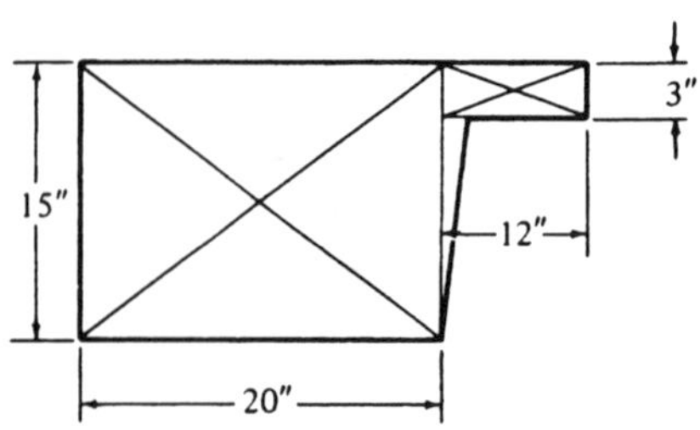

Figure 9-14 Torsion section, *B*-3.

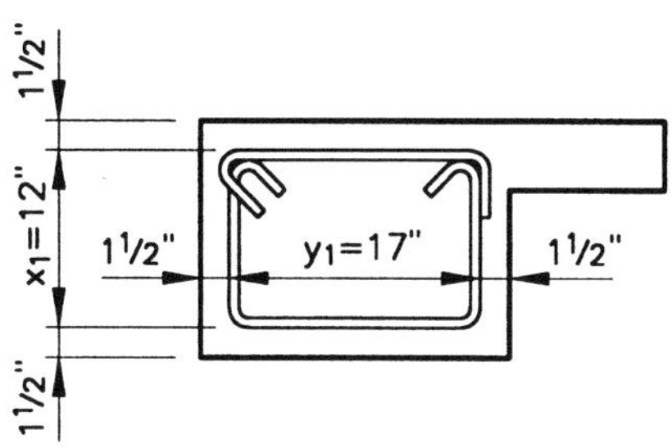

Figure 9-15 Closed stirrup dimensions.

Maximum Combined Torsion-Shear Stress Allowed

The maximum permitted,

$$\sqrt{\left(\frac{V_u}{b_w d}\right)^2 + \left(\frac{T_u p_h}{1.7A_{oh}^2}\right)^2} = \phi\left(\frac{V_c}{b_w d} + 8\sqrt{f_c'}\right) \quad \text{(Section 11.6.3.1)}$$

$$\text{Allowable} \quad \frac{T_u p_h}{\phi 1.7A_{oh}^2} = \sqrt{(2\sqrt{f_c'} + 8\sqrt{f_c'})^2 - 32^2} = 632 \text{ psi}$$

$$A_{oh} = (20 - 3.375)(15 - 3.375) = 193 \text{ in.}$$

$$p_h = 2(16.625 + 11.625) = 56.5 \text{ in.}$$

$$\frac{T_u p_h}{\phi 1.7A_{oh}^2} = \frac{23(12{,}000)(56.5)}{(0.85 \times 1.7)(193^2)} = 291 \text{ psi} < 632 \text{ psi}$$

Since the maximum torsional stress, v_{tu}, is less than that allowed, the trial width of 20 in. need not be increased. Additional trials are unnecessary (check solution above on chart, Fig. 9-25).

Reinforcement for the Shear

The area of (open) stirrups required for shear,

$$A_v = \frac{V_u/(\phi b_w d) - V_c/(b_w d)}{f_y} \quad \text{(Section 11.5.6)}$$

$V_u/(\phi b_w d) - V_c/(b_w d) = 169 - 2\sqrt{4000} = 43$ psi $< 4\sqrt{f_c'} = 253$ psi; maximum spacing for the vertical stirrups can be used (Section 11.5.4.3). Maximum spacing, $s = 0.5d$ (Section 11.5.4.1). $s = \frac{1}{2}(12.5) = 6.25$ in. max. At $s = 6$ in.,

$$A_v = \frac{(43)(20)(6)}{(60{,}000)} = 0.086 \text{ in.}^2 \text{ required.}$$

Reinforcement for the Torsion

The area of (closed) stirrups required for torsion,

$$\text{Area of one leg, } A_t = \frac{T_u s}{1.7\phi A_{oh} f_{yv}} \quad \text{(Section 11.6.3.6)}$$

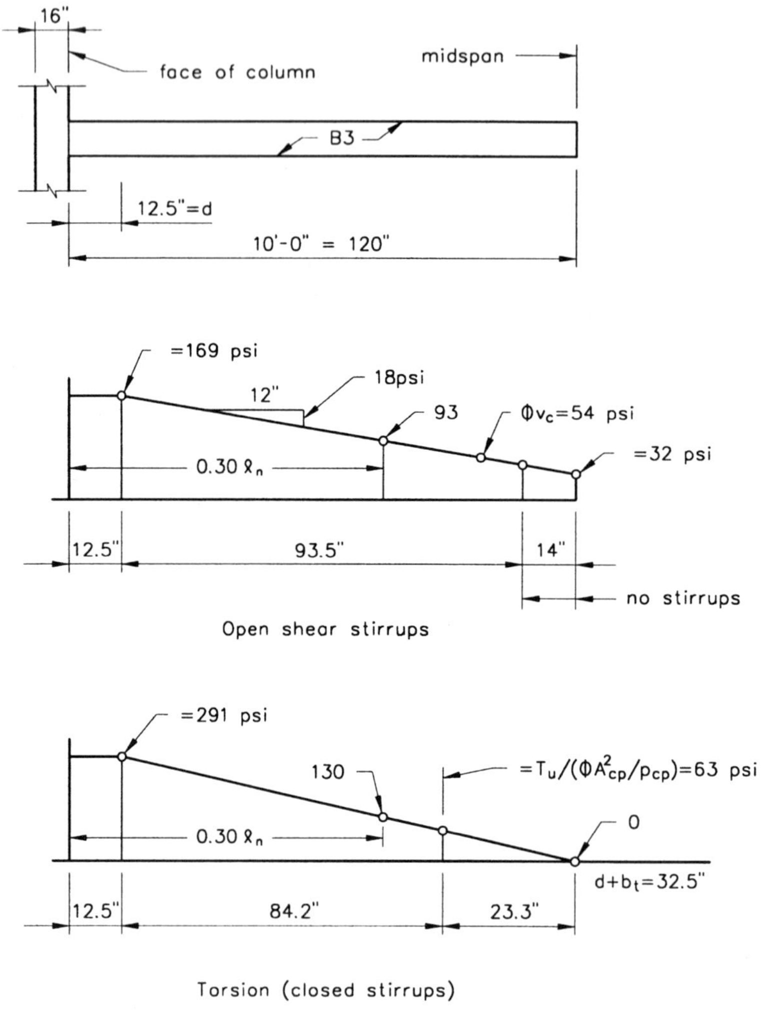

Figure 9-16 Length for spacing stirrups.

where A_{oh} is the area enclosed by the centerline of the closed stirrups. (NOTE: Charts are based on the centerline dimensions of stirrups; see footnote on page 232.)

The minimum area required for one leg of the closed stirrups for torsion with stirrups spaced at 6 in. = A_t:

$$A_t = \frac{(23)(12{,}000)(6)}{(1.7)(0.85)(193)(60{,}000)} = 0.099 \text{ in.}^2$$

The maximum spacing of the closed stirrups,

$$s_{max} = \frac{p_h}{8} = \frac{56.5}{8} \leqslant 7 \text{ in.}$$

Since the maximum spacing for the closed stirrups is 7.0 in. and for open stirrups (for shear) is 0.5 d = 6.25 in., only the closed stirrups will be used, at a spacing of 6 in. or less, in sizes proportioned to resist combined shear and torsion. This arrangement will not only result in a minimum number of stirrups, but also will avoid the complications in design, detailing, and placing that would result from mixing two types of stirrups in tricky placing sequences.

Combine A_v and A_t

The combination $A_v + A_t$, will require more reinforcement than the minimum prescribed for shear alone (Section 11.6.3.8.) If stirrups are required for shear alone, torsion alone, or combined shear and torsion, the minimum amount of combined reinforcement must resist a stress of 50 psi (Sections 11.5.5.3 and 11.6.5.2). The minimum reinforcement is

$$\frac{(50)(20)(6)}{60{,}000} = 0.10 \text{ in.}^2$$

Maximum reinforcement required for combined shear and torsion at the critical section for a spacing, s = 6.0 in. is $2A_t + A_v$ = (2) (0.099) + 0.086 = 0.284 in.^{2}

#4 closed stirrups @6 in. would provide an area = 0.40 in.^{2}

Compute the total number of stirrups required and any added longitudinal bars for torsion required in corners of stirrups. In Fig. 9-16 note that the torsion reinforcement is required for a distance of 12.5 + 76.5 + $(d + b_t) > 120$ in. from the face of the support. The Code requires that such reinforcement be provided for a distance $(d + b_t)$ beyond the point theoretically required (Section 11.6.6.3). $(d + b_t)$ = 12.5 in. + 20 in. = 32.5 in. Use 19-#4 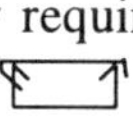 @6 at each end of the span.

Longitudinal Bars for Torsion

Longitudinal bars at least #3 in size and with a diameter at least 1/24 of the stirrup spacing must be provided in each corner of the closed stirrups. The spacing of these bars distributed inside the perimeter of closed stirrups must not exceed 12 in. (Section 11.6.6.2). In this example, three bars are required in both top and bottom. The bottom bars required for flexure will satisfy this requirement, and so only the minimum size of the top bars will be computed here. These top bars, which will also serve as part of the flexural reinforcement, must be extended and lap spliced at midspan.

The required minimum area of all longitudinal bars, A_ℓ, is the larger of the two following expressions (Section 11.6.5.3):

$$(1)\quad A_\ell = \frac{A_t}{s} p_h \left(\frac{f_{yv}}{f_{y\ell}}\right) \cot^2\theta \qquad \text{(Eq. 11-22)}$$

$$(2)\quad A_\ell = \frac{5\sqrt{f'_c}\,A_{cp}}{f_{y\ell}} - \left(\frac{A_t}{s}\right) p_h \left(\frac{f_{yv}}{f_{y\ell}}\right) \qquad \text{(Eq. 11-24)}$$

where A_t/s cannot be taken less than $25b_w/f_{yv}$.

At the usual cutoff for top flexural bars, $0.30\ell_n$, $V_u/(\phi b_w d) = 93$ psi and $T_u/(\phi 1.7A_{oh}^2/p_h) = 130$ psi.

$$(1)\quad A_\ell = \frac{0.099}{6}(56.5)\left(\frac{60}{60}\right) = 0.94 \text{ in.}^2$$

Since $25b_w/f_{yv} = 25(20)/60{,}000 = 0.008$ is less than $0.099/6 = 0.016$, use $A_t/s = 0.016$ in the following equation.

$$(2)\quad A_\ell = \frac{5\sqrt{4000}\,(20)(15)}{60{,}000} - \frac{0.099}{6}(56.5)\left(\frac{60}{60}\right) = 0.65 \text{ in.}^2$$

Use three #4 top bars continuous (lap spliced at midspan). These bars, or equivalent area, must be *added* to flexural requirements.

Shear and Torsion Design Notes

1. Open stirrups may be used wherever torsion is negligible. The anchorage requirements (Section 12.13) are comprehensive including provisions for bar stirrups, deformed WWF, and plain WWF stirrups as well as longitudinal (flexural) reinforcement bent-up to resist shear. For routine use, the single U-shaped stirrup #5 and smaller is normally anchored at the top (on the tension side) by standard 90° stirrup hooks around the longitudinal bars at interior beams (Section 12.13.2.1). At exterior beams, closed stirrups or 135° hooks are required (Section 7.13.2.2). Where multiple open stirrups are required, as for wide beams under high shear stress, the Code permits a one-piece configuration, or various alternative overlapping arrangements of single U-stirrups. See Fig. 9-17(a) for recommended arramgements. For #6, #7, and #8 stirrups with $f_y = 60{,}000$ psi, end anchorage may consist of standard 90° or 135° stirrup hooks enclosing a longitudinal bar plus an embedment length $0.14d_b f_y/\sqrt{f'_c}$ between the member midheight and the outside end of the hook (Section 12.3.2.2.). Finally, for all open stirrups, all bend points *between* anchored ends must enclose a longitudinal bar (Section 12.13.3).

2. In Examples 1 and 2, two-piece closed stirrups for torsion were shown as the general design solution to satisfy the Code requirement of closed stirrups (Section 11.6.4.1). One-piece closed stirrups require either prefabrication of the entire beam cage and placing same as a unit—often impossible because top bars must pass between column verticals—or time-consuming assembly in place,

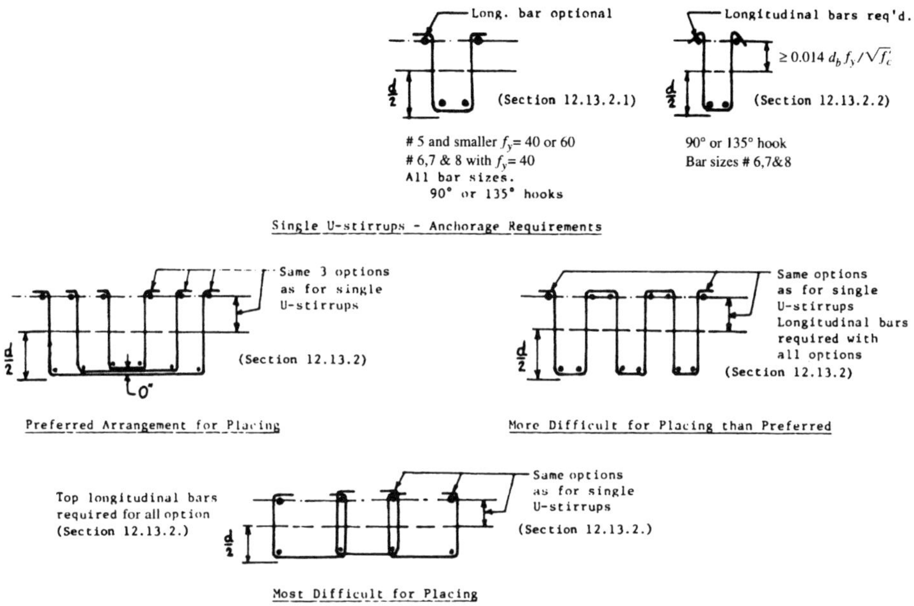

(a) Code Requirements for Anchorage of Open Stirrups

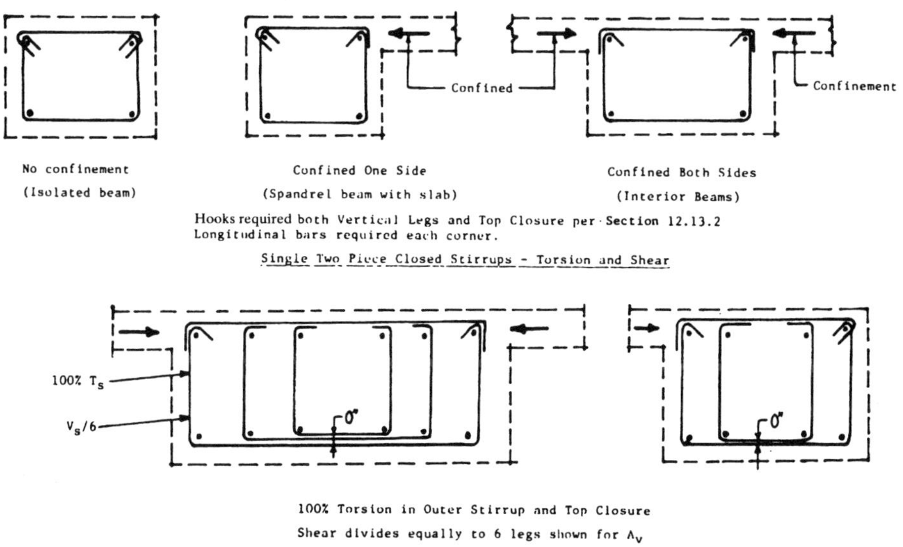

(b) Recommended Two-Piece Closed Single and Multiple U-Stirrups

Figure 9-17 Various closed stirrups and recommendations for their use.

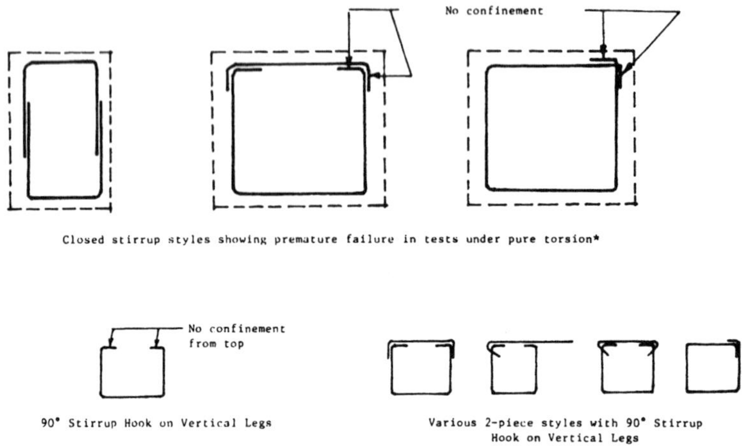

(c) Closed Stirrup Styles Considered Ineffective for High Torsion Stress

Figure 9-17 *(continued)*

perhaps by threading individual flexural bars through stirrups and between column verticals. For practical reasons, the one-piece closed stirrup, spliced by overlapping hooks (Section 7.11.3) should be used only as a last resort. Two-piece closed stirrups should be used wherever possible.

Two-piece closed stirrups must conform to specific Code provisions (Sections 7.11.3 and 12.13.5). The Code provisions are general enough to permit many permutations (two types of end hooks plus full or partial straight embedment). Recommended styles are shown in Fig. 9-17 (b). Tests in pure torsion to failure indicate that some (commonly used) arrangements perform poorly when no concrete confinement is available.* See Fig. 9-17 (c) for unsuitable examples.

Where the straight bar embedment begins at the point of tangency for a hook at the top, the maximum length available to develop the remaining leg is the distance to the mid-depth of the beam, $\frac{1}{2}d$, (Section 12.13.2.2). Fig. 9-17(b) shows some of the special details for two-piece closed stirrups in beams of various shapes as suggested in the ACI Detailing Manual. Design limitations upon these arrangements imposed by the Code (Section 12.13) are also shown.

3. To reduce design time for shear, simple charts of required stirrup size and spacing are shown for the equation

$$\frac{V_u}{\phi b_w d} - \frac{V_c}{b_w d} = \frac{V_s}{b_w d} = \frac{60{,}000 A_v}{b_w s} \qquad \text{(Section 11.5.6.2)}$$

* "Detailing for Torsion," D. Mitchell and M. P. Collins, *ACI Journal*, September 1976.

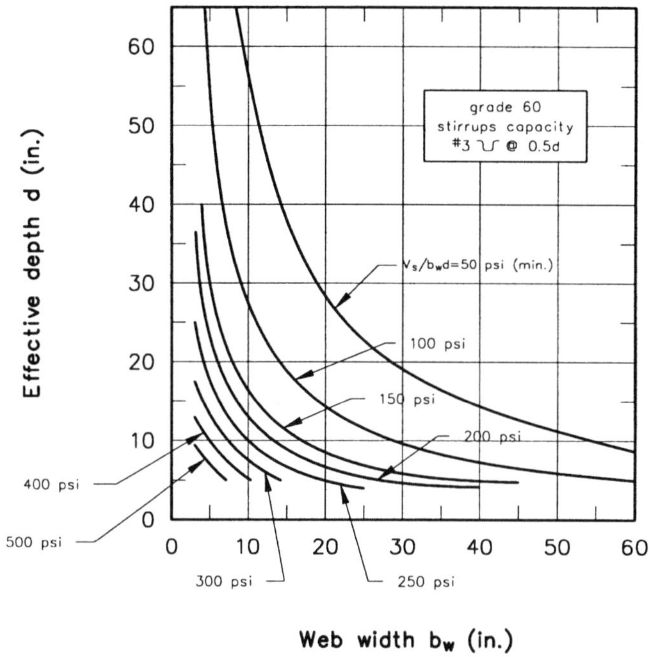

Figure 9-18 #3 stirrup capacity.

See Figs. 9-18, 9-19 and 9-20 for #3, #4, and #5 open ˥⎽⎽⎽Γ stirrups (2 legs effective). The charts here present capacity for Grade 60 stirrups, in terms of $\left(\frac{V_u}{\phi b_w d} - \frac{V_c}{b_w d} = \frac{V_s}{b_w d}\right)$ to be carried by stirrups, applicable with any strength concrete. The charts were prepared for stirrups spaced at 0.50d, the maximum spacing permitted (Section 11.5.4). The range of shear, $V_s/(b_w d)$, is 250 psi (maximum for $\sqrt{f'_c}$ = 4,000 psi) and spacing 0.50d to the minimum, 50 psi, for any f'_c (Section 11.5.5.3). The user can easily interpolate readings for any excess shear, $V_s/(b_w d)$, between values given, select appropriate size of U-stirrup, spaced at 0.50d. Alternatively, the designer can begin with a $V_s/(b_w d)$ for any beam and a desired size of stirrup and reduce the spacing so that:

$$s = \left[\frac{V_s/(b_w d) \text{ tabulated}}{V_s/(b_w d) \text{ actual}}\right](0.50d)$$

If beam width makes multiple U-stirrups feasible as shown in Fig. 9-17(b) use a fraction of the total beam width for two stirrup legs as b_w, and increase capacity proportionally as follows:

$$\text{Capacity}\left(\frac{V_s}{b_w d}\right) = \left(\frac{\text{No. of stirrup legs furnished}}{2}\right)\left[\left(\frac{V_s}{b_w d}\right) \text{tabulated}\right]$$

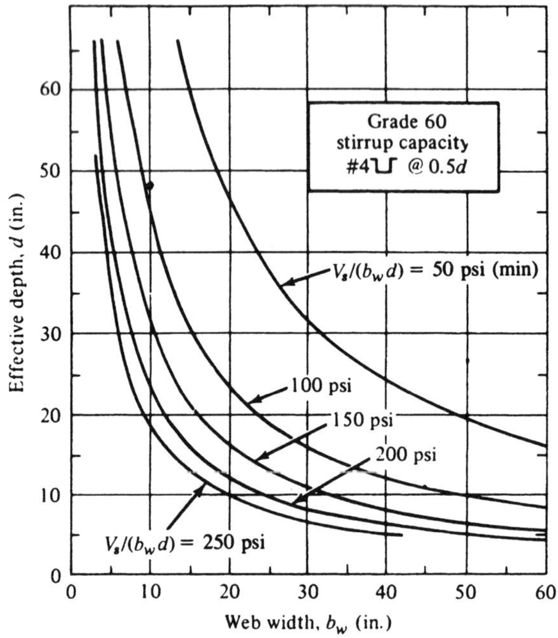

Figure 9-19 #4 stirrup capacity.

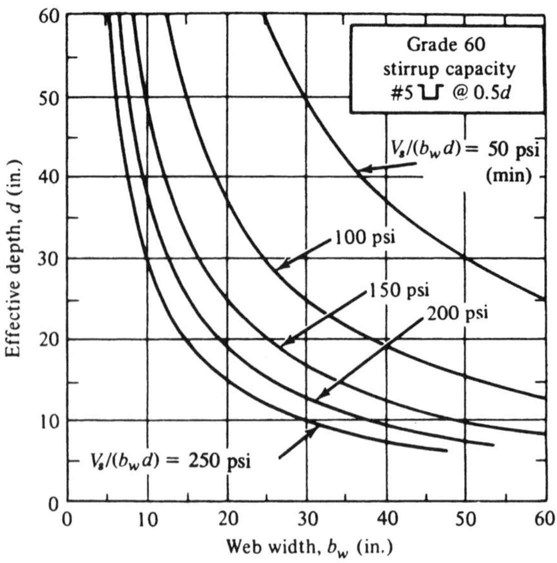

Figure 9-20 #5 stirrup capacity.

Note that a longitudinal bar must be enclosed within each bend top and bottom in multiple stirrups (Section 12.13.3).

4. Beams subject to torsion involve lengthy calculations if performed manually to satisfy a number of interdependent Code requirements as in manual Example 2. Since all edge beams and many interior beams with possible heavy unbalanced live loadings on either side must be investigated, rapid answers to the following design questions in the order listed are essential to the practical designer:

(*a*) Whether the effect of torsion can be neglected (Section 11.6.1); is

$$\frac{T_u}{\phi A_{cp}^2/p_{cp}} \leqq 1.0\sqrt{f_c'}$$

(*b*) Whether the combined torsional stress plus shear stress is excessive requiring selection of a larger size section (Section 11.6.3.1.)

(*c*) Whether torsion or shear will control the maximum stirrup spacing and whether this spacing involves an impracticable number of stirrups, requiring in most cases selection of a larger size section.

Design aids to provide the answers to the above questions and to permit a rapid completion of the design where stirrups are required including the selection of stirrups have been included in this chapter (see Figs. 9-18 through 9-25). The procedure for the use of these aids is:

a. Compute the applied factored shear, V_u, and torsion, T_u, at the same critical section, a distance d from the face of the support.

b. Compute the unit shear and torsional stresses for the beam section selected as a first trial.

$$\text{shear stress} = \frac{V_u}{\phi b_w d} \qquad \text{torsion stress} = \frac{T_u}{\phi 1.7 A_{oh}^2/p_h}$$

c. Enter Fig. 9-25(a) with values for $\sqrt{f_c'}$, $V_u/(\phi b_w d)$, and $T_u/(\phi 1.7A_{oh}^2/p_h)$. Calculate values for k_1 and k_2. Read whether $T_u/(\phi 1.7A_{oh}^2/p_h)$ exceeds the maximum allowable value. Select a larger section and repeat if $T_u/(\phi 1.7A_{oh}^2/p_h)$ is above the curve.

$$\frac{T_u}{\phi 1.7 A_{oh}^2/p_h} \quad \text{and} \quad \frac{V_s}{b_w d} = \frac{V_u}{\phi b_w d} - \frac{V_c}{b_w d}$$

torsional and shear stresses, respectively, which must be carried by the stirrups. At this point, for discontinuous ends only, it will usually be possible to reduce the number of stirrups substantially by the use of the "long" formula for $V_c/b_w d$.

$$\frac{V_c}{b_w d} = 1.9\sqrt{f_c'} + 2500\rho_w \frac{V_u d}{M_u} \qquad \text{(Section 11.3.2.1)}$$

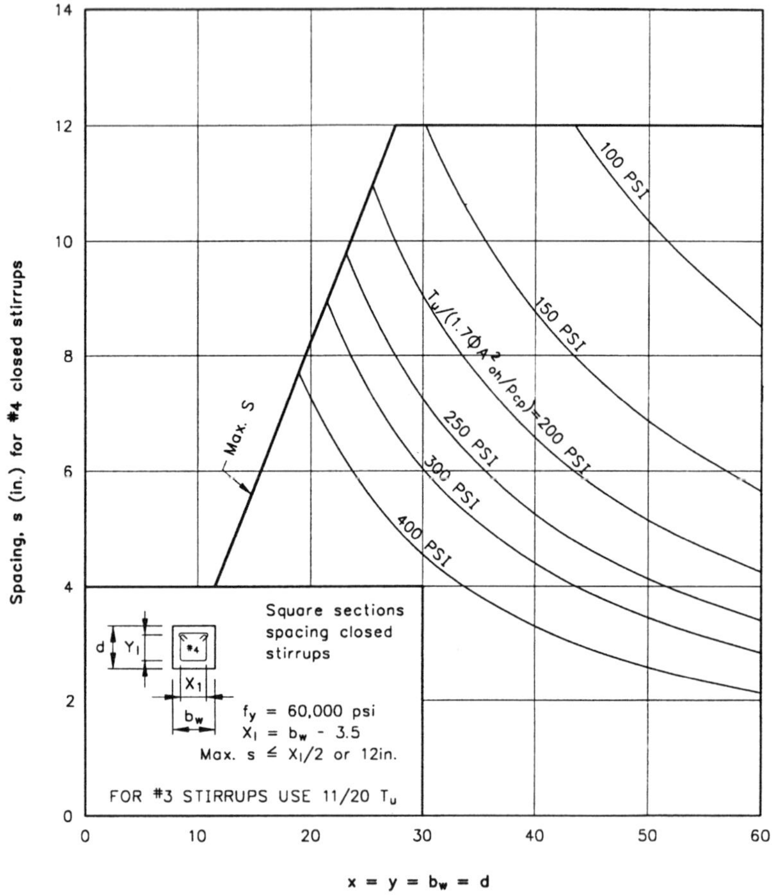

Figure 9-21 Square sections—closed stirrup spacing chart.

Read the maximum spacing limits for open (shear) stirrups. Increase the beam size and repeat if it is not desirable to reduce maximum stirrup spacing from 0.5 *d* to 0.25 *d*.

d. Enter Figs. 9-21 to 9-24 with value, $T_u/(\phi 1.7A_{oh}^2/p_h)$. Read maximum spacing of closed stirrups for torsion, in inches for #3 and #4 stirrup sizes. For very large beams only, it may be desirable to compute spacings for #5.

e. Enter Figs. 9-18 to 9-20 with the value, $V_s/(b_w d)$. Read size of stirrups to permit spacing at 0.50 *d*. Compute size for spacing at 0.25 *d*.

f. Use judgment to select the simplest practicable arrangement of stirrups in minimum numbers to satisfy the Code. A sketch similar to that in Fig. 9-16 will usually save time. The authors recommend that the designer *not mix open and closed stirrups in an alternate placing sequence.* If torsion controls the spacing throughout, use only closed stirrups; if torsion con-

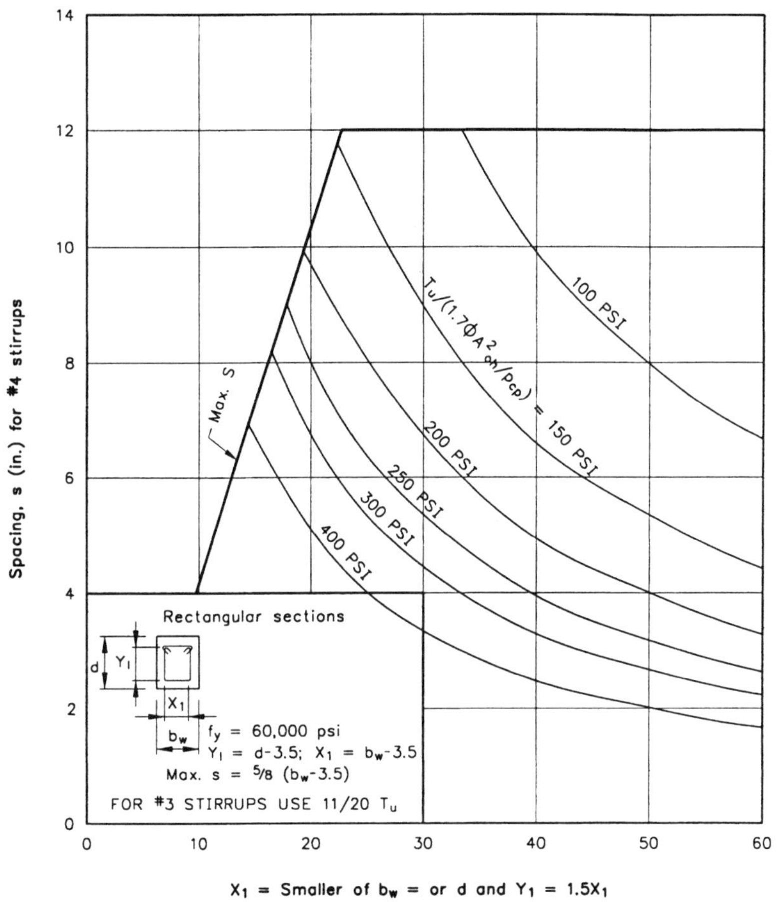

Figure 9-22 Rectangular sections (1.5:1)—closed stirrup spacing chart.

trols for only a short distance at each end, it is better to use closed stirrups until torsion can be neglected, and then open stirrups beyond. The simple procedure illustrated in Example 1 can be adopted to determine various stirrup size-spacing combinations acceptable under the Code for comparison.

g. If the final design requires closed stirrups for torsion, longitudinal bars are also required as part of the torsional reinforcement. At the critical section for torsion, the longitudinal bars required for flexure can be utilized. These bars will usually suffice for interior beams, but must be increased in spandrel beams where the maximum flexure and torsion occur under the same load combination.

The design charts (Figs. 9-26 and 9-27) will permit a rapid selection of the minimum size longitudinal bars for extension beyond the normal cutoff point of

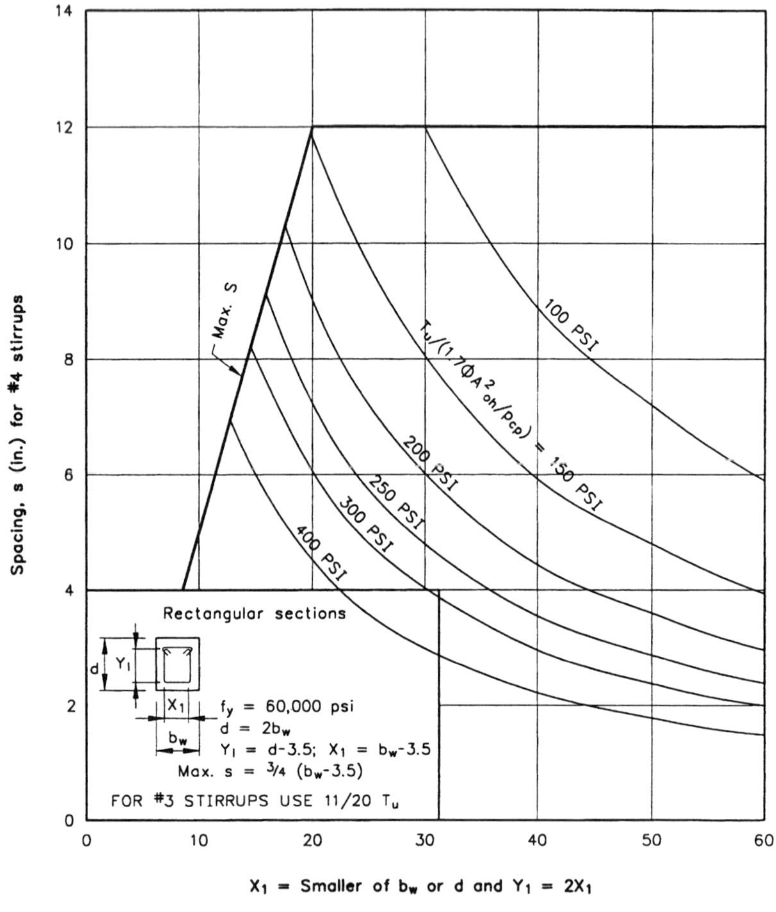

Figure 9-23 Rectangular sections (2:1)—closed stirrup spacing chart.

the top bars for flexure. Shear and torsion at this point are both low, and both charts are, based on

$$\frac{T_u}{\phi 1.7A_{oh}^2/p_h} = 50 \text{ psi}$$

Enter Fig. 9-26 with the actual $T_u/(\phi 1.7A_{oh}^2/p_h)$; read A_t = minimum area of one leg of closed stirrups at maximum spacing, $s = \frac{1}{2}x$; The total area of all longitudinal bars, $A_\ell \geqslant 8A_t$, and $A_\ell \geqslant 5\sqrt{f_c'}\, b_w^2/f_{y\ell} - 8A_t$, but not less than four #3 bars. Enter Fig. 9-27 with the value of $b_w = h$ and the value $T_u/(\phi 1.7\sqrt{f_c'}\, A_{oh}^2/p_h)$ which is the ratio of torsion stress to $\sqrt{f_c'}$ (psi); read A_ℓ/A_t. Calculate the number of bars required, one per corner and spaced not more than 12 in. around the inside perimeter of the closed stirrups. The area of each bar is *not* required to be equal, which will aid matching the total to that

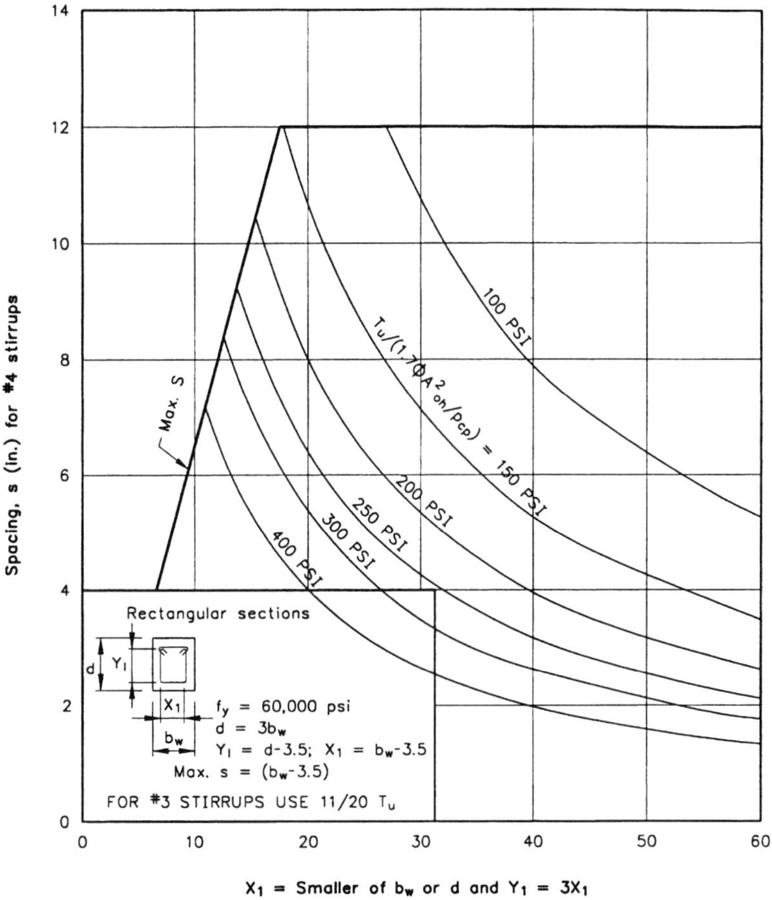

Figure 9-24 Rectangular sections (3:1)—closed stirrup spacing chart.

required. The centroid of the additional longitudinal bars should approximately coincide with the centroid of the section.

Figures 9-26 and 9-27 were prepared for square sections. For office use, the authors suggest that readers prepare similar charts for various ratios of b_w/h to cover the usual range of rectangular sections. The sample charts are merely applications of the Code (Sections 11.6.3.5, 11.6.3.6, 11.6.3.7, and 11.6.5.3).

Flexural Design Notes

(1) In most of the manual examples, the flexural computations for the required areas of tension reinforcement, A_s, were carried out by trial and review. The flexural design equations are not prescribed directly in the Code, which merely lists the necessary assumptions for design. It has been assumed that the

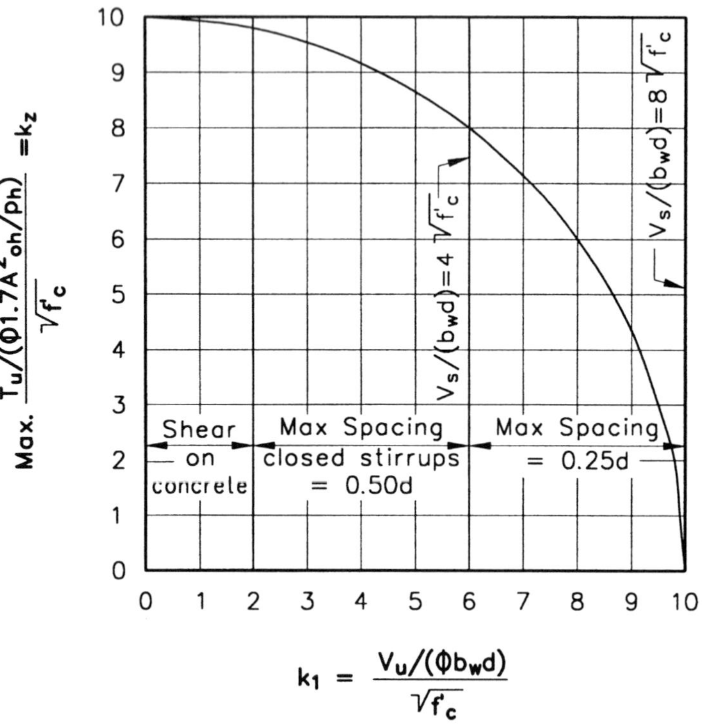

Figure 9-25 Maximum torsion with shear.

necessary design equations resulting from these assumptions are well known and accepted in the form shown below. These equations are established in simple terms for *review* of trial design sections. A direct solution for rectangular sections with tension reinforcement only is

$$A_s = \frac{1.7f'_c bd}{2f_y} - \frac{1}{2}\sqrt{\frac{2.89(f'_c bd)^2}{f_y^2} - \frac{6.8f'_c bM_u}{\phi f_y^2}}, \quad \text{or}$$

$$A_s = 0.85\left(1 - \sqrt{1 - \frac{2.35M_u}{\phi f'_c bd^2}}\right)\frac{bdf'_c}{f_y}$$

Figure 9-28 solves this equation for the flexural constants in the familiar forms which have been used in handbooks for the straight-line flexural design theory for many years. These constants provide quick *direct* solutions for area reinforcement, A_s, or size of section, bd, to satisfy the 1995 Code.

$$(1)\ A_s = \frac{M_u}{Fd} \qquad (2)\ M_u = Rbd^2$$

in which M_u is in ft-kips; A_s in in.^{2}; and d in in.

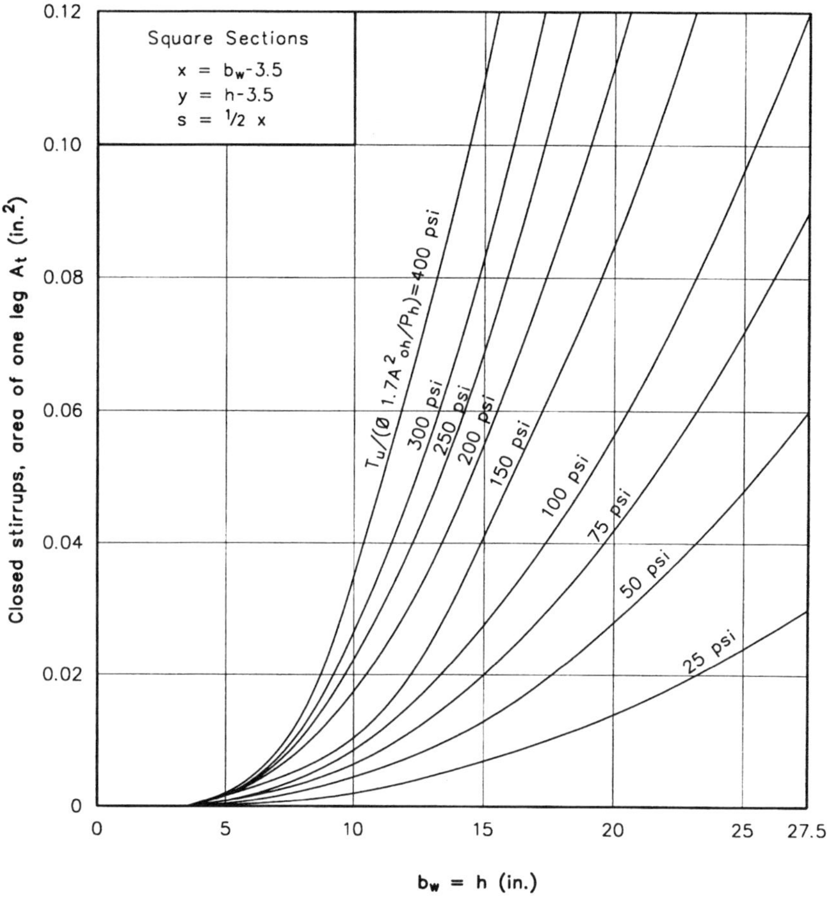

Figure 9-26 Stirrups for torsion—square sections.

(2) Curves given in Fig. 9-28(a) are for f_y = 60,000 psi. Derivations are given for quick adjustments of the curves for any concrete strength, f'_c, or yield strength, f_y. Note that the relations are linear. Three curves are given for the tension steel area constant, F, with concrete strengths, f'_c = 3,000 psi, 4,000 psi and 6,000 psi .

(3) In Fig. 9-28 (*b*), three curves are presented for the compression constant, R, with concrete strengths, f'_c = 3,000 psi, 4,000 psi, and 6,000 psi using f_y = 60,000 psi. For other values of f_y, multiply the reinforcement ratio, ρ, by the ratio 60,000/f_y. It will be noted that the value of R is not sensitive to f'_c, except as to the limiting maximum values of reinforcement ratio, ρ. For most designs where 3,000 psi $\leqq f'_c \leqq$ = 6,000 psi, the values of R can be read directly with sufficient accuracy for practical design.

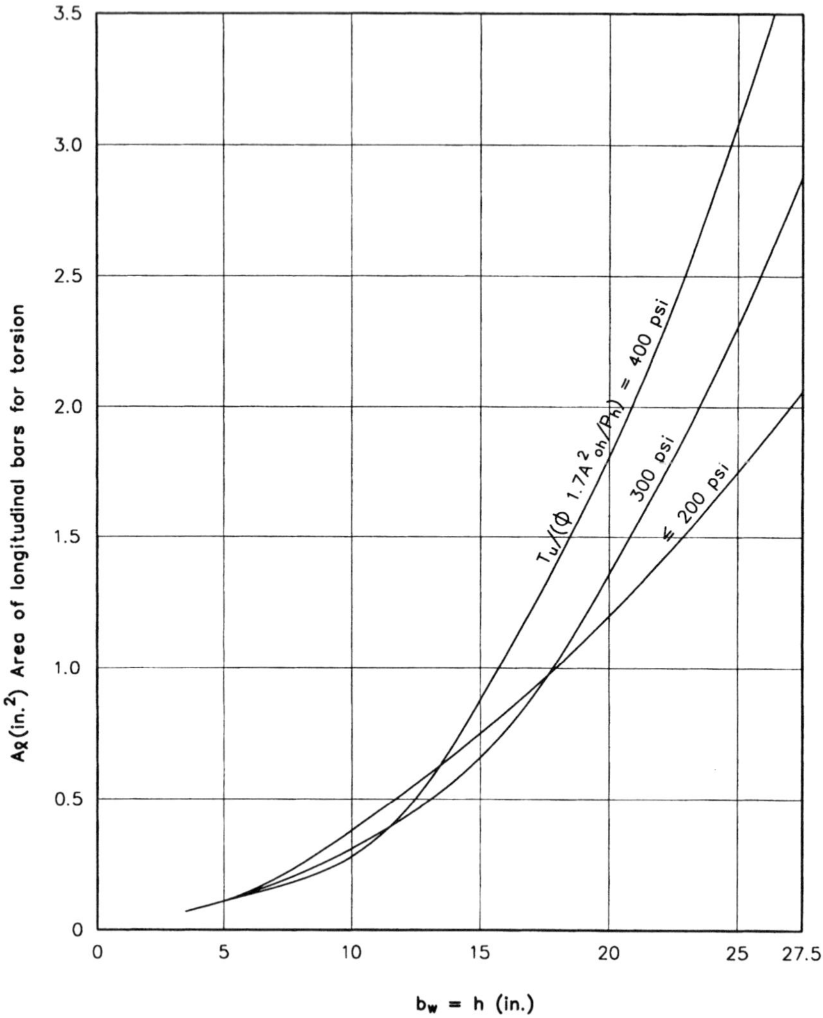

Figure 9-27 Longitudinal torsion reinforcement for square sections.

MINIMUM REINFORCEMENT FOR FLEXURAL MEMBERS

The Code requires minimum reinforcement for flexural members (Section 10.5) for every section of a flexural member where tensile reinforcement is required by analysis. The minimum reinforcement must be provided in both positive and negative moment regions. An alternative Code provision (Section 10.5.3) would require at least one-third more reinforcement than required by analysis.

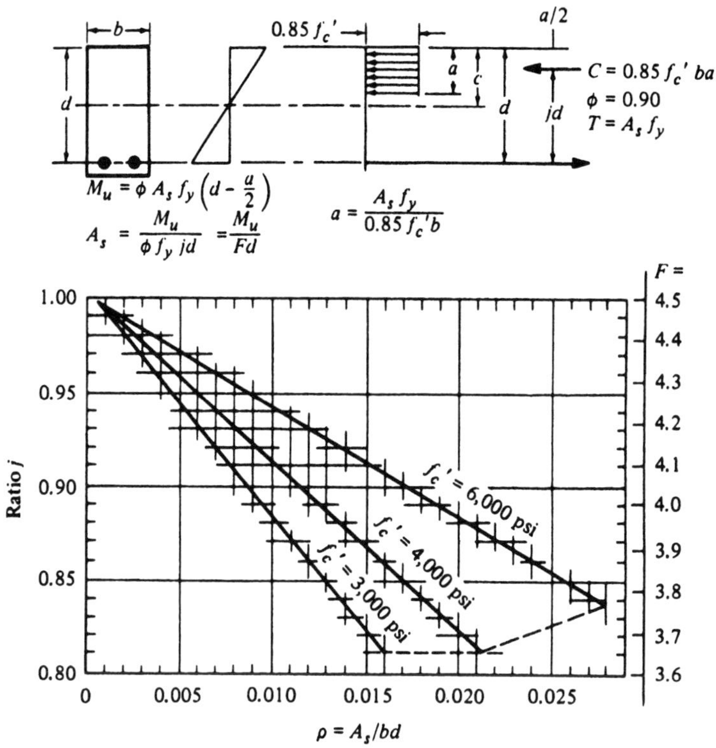

f_y = 60 ksi

f_c' =	a =	j =
3	23.53 ρd	$(1-11.76\,\rho)$
4	17.65 ρd	$(1-8.82\,\rho)$
6	11.76 ρd	$(1-5.88\,\rho)$

$A_s = \frac{M_u}{Fd}$ $\quad F = \frac{\phi f_y j}{12}$

For f_y = 60, M_u in ft-kips

$F = 4.5j$

$$j = (1 - \tfrac{1}{2}\tfrac{a}{d});\; y = \left(1 - \frac{1}{2}\frac{A_s f_y}{0.85 f_c' bd}\right) = \left(1 - \frac{\rho f_y}{1.7 f_c'}\right)$$

Figure 9-28(a) Rectangular section—flexural constants.

The purpose for a minimum ρ is quite simply to avoid a sudden (brittle) failure under any overload that cracks the concrete by providing sufficient reinforcement to ensure a reserve strength after cracking (Commentary Section R10.5.1). The formula for minimum flexural reinforcement $A_s = 3\sqrt{f_c'}\, bd/f_y$ provides a greater amount of reinforcement for higher strength concrete. For a very low strength concrete, the minimum area of reinforcement cannot be less that than $200\, b_w d/f_y$.

For a rectangular section, the average cracking moment calculated according to Code Section 9.5.2.3 is, $M_{cr} = 7.5\sqrt{f_c'}\,(bh^2/6) = 1.25\sqrt{f_c'}\, bh^2$. Moment capacity provided by the minimum area of reinforcement is $\phi M_n = \phi A_s f_y jd =$

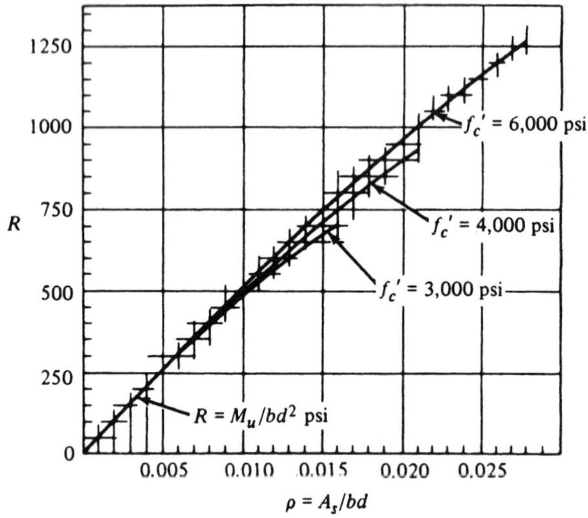

$$M_u = \frac{\phi A_s f_y (d - a/2)}{12{,}000} \text{ ft-kips} \qquad M_u = Rbd^2$$

$$\frac{M_u}{bd^2} = \frac{\phi f_y \rho j}{12{,}000} \qquad \text{For } f_y = 60{,}000 \text{ psi}, \quad R = 4.5 \rho j$$

Figure 9-28(b) *(continued)*

$3\phi\sqrt{f_c'}\,jbd^2$. For $d = 0.9h$ and assuming $j \geqslant 0.96$, the ratio of the moment capacity of the reinforced section to the cracking moment of the unreinforced section is $3\phi\sqrt{f_c'}\,jbd^2/(1.25\sqrt{f_c'}\,bh^2 = 3(0.9)(0.96)(0.9h)^2/(1.25h^2) = 1.68$. The specified minimum reinforcement results in a not unreasonable "safety factor", SF = 1.68 in both the positive and negative moment regions of the members.

Obviously, the alternate minimum provision for minimum reinforcement in beams (Section 10.5.3) is more rational which is reflected in a lower safety factor, $SF = 1.33$. In this case, this more rational approach (adopted in 1963) involves no more real design time since it is necessary in design of any flexural member to calculate required A_s.

The provisions for minimum flexural reinforcement in "structural slabs" (Section 10.5.4) are, however, really not rational. The basic purpose is unchanged—to avoid brittle failure. The values prescribed for convenience were simply picked up by reference to "temperature and shrinkage" minimum ratios (Section 7.12). These values were revised from 0.0025*bd* and 0.002*bd* for Grades 40 and 60 to 0.002*bh* and 0.0018*bh* in 1963, which is proper for shrinkage and temperature (functions of total thickness, *h*) but vary wildly as flexural minima (functions of *d*). The Commentary explains the reduced need for proper safety factors in structural slabs on the basis that slabs permit lateral distribution of (concentrated?) overloads (Commentary R10.5.4). Since "structural slabs of uniform thickness" include two-way flat plates where brittle failure is likely to be quite serious, the authors would recommend a more rational approach. It would be prudent to provide at least a 1.33 reserve reinforcement area (Section 10.5.3); or a reinforced design say at least 1.68 times the cracking strength; whichever is larger. Where the suspended one- or two-way slabs are significantly restrained (Section 7.12.1.2) as in multistory construction, by large columns or shear walls, residual shrinkage stresses are likely to reduce computed cracking load and the second criterion can become more important.

A second difficulty arising from Code requirements for minimum reinforcement in flexural slabs is purely practical. The user finds maximum spacing requirements (Section 7.12.2.2) based on shrinkage and temperature (essentially uniform stress) rather than flexure. For shrinkage and temperature, the reinforcement is properly arranged with its centroid at mid-depth and the minimum spacing can properly be measured on the diagonal as now permitted by Section 7.12.2.2. For flexure, minimum reinforcement is located almost always in a plane as far from mid-depth as the minimum cover requirements of the Code permit (Section 7.7). For flexure, the maximum spacing should be measured in the plane of the reinforcement, top or bottom separately. See Figure 9-29.

SHRINKAGE AND TEMPERATURE (TENSION)

Grade 60
$T_s = (0.0018bh)\,(60{,}000) = 108bh$

Grade 40
$T_s = (0.002bh)\,(40{,}000) = 80bh$

Equivalent stresses in concrete
Grade 60: 108 psi
Grade 40: 80 psi

These low values ensure that shrinkage stress over-all (carried *across* cracks from joints, restrained, or free ends) is never higher than 108 psi or 80 psi, respectively.

The initial crack pattern is set at an early age when f'_c is low enough so that $f_{ct} = 108$ or 80 psi. Further cracking will be avoided as shrinkage reinforcement yields at cracks and delivers the low stress (108 or 80) to restraining elements or opening joints. At these low strengths, bond is an uncertain value, and the concrete will crush at deformations near a crack to allow a finite length for yielding to a total strain equal to the crack width.

Reinforcement for shrinkage and temperature tension force, T_s, must be provided in a one-way slab in a direction normal to the flexural reinforcement as shown in Fig. 9-30.

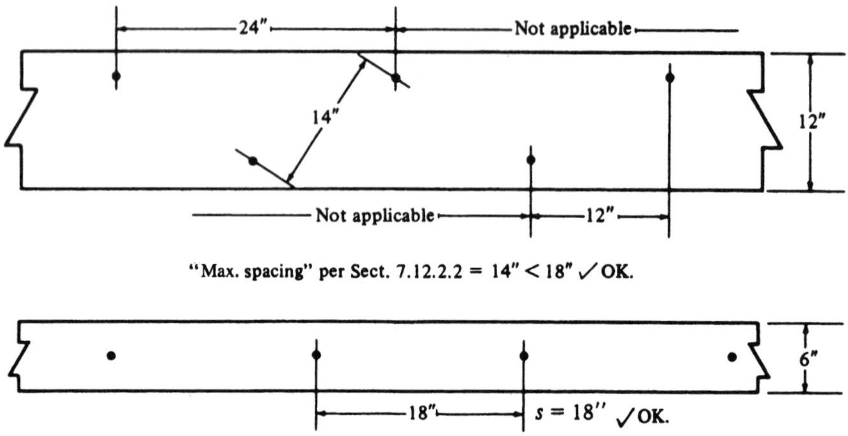

(a) **Shrinkage & Temperature Max. Spacings per Section 7.12.2.2**

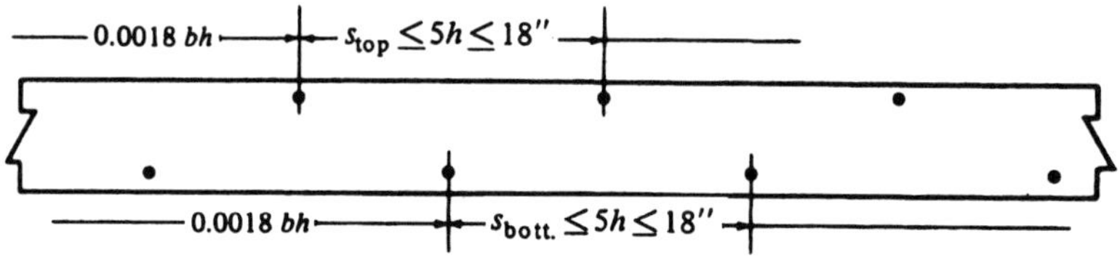

(b) **Suggested maximum spacing of flexural steel**

Figure 9-29 Maximum bar spacings.

The design rationale is that the initial crack pattern is at a sufficiently close spacing for the growth in concrete tensile strength to resist later shrinkage.

If the designer has provided "control joints" for stress relief at the proper interval which is dependent upon curing conditions (luck?) the initial crack pattern will coincide with the control joints, and no further (uncontrolled) cracking will develop. Since control over curing and indirectly rate of shrinkage is less than complete, it is prudent to regard the minimum percentages prescribed in the Code as truly minimum—for mild exposure conditions and closely spaced construction or control joints, say about 20 ft. Use larger amounts of shrinkage reinforcement for larger joint spacings, severe exposures, or full

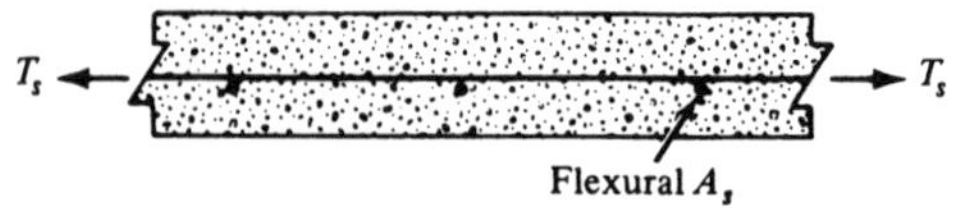

Figure 9-30 Temperature-shrinkage reinforcement in one-way slabs.

restraint, up to about 0.006*bh* Grade 60 as for continuously reinforced concrete pavement with no joints.

$$T_c = f_{ct}bh = 6.7\sqrt{f_c'}\,bh$$

$$\text{Stress in concrete: } f_{ct} = 6.7\sqrt{f_c'}$$

$$\text{For } f_c' = 3{,}000 \text{ psi: } f_{ct} = 6.7\sqrt{3{,}000} = 367 \text{ psi}$$

$$\text{Required Grade 60 reinforcement: } pbh = 367bh/60{,}000 = 0.006bh$$

10 Columns

1995 ACI BUILDING CODE

The most notable change in column design provisions of the ACI 318-95 is a revision to slenderness effects to better recognize the use of second-order analyses and to improve the arrangement of the provisions dealing with nonsway (braced) and sway frames. Separate moment magnification factors and separate moment magnification equations are given for nonsway frames and for sway frames.

DEFINITIONS: COLUMN, WALL, AND PEDESTAL

One of the ambiguous areas in the 1995 Code is the distinction between a load-bearing wall (see Chapter 11) and a column. The legal definitions emphasize that a column is "used *primarily* to support axial compressive loads" and that a wall is "used to enclose or separate spaces" (Section 2.1). Columns and pedestals are distinguished by the ratio of their height, ℓ_c, to their least dimension, c_1 or c_2; for columns, which must be reinforced (Section 10.9), the minimum ratio is 3; for pedestals, which may be plain concrete, the maximum ratio is 3 (Section 2.1). When a pedestal of any height is loaded beyond the capacity of plain concrete, it must be reinforced *and* designed as a column (Section 22.4.3), and so the overlap of these definitions at the height-to-least dimension ratio equals 3 will seldom cause confusion.

MINIMUM SIZES

There are no limits on the ratio of column cross-section dimensions, c_1/c_2; minimum column dimensions, c_1, c_2; or minimum column cross-sectional areas, A_g; specifically prescribed in the 1995 Code. Indirect limits, however, are imposed by minimum eccentricity $e = (0.6 + 0.03h)$ in. (Section 10.12.3.2) and the moment amplification prescribed for slender columns (Sections 10.10 through 10.12) which limit the minimum size of practicable columns.

DESIGN-CODE REQUIREMENTS

Reinforced concrete column design can be performed manually since the prescribed design assumptions result in simple geometric relationships for strain compatibility, stress in concrete, and stress in reinforcement, but practical economics make the cost of design prohibitive for large numbers of columns as in routine work. Since the basic assumptions for design (Section 10.2) are unchanged from previous codes, the designer can still use design aids based on the minimum eccentricity requirements of the 1963 and 1971 codes, at least for "short" column capacity. These design aids will be applicable through the full range of the load-moment interaction curve.

DESIGN AIDS

Load capacities for short columns with specific sizes, shapes, reinforcement, concrete strengths, and eccentricities are tabulated in the CRSI *Design Handbook**. Column interaction curves showing the full range of load and moment capacities for a variety of concrete and reinforcing steel strength combinations are available in the ACI Design Handbook. Computer programs for column design are widely available.

SLENDERNESS EFFECT

Design aids have been developed for slender column design to conform to the requirements (Sections 10.10 and 10.11) in this chapter. The slenderness effect may be neglected for columns braced against sidesway (Section 10.12.2) when $k\ell_u/r < (34 - 12\,M_1/M_2)$. For columns not braced against sidesway slenderness may be neglected when $k\ell_u/r < 22$ (Section 10.13.2) or when $Q \leqslant 0.05$ (Section 10.11.4.2). Where columns are braced, the effective length ratio, k, is 1.0 (Section 10.12.1); for columns not braced against sidewsay, $k \geqslant 1.0$ (Section 10.13.1). Other terms are defined as follows: ℓ_u = unsupported length; r = radius of gyration for cross sections ($r = 0.30h$ for rectangular shapes and $r = 0.25h$ for round columns); M_1 and M_2 are the end moments of the column, where $M_1 < M_2$. M_2 is always positive, and M_1 is positive for single curvature, negative for double curvature (Section 10.0).

ALTERNATE DESIGN METHOD

For design by the alternate method (Section 8.1.2) using unity load and strength-reduction (ϕ) factors, column design capacity is taken as simply 40 percent of that computed by the strength design method. In Equations (10-10) and (10-19), the term P_u must be replaced by 2.5 times the design axial load (Section A.6.2) and the stiffness-reduction factor, $\phi_K = 0.75$, must be replaced with 1.0.

* CRSI *Design Handbook*, 1996.

UNIFIED DESIGN PROVISIONS

Design is also permitted by using the Unified Design Provisions (Section 8.1.3). Transition of the ϕ-factor for the Unified Design Provisions occurs in the range of tensile strain ϵ_t between 0.002 and 0.005.

DESIGN ASSUMPTIONS FOR COLUMNS (SECTION 10.2)

1. Strain compatibility and equilibrium of forces must be satisfied.
2. Strains in the reinforcement and the concrete are proportional to the distance from the neutral axis.
3. The maximum compressive strain in the concrete is 0.003.
4. The stress in the reinforcing steel is equal to the strain times E_s which must be equal to or less than f_y.
5. The tensile strength of concrete equals zero.
6. The concrete compressive stress block may be taken as any shape that can be justified by tests (i.e., give the same answer as assumption 7.)
7. A rectangular stress block with concrete stress equal to 0.85 f_c' and extending from the extreme fiber in compression to a straight line parallel to the neutral axis, located at a distance of $a = \beta_1 c$, may be used; $c =$ distance from the outer fiber to the neutral axis; $\beta_1 = 0.85$ when $f_c' \leqslant 4{,}000$ psi and 0.05 less for each 1,000 psi over 4,000 psi but not less than 0.65 (see Fig. 2-7).

DESIGN REQUIREMENTS FOR COLUMNS (SECTION 10.3)

The basic minimum requirements for the design of short columns are

1. When the design axial load strength, ϕP_n, is less than the smaller of $0.10 f_c' A_g$ or ϕP_b, the tensile reinforcement ratio, $\rho = A_s / bd$, must not exceed 0.75 of the ratio producing balanced strain conditions for a singly-reinforced section under pure flexure plus balancing reinforcement for the force in compressive reinforcement, if any, (Section 10.3.3).
2. Balanced conditions exist when the tensile reinforcement strain is at its yield strain and the assumed ultimate strain of concrete in compression equals 0.003 (Section 10.3.2)
3. that maximum factored load, $P_u \leqslant 0.85 \phi P_0$ for spirally-reinforced columns or $P_u \leqslant 0.80 \phi P_0$ for tied columns (Section 10.3.5), where

$$P_0 = 0.85 f_c' (A_g - A_{st}) + f_y A_{st}.$$

APPLICATIONS

Consider designs of short tied columns with $e = 0.10h$. Use longhand application of the design assumptions and design requirements listed. Compare wall-like columns and walls. Use $f'_c = 4{,}000$ psi; $f_y = 60{,}000$ psi; try minimum reinforcement ratios, $(\rho + \rho') = 0.005$, (Section 10.8.4) and maximum, $(\rho + \rho') = 0.08$ (Section 10.9.1).

Direct design using the basic assumptions is almost impossible by longhand methods of calculation for a given factored axial load, P_u, and factored moment, M_u, but an indirect design by computations of capacity for assumed *positions of the neutral axis* is simple. Computation of several control points for a load-moment interaction curve and plotting such a curve will permit the capacity for any given eccentricity to be read with an accuracy appropriate for design. The upper portion of the interaction curve, where compression controls the design, is very nearly a straight line. Straight line interpolation gives a very close approximation between axial maximum load capacity and an eccentricity, $e = 0.10h$. See the typical interaction curve for the full range of e (Fig. 10-16).

EXAMPLE 1: Consider a wall-like column with interior exposure, 10 in. × 60 in., with minimum reinforcement. Use 10-#5 vertical bars placed equally in the long faces. $A_g = 10 \times 60 = 600$ in.^{2}, $(\rho + \rho') = 3.10/600 \approx 0.005$. Since the same basic assumptions and requirements for design apply for design of wall-like columns and for design of walls as columns (Section 14.4), the design procedures will be identical, but the capacities computed will vary slightly depending upon the cover of the bars. (With the reinforcement ratio used in this example and the bar sizes indicated, the choice of terminology is the designer's option.)

For columns, the minimum cover required is $1\frac{1}{2}$ in. to the *ties* for cast-in-place construction (Section 7.7.1-c); the minimum size of ties is #3 (Section 7.10.5.1) (see Fig. 10-1).

For walls designed as columns, if ties are provided, minimum size #3, the minimum cover required is $\frac{3}{4}$ in. to the *ties* (Section 7.7.1-c) (see Fig. 10-2).

For walls designed as columns, lateral ties may be omitted (1) if the area of the vertical bars is $0.01A_g$ or less, or (2) where the vertical bars are not required as compression reinforcement (Section 14.3.6) (see Fig. 10-3). The first provision permits the use of vertical bars in one or two faces without ties for tension or compression stress, provided that the area is less than $0.01A_g$. It would permit the vertical reinforcement to be concentrated in the compression face of the wall in amounts between $0.0012A_g$ minimum (Section 14.3.2-a) and $0.01A_g$ maximum without ties (Section 14.3.6). If the wall were subjected to loadings causing tension, additional vertical bars as required can be provided in the tension face without ties. *Ties are required only where vertical bars in compression exceed in area $0.01A_g$.* The above interpretation is the most liberal and is recommended as such by the authors, although the Code provisions are sufficiently ambiguous to permit other interpretations.

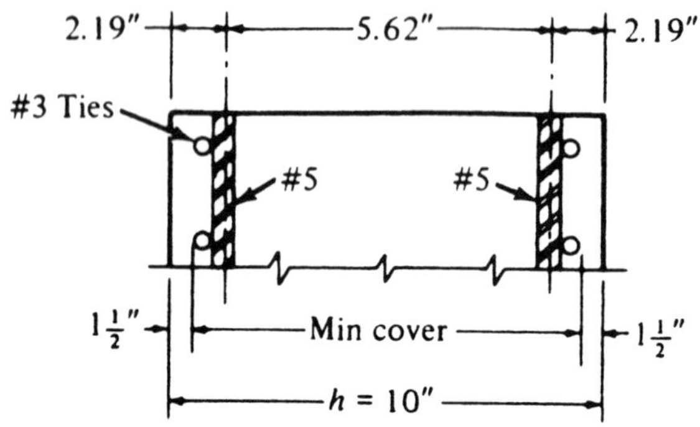

Figure 10-1 Column cover.

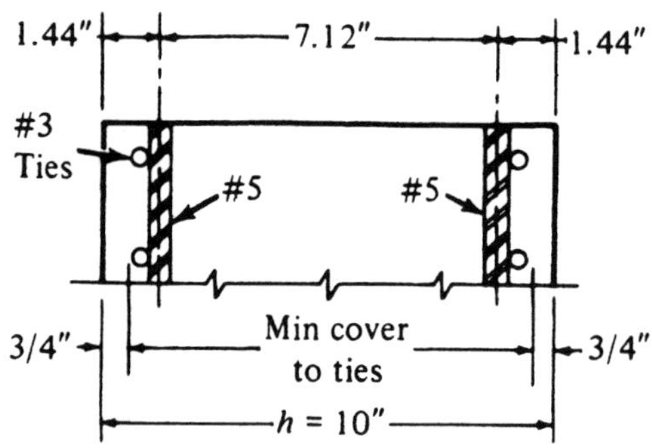

Figure 10-2 Cover for wall with ties.

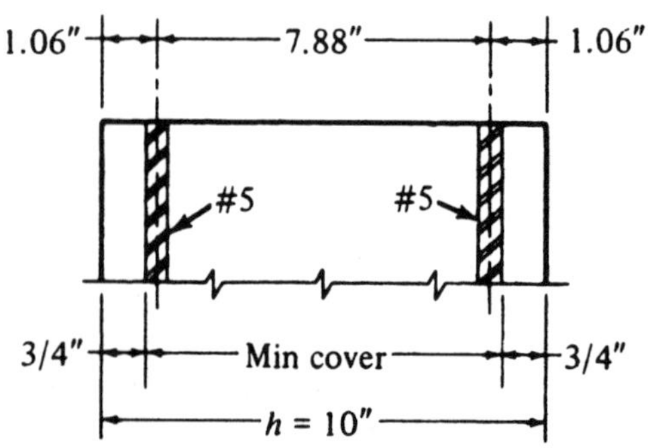

Figure 10-3 Doubly-reinforced wall without ties.

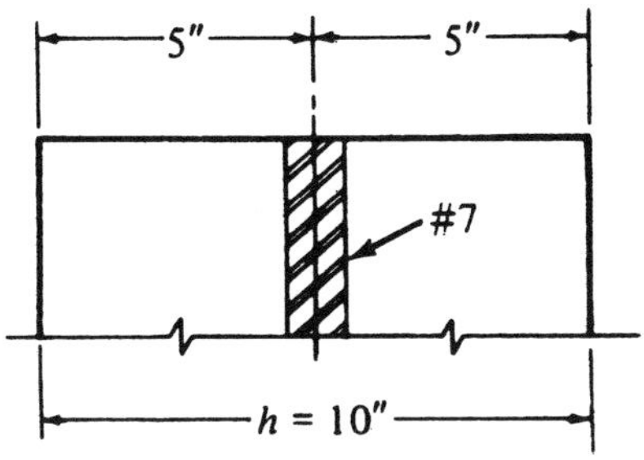

Figure 10-4 Singly-reinforced wall with ties.

The previous three arrangements are for a doubly-reinforced cross-section. Walls may also be singly-reinforced if not more than 10 in. thick (Section 14.3.4) and designed by the"wall formula"(Code Eq. 14-1) under the empirical method, provided the eccentricity, $e \leqq h/6$ (Section 14.5.1). For larger eccentricities, such walls must be designed as columns (Section 14.4). See Fig. 10-4 for an equivalent singly-reinforced simple arrangement of 5-#7 verticals without ties, $\rho = 0.005$.

For the wall-like column of Fig. 10-1, compute the maximum capacity at $e = 1$ in. by constructing the upper portion of its load-moment interaction curve and interpolating along the curve to the point where $e = 1$ in. (see Fig. 10-5). Take $f_c' = 4$ ksi.

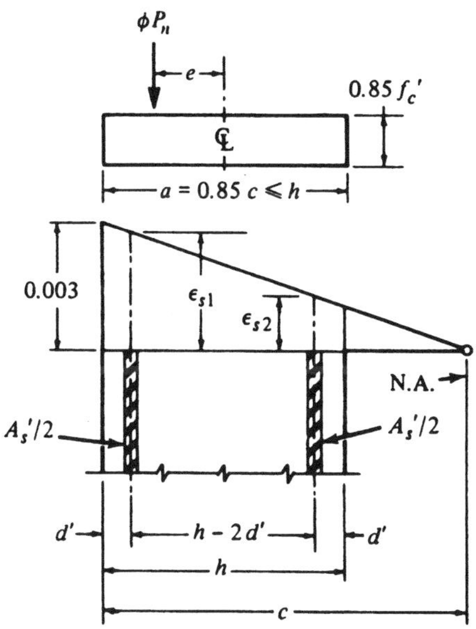

$a = 0.85\,c$

Max $\epsilon_c = 0.003$

$\epsilon_{sf} = 0.00207$ for $f_y = 60$

ϵ_{s1} = larger steel strain (+)

ϵ_{s2} = lesser steel strain; may be (−) for tension

b = width

$$\phi P_n = \phi \left[0.85 f_c'(ba - A_s') + \tfrac{1}{2} A_s' f_y \frac{(\epsilon_{s1} + \epsilon_{s2})}{\epsilon_{sf}} \right]$$

where $\epsilon_{s1} > \epsilon_{sf}$ is taken as ϵ_{sf}
and where $\epsilon_{s2} > \epsilon_{sf}$ is taken as ϵ_{sf}

$$e = \frac{\phi M_n}{\phi P_n}$$

Figure 10-5 Notation for stress and strain (example 1).

Case (a)

$$e = 0;\ c = \infty;\ a = 10\text{ in.};\ \varepsilon_c = \varepsilon_{s2} = \varepsilon_{s1} = 0.003;\quad 10\text{-\#5}, A_s = 3.10\text{ in.}^2$$

$$\phi P_0 = (0.70)[(0.85)(4)((10)(60) - 3.10) + (3.10)(60)]$$

$$= (0.70)(2029 + 186)$$

$$\phi P_0 = 1551\text{ kips}\quad \phi M_n = 0$$

Maximum axial allowable capacity = 0.8(1551) = 1241 kips

Case (b)
(See Fig. 10-6.) Take $c = 10/0.85 = 11.76$ in.; $a = 10$ in.

$$\varepsilon_{s1} > 0.00207, f_{y1} = 60\text{ ksi};\ \varepsilon_{s2} = \left[\frac{3.95}{11.76}\right](0.003) = 0.001008$$

$$\phi P_n = 0.70\left[2029 + \frac{3.10}{2}(60) + \frac{3.10}{2}\left(60 \times \frac{0.001008}{0.00207}\right)\right]$$

$$= (0.70)(2029 + 93 + 45.3)$$

$$\phi P_n = 1517\text{ kips}\quad \phi M_n = \phi P_n e = (0.70)(93 - 45.3)\tfrac{1}{2}(5.62) = 93.8\text{ in.-kips}$$

$$e = \phi M_n/\phi P_n = 0.0618\text{ in.}$$

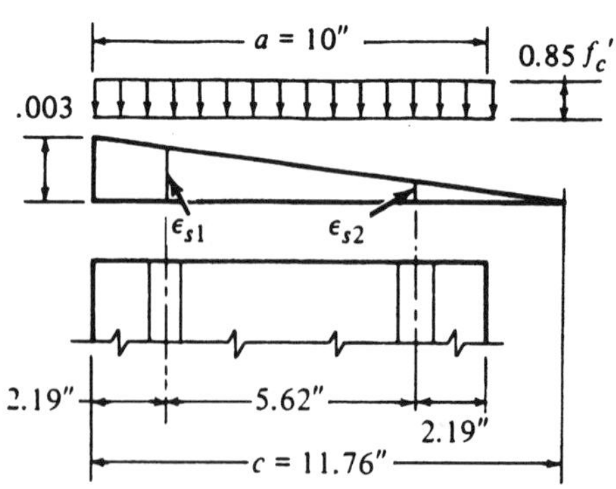

Figure 10-6 Case (*b*).

Case (c)
(See Fig. 10-7.) Take $c = 10$ in.; $a = 8.5$ in.

$$\varepsilon_{s1} = \frac{7.81}{10.00}(0.003) > 0.00207; f_{y1} = 60\text{ ksi}$$

$$\varepsilon_{s2} = \frac{2.19}{10.00}(0.003) = 0.000657;$$

$$\phi P_n = 0.70\left[(0.85)(4)((8.5)(60) - 3.10) + (1.55)(60) + (1.55)(60)\left(\frac{657}{2070}\right)\right]$$

$$= 0.70(1723 + 93 + 29.5) = 1206 + 65 + 20.7$$

$$\phi P_n = 1292 \text{ kips} \qquad \phi M_n = (1206)(5.00 - 4.25) + (65.0 - 20.7)(2.81)$$

$$\phi M_n = 1029 \text{ in.-kips}; \; e = 1029/1292 = 0.796 \text{ in.}$$

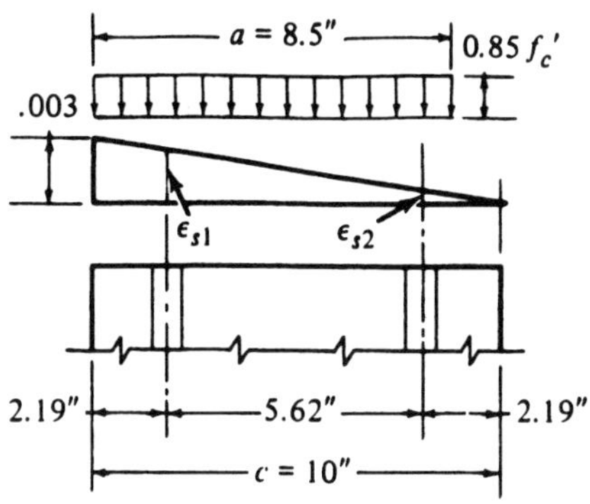

Figure 10-7 Case (*c*).

Case (d)
(See Fig. 10-8.) Take $c = 9.41$ in.; $a = 8.0$ in.

$$\varepsilon_{s1} = \frac{7.22}{9.41}(0.003) > 0.00207, f_{y1} = 60 \text{ ksi}$$

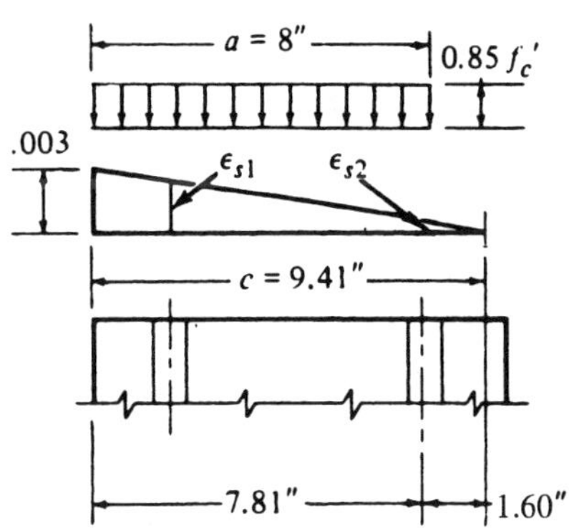

Figure 10-8 Case (*d*).

$$\varepsilon_{s2} = \frac{1.60}{9.41}(0.003) = 0.000510, f_{y2} = \frac{510}{2070}(60) = 14.8 \text{ ksi}$$

$$\phi P_n = (0.70)[(3.4)((8)(60) - 1.55) + 1.55(60 + 14.8)]$$

$$\phi P_n = (0.70)(1627 + 93 + 22.9) = 1139 + 65 + 16 = 1220 \text{ kips}$$

$$\phi M_n = (1139)(5 - 4) + (65 - 16)(2.81) = 1277 \text{ in.-kips}$$

$$e = \phi M_n/\phi P_n = 1277/1220 = 1.047 \text{ in.}$$

Case (e)
(See Fig. 10-9.) Take $c = 7.81$ in.; $a = 6.64$ in.

$$\varepsilon_{s1} = \frac{5.62}{7.81}(0.003) > 0.00207, f_{y1} = 60 \text{ ksi}$$

$$\varepsilon_{s2} = 0, \quad f_{y2} = 0$$

$$\phi P_n = (0.70)[(3.4)((6.64)(60) - 1.55) + (1.55)(60)]$$

$$= (0.70)(1349 + 93) = 944 + 65.1$$

$$\phi P_n = 1009 \text{ kips}$$

$$\phi M_n = (944)\left(5.00 - \frac{6.64}{2}\right) + (65.1)(2.81) = 1586 + 183$$

$$\phi M_n = 1769 \text{ in.-kips} \quad e = 1769/1009 = 1.75 \text{ in.}$$

If the five points for (ϕP_n, ϕM_n) computed in (a), (b), (c), (d), and (e) are plotted as the upper portion of a load-moment interaction curve, the strength at $e = 1.0$ in. can be determined by simple straight line interpolation since this portion of the curve is very nearly a straight line (see Fig. 10-10). Interpolating

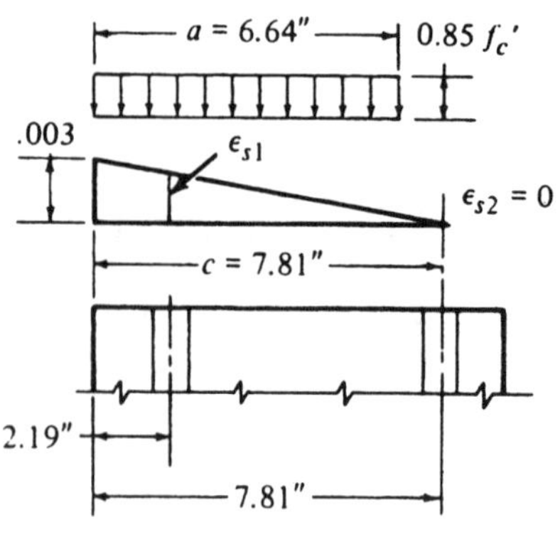

Figure 10-9 Case (*e*).

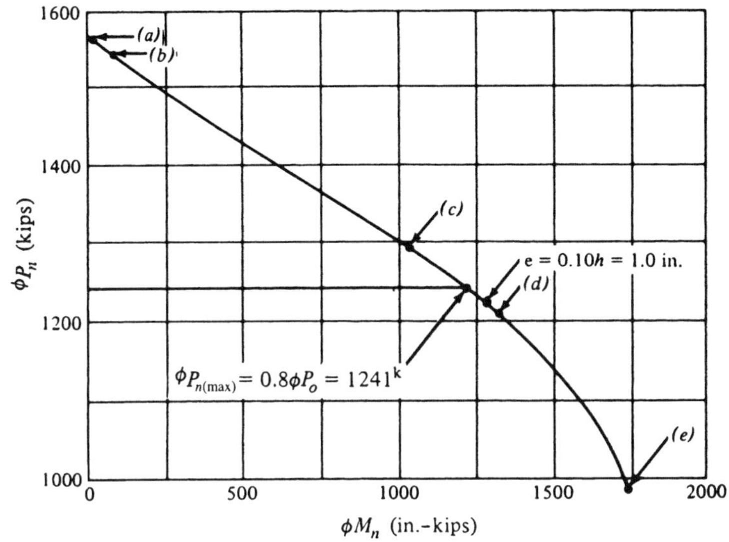

Figure 10-10 Capacity at $e = 0.10h$ about minor axis.

between (c), $e = 0.796$ in., and (d), $e = 1.047$ in., for $e = 1.0$ in., $\phi P_n = 1233$ kips and $\phi M_n = 1233$ in.-kips.

The requirement for a minimum eccentricity had the effect of cutting off the top of the interaction curve since all usable combinations of load and moment lie below this line. Note that the column of Example 1 has been investigated for short column capacity with bending about the minor principal axis, that is, parallel to the long dimension of the cross section. The maximum allowable load, $\phi P_{n(\max)} = 0.8\ \phi P_0 = 1241$ kips. The current Code provisions for maximum axial load have the same effects as former Code provisions for minimum eccentricity. Interpolate for moment capacity at $\phi P_{n(\max)} = 1241$ k. $\phi M_n = 1206$ in.-kips, and $e = 0.97$ in. ≈ 1.0 in. O.K.

Capacity About Major Axis

To complete the investigation of the maximum short-column capacity for the column of Example 1, it is necessary to determine the capacity about the major axis. For the wall-like column of Example 1 with bending about the major axis, usual practice in detailing a 10 in. × 60 in. column for 10-#5 bars in two faces would be to space them equally, beginning with a half-space, unless the designer indicated some other specific arrangement desired (see Fig. 10-11). The capacity at $e = 0$ for bending in the longer direction is the same as case (a) previously calculated, $\phi P_n = 1551$ kips; $\phi M_n = 0$; $e = 0$. Also the maximum allowable load, $\phi P_{n(\max)} = 1241$ kips is unchanged.

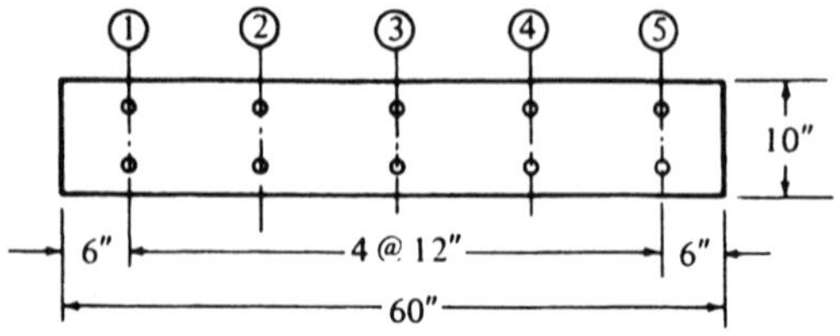

Figure 10-11 Bar layout plan (Example 1).

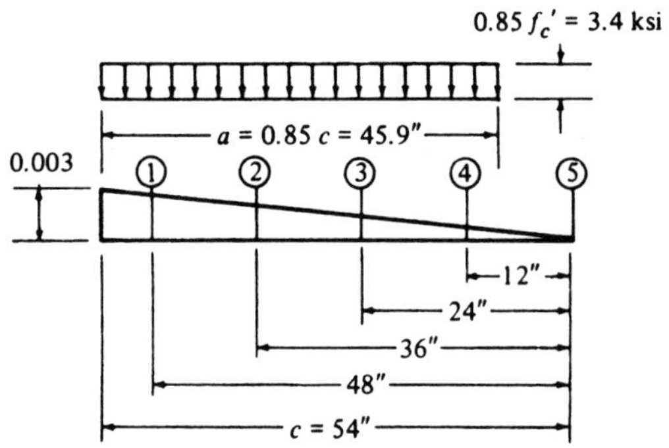

Figure 10-12 Case (*f*).

Case (f)
(See Fig. 10-12.) Take $c = 54$ in. so that $f_{s5} = 0$.

$$f_{s1} = (48)\frac{(0.003)}{(54)}\frac{(60)}{(0.00207)} = (48)(1.61) > 60;\ \text{use } f_{s1} = 60 \text{ ksi}$$

$$f_{s2} = (36)(1.61) = 58 \text{ ksi}$$

$$f_{s3} = (24)(1.61) = 38.6 \text{ ksi}$$

$$f_{s4} = (12)(1.61) = 19.3 \text{ ksi}$$

$$f_{s5} = (0)(1.61) = 0 \text{ ksi}$$

$$\begin{aligned}\phi P_n &= (0.70)[(0.85)(4)((45.9)(10) - (.31)(8)) \\ &\qquad + (0.62)(60 + 58 + 38.6 + 19.3 + 0)] \\ &= (0.70)(1552 + 37.2 + 35.9 + 23.9 + 12.0 + 0) \\ &= (1086 + 26.0 + 25.1 + 16.7 + 8.4) \\ &= 1162 \text{ kips}\end{aligned}$$

$$\begin{aligned}\phi M_n &= (1086)(30.00 - 22.95) + (26.0)(24) + (25.1)(12) - (8.4)(12) \\ &= 8481 \text{ in.-kips}\end{aligned}$$

$$e = 8481/1162 = 7.30 \text{ in.}$$

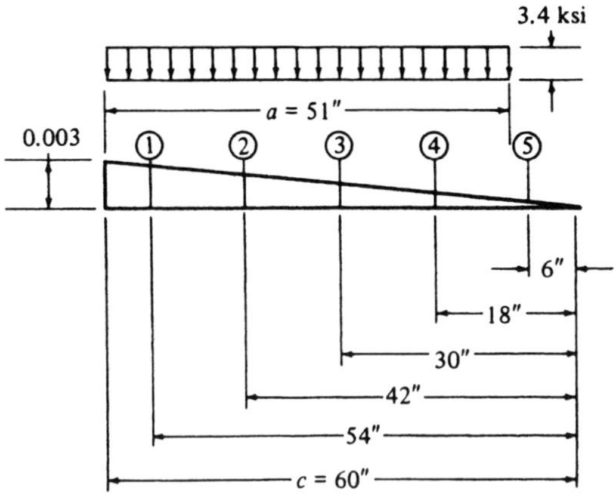

Figure 10-13 Case (g).

Case (g)
(See Fig. 10-13.) Take $c = 60$ in.; $a = 51$ in.

$$f_{s1} = (54)\frac{(0.003)}{(60)}\frac{(60)}{(0.00207)} = 54(1.45) > 60; \text{ use } f_{s1} = 60 \text{ ksi}$$

$$f_{s2} = (42)(1.45) > 60; \text{ use } f_{s2} = 60 \text{ ksi}$$

$$f_{s3} = (30)(1.45) = 43.5 \text{ ksi}$$

$$f_{s4} = (18)(1.45) = 26.1 \text{ ksi}$$

$$f_{s5} = (6)(1.45) = 8.7 \text{ ksi}$$

$$\phi P_n = (0.70)[(3.4)((51(10) - 3.10) + (0.62)(60 + 60 + 43.5 + 26.1 + 8.7)]$$

$$\phi P_n = (1206 + 26.0 + 26 + 18.9 + 11.3 + 3.8) = 1292 \text{ kips}$$

$$\phi M_n = (1206)(30 - 25.5) + (26)(24) + (26)(12) + (18.9)(0) - (11.3)(12) - (3.8)(24) = 6248 \text{ in.-kips}$$

$$e = 6248/1292 = 4.83 \text{ in.}$$

Plotting the values for capacity from cases (a), (f), and (g) develops the straight line top portion of the interaction curve for short-column capacity bending about the major axis. Interpolation shows approximately $\phi P_n = 1241$ kips and $\phi M_n = 7214$ in.-kips at $e = 5.74$ in. (see Fig. 10-14).

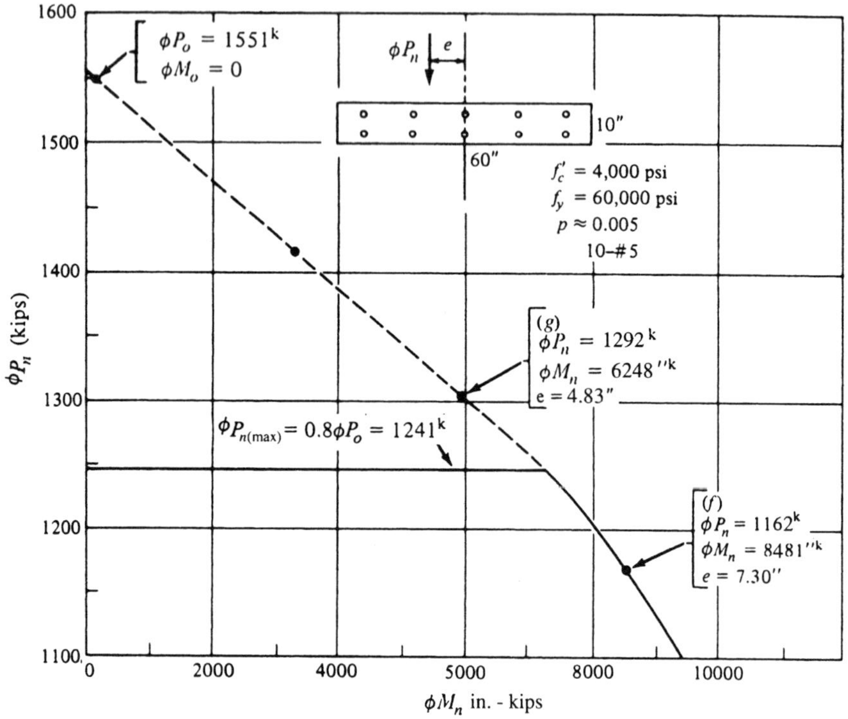

Figure 10-14 Capacity at $e = 0.10h$ about major axis (example 1).

Short columns have the maximum axial load controlled by $\phi P_{n(\max)}$ calculated by Section 10.3.5. This code section has the same effect as the minimum eccentricity provisions of previous editions of the Code. A 10 × 60 long column will have its axial load capacity drastically reduced in the 10 in. dimension due to the effects of slenderness (Sections 10.10 and 10.11) if it is not braced against sidesway. See the discussion of long columns following in this chapter.

SHORT CUT TO DIRECT DESIGN OF COLUMNS

The capacity of columns at the minimum eccentricity (implied by the maximum axial load provisions of Section 10.3.5.1 of the Code) is an important practical consideration because most routine interior building columns are in this class. Where walls or other bracing provide lateral resistance to sidesway of the structure, the design of most if not all columns will be controlled by this Code provision.

The major variables are $\phi = 0.70$ for walls and tied columns, f'_c, f_y, and the total reinforcement ratio, $\rho = A_s/A_g$. Based on these variables, a simple design aid for columns can be constructed, (see Fig. 10-15). The chart will enable the designer to make a fast selection of a preliminary size column with a low rein-

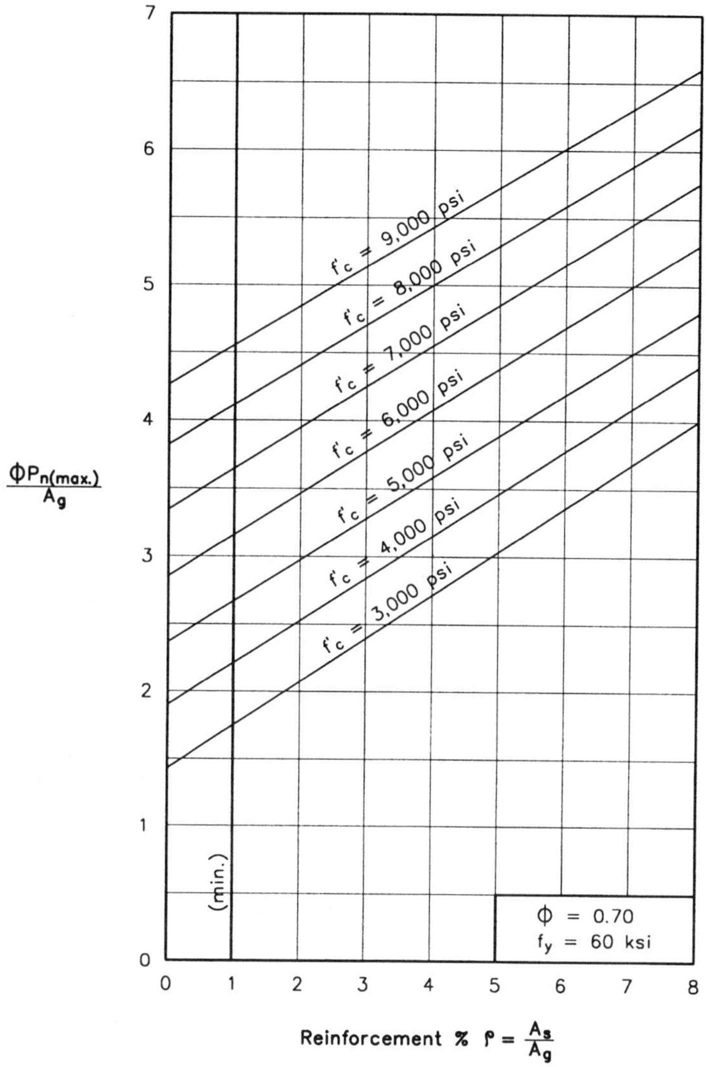

Figure 10-15 Preliminary design—tied columns for maximum design axial load strength, $\phi P_{n(\max)}$.

forcement ratio that can be used for elastic analyses and easily adjusted if the factored moment exceeds the design moment strength at maximum design axial load strength by adding reinforcement, usually without changing the column size, which would require a second trial analysis. The results are usually accurate within acceptable limits for final design where the actual eccentricity computed is small.

APPLICATION OF PRELIMINARY DESIGN CHART TO WALLS AND LIGHTLY REINFORCED COLUMNS

The chart in Fig. 10-15 covers the range of reinforcement ratios from $\rho = 0.01$ to $\rho = 0.08$ because these are the limits established in the Code for "compression members" (Section 10.9.1). The designer is advised to interpret this requirement as intended for principal columns loaded to full capacity. The ratio $\rho = 0.005$ becomes particularly significant since one of the exceptions provided to the requirement for a minimum $\rho = 0.01$ is that " ... a reduced effective area A_g not less than one-half of the total area may be used for determining minimum reinforcement and design strength" (Section 10.8.4). This provision permits retaining the same size column from footing to roof for most buildings to save the cost due to formwork changing and at the same time reducing the minimum area of vertical bars to a more economical minimum of 0.005 times A_g in the upper stories. For bearing walls designed as columns, the minimum ratio of vertical reinforcement to gross area is 0.0015 for bars larger than #5 (Section 14.3.2). Since walls may be designed without ties (Section 14.3.6) for ratios of vertical reinforcement, $\rho \leqslant 0.01$, by the same capacity formulas as columns, it seems reasonable to interpret the intent of the Code that the minimum area of vertical bars in tied columns, $\rho = 0.005$, also permits full utilization of reinforcement and concrete capacities.

LOAD-MOMENT INTERACTION CURVES

From the preceding explanations and examples, it is evident that longhand calculations to apply the basic assumptions and requirements for routine column design are not economically practicable. The designer should have access to design aids for short-column capacities. Offices with personal computers can choose from many readily available programs. ACI Design Handbooks present short-column design aids in the form of interaction curves; also some computer output is in this form. It behooves the designer to become familiar with this form of design data.

Key Points on Interaction Curves

See Fig. 10-16 for a typical load-moment interaction curve for a "short tied column" with the key points for plotting of same identified. Note that it was not necessary for Code purposes to provide standard nomenclature for all of these key points. The nomenclature in Fig. 10-16, followed elsewhere in this Guide, has been used informally by many writers and should either be familiar to the designer or self-explanatory. Briefly, the key points, in order from the top down, are identified and defined as follows:

Note: All points on the interaction curve of Fig. 10-16 are design capacities, including the ϕ factor $= 0.70$ for tied columns.

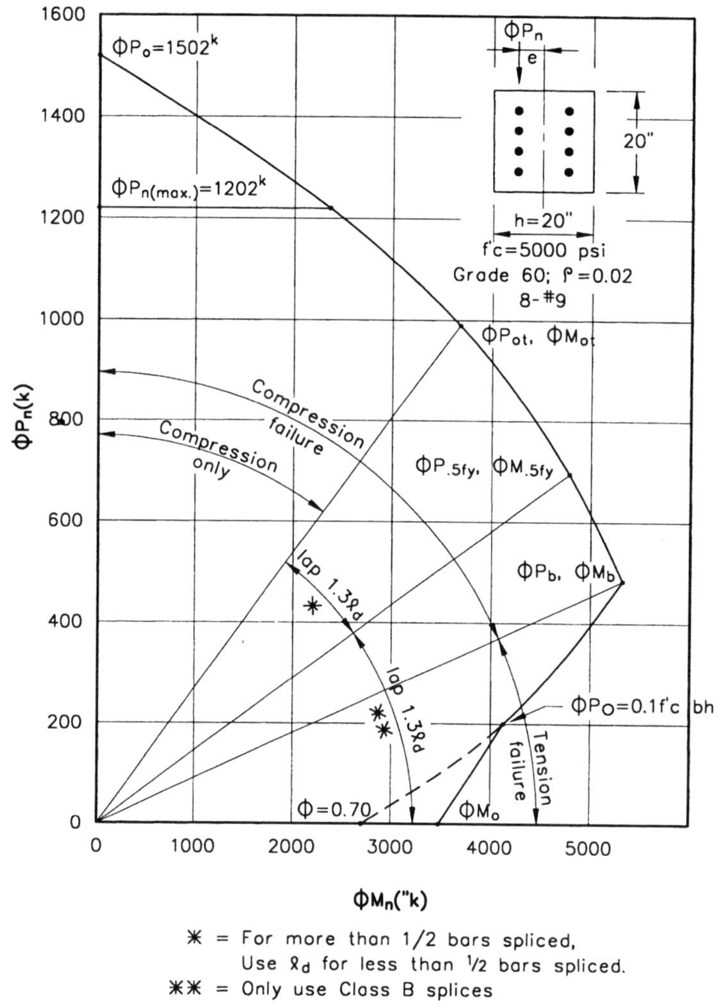

Figure 10-16 Typical interaction curve.

1. ϕP_0 – axial load capacity at $e = 0$ (not permitted for use in design; useful only for drawing the curve).
2. $\phi P_{n(\max)}$ = maximum axial load capacity, $\phi M_{n(\max)}$ is the associated moment capacity. $\phi P_{n(\max)} = 0.80\ \phi P_0$ for tied columns, and $\phi P_{n(\max)} = 0.85\ \phi P_0$ for spirally-reinforced columns (Sections 10.3.5.1 and 10.3.5.2).
3. ϕP_{0t}—axial load capacity at the point of zero tension in the reinforcement nearest to the "tension face." Compression lap splices used for lesser eccentricities will satisfy the Code requirements for minimum tensile capacity (Section 12.17.2.1). End-bearing splices with a minimum stagger

of 50 percent A_g in each face will satisfy the Code for somewhat greater eccentricities.*

4. $\phi P_{0.5 f_y}$—axial load capacity at the point where tensile stress equals $0.5 f_y$ in the reinforcement nearest to the "tension face." If less than one-half of the bars are spliced at any section, Class A tension lap splices may be used (Section 12.17.2.2).
5. ϕP_b—axial load capacity at the point of balanced conditions; the compression strain in the concrete reaches 0.003 and the stress in the tensile reinforcement reaches f_y simultaneously; the point of transition between a failure by compression in concrete or yielding of the reinforcement controlling. ϕM_b is the accompanying moment capacity.
6. ϕP_ϕ = the axial load capacity at the beginning point for the transition from a ϕ-factor of 0.70 for compression members (tied columns) to 0.90 for flexural members (beams). ϕM_ϕ is the accompanying moment capacity.
7. ϕM_0—moment capacity at the point of "pure flexure," where $\phi P_n = 0$; $e = \infty$, and $\phi = 0.90$.

Major and Minor Axes–Double Interaction Curves

Several features of the typical load-moment relationship may be observed in Fig. 10-16. The identifying sketch showing axis of bending is necessary for any reinforcement arrangement except the standard 4-bars with one in each corner and those with reinforcement symmetrical about both axes of bending, square or round columns with bars in a circular pattern, and multiples of four bars in all faces of square columns. Except for these doubly-symmetrical reinforcement arrangements in square columns, round columns, or special doubly-symmetrical column shapes, two load-moment curves are required for bending about the major and the minor axes (see Figs. 10-10 and 10-14). Note the effect of the transition in Fig. 10-16 from $\phi = 0.70$ for the tied column to $\phi = 0.90$ for pure flexure. This transition typically begins at $\phi P_\phi = 0.10 f'_c A_g$, but for unusual columns of low f'_c, high f_y, high reinforcement ratios, unsymmetrical layout ($A'_s < A_s$), and high ratios of cover to total depth, where $\phi P_b \leqq 0.10 f'_c A_g$, the transition begins at ϕP_b (Section 9.3.2.2).

END-BEARING SPLICES

More important features of the Code for practical everyday design are shown in Fig. 10-16. The point of zero tension in the vertical bars on the tension face is easily located by longhand calculations, and is identified in most handbooks in either load tables or interaction curves and in most computer design programs. Under the Code, columns for all factored load-moment combinations above this

* See Fig. 10-17.

point can be designed for end-bearing splices. Splices must be either staggered or, if not staggered, must have additional bars provided at the splice locations. The continuing bars in each face of the column must have a tensile capacity not less than 0.25 times the yield strength of all the bars in that face (Sections 12.16.1, 12.16.4 and 12.17.4). Since column splice design for multistory columns is a most important problem not yet susceptible to mechanical solutions and practical columns commonly lie above the point (ϕP_{ot}, ϕM_{ot}), this point becomes essential to economical design. Note double interaction curves such as these shown in Fig. 10-17 may be constructed for values of $\phi P_n < \phi P_{ot}$ showing reduced bending capacity based on the ability of only 50 percent of the bars to resist tension

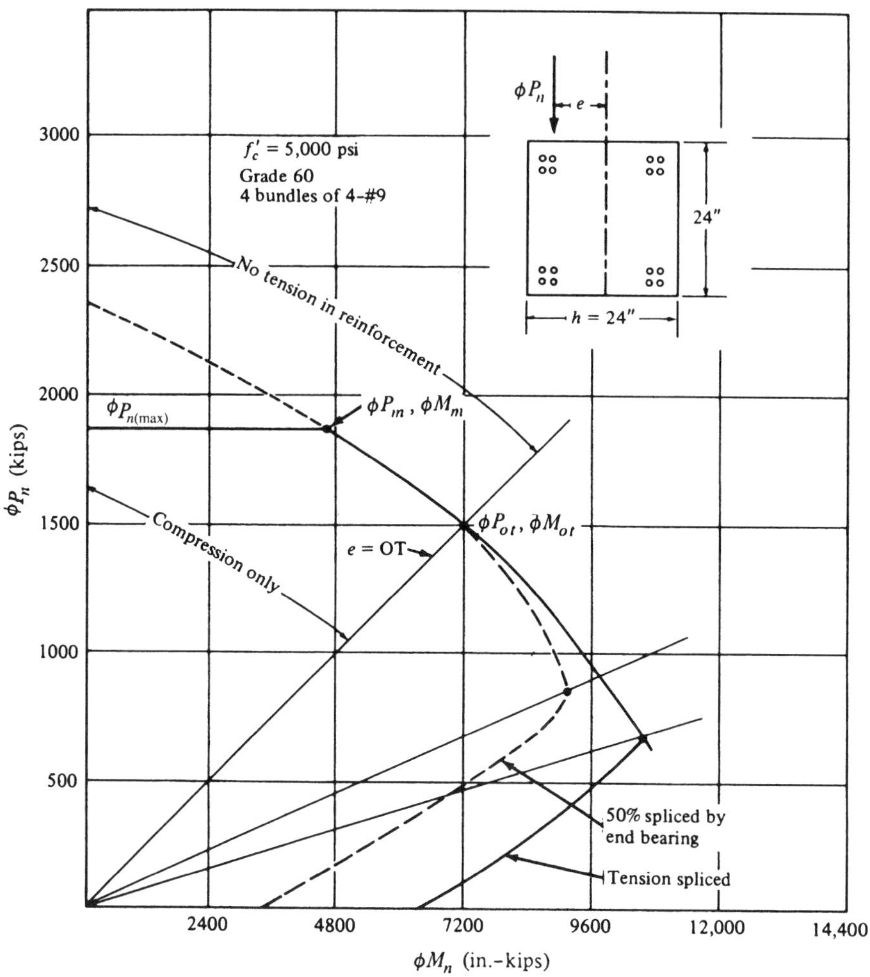

Figure 10-17 Interaction curve—end-bearing splices.

in a design where end-bearing splices are used for 50 percent of the bars (stagger 50 percent). The curve for 50 percent end-bearing splices (dashed line) must be used for factored axial loads less than ϕP_{ot}. End-bearing splices must be staggered at least $1.3\ell_d$ to allow a Class B tension lap splice between bars.

BUNDLED VERSUS SEPARATE BARS

For smaller bars, the most economical design is to use compression lap splices of 100 percent of A_s at the splice section, using bars one story in height. For the large bars, #14 and #18, for which lap splices are not allowed (Section 12.14.2.1), the end-bearing splice with 50 percent stagger using bars of two story height is the economical solution. Tension lap splices can become quite costly for larger size bars (see Chapter 13 for required lap lengths). In Fig. 10-17, note the reduction in capacity in the zone of high moments with 50 percent A_s spliced by end bearing; few practical columns lie in this zone. Note also, this column has greater efficiency to resist bending in either direction with four 4-bar bundles in the corners of the column than it would have for 16 separate bars in all four faces. The bundled arrangement requires one tie per spacing instead of three per set required for separate vertical bars (Section 7.10.5.3). See the following section in this chapter on column details.

COMPARISON OF TYPE OF COLUMN–SQUARE, ROUND, TIED, AND SPIRAL

Figure 10-18 contains a number of interaction curves for comparable column designs. Comparisons are given for:

1. Efficiency of locating 6-#18 vertical bars, $\rho = 0.06$ in two faces to resist bending in one direction. See Fig. 10.18(a) for bending capacity about the major and minor axis and for capacity of a round spiral column with the same area.
2. Efficiency of square tied columns versus round columns with ties for moment capacity wherever eccentricity exceeds about $M_{n(\max)}/P_{n(\max)} \approx 0.10h$ [see Fig. 10-18(b)].
3. The small gain in capacity for spirals versus circular ties in round columns [see Fig. 10-18(c)].

In cases (a) and (c), the concrete areas and vertical reinforcement are identical. The spiral column requires a 3/8 in. diameter spiral at 1-3/4 in. pitch (Sections 10.9.3 and 7.10.4), weighing 13.4 lb per ft, plus two spacers weighing 1.4 lb per ft, plus finishing turns, say total weight equals 15 lb per ft. The tied column requires one #4 tie per set at 20 in. spacings (Section 7.10.5) weighing 2.4 lb per ft. The unit in-place cost of reinforcing steel for light ties and spirals is two to three times that for the heavy vertical bars. The efficiency per pound in terms of added column capacity is far greater for vertical reinforcement, and thus more capacity at less cost for any loading can be provided by adding up to

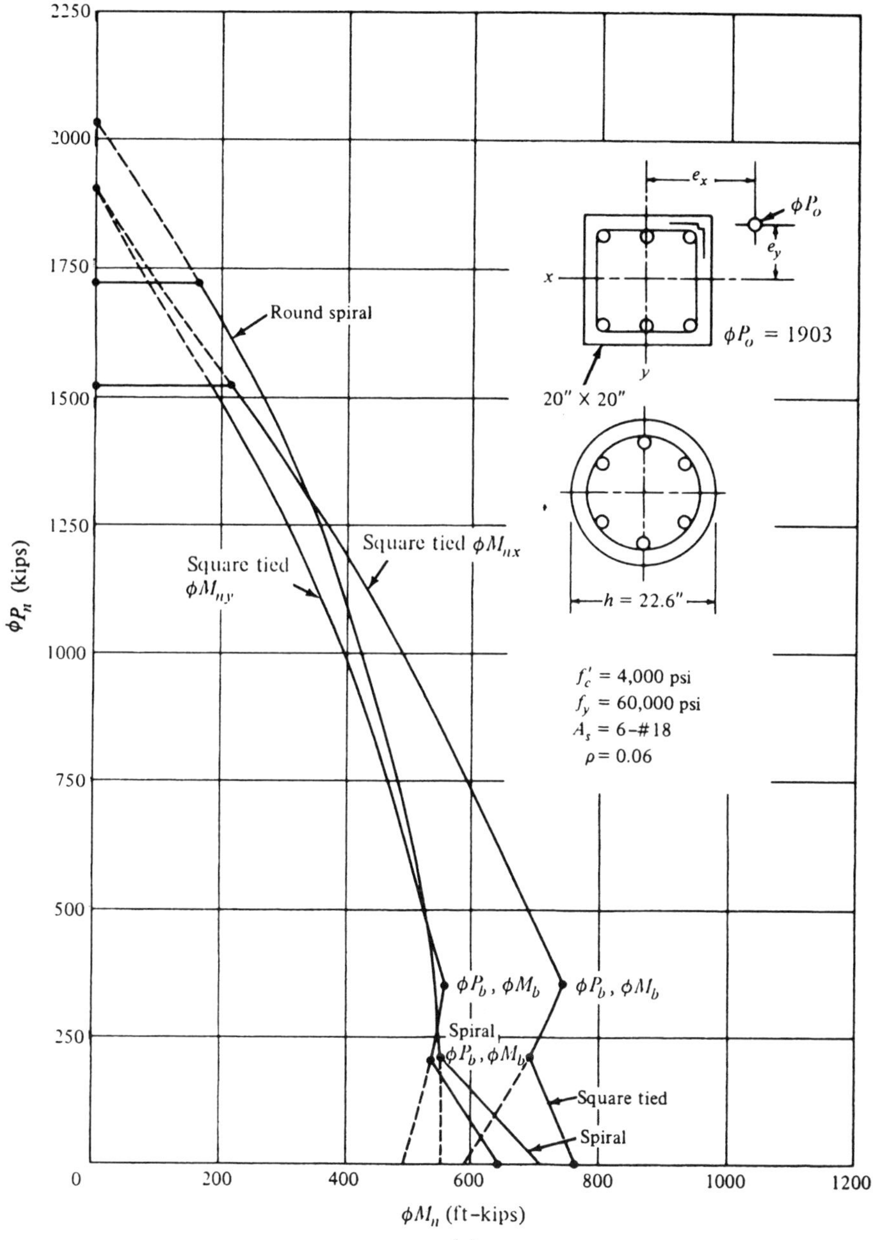

(*a*)

Figure 10-18(a) Square tied vs. round spiral columns.

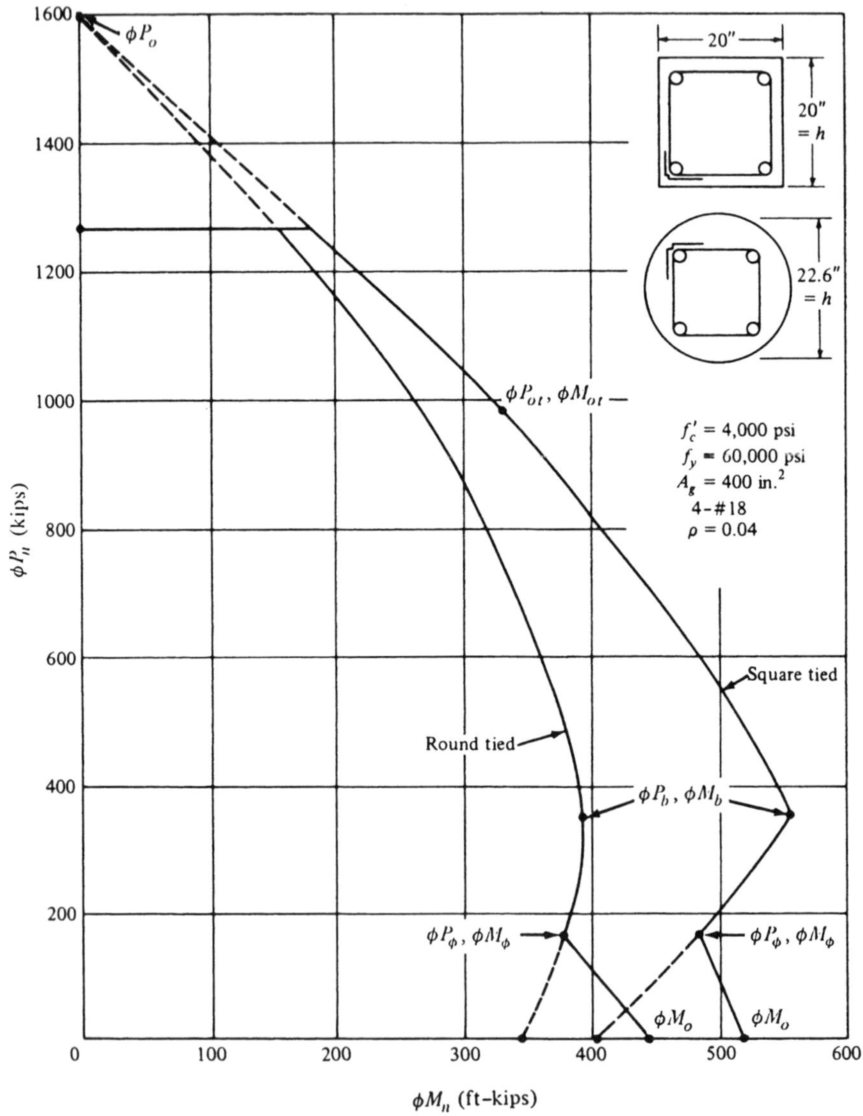

Figure 10-18(b) Square vs. round tied columns.

two more #18 vertical bars to the tied columns for Figs. 10-18(a) and (c) than by the use of spirals.

Similar comparisons lead to the conclusion that the cost of spiral reinforcement under the 1995 Code can seldom be justified (in nonseismic areas) unless three conditions exist: (1) the maximum axial load capacity may be used,

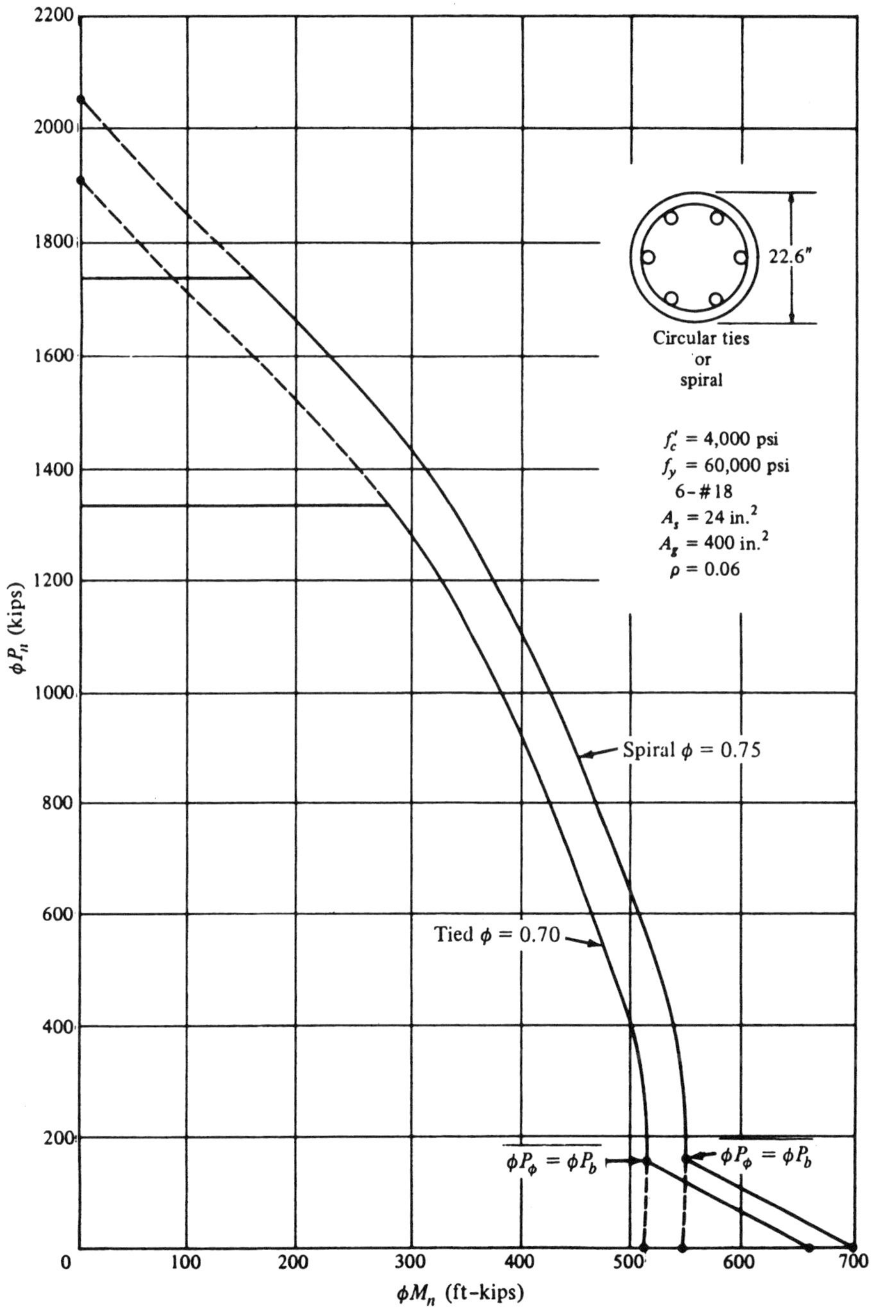

Figure 10-18(c) Round columns—spiral vs. ties.

(0.85/0.80)(0.75/0.70) = (spiral/tied) capacity, (2) the maximum reinforcement ratio, $A_s'/A_g = 0.08$, is insufficient for the maximum tied column desirable, and (3) the maximum practicable concrete strength, f_c' is specified. When these three conditions exist, the only reinforced concrete design recourse under the Code is to utilize a spiral. Even then for a square column, it will always be more difficult to provide the vertical bars required for a reinforcement ratio of 0.08 in a circular pattern in one circle. ("Square" spirals, sometimes specified, are usually impracticable due to fabricating limitations, and, if attempted, will increase costs.)

Where any sizable factored moments exist, the use of spirals even with the larger ϕ-factor, becomes very inefficient due to the inefficient circular pattern required for the vertical bars. In Fig. 10.18(*b*), the moment capacity is almost 30 percent greater for the square column than for the round with equal ϕ-factors. Note that the four-bar round tied column, is permitted under the Code (Section 10.9.2). Four bars in a circular column can be tied by a standard square tie. A provision of the Code (Section 10.9.2) allows three-bar tied columns. Care must be taken to orient the three bars properly if this provision is used. Alternatively, the column capacity may be based on the axis of bending having the least moment resistance, and the orientation of bars during construction may be arbitrary.

In Fig. 10-18(*c*), observe the very slight additional capacity, 0.75/0.70, for most of the load-moment interaction range developed by the spiral versus circular ties. Separate circular ties are specifically permitted and, of course, only one per tie spacing is required (Section 7.10.5.3). This requirement can also be satisfied by a "continuous circular tie" at a pitch equal to the required tie spacing (Commentary R7.10.5).

SPECIAL PROBLEMS IN COLUMN DESIGN

Biaxial Bending

Minimum eccentricity must be considered for each axis separately, but the Code does not require that the minimum eccentricity should be considered for both axes simultaneously. Where factored moments larger than minimum exist about both x and y axes simultaneously with the maximum factored axial load, P_u, the biaxial bending conditions must be considered. This design condition is often encountered in the practical design of corner columns, wall columns with heavy spandrel beams, and columns supporting two-way construction, although possibly for a minority of the columns in a particular structure. To illustrate biaxial bending, a three-dimensional load-moment curve interaction surface is required (see Fig. 10-19).

The interaction surface for a round column with six or more bars in a circular pattern is formed simply by rotating the planar load-moment interaction

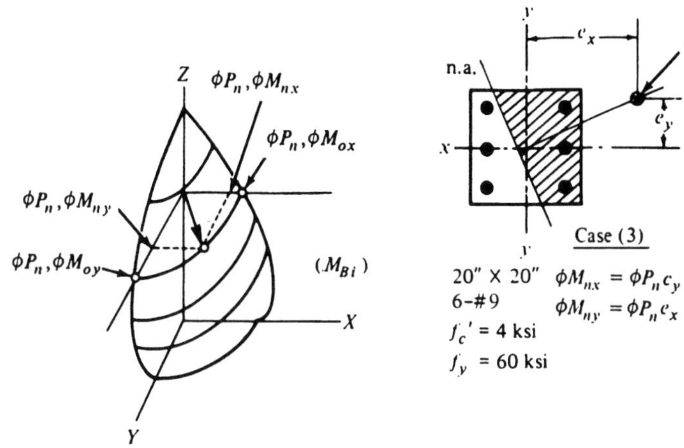

Figure 10-19 Biaxial load-moment interation surface.

curve through 90°. All combinations of design moment strength, $\phi P_n e_x$ and $\phi P_n e_y$ at a given design strength ϕP_n, constitute a horizontal slice through the interaction surface (see Fig. 10-20). For the special case of a polar symmetrical (round) column, an exact solution is simple because the curves are circular at each slice or level for the design strength ϕP_n.

CASE 1: POLAR SYMMETRY—ROUND COLUMNS WITH VERTICAL BARS IN A CIRCULAR PATTERN. Biaxial moment design strength is represented by a circular curve with a radius equal to ϕM_n. Uniaxial design moment strengths, $\phi M_{nx} = \phi P_n e_x$ and $\phi M_{ny} = \phi P_n e_y$, occurring simultaneously about both axes are simply combined for the biaxial design moment strength (see Fig. 10-20).

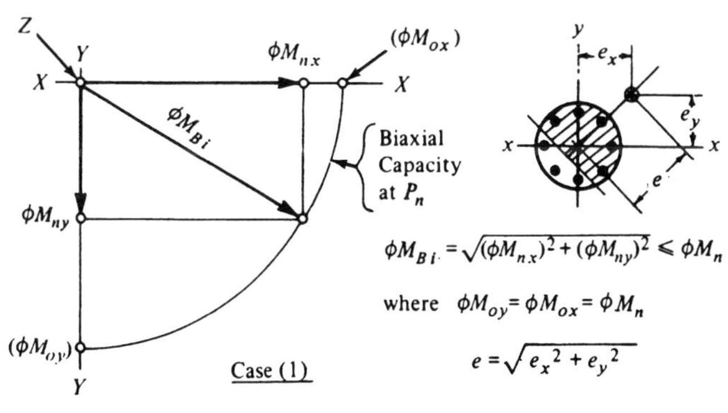

Figure 10-20 Biaxial bending—polar symmetrical section—case 1.

$$\phi M_{Bi} = \sqrt{(\phi M_{nx})^2 + (\phi M_{ny})^2} \leqslant \phi M_n \tag{10-1}$$

where $\phi M_n = \phi M_{nx} = \phi M_{ny}$ is the design strength for uniaxial bending moment, $\phi P_n e$, and $e = \sqrt{(e_x)^2 + (e_y)^2}$. (This analysis is based on the assumption that the vertical bars can be replaced by an equivalent circular tube area. This assumption is sufficiently accurate when six or more bars are present.)

CASE 2: ROUND AND SQUARE COLUMNS WITH EQUAL BENDING DESIGN STRENGTHS ABOUT BOTH PRINCIPAL AXES. ($M_{ox} = M_{oy}$) For round columns with four bars, square columns with four bars (or bundles of bars) one in each corner, or square columns with equal reinforcement in all four faces; the uniaxial bending design strength about the X-axis, ϕM_{ox}, is equal to the uniaxial bending design strength about the Y-axis, ϕM_{oy}. The curve of biaxial design strength is the horizontal slice at any load level, ϕP_n, through the biaxial interaction surface. This curve is nearly circular, even though the biaxial moment design strength is not exactly constant. For a rapid practical solution note that the straight line connecting ϕM_{ox} and ϕM_{oy} always lies *inside* the biaxial design strength curve. When using longhand solutions or a handbook giving the uniaxial design strength, $\phi M_{ox}(\phi M_{ox} = \phi M_{oy})$, simply add the design moments $\phi M_{nx} + \phi M_{ny}$. Use the sum as the design moment, ϕM_n. Select a column with a design strength for uniaxial moment, ϕM_{ox}, equal to or greater than ϕM_n.

$$\phi M_n = \phi M_{nx} + \phi M_{ny} \leqslant \phi M_{ox}$$

(see Fig. 10-21.)

CASE 3: COLUMNS WITH UNEQUAL BENDING DESIGN STRENGTHS ($M_{ox} \neq M_{oy}$) Rectangular columns or square columns with unequal amounts of reinforcement about X or Y axes have more complicated biaxial load-moment interaction surfaces. Horizontal slices for three values of axial design strength of an unsymmetrical square column are shown in Fig. 10-22. The solution proposed for case 2 becomes the more general form:

$$\frac{\phi M_{nx}}{\phi M_{ox}} + \frac{\phi M_{ny}}{\phi M_{oy}} \leqslant 1 \tag{10-3}$$

which is merely a linear interpolation of the column design strength between two axes, and where ϕM_{nx} and ϕM_{ny} are the design moments occurring simultaneously, and ϕM_{ox} and ϕM_{oy} are the uniaxial design moment strengths for bending only about either the x- or y-axes. Since ϕM_{ox} is not equal to ϕM_{oy}, the equation cannot be reduced to the simple form of Eq. 10-2. To illustrate the application of Eq. 10-3, consider the column of Fig. 10-22. The computer solution at an axial load strength, ϕP_n, of 200 kips gives uniaxial moment strengths, $\phi M_{ox} = 285$ ft-kips and $\phi M_{oy} = 260$ ft-kips. For the simultaneously

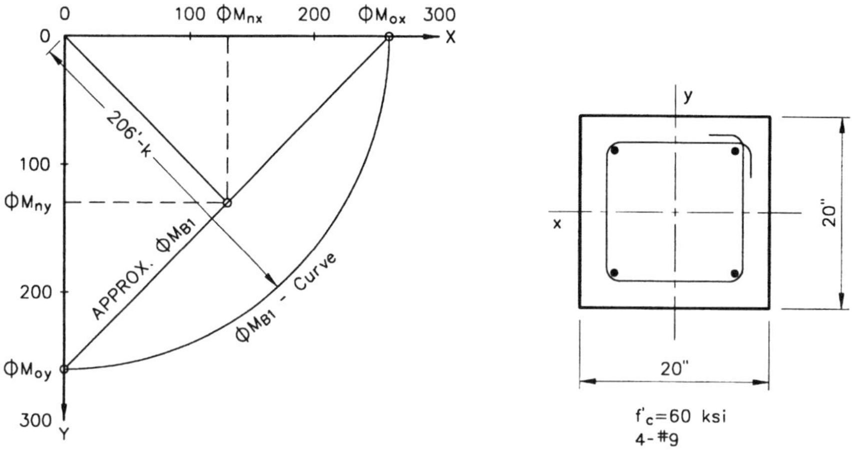

Figure 10-21 Capacity design strength approximation for columns with symmetry about *X*- and *Y*-axes—case 2.

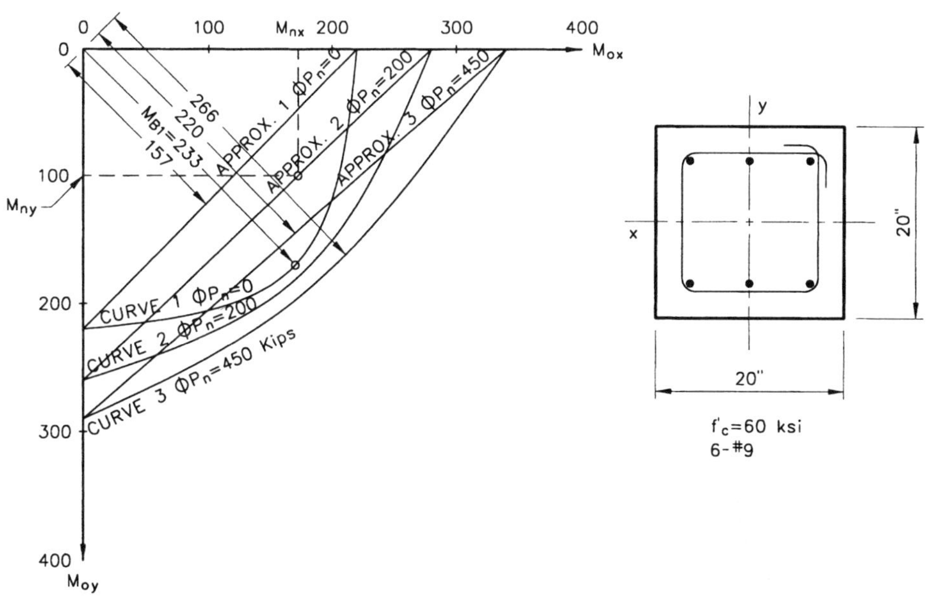

Figure 10-22 Biaxial capacity—unsymmetric section—case 3.

occurring moment strengths of $\phi M_{nx} = 175$ ft-kips; $\phi M_{ny} = 100$ ft-kips. Eq. 10-3 results in

$$\frac{\phi M_{nx}}{\phi M_{ox}} + \frac{\phi M_{ny}}{\phi M_{oy}} = \frac{175}{285} + \frac{100}{260} = 0.614 + 0.385 = 0.999 < 1.00$$

It will be observed that this solution lies at the straight line of a graphic solution of Curve 2 in Fig. 10-22.

To determine the maximum error conservatively underestimating the biaxial design strength of a column by the use of Eq. 10-3, see Curve 1 of Fig. 10-22. The design axial load strength, ϕP_n = zero is drawn from a computer solution. The biaxial design moment strength, ϕM_{Bi}, is the combination of moments at which the column bends about a neutral axis at 45° to the principal axis. The maximum difference between the exact solution (Curve 1) and its straight line approximation is greater than the difference caused by the approximations to Curves 2 and 3. $\phi M_{Bi} = 233$ ft-kips; the straight line is about 157 ft-kips. The conservative error $= 76/233 = 33$ percent.

Equation 10-3 is most accurate for axial loads near the balance point. Design axial strength at the balance point for the column in Fig. 10-22 is about 416 kips. For a design axial strength of 450 kips, the error of Equation 10-3 can be calculated based on the difference between the design moment strengths of Curve 3 and its straight line approximation as $(266 - 220)/266 = 17$ percent.

A second method of solution for biaxial bending problems is for use with nominal axial strengths $P_n > P_b$. This method which is suitable for longhand calculations is based on the equation

$$\frac{1}{\phi P_n} = \frac{1}{\phi P_{nx}} + \frac{1}{\phi P_{ny}} - \frac{1}{\phi P_0}$$

which was proposed by Bresler.* ϕP_n is the design strength under biaxial bending; ϕP_{nx} and ϕP_{ny} are the uniaxial design strengths with only e_y and e_x, respectively, and ϕP_o is the axial load design strength at $e = 0$ (see Fig. 10-23). The solution in Fig. 10-23 using this equation was developed for a single value, $P_n =$ 450 kips, with various uniaxial load capacities from a computer program reversing the normal procedure which would solve for the design strength, ϕP_n. The solution for all combinations at the single load level is presented so that the reader may compare the accuracy of this equation to other solutions. At a design axial load of 450 kips and with large bending moments about both axes, the second solution results in a large unconservative error of $(383 - 270)/270 = 42$ percent. The authors do not recommend the use of the second solution except for columns with large axial loads such that little or no tensile stress exists at any point of the column section.

* Bresler, Boris, "Design Criteria for Reinforced Concrete Columns Under Axial Load and Biaxial Bending," ACI *Journal Proceedings* V.57, Nov. 1960, pp. 481–490.

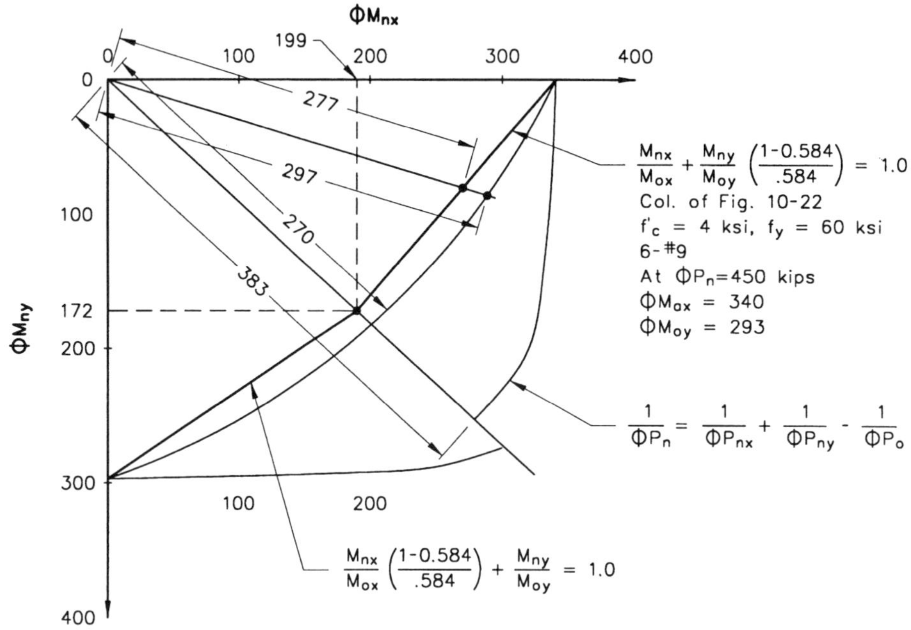

Figure 10-23 Biaxial capacity—manual solution.

It will be noted that the last equation, although simple, requires the complete load-moment interaction curve data to solve. For sections with unequal capacities in bending about the principal axes—the general case—both interaction curves or complete tabulations of capacities about both axes are required. Thus, the second solution is convenient only if the user has a computer-prepared handbook available. The straight line approximation or the third solution presented below are the only solutions convenient for use requiring only the data shown for Figs. 10-20 through 10-22.

The equation of the second solution substantially overestimates the strength for lightly reinforced columns when the axial loads are near balanced conditions. The equation is only applicable wherever values of $P_n \geqslant P_b$. For lesser axial loads, or for more accurate and conservative results, use Eq. (10-2) or Eq. (10-3) or the third solution presented in the following paragraph.

A third solution* can be used for axial load $0 \leqslant P_n \leqslant P_o$. It is suitable for simple calculation and is based on a bilinear interaction approximation between the design moment strengths at each of the two axes. The design moment strengths are expressed in nondimensionalized form and the equations for the design moment strength are given in Fig. 10-24 where β_b is a function of $P_n/f_c'A_g$ and $\rho f_y/f_c'$ can be obtained from Fig. 10-25.

* "Biaxial Bending Simplified" by A. J. Gouwens, 1975 ACI *SP-50 Reinforced Concrete Columns.*

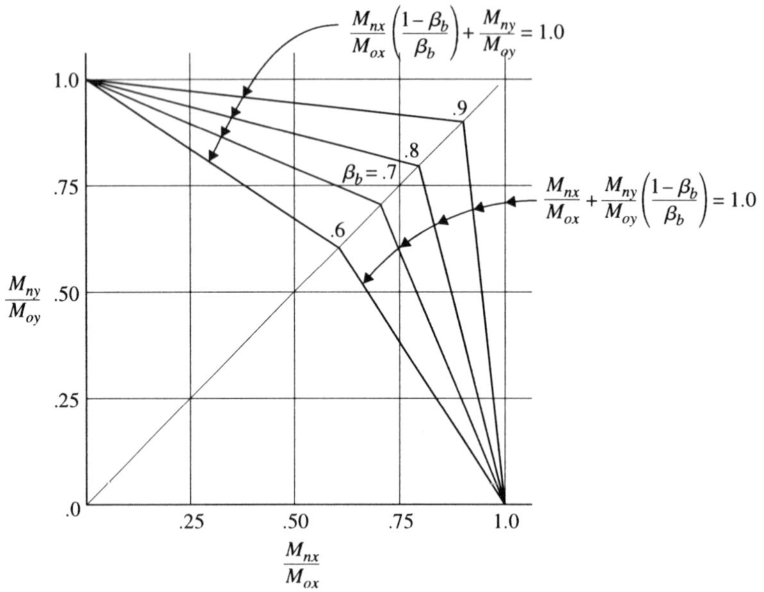

Figure 10-24 Straight-line interaction approximation.

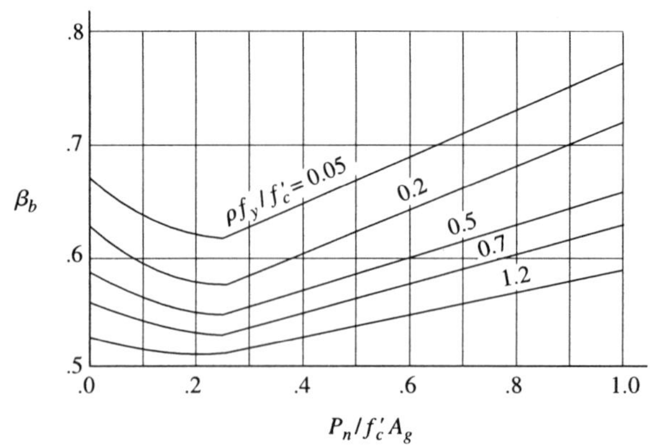

Figure 10-25 Simplified β_b as given by Eq. (6) to (10).

Fig. 10-25 and the equations in the next paragraph are empirical equations which form a lower bound estimate of the β_b value for columns with length-to-width ratios up to 3.0 and for various arrangements of reinforcement.

When $P_n \geqslant 0.25\, f'_c A_g$

$$\beta_b = \beta_{25} + 0.2\,\frac{P_n - 0.25 f'_c A_g}{0.85 f'_c A_g + A_s f_y}$$

or when $P_n < 0.25\, f'_c A_g$

$$\beta_b = \beta_{25} + \left(0.25 - \frac{P_n}{f_c' A_g}\right)^2 \left(0.85 - \frac{0.5\rho f_y}{f_c'}\right)$$

β_{25} is the value of β_b when $P_n = 0.25 f_c' A_g$. When $\rho f_y / f_c' \geqslant 0.5$

$$\beta_{25} = 0.485 + \frac{0.03\rho f_y}{f_c'}$$

and when $\rho f_y / f_c' < 0.5$

$$\beta_{25} = 0.545 + 0.35\left(0.5 - \frac{\rho f_y}{f_c'}\right)^2$$

For the column shown in Fig. 10-23 with $\rho f_y / f_c' = (6)(60)/((4)(400)) = 0.225$

$$\beta_{25} = 0.545 + 0.35(0.5 - 0.225)^2 = 0.572$$

and with $\phi P_n = 450$ kips

$$\beta_b = 0.572 + 0.2\,\frac{450/0.9 - 0.25(4)(400)}{0.85(4)(400) + 6(60)} = 0.584$$

$$M_{nx} = 0.584 \times 340 = 199 \text{ ft-kips}$$

$$M_{ny} = 0.584 \times 295 = 172 \text{ ft-kips}$$

The maximum conservative error occurs about at the point where $M_{nx}/M_{ny} = 3$ and is $(297 - 277)/297 = 7$ percent.

The designer will find curves for the "exact" solution of biaxial bending load capacities most helpful for further study, if desired. Not all curves will give exactly the same answers for biaxial capacity depending upon which form of stress block (rectangular or parabolic) was used; whether the bar areas were deducted from the concrete in the stress block; and, whether the exact depth for each bar size with minimum cover and ties were used or an average value.

Slender Columns

The Code requires column moments to be magnified for slenderness effects. Slenderness effects are considered to include buckling due to the secondary moment caused by lateral deflection. Two procedures for calculating the required strength to allow for slenderness effects are permitted. The preferred method is a comprehensive second-order analysis of the structure, and an application of the appropriate increases in factored moments resulting from lateral deflections considering material nonlinearity and cracking, as well as the effects of member curvature and lateral drift, duration of the loads, shrinkage and creep, and interaction with the supporting foundation (Section 10.10.1). An alternate method is to calculate the column moment magnification based upon a first-order frame analysis using the stiffnesses of the columns and beams given in Section 10.11.1 as

Beams . . . . $0.35I_g$
Columns . . . . $0.70I_g$
Walls
  Uncracked . . . . $0.70I_g$
  Cracked . . . . $0.35I_g$
Flat plates and flat slabs . . . . . $0.25I_g$

Columns must be checked for slenderness as nonsway (braced) frames subjected to gravity loads and as sway (unbraced) frames subjected to wind or seismic loads. Nonsway frames are covered in Section 10.12 and sway frames are covered in Section 10.13. Each section has unique equations for the moment magnification factor, and effective length. A minimum factored moment is specified for nonsway frames (Section 10.12.3.2), but no minimum factored moment is specified for sway frames.

For purposes of applying Code requirements to design columns for slenderness, one might consider structures classified as follows: (1) very tall, where the slenderness provisions determine the need for shear walls or other special lateral bracing, (2) tall enough so that slenderness effects are a major factor but special lateral bracing is not required, (3) short, where lateral load effects are minor, and (4) special, such as two-column multistory frames. The vast majority of buildings fall into class 3 where the approximate evaluation of slenderness effects (Section 10.11) is appropriate. Since most tall structures involve a number of equal size, equal height columns and some lateral bracing, the only practical approach for classes 1 and 2, as well as special structures, is a second-order analysis, the Code-preferred method (Section 10.10.1). The requirements of Section 10.10 of ACI 318-95 can be met by using an iterative P-Δ Analysis that is suitable for computer solution[†] or by Direct P-Δ Analysis by manual computation or by computer solution[‡].

For the common structures, with height-to-base width ratios less than two, the alternate evaluation of slenderness can be quite simple. In such structures, large bending moments in columns are unusual. With ordinary story heights, column sizes, and spans, each lateral force resisting frame will usually have a

[†] ICES-STRUDL program for IBM 360 by CRSI-PCA-MIT; and IBM STRESS for IBM 1130.

[‡] MacGregor, J. G., and Hage, S. E., "Stability Analysis and Design of Concrete," *Proceedings, ASCE*, Vol. 103, No. ST 10, October 1977; and Gouwens, A. J., "Lateral Load Analysis of Multistory Frames with Shear Walls," Bulletin AEC 2, Portland Cement Association Computer Program, 1968.

sufficient number of columns so that the wind moment is small, and only in edge columns with no cantilever will gravity load moment become an important factor. Most columns in such frames can be designed at a low reinforcement ratio for the maximum factored axial load. See preliminary selection chart, Fig. 10-15.

The alternate evaluation of slenderness effects and design to resist same then consists essentially of seven steps:

1. Determine if each story in the frame is "braced" against sidesway.
2. Determine the column flexural stiffness.
3. Determine effective length factor, k.
4. Determine required factored moment, M_2, and moment magnification factor for the nonsway (braced) frame, δ_{ns}, about both axes separately.
5. Determine factored moments, M_1, and M_2, and moment magnification factor for sidesway, δ_s, about both axes separately.
6. Increase reinforcement ratio as required for the factored axial load, P_u, and magnified factored moment, $M_c = \delta_{ns} M_2$ for braced frames, or for unbraced frames, use the larger magnified moment resulting from either

$$M_1 = M_{1ns} + \delta_s M_{1s}$$

$$M_2 = M_{2ns} + \delta_s M_{2s}$$

7. Check for extraordinarily slender columns in an unbraced frame, and check for overall frame stability.

STEP 1. A story may be permitted to be assumed as a nonsway story if it satisfies either of two conditions. If the increase in the column end moments due to second-order effects does not exceed 5 percent of the first-order end moments (Section 10.11.4.1), the story is designated as nonsway. Since this method can only be used with the benefit of a second-order analysis, the second code method is of more interest when designing by the alternate method of slenderness evaluation. In the second method, the code uses the stability index $Q = \Sigma P_u \Delta_o / V_u \ell_c$. When $Q \leq 0.05$, the story may be considered braced (Section 10.11.4.2).

V_u = total factored lateral load resisted by the story

Δ_o = first-order lateral deflection by elastic analysis with factored loads

ℓ_c = center-to-center height of the story

STEP 2. Compute column stiffness, with consideration of sustained load effects and reinforcement ratio (Section 10.12.3). Two formulas are prescribed:

$$EI = \frac{0.2E_c I_g + E_s I_{se}}{1 + \beta_d} \quad \text{Code Eq. (10-12)}$$

and

$$EI = \frac{0.4E_cI_g}{1 + \beta_d} \quad \text{Code Eq. (10-13)}$$

where subscripts c = concrete; se = reinforcement ; g = gross; and β_d = ratio of the maximum factored axial dead load to total factored axial load, always positive.

Equation (10-13) is a simplified approximation of Equation (10-12) (Commentary R10.12.3). The maximum value of EI as obtained from either equation can be used, but Equation (10-12) is generally more accurate.

STEP 3. For braced frames, k = 1.0 (Section 10.12.1). For unbraced frames, k must be greater than 1.0 (Section 10.13.1) (see chart, Fig. 10-26, and the section of this chapter titled "Determination of the Effective Length, Unbraced

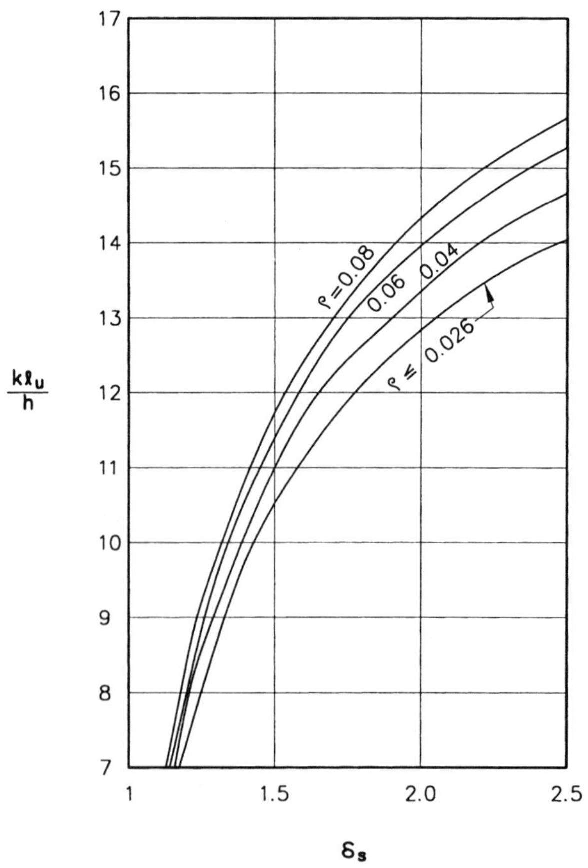

Figure 10-26 Slender columns—Unbraced.

Frames"). In determining the effective length factor k, it is recommended that the stiffness of the flexural members be made on the basis of $0.35I_g$ and the stiffness of the columns on the basis of $0.70I_g$ (Commentary R10.12.1).

STEP 4. Using the value of EI, the critical buckling load, P_c is computed as

$$P_c = \frac{\pi^2 EI}{(k\ell_u)^2} \quad \text{Code Eq. (10-11)}$$

The nonsway frame moment magnification factor, δ_{ns}, is defined as

$$\delta_{ns} = \frac{C_m}{1 - P_u/0.75P_c} \geqq 1.0 \quad \text{Code Eq. (10-10)}$$

in which $C_m = 0.6 + 0.4\,(M_1/M_2) \geqslant 0.4$ by Code Eq. 10-14 where M_2 is the larger factored end moment (+) and M_1 is the smaller factored end moment, (+) for single curvature and (−) for double curvature.

When the maximum factored end moment is less than the minimum specified factored end moment, the moment M_2 of Code Eq. (10-9) must be based on the specified minimum factored end moment $M_{2,\min}$ of Code Eq. 10-15. Where computations show no eccentricity at both ends of the column, conditions of curvature must be based on a ratio of $M_1/M_2 = 1.0$. Where the eccentricities are calculated and are less than the minimum, $0.6 + 0.03h$, for design, computed moments may be used to compute C_m (Commentary Section R10.12.3.2).

Short-cut solutions to determine δ_{ns} (Step 4) at $P_u = \phi P_{n(\max)}$*; reinforcement concentrated in the two faces parallel to the neutral axis or neglected in other faces, for given f_y and f_c' are easily prepared. The charts in Fig. 10-27 present direct solutions for δ_{ns}, using normal weight concrete, $f_c' = 4{,}000$ psi, and $f_y = 60{,}000$ psi and

$$\beta_d = \frac{1.4D}{1.4D + 1.7L} = 0.6$$

for square sections for sustained loads.

Step 4, determination of δ_{ns}, becomes very simple with charts such as these. For braced frames, enter Fig. 10-27 appropriate Chart A, B, C, D, or E with ρ, $k\ell_u/h$, and C_m and read δ_{ns} directly.

STEP 5. Using the EI values of Section 10.11.1, the ΣP_c is calculated for all the columns in the story. The magnified sway moment produced by sidesway, including lateral and gravity loads, is given by:

$$\delta_s M_s = \frac{M_s}{1 - \Sigma P_u/0.75\Sigma P_c} \geqslant M_s \quad \text{(Code Eq. 10-19)}$$

The value of $\delta_s M_s$ must be less than 2.5 M_s or the structure as a whole is considered to be unstable.

---

* δ_{ns} at $P_u = \phi P_{n(\max)}$ will be conservative for smaller values of P_u, and correct at the maximum factored load $P_u = \phi P_{n(\max)}$.

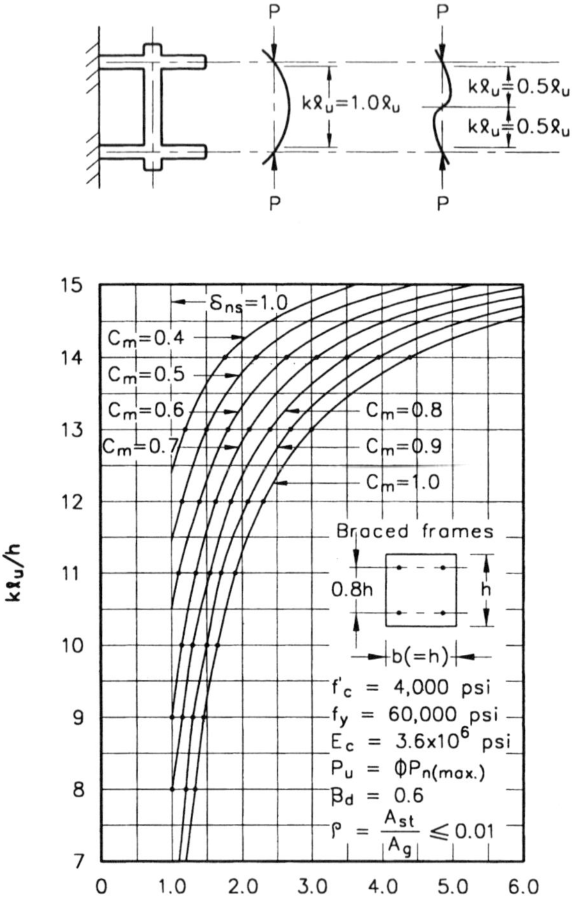

δ_{ns} = Moment magnification factor

(A)

Figure 10-27 Effective heights limits—braced frame and slender columns, charts A to E.

No minimum factored moment is required for sway frames.

An alternate way to calculate the magnitude of the sway moment is based on the stability index and is given by

$$\delta_s M_s = \frac{M_s}{1 - Q} \geqslant M_s \quad \text{(Code Eq. 10-18)}$$

A shortcut solution to determine δ_s (Step 5) for $P_u = \phi P_{n(\max)}$ for given f_y of reinforcement and f_c' is easily prepared.

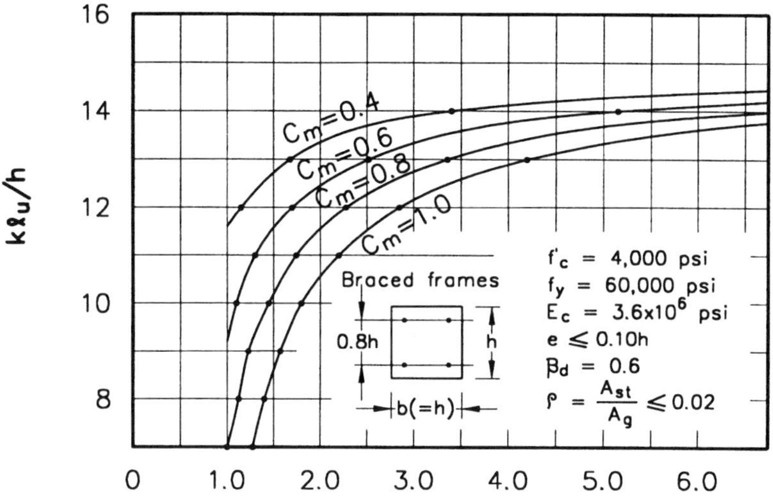

δ_{ns} = Moment magnification factor

(B)

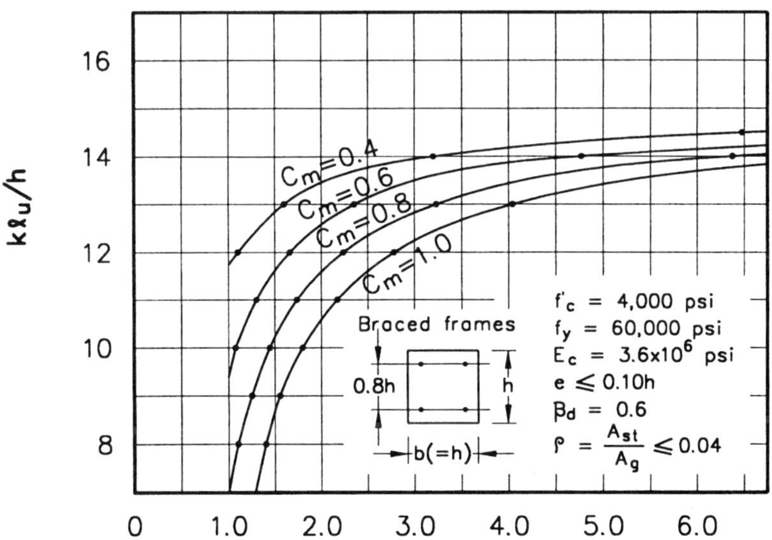

δ_{ns} = Moment magnification factor

(C)

Figure 10-27 *(continued).*

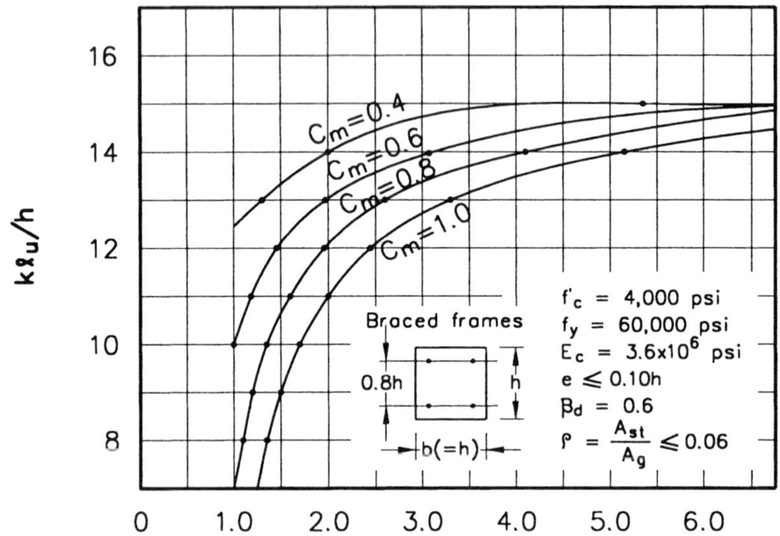

δ_{ns} = Moment magnification factor

(D)

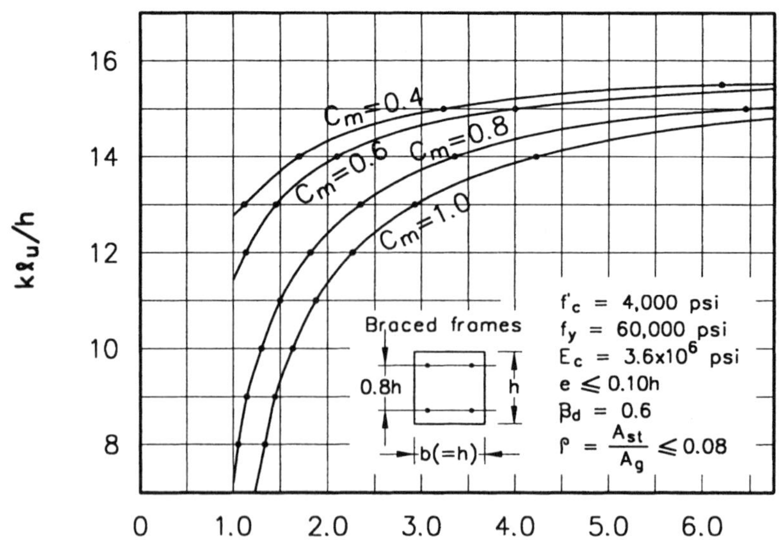

δ_{ns} = Moment magnification factor

(E)

Figure 10-27 *(continued).*

Figure 10-26 was prepared using values of f'_c, f_y, and E_c similar to those used in Fig. 10-27 of Step 4. β_d for lateral wind load was taken as zero and the column moment of inertia was taken as $0.70 I_g$.

STEP 6. This consists of selecting a higher reinforcement ratio to satisfy load-moment conditions of P_u, and the magnified moment.

Example: See page 305.

STEP 7. Individual columns which are very slender should be checked for nonsway magnification of the moment resulting from gravity loads plus the sway magnified lateral load. The code requires that when

$$\ell_u > \frac{35}{\sqrt{P_u / f'_c A_g}}$$

the column must be designed (Section 10.13.5) for the larger factored moment calculated from

$$M_c = \delta_{ns}(M_{1ns} + \delta_s M_{1s})$$

or

$$M_c = \delta_{ns}(M_{2ns} + \delta_s M_{2s})$$

where δ_{ns} is calculated using β_d based on the load combination under consideration. For structures where the sway moments are caused by wind or seismic loads, $\beta_d = 0$.

All sway frames must be checked for stability of the structure as a whole under factored gravity loads considered. The β_d value to be used for the stability check is specified in Section 10.0. The value of δ_s must not exceed 2.5 (Section 10.13.6). Structures with δ_s greater than 2.5 may have large increases in δ_s due to small variations of EI or of factored axial loads. Such structures should be stiffened.

DETERMINATION OF THE EFFECTIVE LENGTH

Braced Frames

In braced frames, neglecting axial shortening of members, the intersections of the column and beam centerlines are considered as braced points by definition. The effective length of a member in a buckling curve about its centerline, defined as ℓ_u, is the length between points of contraflexure, reversals of curvature (hinges). The column is bent in single curvature between the braced points for $k\ell_u = \ell_u$, or in double curvature for $0.5\ell_u \leq k\ell_u \leq 1.0\ell_u$. (See Fig. 10-27, Chart A).

In reinforced concrete frames consisting of beams and columns with average ratios of column to beam stiffness varying from 0.2 to 4.0, the effective length,

$k\ell_u$, in braced frames can vary from about 0.6 ℓ_u to 0.9 ℓ_u. The same column under different patterns of live load can have different effective lengths as it bends from single to partial or full double curvature. As a simple, quick, and safe design procedure, the authors recommend using $k\ell_u = 1.0$ for all columns in braced frames. Note that ℓ_u is to be taken as the *clear* distance between lateral supports.

Unbraced Frames

In unbraced frames, the column-beam intersections can deflect laterally and buckle as a unit so that $k\ell_u$ may extend several stories in height, particularly in frames with flexible beams and stiff columns or in lower stories of frames anchored to a rigid foundation. The following formulas provide sufficient accuracy for design and are appropriate for practical use*:

$$k = \frac{20 - \Psi_m}{20}\sqrt{1 + \Psi_m} \quad \text{for } \Psi_m < 2.0$$

$$k = 0.9\sqrt{1 + \Psi_m} \quad \text{for } \Psi_m \geqslant 2.0$$

where Ψ_m is the average ratio at top and bottom of the column,

$$\Psi = \frac{\Sigma K_{\text{col}}}{\Sigma K_{\text{beam}}} = \frac{\Sigma E_c I_c/\ell_c}{\Sigma E_c I_b/\ell_b}$$

Use $E_c I_c$ and $E_c I_b$ computed per Section 10.11.1 for columns and beams. For frames consisting of columns and flat slabs, use $\Psi = \Sigma K_c/K_{es}$ (see page 30). Use the lower $E_c I_b$ values also given for flat plates and slabs in Section 10.11.1.

DERIVATIONS AND BASIS OF CHARTS

As aid to readers who may wish to extend the charts in Fig. 10-26 and Fig. 10-27 for other concrete strengths, D/L ratios, or lightweight aggregate concrete (lower E_c values), the derivation is as follows:

1. For nonsway columns with $\rho \leqslant 0.01$, use Code Eq. (10-13); $EI = E_c I_g/0.4$.
   For nonsway columns with $\rho \geqslant 0.02$, use Code Eq. 10-12; assume an average value for the ratio of the center-to-center of bars to overall column dimension. Take $gh = 0.8h$, which is closely applicable to a wide variety of column sizes, cover depths, and bar sizes.
2. For sway columns, use EI based on Section 10.11.1.
3. For P_u at $\phi P_{n\,(\max)}$ take the value of P_u/A_g for $f_c' = 4{,}000$ psi from the preliminary design chart (Fig. 10-15) which is based on Code Eq. (10-2). For

---

* Richard W. Furlong, "Column Slenderness and Charts for Design," *ACI Journal* (Jan. 1971), pp. 9-18.

a square tied column, $P_u = (P_u/A_g)(h)^2$. Alternatively, P_u for spiral columns. (NOTE: Fig. 10-15 is based on $\phi = 0.70$.)

4. Consider all reinforcement concentrated in the corners of a square column.
5. Compute coefficients for $E_c = 3.6 \times 10^6$ psi; $\phi = 0.70$; and

$$\beta_d = \frac{1.4D}{1.4D + 1.7L} = 0.6$$

for sustained loads in nonsway frames and $\beta_d = 0$ for sway frames.

It will be noted that other values for E_c can be substituted directly into the tabulated coefficient for *EI* only for $\rho \leqq 0.01$, which utilizes Code Eq. (10-13). The coefficient itself will change when based on Code Eq. 10-12. Coefficients for P_c which include the term *EI* will require revision proportional to changes in the coefficient for *EI*.

6. Points in the charts of Figs. 10-26 and 10-27 are located using the tabulated coefficients, and will form a family of similar curves in each case. As an example, consider a 12 in. square column, $f'_c = 4$ ksi; $\rho = 0.08$; braced so that $k = 1.0$; with unsupported height = 17 ft; where $C_m = 1.0$ (conservative for small values of eccentricity at each end).

$$\left(\frac{h}{k\ell_u}\right)^2 = \left(\frac{1}{17}\right)^2$$

From Table 10-1 coefficients,

$$P_c = 2700\left(\frac{h}{k\ell_u}\right)^2; \; P_u/0.75 = 5.86.$$

$$\delta_{ns} = \frac{C_m}{1 - \dfrac{P_u/0.75}{P_c}} = \frac{1.0}{1 - \dfrac{5.86}{2700}(17)^2}$$

$$\delta_{ns} = \frac{1}{1 - 0.628} = 2.69$$

See Chart E, Fig. 10-27. For $k\ell_u/h = 17$ and $C_m = 1.0$, read $\delta_{ns} = 2.7$.

For a square column, radius of gyration, *r*, equals $0.3h$ (Section 10.11.2). Slenderness effects may be neglected for nonsway columns with $M_1/M_2 = 1.0$ and with ratios of $k\ell_u/r < 22$ in all frames (Section 10.12.2). Note charts begin with ratios of $k\ell_u/h = 7$.

EXAMPLE 1: SLENDERNESS EFFECTS—UNBRACED FRAME—APPROXIMATE P-Δ ANALYSIS. Consider an interior column in a low building or the upper stories of a high-rise building supporting a flat plate floor. The flat plate design requires a 12 × 12 in. minimum size square column. For a 12 × 12 square column, the ratio Ψ, of the column stiffness to floor system

TABLE 10-1 Square Column Slenderness Coefficients

$f'_c = 4$ ksi; Grade 60 Bars; $E = 3.6 \times 10^3$ ksi; $b = h$; $P_u = \phi P_{n(\max)}$

$\rho = A_{st}/A_g$	EI[a]	P_c (ksi)	$P_u/0.75$ (ksi)
$\leqslant 0.01$	$0.0209\ E_c h^4$	$750\ (h/k\ell_u)^2$	3.15
0.02	$0.0264\ E_c h^4$	$900\ (h/k\ell_u)^2$	3.57
0.03	$0.0344\ E_c h^4$	$1{,}200\ (h/k\ell_u)^2$	3.93
0.04	$0.0425\ E_c h^4$	$1{,}500\ (h/k\ell_u)^2$	4.29
0.05	$0.0510\ E_c h^4$	$1{,}800\ (h/k\ell_u)^2$	4.72
0.06	$0.0585\ E_c h^4$	$2{,}100\ (h/k\ell_u)^2$	5.07
0.07	$0.0665\ E_c h^4$	$2{,}400\ (h/k\ell_u)^2$	5.50
0.08	$0.0745\ E_c h^4$	$2{,}700\ (h/k\ell_u)^2$	5.86

[a] For $\rho \leqq 0.01$, EI per Code Eq. 10-13; for $\rho \geqq 0.02$, EI per Code Eq. 10-12.

stiffness at the ground floor and second floor is 0.45. The columns are not braced against sidesway. The total factored axial load after live load reductions is 372 kips; and the unbraced first floor height is 9′-4″. There are a sufficient number of columns per frame in both directions so that the factored wind moment is only 270 in.-kips. Since the factored wind moment is less than the minimum braced frame factored moment of $M_{2,\min} = P_u(0.6 + 0.03h) = 357$ in.-kips, the interior columns will be designed for the minimum braced frame moment about both axes. Grade 60 bars and, if feasible, concrete $f'_c = 4{,}000$ psi, will be used.

From any handbook, select a short column design at lowest reinforcement ratio: 12 × 12 in. square tied column, four #9; $P_u = 372$ k. For the slenderness effect compute effective length factor, k, as follows:

$$k = \frac{20 - \Psi}{20}\sqrt{1 + \Psi}; \quad \text{take } \Psi = \frac{K_{\text{col}}}{K_{\text{floor}}} = 0.45$$

$$k = 1.174; \quad k\ell_u = (1.174)(9.33) = 11.0$$

(See Fig. 10-26, Unbraced.) Enter with $k\ell_u/h = 11.0$, and $\rho = 0.0278$.

Read $\delta_s = 1.7$. The magnified factored moment is thus $1.7 \times 270 = 459$ in.-kips. From a handbook, the reinforcement required for this moment is four #10 bars. Since a square column with four vertical bars has equal bending capacity about both principal axes, separate slenderness reduction for each is unnecessary in this example.

EXAMPLE 2: SLENDERNESS EFFECTS—UNBRACED FRAME—COMPARISON OF THE APPROXIMATE METHOD (SECTION 10.11) AND A DIRECT P-Δ ANALYSIS (Section 10.10). Slenderness effects will be considered for gravity plus wind loads for a three-bay wide five-story building with 20 ft square bays, 8 in. thick flat plate floors, 16 in. square columns, 12′-0″ story heights, $f'_c = 4{,}000$ psi, $f_y = 60{,}000$ psi, and the outer face of the edge columns flush with the edge of the slab. This construction is the same as that used in the two-way solid flat plate braced frame examples of Chapter 5 with both the Direct Design Method (Section 13.6) and the Equivalent Frame Method (Section 13.7). See Fig. 10-28.

In both the Approximate Method and the Direct P-Δ Method, only the column strip will be considered effective as a part of the unbraced frame.

Approximate Method (Section 10.11)

In determining the effective length factor, k, the effect of cracking and reinforcement must be considered in computing relative stiffness (Section 10.11.1).

Prior to application of wind load, the areas of the slab adjacent to columns will be subjected to negative moments caused by dead and live load. Cracking around the columns causes a reduction of the column strip stiffness for any increased wind load moments to 0.25 (0.50 I_g). While wind load increases the moment on one side of the column, wind load decreases the moment on the opposite side of the column. The column strip stiffness for elastic reduction of moment due to wind load is 1.0 (0.50 I_g). The average column strip stiffness where one-half the column strip length has increasing curvature and one-half the column strip length has decreasing curvature is $\frac{1}{2}(0.25 + 1.0)(0.50\, I_g) = 0.31\, I_g$. In structures where the moment due to 0.75(1.4D + 1.7L + 1.7W) exceeds 2(1.4D + 1.7L), the column strip stiffness would be reduced to $\frac{1}{2}(0.25 + 0.25)(0.50\, I_g) = 0.12\, I_g$. In this example, the wind load moments are less than one-half the gravity load moments and the average column strip stiffness is conservatively assumed to be 0.25 I_g. Column stiffness is assumed as 0.70 I_g with $\beta_d = 0$.

Slab column strip EI;

$$EI = 0.25 I_g E_c = (0.25)(1/12)(240)(8)^3(3{,}600{,}000)$$
$$= 9.22 \times 10^9 \text{ lb. in.}^2.$$

Column (assume 4-#9)

$$EI = E(0.70 I_g)$$
$$= (3{,}600{,}000)(.70)(16^4/12) = 13.76 \times 10^9 \text{ lb. in.}^2.$$

ΣEI of columns/ΣEI of flexural members

First Story

$$\Psi \text{ (Bottom)} = 0 \text{ (Fixed base)}$$

$$\Psi \text{ (Top Ext)} = \frac{2(13.76)}{12 \times 12} \times \frac{12 \times 20}{9.22}$$
$$= 4.97$$

$$k \text{ (Ext)} = 1.50 \text{ (Commentary Section R10.12.1)}$$

$$\Psi \text{ (Top Int)} = \frac{2(13.76)}{12 \times 12} \times \frac{12 \times 20}{2(9.22)}$$
$$= 2.49$$

$$k \text{ (Int)} = 1.30$$

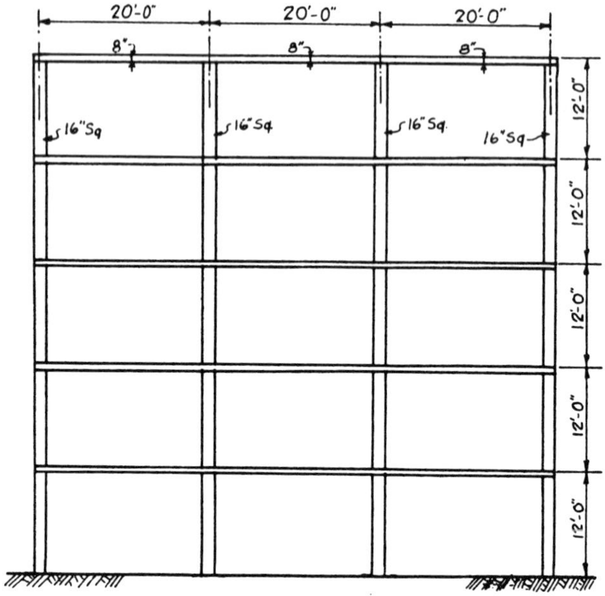

Five Story Frame under Lateral Load

Frames spaced at 20′–0″; story height, ℓ_u = 12′–0″

f'_c = 4,000 psi for all columns and slabs; f_y = 60,000 psi

Floors and roof are 8″ thick flat plates; all columns are 16″ sq.

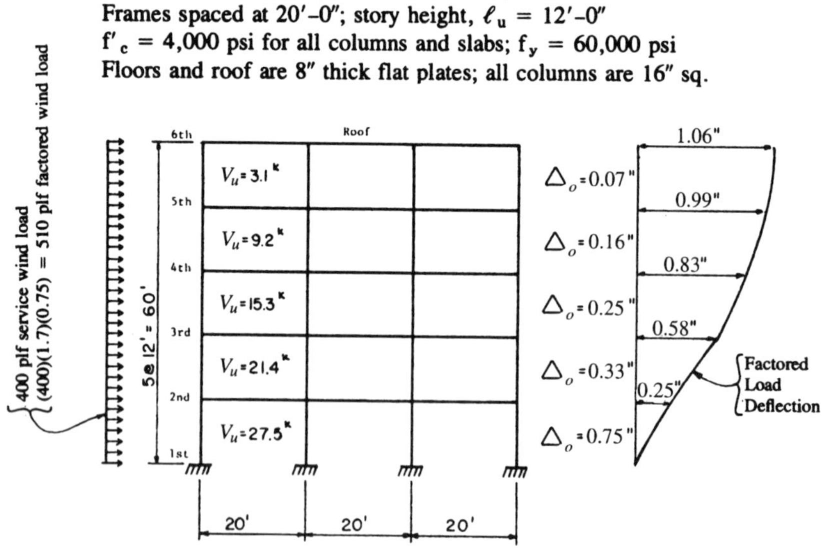

Service Loads:

8″ slab	100 psf	
Partition	18 psf	
	118 psf × 1.4 =	165 psf
Live	50 psf × 1.7 =	85 psf
	168 psf	250 psf
		× 60 × 20 = 300 kips/floor
Exterior wall - 30 psf × 1.4 = 42 psf		
		× 12 × 20 × 2 = 20 kips/story

Figure 10-28 Unbraced Frame.

Second, Third, and Fourth Stories

$$\Psi \text{ (Bottom Ext)} = \text{(Top Ext.)} = 4.97$$

$$k \text{ (Ext)} = 2.2$$

$$\Psi \text{ (Bottom Int)} = \text{(Top Int.)} = 2.49$$

$$k \text{ (Int)} = 1.71$$

Fifth Story

$$\Psi \text{ (Bottom Ext)} = 4.97$$

$$\Psi \text{ (Top Ext)} = \frac{13.76}{12 \times 12} \times \frac{12 \times 20}{9.22}$$

$$= 2.49$$

$$k \text{ (Ext)} = 1.93$$

$$\Psi \text{ (Bottom Int)} = 2.49$$

$$\Psi \text{ (Top Int)} = \frac{13.76}{12 \times 12} \times \frac{12 \times 20}{2 \times 9.22}$$

$$= 1.24$$

$$k \text{ (Int)} = 1.53$$

For transient wind loads, the sustained shear load is zero and thus $\beta_d = 0$ (Section 10.0)

The values of EI are based on the previously determined value of the stiffness of the columns, $EI = 13.76 \times 10^9$. No adjustment is necessary for $\beta_d = 0$ and all stories of the building $EI = 13.76 \times 10^9$ lb. in.^{2}.

$$P_{cr} = \pi^2 EI/(k\ell_u)^2 \qquad \text{(Section 10.12.3, Eq. 10-11)}$$

In this expression for the critical load the previously determined values of k for exterior and interior columns must be used along with the EI values. Computed values of ΣP_{cr} for the different stories are as follows:

$$P_{cr} \text{ (1st story)} = \frac{\pi^2(13.76 \times 10^9)(2)}{[(1.50)(122)]^2(1{,}000)} + \frac{\pi^2(13.76 \times 10^9)(2)}{[(1.30)(112)]^2(1{,}000)}$$

$$= 20{,}920 \text{ kips}$$

$$P_{cr} \text{ (2nd story)} = \frac{\pi^2(13.76 \times 10^9)(2)}{[(2.20)(122)]^2(1{,}000)} + \frac{\pi^2(13.76 \times 10^9)(2)}{[(1.71)(112)]^2(1{,}000)}$$

$$= 11{,}180 \text{ kips}$$

$$P_{cr} \text{ (3rd story)} = 11{,}180 \text{ kips}$$

$$P_{cr} \text{ (4th story)} = 11{,}180 \text{ kips}$$

$$P_{cr}\text{ (5th story)} = \frac{\pi^2(13.76 \times 10^9)(2)}{[(1.93)(122)]^2(1{,}000)} + \frac{\pi^2(13.76 \times 10^9)(2)}{[(1.53)(112)]^2(1{,}000)}$$

$$= 14{,}150 \text{ kips}$$

The moment magnifier, δ_s, (Section 10.11.5.1, Eq. 10-8) can now be computed.

$$\delta_s = \frac{1}{1 - \Sigma P_u/(0.75\Sigma P_c)}$$

Story	ΣP_u kips	ΣP_c kips	δ_s
5th	300	14,150	1.03
4th	620	11,180	1.08
3rd	940	11,180	1.13
2nd	1260	11,180	1.18
1st	1590	20,920	1.11

DIRECT P-Δ METHOD

The requirements of Section 10.10 of ACI 318-95 can be met by a direct *P*-Δ analysis.* The direct *P*-Δ method of Section 10.13.4.2 which was presented by MacGregor and Hage is based on an equation for a geometric series that is derived for the final second-order deflection as a function of the first-order elastic deflection. This direct *P*-Δ analysis provides the following simple method of computing the moment magnifier, δ_s, when the stability index, Q, is greater than 0.05 but equal to or less than 0.33 ($0.05 < Q \leqslant 0.33$).

$$\delta_s = 1/(1 - Q)$$

this will obviously give the same results as Code Equation 10-16, where

$$Q = P_u \Delta_o / V_u \ell_c$$

and

P_u = Summation of factored loads in a given story

Δ_o = Elastically computed first-order lateral deflection due to V_u (neglecting P-Δ effects) at the top of the story relative to the bottom of the story

V_u = Total factored lateral force within the story

ℓ_c = Height of story, center to center of floors or roof

* MacGregor, J. G., and Hage, S. E., "Stability Analysis and Design of Concrete," *Proceedings, ASCE*, Vol. 103, No. ST 10, October, 1977.

Computations for the stability index, Q, and the moment magnifier, δ_s, for each story are as follows:

Fifth Story (See Fig. 10-28, $\Delta_o = 0.07$

$$Q = \frac{P_u \Delta_o}{V_u \ell_c} = \frac{(300)(0.07)}{(3.1)(144)} = 0.047$$

$$\delta_s = 1/(1 - Q) = 1/(1 - 0.047) = 1.05$$

Fourth Story

$$Q = \frac{(620)(0.16)}{(9.2)(144)} = 0.075$$

$$\delta_s = 1/(1 - 0.075) = 1.08$$

Third Story

$$Q = \frac{(940)(0.25)}{(15.3)(144)} = 0.107$$

$$\delta_s = 1/(1 - 0.107) = 1.12$$

Second Story

$$Q = \frac{(1260)(0.33)}{(21.4)(144)} = 0.135$$

$$\delta_s = 1/(1 - 0.135) = 1.16$$

First Story

$$Q = \frac{(1590)(0.25)}{(27.5)(144)} = 0.100$$

$$\delta_s = 1/(1 - 0.100) = 1.11$$

A summary of the moment magnifier obtained by using the Approximate Method (Section 10.11) and the Direct P-Δ Analysis (Section 10.13.4.2) of MacGregor and Hage and a second-order frame analysis computer program by Gouwens* is shown in Table 10-2.

---

* Gouwens, A. J. "Lateral Load Analysis of Multistory Frames with Shear Walls," Bulletin AEC2, Portland Cement Association, 1968.

TABLE 10-2 Comparison of Slenderness Effects by the Approximate Methods (Section 10.13) and the Direct P-Δ Method (Section 10.10)

	Moment Magnifier		
Story	ACI 318-95 Section 10.13.4.3 Eq. 10-19 $= 1/(1 - \Sigma P_u/0.75P_o)$	ACI 318-95 Section 10.13.4.2 $= 1/(1 - Q)$	Gouwens Computer
5th	1.03	1.05	1.07
4th	1.08	1.08	1.10
3rd	1.13	1.12	1.13
2nd	1.18	1.16	1.15
1st	1.11	1.11	1.13

EXAMPLE 3: SLENDERNESS EFFECTS—BRACED FRAME. Consider an edge column in a low building with end walls and stairwell walls (shear walls) which provide lateral bracing in the E–W direction (Fig. 10-29). Factored axial load, P_u = 390 kips for the first floor columns which have a clear height, h, of 14 ft 6 in. The bending moment in the E–W direction computed for a preliminary column size of 10 in. × 14 in. was 98 ft-kips at the top of the first floor edge column; individual column footings were considered hinged, moment at the bottom equal zero. It is desired to maintain the 10 in. width of these columns in the N–S direction. Use Grade 60 bars; f'_c = 5000 psi for columns. Short column design requirements:

$$P_u = 390 \text{ kips}$$

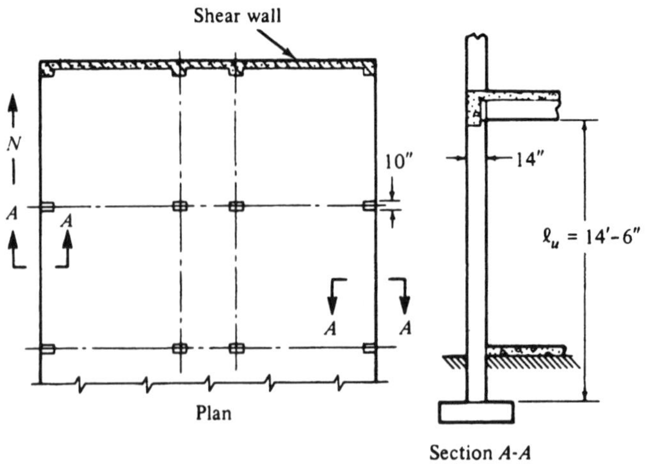

Figure 10-29 Column location.

$$M_{ux} = 98 \text{ ft-kips}$$

$$e_x = 3.03 \text{ in.}$$

$$M_{uy} = (390)\,\frac{(0.6 + 0.03 \times 10)}{12}$$

$$= 29.3 \text{ ft-kips}$$

Select a short column from a design handbook, 10 × 14, 4-#11, $\rho = 0.0445$, capacity, $\phi P_n = 393$ kips at $e = 3$ in. Compute ratio, M_1/M_2, smaller-to-larger end moment. $M_1 = 0$, $M_2 = 98$ ft-kips; $M_1/M_2 = 0$. Take $k = 1.0$.

$$\frac{k\ell_u}{h} = \frac{14.5 \times 12}{14} = 12.4 \quad \frac{k\ell_u}{r} = \frac{(1.0)(12.4)}{0.3} = 41.4 > 34 - 12(M_1/M_2),$$

and so slenderness effects must be considered. Compute the effect of end restraints,

$$C_m = 0.6 + 0.4\,\frac{M_1}{M_2} = 0.6$$

In Fig. 10-27, select Chart C for braced frames with $\rho = 0.04$, enter with $C_m = 0.6$ and $k\ell_u/h = 12.4$ and read $\delta_{ns} = 2.2$; repeat with chart for $\rho = 0.06$, and read $\delta_{ns} = 1.8$. Use $\delta_{ns} = 2.1$. $\delta_{ns}\,M_2 = 2.1 \times 98$ ft-kips $= 206$ ft-kips.

Select a 10 × 14 columns, 6-#11 vertical bars, $\rho = 0.0668$ with a design capacity as a short column of $P_u = 390k$ at $e = 6.3$ in. Note that use of slenderness coefficients from charts for $f'_c = 4000$ psi is conservative for designs using $f'_c = 5000$ psi since δ_{ns} is reduced slightly with increased E_c.

Again as in the preceding example, by inspection it is unnecessary to compute biaxial slenderness reductions in strength about each axis and biaxial bending strength. The N–S moments are small (29.3 ft-kips), the column strength for bending about the minor axis is $P_u = 586$ kips at $M_u = 48.8$ ft-kips, and by inspection, the full factored load can occur only when panels on both sides of the E–W (minor) axis are fully loaded, and full factored axial load and factored biaxial bending moment cannot occur simultaneously.

Variation From Maximum Load, $P_u = \phi P_{n(\max)}$

The question of error involved in applying the charts of Fig. 10-27 to conditions where $P_u < \phi P_{n(\max)}$ as in Example 3 can be readily determined, and corrections applied. The moment magnification factor, δ_{ns}, for slenderness effects increases with larger values of factored axial load, P_u

$$\delta_{ns} = \frac{C_m}{1 - P_u/0.75P_c} > 1.0 \quad \text{Code Eq. (10-10)}$$

Since the factored axial load, P_u strength, is less than $\phi P_{n(\max)}$, it is conservative to use the charts for δ_{ns} with $P_u \leqslant \phi P_{n(\max)}$, but, as the factored axial load becomes much smaller than $\phi P_{n(\max)}$, it obviously may become overconservative and wasteful.

Check the case ($\delta_{ns} = 2.1$) of Example 3. Determine a more refined value of δ_{ns} when $P_u < \phi P_{n(\max)}$, $P_u = 390$ kips and $\phi P_{n(\max)} = 625$ kips.

$$\delta_{ns} = 2.1 = \frac{1}{1 - 625/0.75P_c}$$

$$1 - \frac{625}{0.75P_c} = \frac{1}{2.1} = 0.48$$

$$\frac{625}{0.75P_c} = 0.52 \text{ compared to } \frac{393}{0.75P_c} = 0.33$$

$$1 - \frac{393}{0.75P_c} = 0.67$$

$$\text{Corrected for } P_u = 393 \text{ kips, } \delta_{ns} = \frac{1}{0.67} = 1.49 \text{ vs. } 2.10$$

Corrections for such changes in the load, P_u, when using the charts of Fig. 10-27 are not significant. Such refinements in the elaborate analysis for slenderness effects, in view of the much larger uncertainties involved in the assumptions and tests upon which the theory is based, are not warranted. The expenditure of design time for more definite and more significant refinements arising from consideration of reductions in the effective depth, *d*, at splices of reinforcement, routinely neglected by many designers, is much more easily justified. See the following section.

Column Details

Under the 1995 ACI Code, the design of columns has been rationalized to reconcile theory and test results and refined to result in nearly uniform safety against overloads. Computer programs and computer-prepared handbook designs, not only available but necessary for practical design, should not, however, induce such confidence in the apparently authoritative precision of calculations, that the art of design is neglected. None of the design aids to the best of the authors' knowledge consider the effect of practical details such as splices of reinforcement and construction joints, always present in real building columns. With the more precise estimation of strength, consideration of these practical details and appropriate corrections to design capacity, often considered negligible under past codes, becomes more important. Most of the Code requirements for column details have been consolidated into seven sections of the Code (Sec-

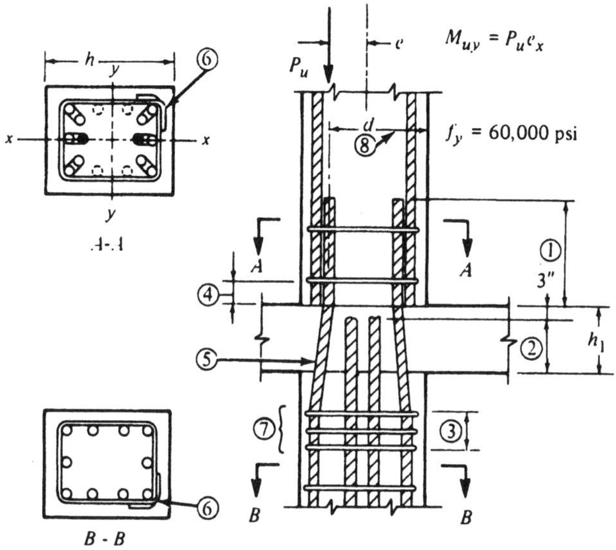

Notes to Fig. 10-30:

(1) Compression lap lengths = 30 bar diam (see Table 13-2). Within the minimum tie area = 0.0015 *hs*, lap length = 25 bar diam. for #11 or smaller bars only (Section 12.17.2.4).

(2) Standard practice detail length is (h_1 − 3 in.) unless designer specifies differently (ACI 315). This dimension must be $1.0\,\ell_d \geqq 8$ in. for compression if cutoff bar is required for compression just below the floor (Section 12.3). For compressive ℓ_d lengths, see Table 13-1.

(3) Extra ties must be located within 6 in. of the lower bend point of off-set verticals. (Section 7.8.1.3).

(4) The first tie must be located no more than one-half tie spacing above floor (Section 7.10.5.4).

(5) The maximum slope from the vertical is 1:6 (Standard detail ACI 315 and Section 7.8.1.1).

(6) Alternate hook position (ACI 315 recommended note).

(7) These ties designed to resist 1.5 times the horizontal component of the compressive force in the offset bent bars = $(1.5/6) f_y A_s'$ (Section 7.8.1.3).

(8) The reduced depth, *d*, at the splice equals the column thickness minus the concrete cover minus the tie dia. and minus one and one-half bar dia. (Section 7.7.1).

Figure 10-30 Compression lap splices—100% A_s—with offset bent vertical bars.

tions 7.6, 7.7, 7.8, 7.10, 12.14, 12.16, and 12.17). Some of these common requirements are shown in Fig. 10-30. Note the reduction in the effective depth, minimum one bar diameter to the reinforcement at the base of the upper story column. Even where the column size is unchanged, it is standard practice to off-set bend the vertical bars from below, but not necessary except in some round columns where spacing limitations permit two sets of verticals in one ring. The location of the offset bars may slightly reduce the design moment strength of columns with compression lap splices. It can have a much greater effect on reducing the design moment strength of columns with tension lap splices.

The art of design is involved in maintaining the design capacity at full value through all critical sections of a column as it is a member constructed in pieces with the joints and splices customarily located at points of maximum moment. Special details may be required with the use of higher strength concrete in the column than in the floor system or footing for overall economy. Where the column concrete, f'_c, is not more than 40 percent greater than f'_c for the floor system, no special details are required (Section 10.15). If it is desired to use a column concrete strength in excess of this value, the designer has three options:

1. Specify that the column strength concrete be placed in the floor system over the column area and in the slab around the column, as shown in Fig. 10-31, for a distance of 2 ft measured at the top surface of the slab. This operation involves the placing of two different mixes of concrete simultaneously and requires continuous dependable field inspection better than will usually be supplied. It has the advantage of simplicity and provides a small bonus shear capacity in the surrounding floor, but the authors recommend it for use only where field inspection is provided (see Fig. 10-31).
2. The second option is to add dowels or spirals through the floor system to increase the capacity by the required amount. Spirals are seldom practicable and generally uneconomical for nonseismic design. If the additional capacity is to be provided by dowels, these dowels must be developed for compression on either side of the weaker floor system depth (see Fig. 10-32). For the usual column designed for $P_u = \phi P_{n(\max)}$ about both principal axes and in either direction, the added dowel area may be located symmetric about both axes. Where a column is required to resist a large bending moment about one axis, the added dowels should be located within the plan of the column near the centroid of the assumed stress block.

   The added dowels are easily overlooked, especially because they will usually be added into the fresh concrete at the conclusion of casting the

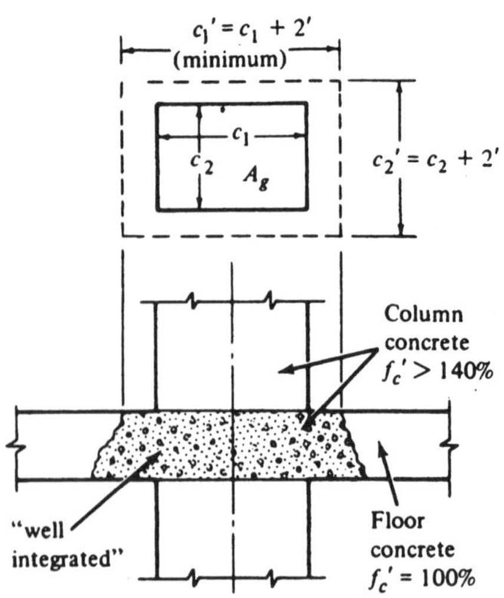

Note: Column concrete in the floor *must* be cast *first*, as shown.

Figure 10-31 Two-strength floor (not recommended).

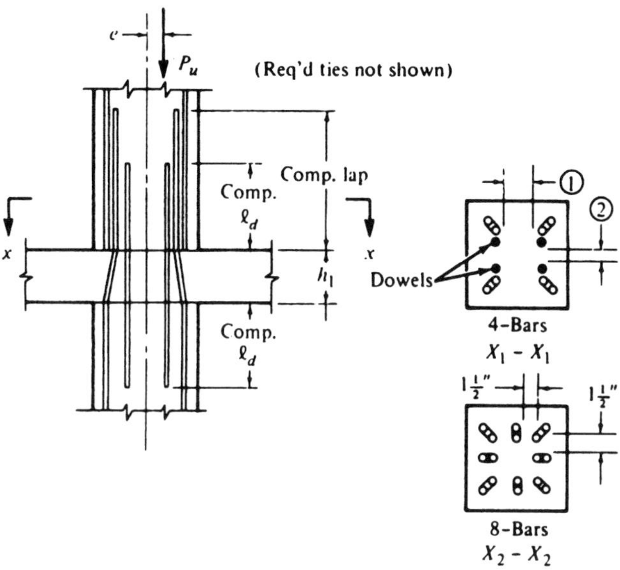

Notes to Fig. 10-32:
(1) Dimensions (1) and (2) must be at least one and one-half bar diam $\geqq 1\frac{1}{2}$ in. clear between any group of dowels or bottom bar and top bar. If the splices were at minimum spacing, *Section* $X_2 - X_2$, dowels are impractical (Section 7.6.3).
(2) If spacing of vertical bars permits the addition of dowels, required dowel length is equal to 2 times ℓ_d plus the floor depth, h_1 (see Table 13-1).

Figure 10-32 Offset verticals plus added dowels (not recommended).

lower column. If they are used to supplement offset verticals from below, the authors recommend provision of complete details showing location both in plan and elevation in the column schedule as well as field inspection before casting the floor. Provision of added capacity under this option is simpler where separate dowels are to be used in any event (see Fig. 10-33). Where separate dowels are provided for the splice, the added capacity can be supplied simply by requiring the use of larger size dowels.

3. The third option is to compute a reduced column capacity based on a composite value of f_c' equal to 75 percent of the column concrete and 35 percent of the floor concrete f_c'. This option is self-defeating for economy, except for columns controlled by tension.

In general for practical design, the authors recommend selection of concrete strengths within the limit of 40 percent so that this problem can be neglected since all three options require special conditions to be effective.

Splices in Column Verticals

Columns bars which are required only to resist compression may be butt spliced, Fig. 10-34 (Section 12.16.4). Splices must be staggered so that the continuing

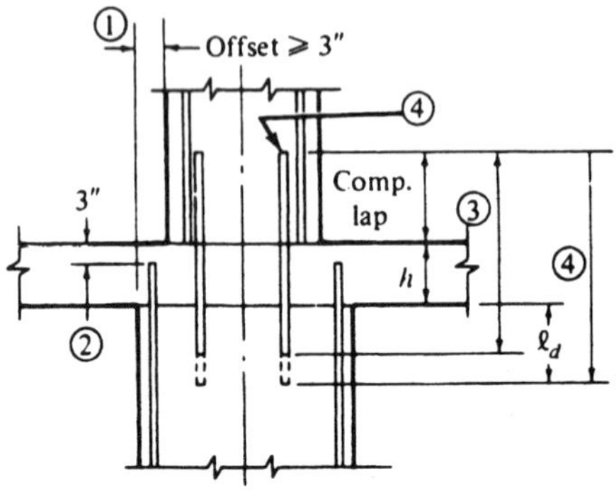

Notes to Fig. 10-33:
(1) Where face of column above is offset 3 in. or more from the face of the column below, offset bars are not permitted; the splice must be made by separate dowels (Section 7.8.1.5).
(2) Cutoff column verticals stop 3 in. below finished floor unless otherwise noted (ACI 315 standard practice).
(3) Length of dowel equals 2 times the required lap for tension or compression plus 3 in.
(4) If f_c' for the column concrete is more than 40 percent greater than for floor concrete, added compression capacity required may be provided by increasing the size of the separate straight dowels. For this purpose, the extension of the dowel below the floor must be $1.0\,\ell_d$ (Section 10.13.2). Dowel length will be as determined above or one compression lap length plus the floor thickness plus $1.0\,\ell_d$, whichever is larger. Note here also that it will be coincidental if midpoint of the dowel length is located at the finish floor line; designer must indicate embedment required; otherwise placing practice is to extend half the length above finished floor.

Figure 10-33 Separate dowels at offset faces.

bars in each face of the column have a tensile strength, based on yield strength f_y, not less than 0.25 times the area of vertical reinforcement in that face (Section 12.17.4). The difficult design requirements for splices of column verticals involve tensile capacity (Sections 7.8, 12.15 and 12.17). In columns controlled by tension, where all splices are located near points of maximum moment, "full" tensile splices are required. Full tensile splices may consist of: (1) Class B tension laps Figs. 10-33 and 10-36; (2) full welding, butt welds developing 125 percent of f_y Fig. 10-35; (3) full mechanical connections, proprietary splice devices developing 125 percent of f_y (Sections 12.15.2, 12.15.3, and 12.15.4). Whenever the calculated tension in a bar exceeds $0.5\,f_y$, a full-tension splice is required (Section 12.17.2) regardless of the percentage spliced. Whenever the calculated tension in a bar is less than $0.5\,f_y$, Class B splices must be used if more than one-half the bars are spliced at any section, and Class A splices may be used if half or fewer of the bars are spliced at any section, and if alternate lap splices are staggered by one development length.

Columns with eccentricity inside the kern ($e \approx 0.17h$ for square and $0.13h$ for round) have no tension, and if the bars are #11 or smaller, compression lap splices for all bars at one section are adequate for sufficient tension to preclude special requirements (Commentary Section R12.17.2.1). If the bars are #14 or #18 in size, end-bearing butt splices can be used, but must be staggered so as not to splice more than 75 percent of the bar area in any face of the column (Section 12.17.4). For larger eccentricities, up to those causing tension of one-

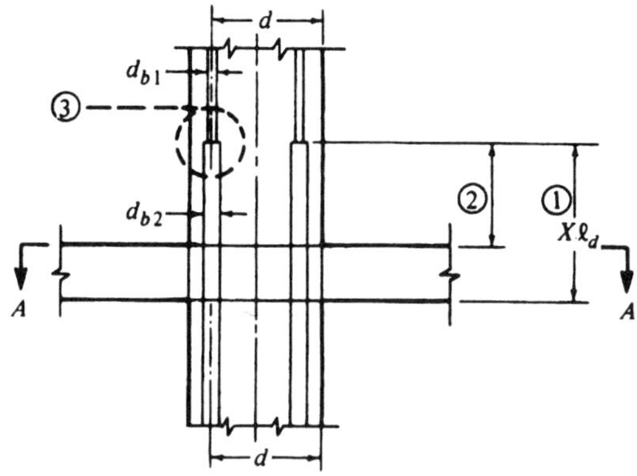

Notes to Fig. 10-34:
(1) *Compression controls.* If the larger bar is required to carry the additional load of the floor slab compression at the top of the lower column, the distance from the bottom of the slab to the splice must be sufficient to develop this additional load. The fraction, $X = [1 - (d_{b1}/d_{b2})^2]$, or ℓ_d for full compression development should be adequate.
Tension controls. Similarly, if the larger bar is required to develop full tensile capacity at the top of the lower column, the extension above the bottom of the floor, must be $X\ell_d$ for full tension development (Section 12.15). See Table 13-3 for ℓ_d.
(2) If column $f'_c > 140$ percent floor f'_c, the difference between the areas of the larger and smaller bars can be considered effective as added dowel area through the floor provided the extension above the floor is $X\ell_d$ for compression (Sections 12.1.1, 12.3, and 10.15.2). See Table 13-1 for compression ℓ_d. The procedure is recommended by the authors in lieu of adding separate dowels when conditions permit its use.
(3) Note that there will be loss of available effective depth, d, for the smaller bar at a concentric butt splice of larger to smaller bars where cover is established by the larger bar. The loss $= 1/2\,(d_{b2} - d_{b1})$. A correction should be considered to computed capacity if tension controls *and* if the splice is located near a point of maximum moment. Location of splices in the center third of the column story height will avoid the point of maximum moment for the usual condition of double curvature. Otherwise, the indicated correction should be considered.

Figure 10-34 Butt splices of larger to smaller bars.

half f_y in the reinforcement, various combinations of staggered end-bearing splices, compression lap splices, tension lap splices, separate dowels, welded splices, or mechanical connections with less than 125 percent f_y capacity may be used.

CONFINEMENT

Vertical bars in columns must be confined by ties (Section 7.10.5) or spirals (Section 7.10.4). In the alternative, beams may provide confinement if they frame into all four sides of a column (Section 7.10.5.5). Typical tie details are shown in Fig. 10-37, and typical tie spacing requirements are shown in Fig. 10-38. Confinement may also be used to shorten lap lengths by a multiplier of 0.83 when the area of the ties throughout the lap splice length have an effective area not less than $0.0015hs$. The area of the tie legs must be satisfied in each direction of the column (Section 12.17.2.4).

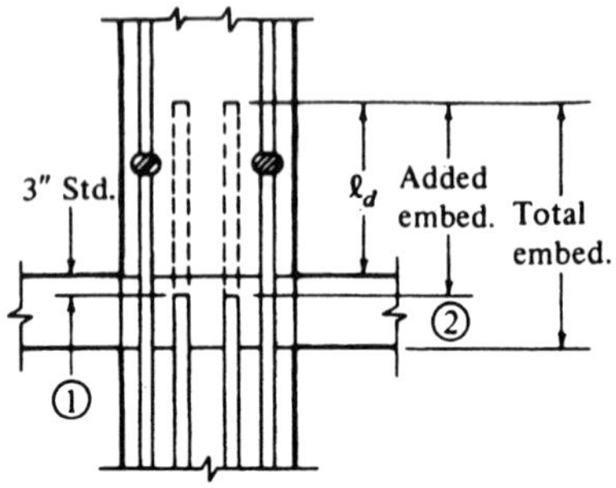

Notes to Fig. 10-35:
(1) Standard detailing practice is to cut off bars as shown when discontinued in the next story unless otherwise dimensioned by the engineer (ACI 315).
(2) If the cutoff bars were required at the top of the lower column for stress added by the floor system, the total extension above the bottom of the floor beam or slab should be $1.0\ell_d$ for tension. See Table 13-1 for compression; Table 13-3 for tension.

If the column concrete $f'_c > 140$ percent floor concrete f'_c, the cutoff bars can be extended to avoid use of separate dowels. To be considered fully effective at the top of the floor slab, the extension required is 3 in. plus ℓ_d for compression (Sections 10.15.2, and 12.3). This device is recommended by the authors where conditions permit instead of using separate dowels.

Figure 10-35 Butt splice same size bars—reduced quantity.

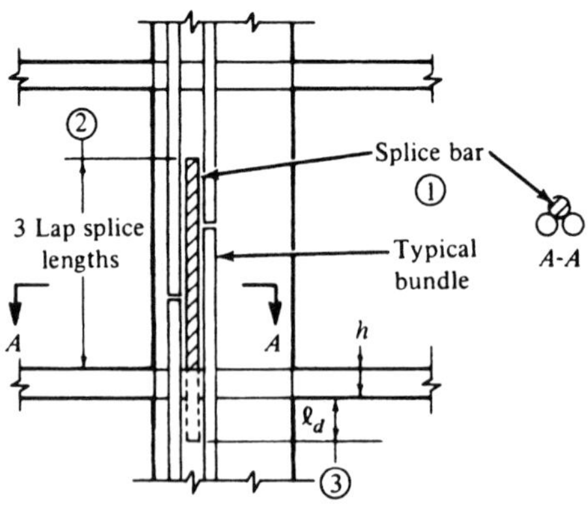

Notes to Fig. 10-36:
(1) The splice bar must be the same size as the bars in the 2-bar bundle if the lap splice is for full development of the bundle, whether for torsion or compression. This method of making a splice technique has been used with sheared ends; or, for partial tension capacity only, with endbearing splices using a smaller size splice bar.
(2) The first splice (left bar of bundle in Figure) is most conveniently located one required lap length above the floor. The splice bar is a minimum of 3 times the lap length required for a bar in a 3-bar bundle which equals 3.6 times the individual bar lap length required (Section 12.4).
(3) If column concrete $f'_c > 140$ percent floor concrete f'_c, extend the splice bar a distance ℓ_d below the floor for compression and it will also serve in place of an added separate dowel. (Section 10.15.2).

Figure 10-36 Bundled bars—lap spliced 2-bar bundle.

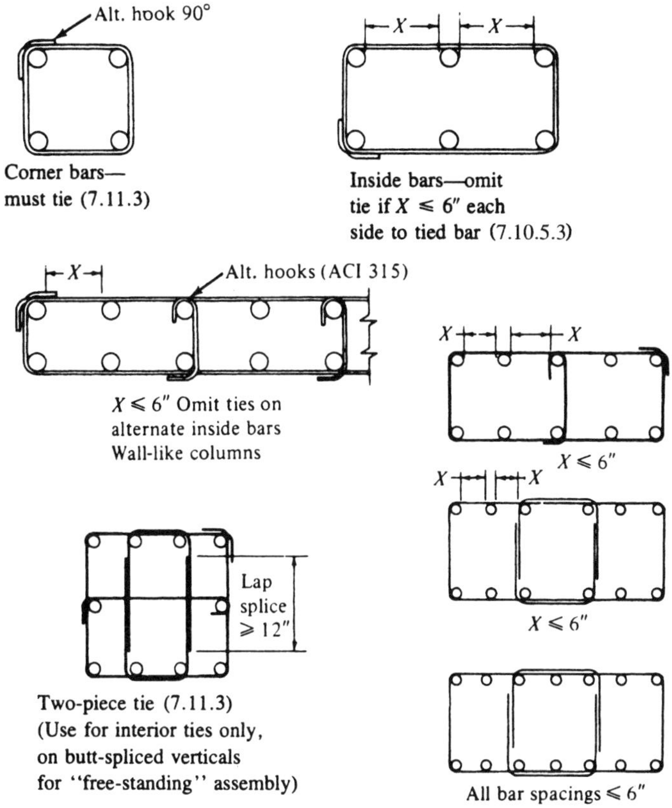

Notes to Fig. 10-37:

(1) For complete detail examples and explanations of accepted industry practice, see *ACI Detailing Manual-*1994 (SP-66).

(2) In any rectangular column the four corner bars are preferably enclosed by a single, one-piece tie to aid in holding the bars in position during erection. For extremely long wall-like columns, a two-piece single tie or overlapping separate rectangular ties may be more convenient to hold the shape of the outer bars in plan. Each bar in the corner of any column must be held in the corner of a tie with an included angle not more than 135° (Section 7.10.5.3).

(3) The 4-bar column detail shows the preferred standard tie with overlapping 90° hooks forming the splice (Section 7.11 and ACI 315). The position of this splice should be staggered at each tie spacing by rotating the tie 90° or 180°. A note to this effect should be added to the detail on the structural drawings and should be repeated by detailers on the placing drawings. (For the confinement area in seismic "ductile frames," overlapping longer 135° hooks are usually required.)

(4) The 6-bar column detail shows the usual standard tie required. The intermediate bars do not require the support of a corner of a tie if the clear spacing (X) on both sides is not more than 6 in. to a bar in the corner of a tie. If $X > 6$ in., the two bars are most easily held by a "candy-stick" tie (the 2-bar tie shown in the "wall-like column" detail). Again, it is recommended that the designer add the note to alternate the position of the 90° and 180° hooks at each tie spacing.

(5) The 2-piece tie and the candy-stick tie are recommended for all interior ties in columns using large, butt-spliced vertical bars, particularly with staggered splices, where in-place assembly on freestanding verticals is required (Section 7.11.3 and ACI 315). The required lap for 2-piece ties is a Class B tension lap (Section 12.15). See tension lap splices, Table 13-6, top bars.

Figure 10-37 Column tie detail requirements (Section 7.11.3).

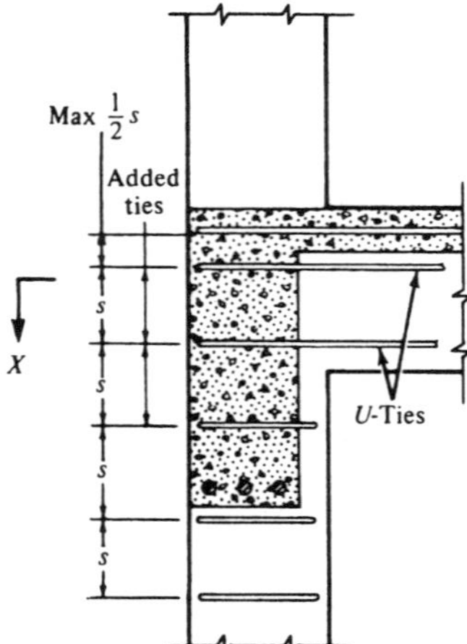

Tied column
(Section 7.10.5.4)

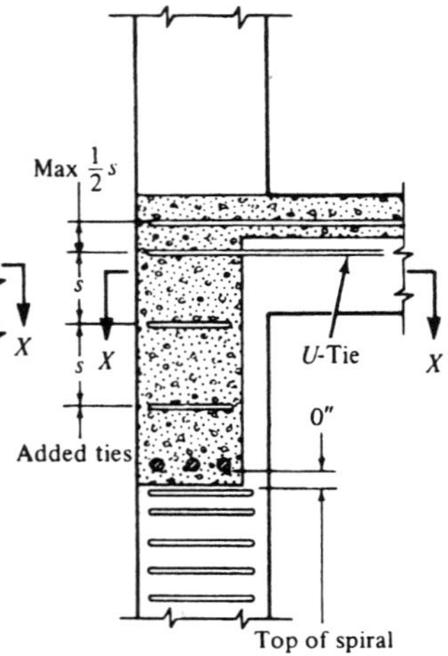

Spiral column
(Section 7.10.4.6)

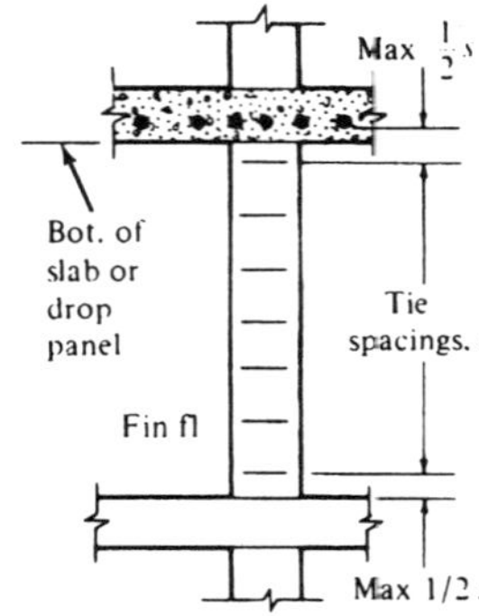

Tie spacing
(Section 7.10.5.4)

Maximum spacing per sets of ties, $s \leqslant$ { 16 Vertical bar diam; 48 Tie diam; Least dimension of col.

Min tie	Vertical bars (largest)	Max spacing *
#3	#5 #6 #7 #8	10 in. 12 in. 14 in. 16 in.
#3	#9 & #10	18 in.
#4	#11	22 in.
#4	#14 & #18	24 in.

*Max spacing not to exceed *least* dimension of the column.

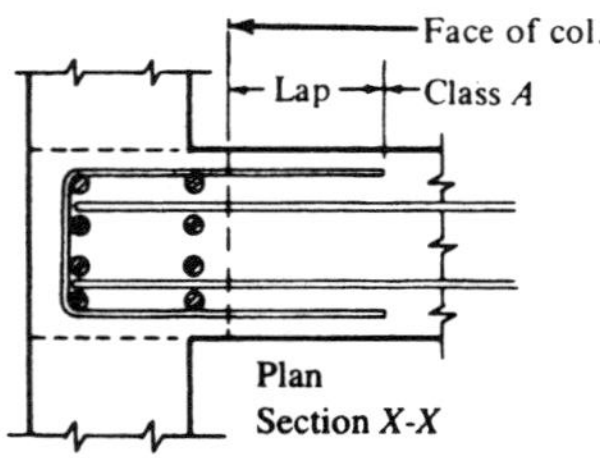

Added *U*-tie used to supplement beam bar embedment
(Sections 7.10.5.3 and 12.2.2)

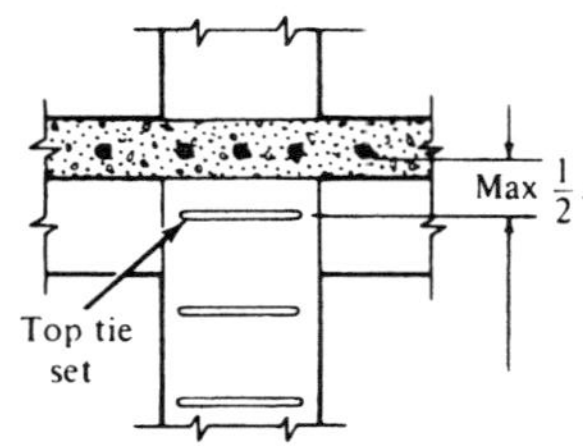

Beams on 1, 2, or 3 sides
(Sections 7.9 and 7.10.5.5)

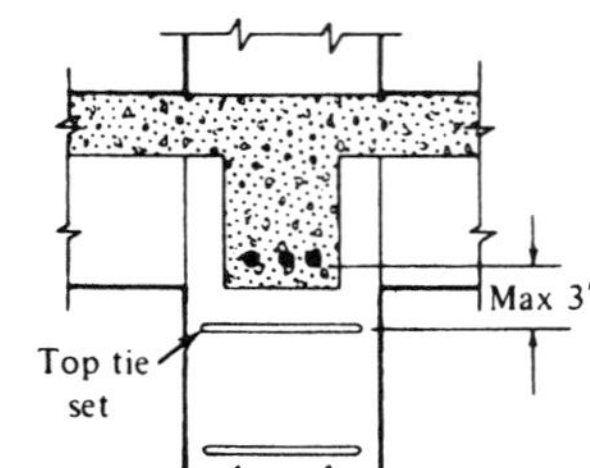

Beam confinement on 4-sides
(Sections 7.9 and 7.10.5.5)

Figure 10-38 Column tie spacing requirements—per set.

Notes to Fig. 10-38:

(1) Maximum tie spacings are governed by three criteria, whichever is least. In no case shall ties or set of ties be spaced further apart than the least dimension of the column. The other two criteria are 16 vertical bar diameters or 48 tie diameters (Section 7.10.5.2). The latter two criteria are easily tabulated for required minimum tie sizes and all practicable sizes of verticals. See tabulation.

(2) The lowest set of ties and the upper set are required to be within one-half tie spacing of the finished floor below and the lowest horizontal reinforcement in the slab or drop panel above, respectively. This requirement is not contingent upon the slab or drop panel providing concrete confinement on all four sides of the column and so it will apply to edge as well as interior columns.

(3) With beams or brackets to provide concrete confinement of the joint area on *all* (four) sides at the top, the top tie set must be within 3 in. of the lowest horizontal reinforcement in the beams or brackets. Note that when one or more sides are not confined by concrete, the ties must be within one-half tie spacing of the lowest steel in the slab.

(4) When only partial confinement by concrete beams is provided, added ties are required up to a maximum of one-half tie spacing, *s*, from the lowest *slab* steel (Sections 7.9 and 7.10.5.5). Above the normal top of the spiral in spiral columns, the same requirement applies (Section 7.10.4.7). Although the Code requires that the spiral extend *to* the level of the level of the lowest horizontal reinforcement in the deepest beam present (dimension 0″), in practice it is usually ended at the construction joint between the lower column and the floor system as the drawing shows. *IMPORTANT:* it is *not* accepted practice to supply ties designated as "added"; if desired, complete details for added ties *must* be prepared by the engineer and typical details shown on design drawings (ACI 315).

(5) A device is illustrated here to satisfy two code purposes with the same added tie. Where the added ties are required only at one unconfined face as shown, such ties need not be closed but may be U-shapes extended into the beam. Properly lapped past the face of the column with the beam bars, they may then be also considered effective additions to either the top or bottom beam bars, depending upon depth position. This device is recommended where depth of the column and/or spandrel is too shallow or congested to permit placing sufficiently long hooked ends on the beam bars required anchorage (Sections 7.9 and 12.12.2).

11 Walls

GENERAL

Concrete walls are defined (Section 2.1) as elements, usually vertical, used to enclose or separate spaces. Minimum reinforcement requirements given for walls in Section 14.3 apply to walls which are used to separate spaces and which function in the structure as compression members. The provisions of Sections 14.4 and 14.5 apply to walls which function as compression members. It should be noted that all minimum reinforcement requirements for walls in Chapter 14 may be disregarded "where structural analysis shows adequate strength and stability" (Section 14.2.7). The required analysis for flexural strength (Section 22.6) can be employed. Temperature and shrinkage reinforcement can be likewise reduced (Section 22.2.1). The design of such walls must be based on lateral forces or any other loads to which they may be subjected (Section 14.2.1). A rational method of design for walls subject to flexure or both flexure and axial compression is provided (Section 14.4). An empirical method of design for bearing walls with small moment (resultant compressive force within the middle-third of the wall thickness) is permitted (Section 14.5.1 and Section 22.6.5.1). The empirical design method is explained in detail (Section 14.5 and Section 22.6.5). Non-loadbearing walls whether precast or cast-in-place, ordinary reinforced or prestressed, must be designed by the rational method, taking into account all loading conditions (Sections 14.4, 16.3, and 18.1.3).

SPECIAL WALLS

Walls which principally resist horizontal shear forces (shear walls in low-rise buildings) in the plane of the wall and parallel to the length of the wall must conform to the requirements for shear (Section 11.10). Shear walls that resist forces from seismic accelerations must conform to the seismic requirements (Section 21.6). Special requirements for walls designed as grade beams are provided (Section 16.4.2). In addition to the general provisions (Section 14.4), precast concrete wall panels must conform to requirements for precast concrete wall panels (Section 16.4.2). Cantilever retaining walls are designed for flexure under Chapter 10 (Section 14.1.2).

COVER

Minimum concrete cover required for the protection of reinforcement (Sections 7.7.1, 7.7.2, and 7.7.3) for purposes other than fire resistance is shown in Fig. 11-1.

FACTORED LOAD COMBINATIONS

Walls must be proportioned for adequate strength using the load factors (Section 9.2) and the strength-reduction factors (Section 9.3). When subjected to lateral earth pressure, H, dead load, D, and/or live load, L, the required strength, U, must not be less than $1.4D + 1.7L + 1.7H$ (Section 9.2.4). When subjected to lateral pressure, F, from liquids, the required strength, U, must not be less than $1.4D + 1.7L + 1.4F$. The vertical pressure from liquids with a well-defined density and controllable height is considered as dead load (Section 9.2.5).

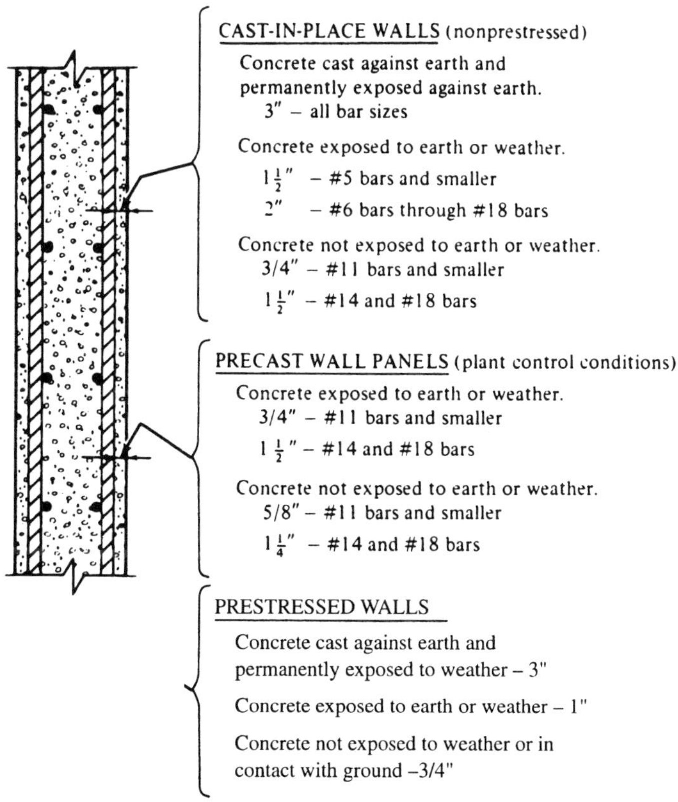

Figure 11-1 Concrete protection for reinforcement.

When D or L reduce the effect of H or F, such as in the design of the heel or toe of a retaining wall, the load factors for strength design must be reduced from 1.4 to 0.9 for D and from 1.7 to zero for L (Section 9.2.4).

NON-LOADBEARING WALLS

Nonloadbearing walls can be defined as vertical elements that support no vertical load other than their own weight. These include basement walls, cantilever retaining walls, area enclosure walls, exterior precast panel walls, tilt-up walls, and cast-in-place walls, for hydraulic structures. Such walls are subjected primarily to flexure or to flexure and relatively small axial compression loads with the resultant compressive forces outside the kern ($e > h/6$) of the section and must be designed by the rational method (Section 14.4). Design for shear forces perpendicular to the face of such walls must conform to the requirements for slabs (Section 11.12).

Minimum vertical reinforcement (Sections 14.3.2, 14.3.5, and 14.3.6) and horizontal reinforcement (Sections 14.3.3 and 14.3.5) for non-loadbearing walls with Grade 60 reinforcement of #5 bars or smaller are shown in Figure 11-2. When bars larger than #5 are used, the minimum vertical and horizontal reinforcement must be increased 25 percent (Sections 14.3.2 b and 14.3.3 b). The Code does not require that minimum reinforcement be placed in the two faces of a wall (Sections 14.1.2, 14.2.7 and 14.4).

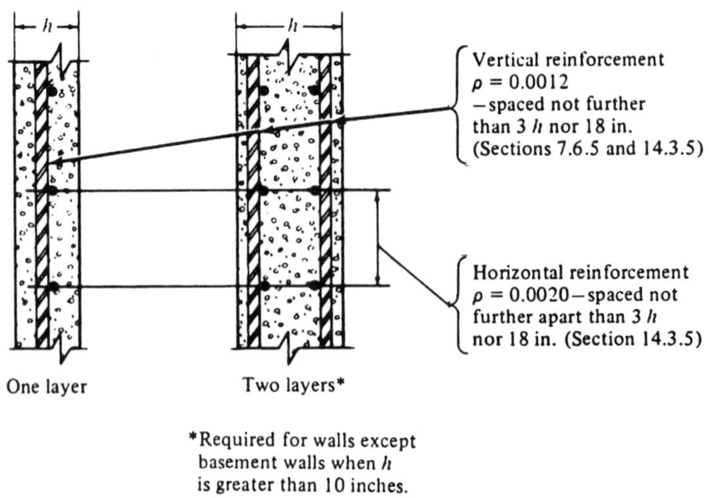

Figure 11-2 Minimum wall reinforcement and maximum spacing.

Thickness *h* (inches)	IN SINGLE LAYER ONLY[a]		IN BOTH FACES[b]	
	Horiz. $\rho = 0.0020$	Vert. $\rho = 0.0012$	Horiz. $\rho = 0.0020$	Vert. $\rho = 0.0012$
6	#3 @9	#3 @18	—	—
8	#4 @12	#3 @11	—	—
10	#5 @15	#4 @16	#4 @18	#3 @18
12	#5 @12	#4 @13	#4 @16	#3 @15
14	#5 @11	#5 @18	#4 @14	#3 @13
16	#5 @10	#5 @16	#4 @12	#3 @11
18	#5 @$8\frac{1}{2}$	#5 @14	#5 @17	#4 @18

[a] Bars arranged in two-way, mat, usually centered in wall.
[b] Bars arranged in identical two-way mats, one near each face of wall.

NOTE: When reinforcement is placed in two layers, maximum spacing limitations apply separately, but minimum reinforcement requirements (ρ) apply to the sum of the areas of reinforcing steel in both faces.

Although a literal application of the requirements for minimum reinforcement (Sections 14.4 and 18.1.3) would apply to prestressed non-loadbearing panel walls, the authors would suggest that nonprestressed reinforcement be added to such wall panels *only* if required for excess tension under a service loading, including temperature and shrinkage effects. One additional exception for non-loadbearing walls is the thin "serpentine"cantilever enclosed wall that may be considered to be a simple thin shell structure (Section 19.1.3). Again a strict application of Code requirements would require minimum reinforcement as in a plane wall (Sections 19.1.2 and 14.4). The authors suggest that vertical reinforcement be provided *only* in such amounts and at such locations as required by structural analysis, considering effects of all factored loads forces without regard to minimum areas of reinforcing steel or spacings prescribed for plane walls (Section 14.2.7).

LOADBEARING WALLS, EMPIRICAL DESIGN

Loadbearing walls are used primarily as compression members. When compression loads are eccentric, the wall must be capable of resisting the compression load and the bending moment resulting from the eccentricity. Loadbearing walls can all be designed by the rational method (Section 14.4), or by the empirical method (Section 14. 5) only if the eccentricity of the load is equal to or less than one-sixth of the wall thickness ($e < h/6$). When designed by the empirical method the ratio of the unsupported height or length whichever is shorter, to thickness ℓ_c/h cannot be more than 25 (Section 14.5.3.1), or 30 for nonloadbearing walls (Section 14.6.1), unless it can be shown by a structural analysis that adequate strength and stability will exist at greater ratios (Section 14.2.7).

Minimum vertical and horizontal Grade 60 reinforcement of #5 bars and smaller for loadbearing walls designed by the empirical method is the same as for loadbearing walls designed by the rational method. It is not required that half of the reinforcement be placed in each layer. If cover requirements are met, a minimum of one-third and a maximum of two-thirds of the reinforcement can be placed in one layer. The minimum quantity of reinforcement for loadbearing walls designed by the empirical method may be waived where structural analysis shows adequate strength (Section 14.2.7).

Wall capacity by the empirical method* (Section 14.5, Eq. 14-1) is based on the average stress, P_{nw}/A_g, as shown in Fig. 11-3 and not on the peak stress that results from eccentricity of the load. In Figure 11-3 the compression load capacity is given for various ℓ_c/h ratios in terms of the average stress and concrete strength as

$$\frac{\phi P_{nw}}{A_g f_c'} = 0.55\phi\left[1 - \left(\frac{k\ell_c}{32h}\right)^2\right] \quad \text{(Section 14.5)}$$

where $k = 0.8$ for walls restrained against rotation at one or both ends, $k = 1.0$ for walls unrestrained at both ends, and $k = 2.0$ for walls not braced against lateral translation (Section 14.5.2)

LOADBEARING WALLS, RATIONAL DESIGN

Wall capacity by the rational method (Section 14.4) uses the general principles and requirements for column design (Section 10.3) to determine the design strength and takes account of the slenderness effects (Section 10.10 or 10.11). Minimum vertical and horizontal reinforcement for loadbearing walls designed by the rational method (Section 14.4) is the same as for nonloadbearing walls, as shown in Figure. 11-2. Lateral ties are not required when the area of the vertical reinforcement is equal to or less than 1 percent of the concrete area, or where it is not required as compression reinforcement (Section 14.3.6) (see also Figs. 10-1, to 10-4, and 10-14).

COMPARISON OF WALL DESIGN METHODS

A comparison of the rational design strength and the empirical design strength of a 12 in.-thick reinforced concrete loadbearing wall for various ℓ_u/h ratios is

* "Interesting Aspects of the Empirical Wall Design Equation," K. M. Kripanarayanan and Discussion; *ACI Journal*, May 1977, p. 204 and November 1977, p. 566. "Strength Investigation of Precast Reinforced Concrete Load Bearing Wall Panels," Oberlander, Garold D. and Everard, Noel, J., University of Texas, Arlington, Texas, 1973.

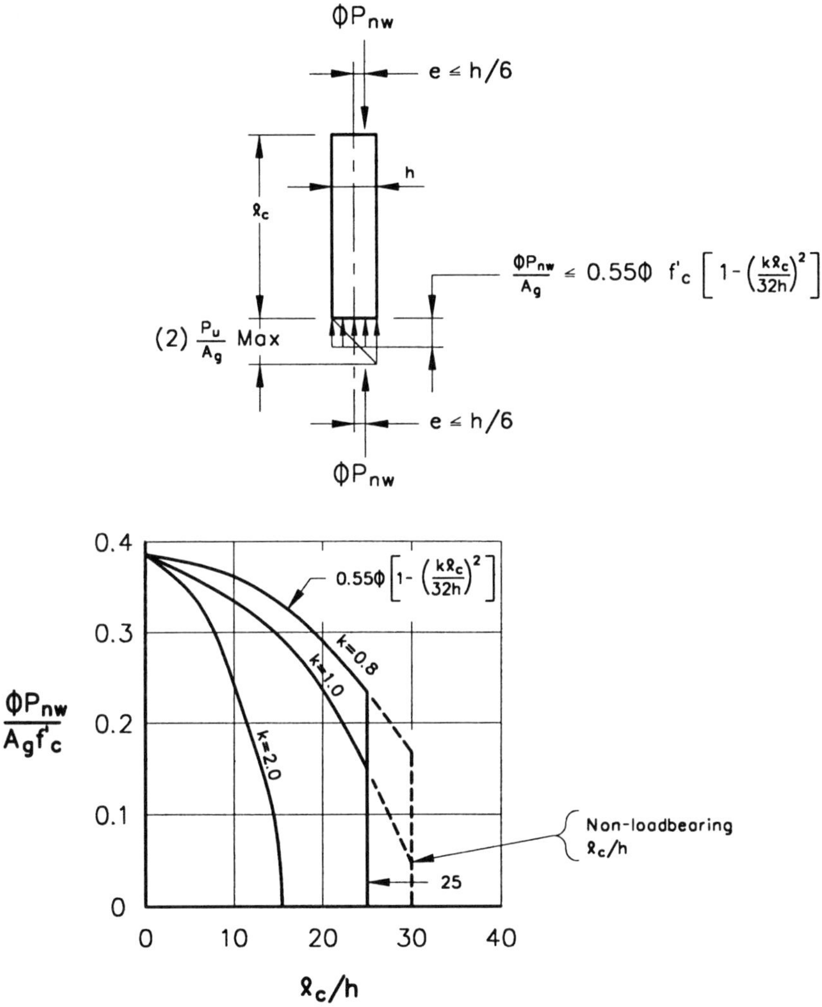

Figure 11-3 Walls by empirical design.

shown in Figure 11-4.* The wall of this example is braced against sidesway and is restrained against rotation at one end. It is reinforced with 1 percent vertical reinforcement and is subjected to a compression load with an eccentricity of 2 in. This ecentricity is the maximum permitted under the empirical design method. Figure 11-4 shows that for single curvature, $k = 1.0$ and $C_m = 1.0$ (Sections 10.11.2 and 10.11.5) the rational method of design provides greater

* See also Chapter 10 for other comparisons. In Chapter 14, $\ell_u = \ell_c$.

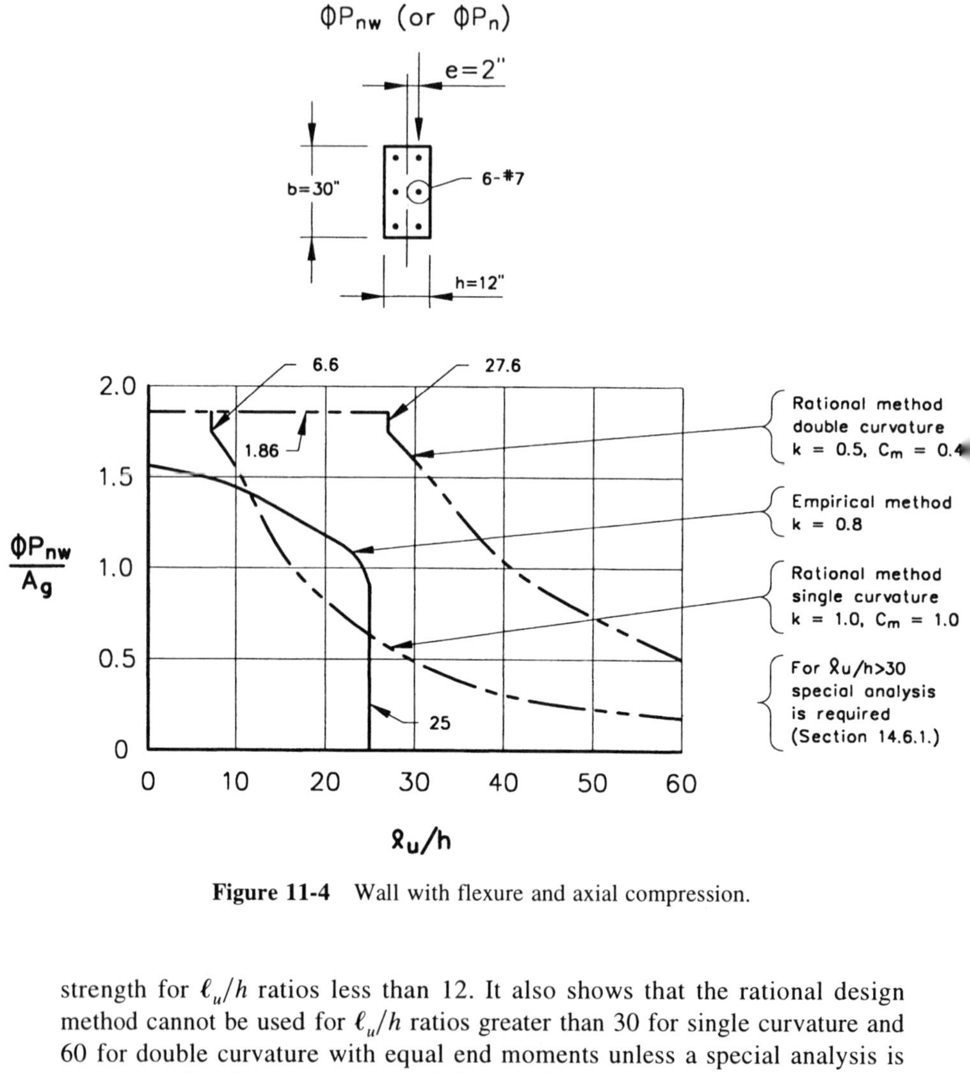

Figure 11-4 Wall with flexure and axial compression.

strength for ℓ_u/h ratios less than 12. It also shows that the rational design method cannot be used for ℓ_u/h ratios greater than 30 for single curvature and 60 for double curvature with equal end moments unless a special analysis is made. For double curvature, $k = 0.50$ and $C_m = 0.40$, the rational method of design provides greater strength for all ℓ_u/h ratios.

The rational design strength of the wall shown in Figure 11-4 was determined from a 12 in. × 30 in. section of wall with $f'_c = 4{,}000$ psi and reinforced with six #7 vertical bars as shown in Fig. 11-4. The radius of gyration was assumed equal to 0.3 h (Section 10.11.2) and the modulus of elasticity of the normal weight concrete was taken equal to 57,000 $\sqrt{f'_c}$ (Section 8.5.1). A load-moment interaction diagram was generated for this section of wall as a short column about the weak axis neglecting slenderness effects. This envelope is shown in Figure 11-5.

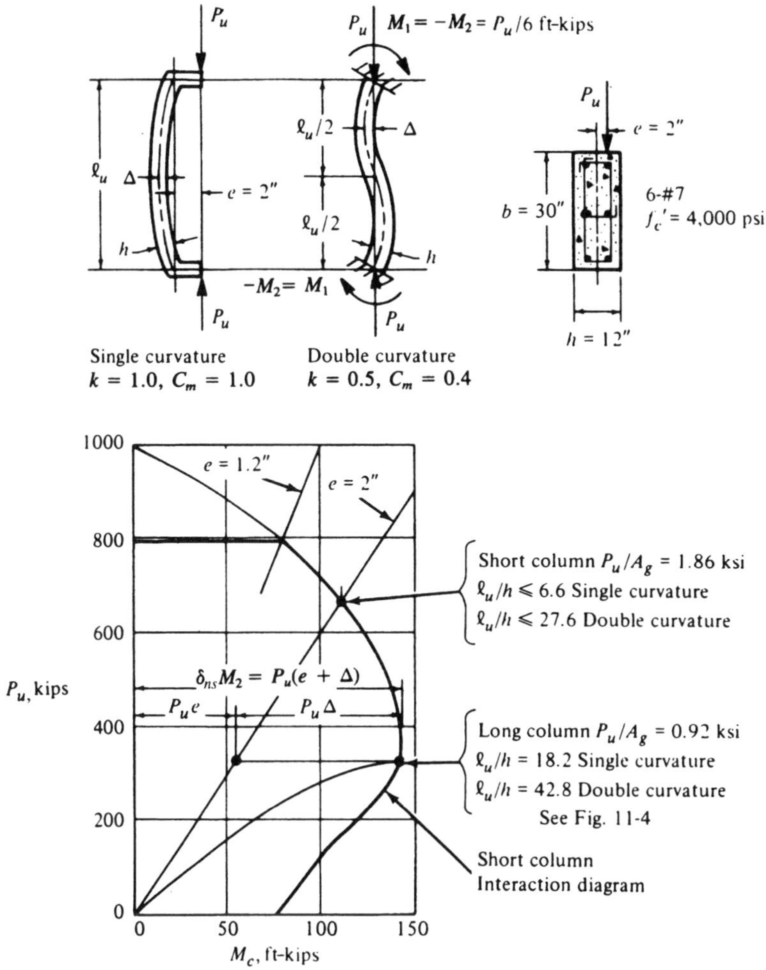

Figure 11-5 Slenderness effects.

SLENDERNESS EFFECTS

An approximate evaluation of slenderness effects was made (Section 10.11) for the wall examples designed by the rational method, but $k\ell_u/r$ values greater than 100 were not considered to avoid making a special analysis (Section 10.11.5). Because the wall of the example is braced against sidesway in the weak direction, slenderness effects can be neglected in this direction for $k\ell_u/r$ values less than $34 - 12\ M_1/M_2$ (Section 10.12.2). For single curvature and $M_1 = M_2$, $k = 1.0$ and $C_m = 1.0$, slenderness effects in the weak direction of the wall can be neglected for $k\ell_u/r$ values less than 22 (ℓ_u/h values less than 6.6). For double curvature and $M_1 = -\ M_2$, $k = 0.5$ and $C_m = 0.4$, slenderness effects in the weak direction can be neglected for $k\ell_u/r$ values less than 46 (ℓ_u/h values less than 27.6).

The load-moment interaction curve for the example wall section as a short column is shown in Figure 11-5 and is applicable for ℓ_u/h ratios less than 6.6 for single curvature and 27.6 for double curvature. As shown in Figure 11-5, for ℓ_u/h ratios greater than 6.6 for single curvature and 27.6 for double curvature, factored moments, $P_u e$, must be increased by the magnification factor, δ_{ns} (Section 10.12.3).

$$\delta_{ns} = \frac{C_m}{1 - (P_u/0.75P_c)} \quad \text{(Code Eq. 10-10)}$$

The critical load P_c, for Eq. 10-10 is determined from the Euler formula,

$$P_c = \frac{(\pi)^2 EI}{(k\ell_u)^2} \quad \text{(Code Eq. 10-11)}$$

EI for use in Eq. 10-11 will be taken in accordance with the equation for columns with low percentages of reinforcement as

$$EI = \frac{0.4E_c I_g}{(1 + \beta_d)} \quad \text{(Code Eq. 10-13)}$$

If dead load, D, is equal to $2.43L$, $\beta_d = 1.4D/(1.4D + 1.7L)$. $\beta_d = 0.667$ and EI for the example wall becomes

$$EI = \frac{(0.4)(57{,}000)(\sqrt{4{,}000})\,(1/12)(30)(12)^3}{(1 + 0.667)}$$

$$= 3{,}740{,}000{,}000 \text{ lb in.}^2$$

with EI known, $k\ell_u$ can be solved for in terms of P_c

$$(k\ell_u)^2 = 3{,}740{,}000{,}000\left(\frac{\pi^2}{P_c}\right)$$

The maximum possible moment magnification factor at any particular eccentricity can be obtained directly from the load-moment interaction curve. Determine the ratio of the maximum moment capacity for any given load to the moment of that load at the particular eccentricity. Figure 11-5 shows that the maximum possible moment magnification factor at eccentricity of 2 in. for balanced load, $\phi P_b = 333$ kips (Section 10.3.2) is equal to $145/55 = 2.64$. The maximum possible moment magnification factor at load $P_u = 536$ kips for $e =$ 2 in. is $134/82 = 1.63$. The maximum possible moment magnification factors for other axial loads can be similarly calculated. Critical loads, P_c, and corresponding ℓ_u/h ratios and moment magnification factors for both single and double curvature conditions were calculated for specific factored loads, P_u.

Compressive stress ratios, P_u/A_g were also calculated for each specific factored load and compared in Fig. 11-4 with stress ratios for walls designed by the empirical method. Two such sample calculations for single curvature follow.

Single Curvature Calculations

1. ($P_u = 333$ kips, $\delta_{ns} = 2.64$, $k = 1.0$, and $C_m = 1.0$)

   Calculate P_c

$$\delta_{ns} = \frac{C_m}{1 - (P_u/0.75P_c)}$$

$$P_c = 715 \text{ kips}$$

   Calculate ℓ_u

$$(k\ell_u)^2 = \pi^2 EI/P_c$$

$$\ell_u^2 = 3{,}740{,}000{,}000\,(\pi)^2/715{,}000$$

$$\ell_u = 227 \text{ in.}$$

   Calculate ℓ_u/h

$$\ell_u/h = 227/12 = 18.9$$

   Calculate P_u/A_g

$$P_u/A_g = 333/(12)(30) = 0.92 \text{ ksi}$$

2. ($P_u = 536$ kips, $\delta_{ns} = 1.505$, $k = 1.0$, and $C_m = 1.0$)

   Calculate P_c

$$1.505 = \frac{1}{1 - (536/0.75P_c)}$$

$$P_c = 2{,}130 \text{ kips}$$

   Calculate ℓ_u

$$\ell_u^2 = 3{,}740{,}000{,}000\,(\pi)^2/2{,}130{,}000$$

$$\ell_u = 132 \text{ in.}$$

   Calculate ℓ_u/h

$$\ell_u/h = 132/12 = 11.0$$

   Calculate P_u/A_g

$$\frac{P_u}{A_g} = \frac{536}{(12)(30)} = 1.49 \text{ ksi}$$

Note that the limiting loads for walls designed by the empirical design method for the usual ℓ_u/h range lie between the limits for walls designed by the rational method. The empirical design method is applicable to both single and double curvature eonditions, but the limiting moment at $e = h/6$ effectively precludes large sidesway or large ℓ_u/h ratios, exeept for nearly concentric loads. The limiting ℓ_u/h ratio for use with the empirical method is 25 (Section 14.5.3.1). For non-loadbearing walls the limiting ℓ_u/h ratio is increased to 30 (Section 14.6), which is also the limit for rational design of bearing walls in single curvature without a special analysis.

SPECIAL WALLS

Plain Concrete

Structural plain concrete (Section 22.6) may be used for continuously supported walls of all types. If closely spaced "control joints" or other provisions for unrestrained shrinkage are not provided or if shrinkage cracking is objectionable, horizontal temperature reinforcement at least equal to the minimum required should be used (Section 7.12). For strength design, the prescribed strength-reduction factor, ϕ, equals 0.65 (Section 9.3.5). The flexural tensile and shear stress permitted for plain concrete should not be exceeded (Section 22.5.3).

Precast Concrete Panels or Precast Crib Retaining Walls

A variety of precast concrete walls are employed, the design of which is outside the scope of the specific Code provisions for walls (Chapter 14 and Section 14.4). Such designs can be approved when the design is submitted under the procedures established by the Code (Section 1.4) or when it conforms to aceepted structural analysis.

Deep Beam Walls

A number of concrete wall applications behave as deep beams and must be designed accordingly (Section 10.7). One common application in this category is deep beam grade walls spanning as simple beams between end supports (footings) and carrying their self-weight plus other wall or column loads above Another is deep spandrel walls supported full depth at each end and resisting small lateral wind forces, as well as carrying their self-weight. The Code provisions for deep beams ensure that reinforcement for wall-like deep beams will be at least equal to that required for walls (Section 10.7.4).

Two-Way Reinforced Walls

Walls for underground enclosures to resist lateral forces due to earth or liquid pressure frequently are supported on three or four sides (basements, tanks and other liquid retaining structures). A realistic two-way elastic analysis will usually reduce overall reinforcement requirements and cracking since two-way reinforcement will conform closely to the elastic analysis of bending. The specific requirements for two-way slab systems based on supports such as columns, or beams between columns (Chapter 13), are difficult to extend to walls (see Chapter 8 of this Guide). Within the general Code requirements (Section 14.4 and the remainder of Chapter 10, two-way reinforcement proportioned by various tables* or by computer analysis methods are acceptable under ACI 318-95 as a "rational" method.

* *Roark's Formulas for Stress and Strain* by W. C. Young, McGraw-Hill Book Company. *Rectangular Concrete Tanks* by A. W. Domel, Jr., Portland Cement Association.

12 Footings

GENERAL

Ultimate soil pressure or ultimate pile capacity as shown by tests is normally not used in design, except to establish an allowable soil pressure, q_a, or allowable pile capacity. The allowable soil pressure or pile capacity is selected with an appropriate factor of safety, using principles of soil mechanics considering both ultimate load and settlement measured in short-time load tests. Any long-time settlement records available on existing older structures in the vicinity, as well as short-time test data filed with the local Building Authority, are used to establish the safe bearing pressures for various soils specified in statutory local building codes.

1995 CODE REVISIONS

An important revision of the ACI Building Code for footing design was the transfer of the requirements for plain concrete footings from Building Code Requirements for Structural Plain Concrete (ACI 318.1-83) and Commentary to Chapter 22. Other revisions have been incorporated as required in remaining text.

SOIL PRESSURE AND PILE CAPACITY

The ACI Building Code is prepared for inclusion in a general statutory building code, and it is expected that the safe allowable bearing capacity for soils or piles will be specified therein or established by test, according to the local building ordinances. The provisions of the ACI Building Code regulate design of footings for column factored loads, usually, $U = 1.4D + 1.7L$ (Eq. 9-1) or $U = 0.75 (1.4D + 1.7L + 1.7W)$ (Eq. 9-2). The proportions of the bearing area are selected for the allowable soil pressure, q_a, under actual (unfactored) D, L, W, and E specified in the General Code (Section 15.2.2).

FOOTING DESIGN

The designer is thus required to select the area of the footing for specified loads (unfactored) and to perform the structural design for factored loads. An assumed soil pressure which results from the factored loads will be used for the design of the footings to resist factored loads. This factored soil pressure, q_s, or the factored pile load used to determine shears and moments in the strength design of footings is determined most simply by dividing the factored column load, U, by the area of the footing or by the number of piles. If the designer does not wish to accumulate both factored and unfactored loads for multistory columns or when the ratio of D/L is not known, such as in the preparation of footing tables for gravity loads, an average load factor such as 1.6 can be used. The allowable soil pressure or the allowable pile load multiplied by this factor is used to determine the factored load that the footing can support.

MINIMUM THICKNESS

The minimum thickness above the bottom reinforcement prescribed for concrete footings is as follows:

Reinforced concrete footings on soil	6 in. (Section 15.7)
Reinforced concrete footings on piles	12 in. (Section 15.7)

The minimum thickness of plain concrete footings on soil is 8 in. (Section 22.7.4). Plain concrete footings on piles are not permitted (Section 22.7.3).

SHEAR

One-way (beam) shear and two-way (punching) shear (Section 11.12) are calculated in the conventional manner. In computing the external shear on any section through a footing supported on piles, the portion of the pile reaction assumed as producing shear on the section is based on a straight line interpolation between full value for center of piles located at $d_p/2$ outside the section and zero value at $d_p/2$ inside the section (Section 15.5.3). d_p is the diameter of the pile at the footing. Two-pile caps and high-capacity pile caps subjected to deep beam shear and "truss" action should be investigated for length available to develop reinforcement. See *CRSI Handbook, 1996.*

The critical sections for reinforced concrete footings for shear are shown in Fig. 12-1. For plain concrete footings cast against earth, substitute $h - 2$ in. for the value of d for reinforced concrete footings (Section 22.4.8). For an isolated rectangular footing, one-way shear should be checked in each direction to determine the maximum value. Two-way shear for square and rectangular footings is checked on the total shear periphery *abcd* in Fig. 12-1.

Unless shear reinforcement is provided, the factored shear force, V_u, must be equal to or less than the design shear strength, ϕV_c, of the concrete.

$\phi V_c = \phi 2\sqrt{f'_c}\, b_w d$ for one-way shear (Section 11.3.1.1) or for reinforced sections subjected to shear and flexure only, the alternate "long" formula may be used.

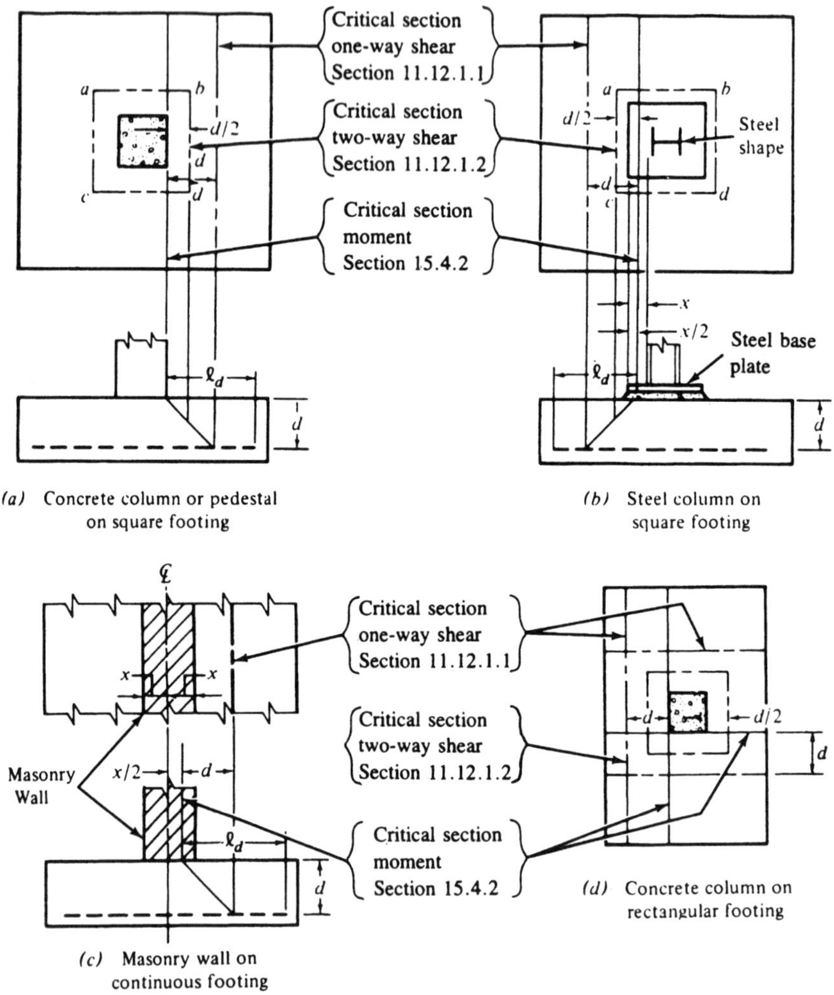

Figure 12-1 Critical sections for shear and moment for reinforced concrete footings—for plain concrete footings cast against earth substitute $h - 2$ in. for d (Section 22.4.8).

$$\phi V_c = \phi[1.9\sqrt{f_c'} + 2500\,\rho_w V_u d/M_u](b_w d) \text{ for one-way shear}$$
(Section 11.3.2.1)

For two-way shear (Section 11.12.2), the smaller of

(a) $$\phi V_c = \phi(2 + 4/\beta_c)\sqrt{f_c'}\,b_o d$$

where β_c = ratio of long side to the short side of the column, concentrated load or reaction area, and

(b)
$$\phi V_c = \phi\left(\frac{\alpha_s d}{b_o} + 2\right)\sqrt{f_c'}\, b_o d$$

where $\alpha_s = 40$ for interior columns, 30 for edge columns, 20 for corner columns, and

(c)
$$\phi V_c = \phi 4\sqrt{f_c'}\, b_o d$$

SHEAR REINFORCEMENT

Shear reinforcement is seldom used in individual reinforced or plain concrete column footings or in two-way mat footings. It is usually more practicable and economical to select footing depths so that the factored shear force is less than the design shear strength of the concrete (see Fig. 12-2). In combined footings

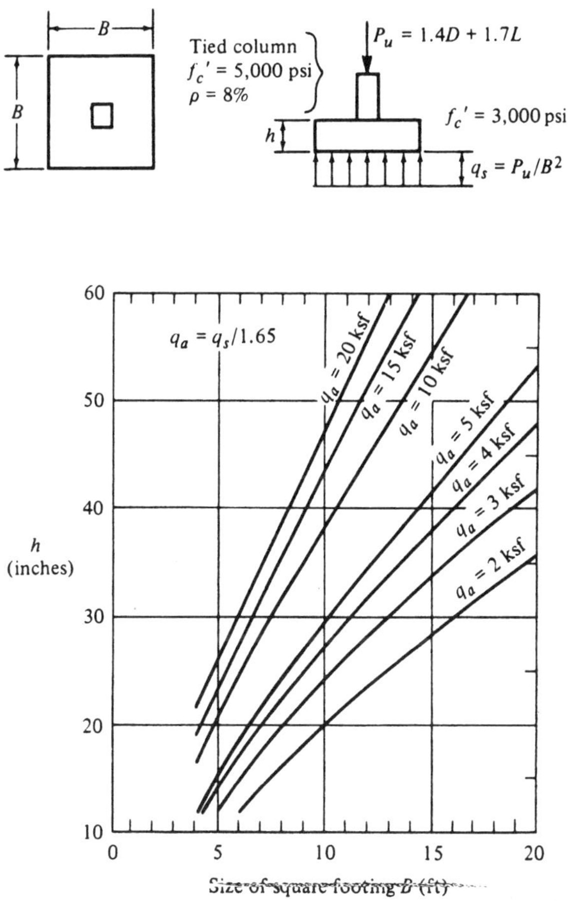

Figure 12-2 Minimum depth without shear reinforcement.

and individual column footings connected by beam-like "straps," shear reinforcement is commonly used, and for special conditions, reinforcement for punching shear may be desirable.

If shear reinforcement, consisting of bars or wires, is provided for one-way beam shear, the value of V_s is limited to a maximum value of $8\sqrt{f'_c}\, b_w d$ (Section 11.5.6.8).

If shear reinforcement, consisting of bars or wires, is provided for two-way shear the shear carried by concrete, V_c, must be reduced from $4\sqrt{f'_c}\, b_o d$ to $2\sqrt{f'_c}\, b_o d$ (Section 11.12.3.1). Nominal shear strength, $V_n = V_c + V_s$, is limited to a maximum value of $6\sqrt{f'_c}\, b_o d$ (Section 11.12.3.2).

For plain concrete footings, it can be shown that the shear stress is not critical. The tension stress due to flexure controls the thickness of plain concrete square or continuous wall footings on soil.

MOMENT

The critical sections for moment for footings are shown in Fig. 12-1. Tension in the reinforcement must be developed on each side of the critical section for moment (Section 12.1). For isolated column footings and continuous wall footings, sufficient length between the critical section (Section 15.6.3) and the nearest end of the bar to develop the bar in tension must be provided (see Fig. 12-1). This development length (ℓ_d) in tension varies directly as the diameter and the yield strength of the bar and inversely as $\sqrt{f'_c}$ (Section 12.2). Concrete cover, spacing of the bars, epoxy coating, bar size, bar position and density of the concrete also affect ℓ_d. The length required (ℓ_d) to develop various size Grade 60 straight bars in tension for different concrete strengths is shown in Table 13-4. If sufficient length is not available to develop a bar size selected for flexure, the required development length may be reduced by using smaller sizes of bars or bars with standard end hooks. The amount of development length required for a standard hook is shown in Table 13-5 (Section 12.5).

Combined footings with short cantilevers, pile caps and deep beam footings sometimes do not provide sufficient length between the critical section and the end of the bar for development of the tension reinforcement. In these cases additional reinforcement, hooks, or other means of anchorage must be used, such as welding a bar of equal size at right angles to the end of the tension bars. The Code does not make any provision to transfer any unbalanced moment in footings by shear. The authors recommend that all unbalanced moment should be transferred by flexure within a distance of $1.5d$ outside opposite faces of the column or pedestal.

BEARING

When the loaded area (area of column pier or base plate) and the supporting area (area at the top of the footing) are equal, the factored unit bearing stress on the loaded area of the footing must be equal to or less than $0.85\,\phi\sqrt{f'_c}$ (Section 10.17.1).

When the supporting area (top of footing *efgh*) is larger than the loaded area (base of column *abcd*) on all sides, the design bearing strength of the supporting area,

$$\phi P_{nb} = [\phi(0.85 f_c' A_1)] \sqrt{\frac{A_2}{A_1}}$$

where

$$\sqrt{\frac{A_2}{A_1}} \leqslant 2$$

and f_c' is the compressive strength of the footing concrete. For this case, the bearing strength of the base of the column,

$$\phi P_{nb} = \phi(0.85 f_c' A_1)$$

where f_c' is the compressive strength of the column concrete. A_1 is the loaded area. A_2, the supporting area, is the lower base of the largest frustum of a right pyramid or cone contained wholly within the footing with upper base the loaded area, A_1, and with side slopes of 2 horizontal to 1 vertical, as shown in Fig. 12-3.

When the bearing strength of the top of the footing or the base of the column is less than the factored axial compressive load, P_u, being delivered by the column to the top of the footing, reinforcement must be provided for transferring

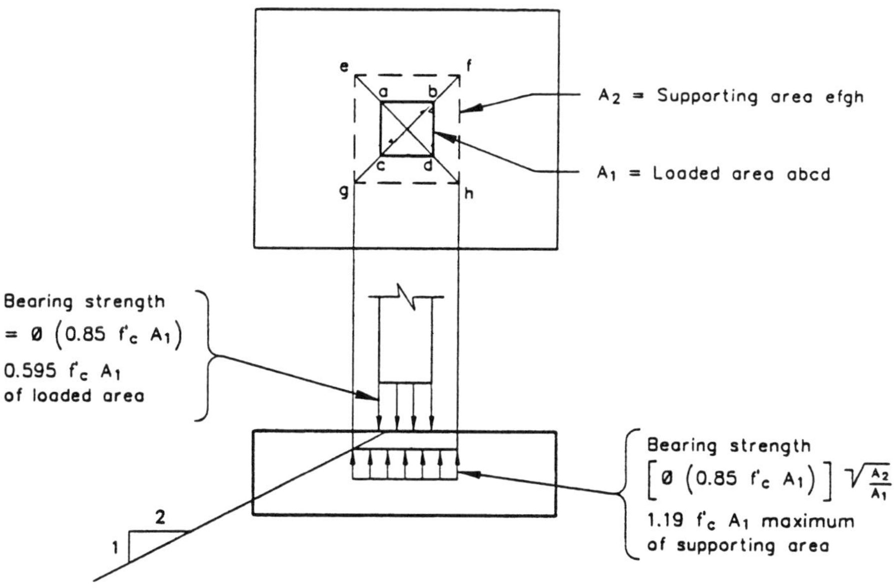

Figure 12-3 Bearing strength of loaded area and supporting area.

the *excess* load by extending the longitudinal bars into the footing or by dowels (Section 15.8.2). Extended longitudinal bars or dowels from a column or pedestal must have a minimum area equal to 0.005 times the loaded area of the column or pedestal (Section 15.8.2.1). There are no limitations on the maximum number of dowels that can be used in a footing other than those due to spacing and cover requirements (Sections 7.6.3 and 7.6.4). The required embedment depth for dowels into the footing must comply with the compression development length provisions (Section 12.3). If a load combination results in a net tensile load on the column, the required embedment depth for dowels must comply with the tension development length provisions (Section 12.2). For shallow footings where tension embedment controls, a 90° end hook can be used to reduce total embedment depth (Section 12.5) (see Table 13-5).

Critical compression dowel embedment cannot be reduced by end hooks (Section 12.5.5) nor with $f_c' > 4444$ psi. If the extended longitudinal bars or dowels of the same size cannot be developed in compression within the footing depth required by strength design, the footing can be made thicker, an equivalent area of smaller size bars can be furnished as shown, or a concrete pedestal as shown in Fig. 12-4 can be provided.

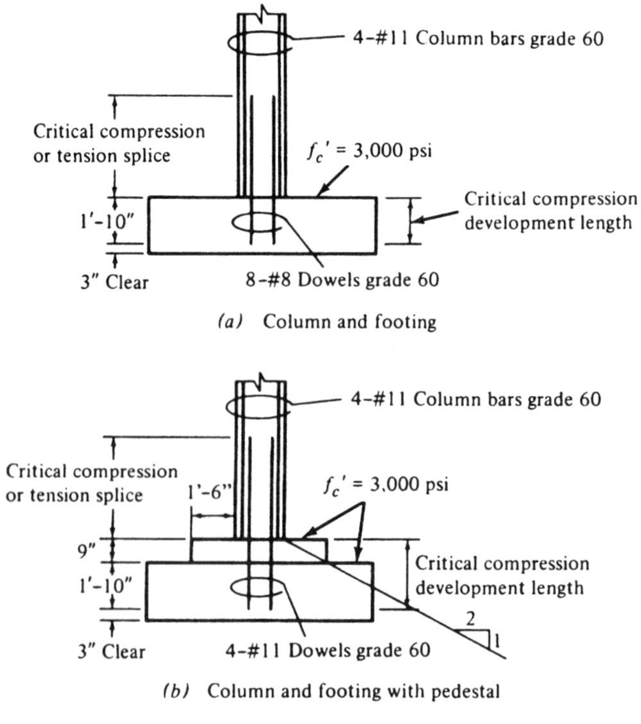

Figure 12-4 Dowel embedment examples.

SPLICES

Lap splices for #14 and #18 bars are not permitted (Section 12.14.2.1) except as column-to-footing compression dowels spliced with bars of a smaller size (Section 15.8.2.3) (See Fig. 12-5). Note the absence of hooks which cannot be considered effective in compression (Section 12.5.5).

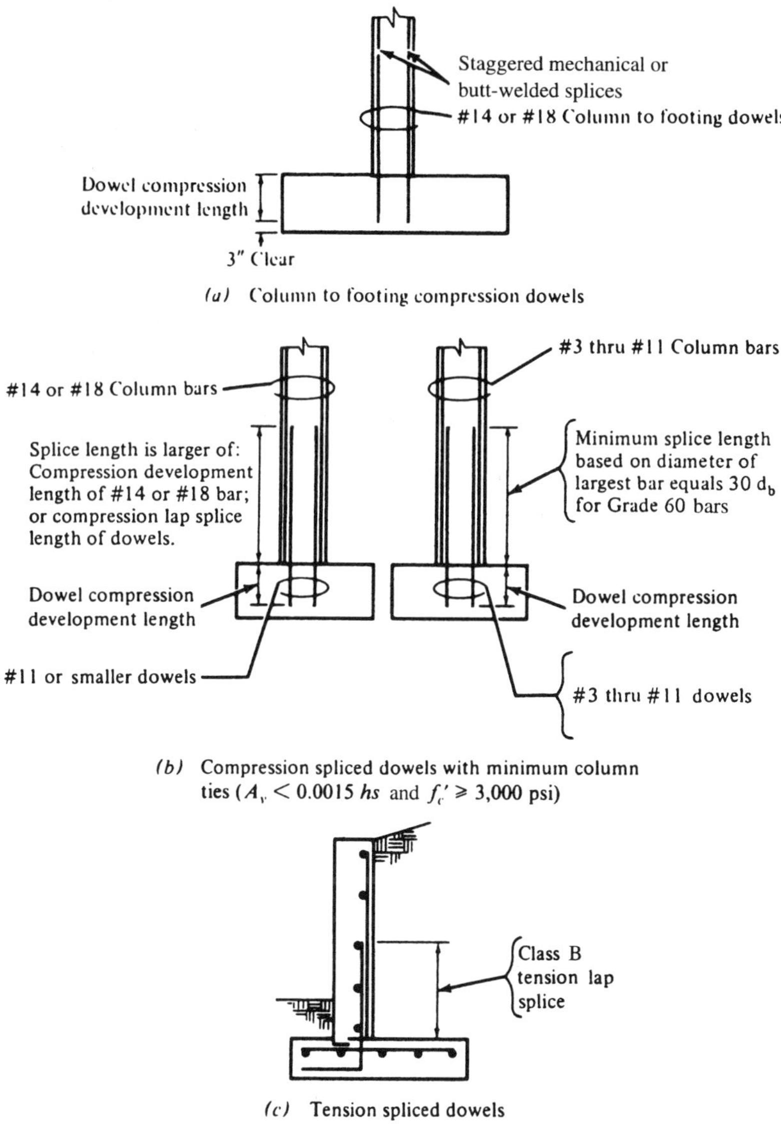

Figure 12-5 Footing dowel splices.

#3 through #11 footing dowels that are spliced in compression with column bars must project into the column the minimum compression splice distance (Section 12.16.2) unless this distance can be reduced because the area of the column ties is greater than 0.0015 *hs* (Section 12.17.2.4) or the splice is made in a spiral column (Section 12.17.2.5). When the size of spliced compression footing dowels and column bars is different, but the total area of each is the same, the splice length must be the larger of the development length of the larger bar or the splice length of the smaller bar.

When footing dowels are designed for critical tension, and spliced with column bars or with wall bars of a retaining wall above, the splice lengths depend on the class of splice required (Section 12.15.1). In the usual case, except for columns controlled by tension or high retaining walls, all footing dowels are spliced at one location at a point of maximum stress (Fig. 12-7) and a Class B tension lap splice is required. (See Table 13-6 for Class A and B tension lap splices with various concrete strengths.)

EXAMPLE. Consider the case of an isolated rectangular footing supporting a 12 in. square tied concrete column with 8-#9 bars, as shown in Fig. 12-6. Because of site conditions, the footing is restricted to a maximum width of 8′-0″ and is supported on soil with a safe allowable soil pressure, q_a, of 3,000 psf. The (unfactored) column service dead load is 163 kips and the column service live load is 100 kips. Grade 60 bars and concrete with f'_c = 3,000 psi have been selected for footing and column.

Determine Required Area of Footing

Actual Column service dead load	D = 163 kips
Actual Column service live load	L = 100 kips
Estimated weight of footing	= 25 kips
	$D + L$ = 288 kips

$$\text{Area of footing } A_f = \frac{D + L}{q_a} = 288/3 = 96.0 \text{ sq ft}$$

A rectangular footing 8′-0″ × 12′-0″ provides the required area without exceeding the allowable soil pressure, and we can now determine the factored soil pressure, q_s, to be used in the strength design of the footing.

$$1.4D = (1.4)(163) = 228 \text{ kips}$$
$$1.7L = (1.7)(100) = \underline{170} \text{ kips}$$
$$1.4D + 1.7L = 398 \text{ kips}$$

$$\text{Factored soil pressure, } q_s = 398/96 = 4.15 \text{ ksf}$$

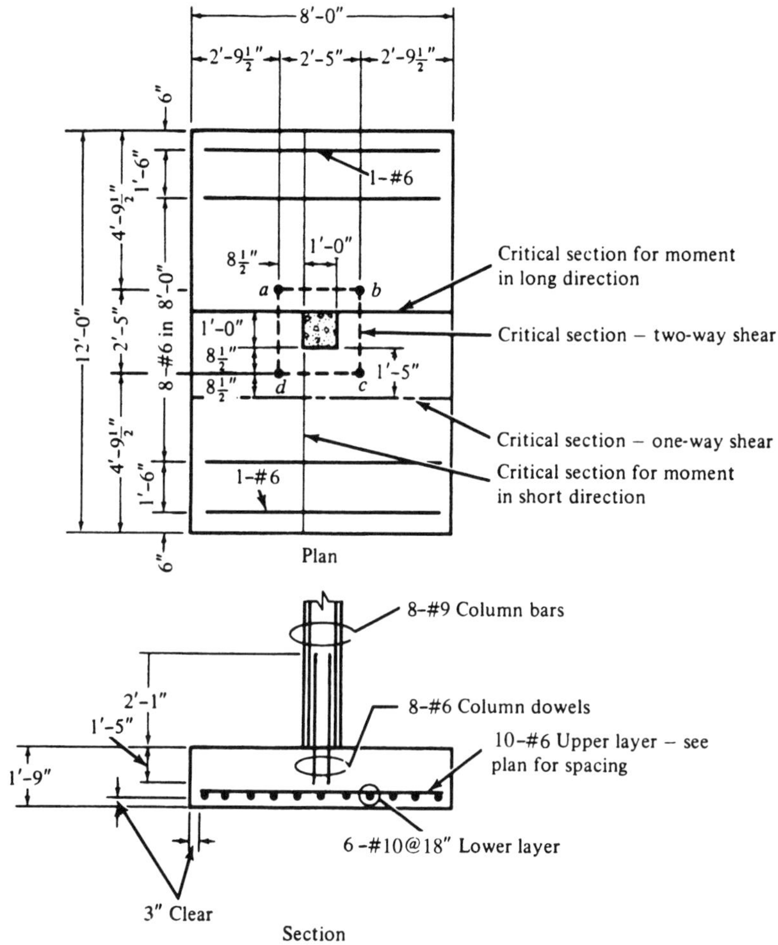

Figure 12-6 Footing design example.

Thickness

Footing depth is usually governed by two-way shear and for this example a trial size depth of 1′-9″ will be assumed and checked. Two-way shear is checked on the shear periphery, *abcd* (Section 11.12.1), with an assumed average depth equal to 4 in. less than the total thickness to allow for cover and one bar diameter.

Two-way shear

$$V_u = (4.15)\left\{(8.0)(12.0) - \left[\frac{(12 + 17)}{12}\right]^2\right\} = 374 \text{ kips}$$

The design shear strength of the concrete, ϕV_c, is based on the smaller of the coefficients of $\sqrt{f_c'}\, b_o d$ for two-way shear (Section 11.12.2.1):

(a) $$2 + \frac{4}{\beta_c} = 2 + \frac{4}{(12/8)} = 4.67$$

(b) $$\frac{\alpha_s d}{b_o} + 2 = \frac{(40)(17)}{(29)(4)} + 2 = 7.86$$

(c) 4 (Use)

$$\phi V_c = \phi 4 \sqrt{f_c'}\, b_o d = (0.85)(4)\sqrt{3{,}000}\,(29)(4)(17)/1000$$
$$= 367 \text{ kips} \approx V_u, \text{ OK}$$

One-way shear is maximum in long direction by inspection.

$$d = 17.5 \text{ and } b_w = 96 \text{ in.}$$
$$V_u = (4.15)(8.00)(4.04) = 134 \text{ kips}$$
$$\phi V_c = \phi 2 \sqrt{f_c'}\, b_w d = (0.85)(2)\sqrt{3{,}000}\,(96)(17.5)/1000$$
$$= 156 \text{ kips} > V_u, \text{ OK}$$

Moment Reinforcement

Determine the area of steel in the long direction with an assumed d of 17.3 in. based on #11 bars in the bottom layer.

$$A_s = \frac{(4.15)(8)(5.5)(2.75)(12)}{(0.90)(60)(0.95)(17.3)} \quad (j \text{ assumed at } 0.95)$$
$$A_s = 6.79 \text{ in.}^2 \text{ (long way)}$$

Select the reinforcing bars in the long direction; the minimum number of bars being based on a maximum spacing of 18 in. or $(96 - 6)/18 = 5$ spaces, using 6 bars.

6-#10 $A_s = 7.62$ in.^{2} (Minimum number)

7-#9 $A_s = 7.00$ in.^{2}

9-#8 $A_s = 7.11$ in.^{2}

11-#7 $A_s = 6.60$ in.^{2} (Minimum area)

Try 6-#10 bars.

Check the required tension development length, ℓ_d, of the #10 bars versus the available embedment length of (5′–6″) − 3″ = 5′–3″ or 63 in. For the #10 bars and referring to Table 13-4; with concrete cover of $3/1.27 = 2.4 d_b$ and center-to-center spacing of $18/1.27 = 14.2\, d_b$, Case 1 applies; from Table 13-4 for $f'_c = 3000$ psi and "other" bars, $\ell_d = 70$ in. > 63 in.; insufficient embedment for #10 bars.

Tension ℓ_d can also be calculated by the more rigorous procedure (Section 12.2.3):

$$\ell_d = \left[\frac{0.075 f_y \alpha\beta\gamma\lambda}{\sqrt{f'_c}\,[(c + K_{tr})/d_b]}\right] d_b \qquad \text{(Eq. 12-1)}$$

For uncoated #10 "other" bars embedded in normal weight concrete, α, β, γ, and λ are equal to unity; $K_{tr} = 0$. The value of c is the smaller of $0.5(18) = 9$ in. or $3 + 1.27/2 = 3.63$ in. Use $c = 3.63$ in. The quantity

$$\frac{c + k_{tr}}{d_b} = \frac{3.63 + 0}{1.27} = 2.85 > 2.5, \quad \text{use } 2.5.$$

Thus,

$$\ell_d = \left[\frac{(0.075)(60{,}000)}{\sqrt{3000}\,(2.5)}\right](1.27) = 41.7 \text{ or } 42 \text{ in.}$$

42 in. < 63 in., the #10 bars are satisfactory.

Determine the area of steel in the short direction with an assumed d of 16.4 in. based on #10 bars in the lower layer and #6 bars in the upper layer.

$$A_s = \frac{(4.15)(12)(3.5)(1.75)(12)}{(0.90)(60)(0.95)(16.4)} = 4.35 \text{ in.}^2 \text{ (short way)}$$

Select the reinforcing bars in the short direction; based on a maximum spacing of 18 in. $= (144 - 6)/18 = 7.7$, say, 8 bars.

8-#7 $A_s = 4.80$ in.^{2}

10-#6 $A_s = 4.40$ in.^{2}

14-#5 $A_s = 4.34$ in.^{2} (Minimum area)

22-#4 $A_s = 4.40$ in.^{2}

Try 10-#6 bars.

In a rectangular footing, a part of the bars in the short direction equal in amount to $\dfrac{2}{\beta + 1}$ times those required for flexure must be placed uniformly across a band width, B, equal to the short dimension of the footing (Section 15.4.4). β is equal to the ratio of the long side to the short side of the footing, or in this case equal to

$$\frac{2}{\beta + 1} = \frac{2}{\dfrac{12}{8} + 1} = 0.80$$

for 10-#6 bars in the short direction, the number to be placed in band-width, B, equals (10)(0.8) = 8 bars. This leaves one bar to be placed on each side of band-width, B. If this bar is placed 18 in. from band-width, B, it will be located within 6 in. of the edge of the footing, which is satisfactory.

Check required tension ℓ_d of the #6 bars versus the available embedment length of (3′-6″) − 3″ = 3′-3″ or 39 in. Referring to Table 13-4 with concrete cover of 4.27/0.75 = 5.7d_b and center-to-center spacing of 96/7 = 13.7 in. or 13.7/0.75 = 18.3 d_b; for $f_c' = 3000$ psi, "other" bars, Case 1, $\ell_d = 33$ in. < 39 in., OK.

Bearing Strength

Column factored axial load, $P_u = 398$ kips

Column concrete bearing strength

$$\phi P_{nb} = \phi(0.85 f_c' A_1) = 0.70(0.85 \times 3 \times 12 \times 12)$$

$$= 257 \text{ kips} < 398 \text{ kips}$$

Dowels are required to transfer the excess load, 398 − 257 = 141 kips. Footing concrete bearing strength

$$\phi P_{nb} = [\phi(0.85 f_c' A_1)] \sqrt{\frac{A_2}{A_1}}$$

where

$$\sqrt{\frac{A_2}{A_1}} \simeq \sqrt{\frac{8 \times 8}{1 \times 1}} = 8 > 2, \text{ use } 2$$

ϕP_{nb} of footing concrete is twice column

$$P_{nb} = 2(257) = 514 \text{ kips} > 398 \text{ kips, OK.}$$

Required area of dowels:

$$A_{s(\text{required})} = \frac{(P_u - \phi P_{nb})}{\phi f_y} = \frac{141}{0.70(60)} = 3.4 \text{ in.}^2$$

Min A_s (Section 15.8.2.1) = 0.005(12 × 12) = 0.72 in.^{2}

Select dowels; 8-#6 (A_s = 3.52 in.^{2}) or 6-#7 (A_s = 3.60 in.^{2}) would provide the required 3.4 in.^{2}

From Table 13-1, compression development length is

#6 bar	17 in.
#7 bar	19 in.

Available embedment length in the footing is 21 − 3 = 18 in. Use 8-#6 dowels.

Lap splice #6 dowels with #9 column bars. Compression lap splice length is larger of (from Table 13-2):

Compression development length of #9 bars	25 in. (Use)
Compression lap splice length of #6 bars	23 in.

COMBINED FOOTINGS AND MATS

The Code forbids use of the "Direct Design Method" for analysis of two-way reinforced footing mats (Section 15.10.2). The "Equivalent Frame" (Section 13.7) likewise is not directly applicable. Mats are not merely inverted two-way slab structures. The "Direct Design Method" (13.6) is an analysis of an indeterminate system, resulting in moments used to compute shears (column reactions) and limited to uniform loads. The footing mat is a *determinate* system with known shears (column loads) even when assumed soil pressure reaction is uniform. More sophisticated assumptions for soil pressure reactions, for instance as a function of mat deflections, require correspondingly more sophisticated analyses.

PILE CAP DESIGN

The principal Code requirement for pile cap design is that the pile reaction is to be assumed as distributed *linearly* across the depth, d_p, in the direction shear is accumulated (Section 15.5.3.3). For moment and shear calculations, the load may be considered concentrated at the center of the pile. Moment calculation is straightforward, equal to the pile reaction times distance from pile center: to face of a concrete column, wall, or pedestal; to a point halfway between middle and edge of a masonry wall; or to a point halfway between face of column and edge of steel base plate (Section 15.4.2). For pile cap layouts where the pile reactions lie either under the column or outside the critical sections, d for one-way caps and $d/2$ for two-way caps (Sections 11.1.3 and 11.12.1.2), design for shear is straightforward.

A great number of the pile caps used in practice, however, fall outside the limits above, including most pile cap layouts with 12 piles or less. These important structural elements might properly be called "orphans" as the Code requirements for shear are either misleading or woefully inadequate for their design. See Figure 12-7. Pile caps are normally designed as two-way slabs, with depth so that shear reinforcement is not required. With some layouts, beam shear may control, certainly for the two-pile cap. Tabulated pile-cap design aids are available, typically reinforced by only a two-way bar mat.*

* *CRSI Handbook, 1996.*

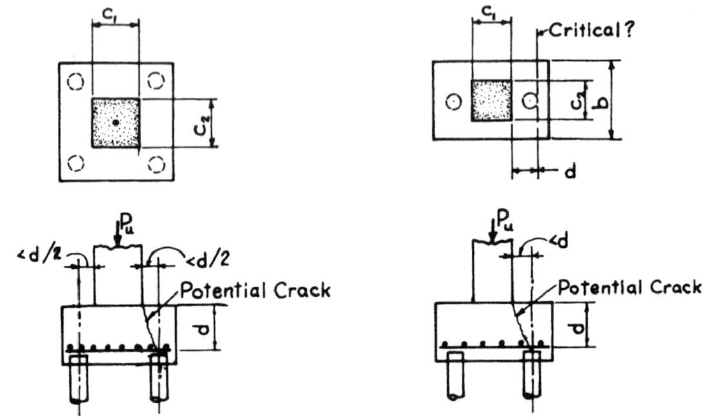

(a) Piles inside critical section at d/2 from face of column (Sections 11.12.1.2; 11.8; 11.9) Two-way Bracket? One-way Bracket?

(b) Piles inside critical section at d from face of column (Sections 11.1.3; 11.8; 11.9) One-way Bracket?

Figure 12-7 Pile caps—orphans of the Code.

For the simplest case, two-pile caps, and for other special layouts where "beam shear" (Section 11.12.1.1) is critical, some guidance for "Deep Beams" is available (Section 11.8). A critical section is established at $0.5a$ from face of support (Section 11.8.5); V_c is established as $2\sqrt{f_c'}\, b_w d \leq V_c \leq 6\sqrt{f_c'}\, b_w d$ (Sections 11.8.6 and 11.8.7); but minimum areas of vertical and horizontal shear reinforcement are required (Sections 11.8.9 and 11.8.10). Finally, the entire Section deals with deep beams using ℓ_n = clear span measured face-to-face of supports (Section 9.0). A comparison of Code limits with test results is shown in Fig. 12-8.

The sections dealing definitely with cantilevers are keyed to design of one-way brackets or corbels: (Section 11.7 Shear-Friction) tailored for precast column brackets; and (Section 11.9—Brackets and Corbels). Again, both require minimum shear reinforcement. None of the provisions for one-way deep elements with concentrated loads less than d from the face of the support provide for design of members where depth or width can economically be increased as necessary to eliminate special shear reinforcement. There are no Code provisions whatever applicable to design of two-way brackets or corbels with depth greater than two times distance "a" (face of support to a single load).

Faced with this problem, most practitioners and most authorities preparing tabulated pile cap designs have simply followed the Code literally—i.e., loads within $d/2$ or d are ignored for two-way shear or one-way shear respectively. In the one exception, the *CRSI Handbook, 1996*, formulas are developed for allow-

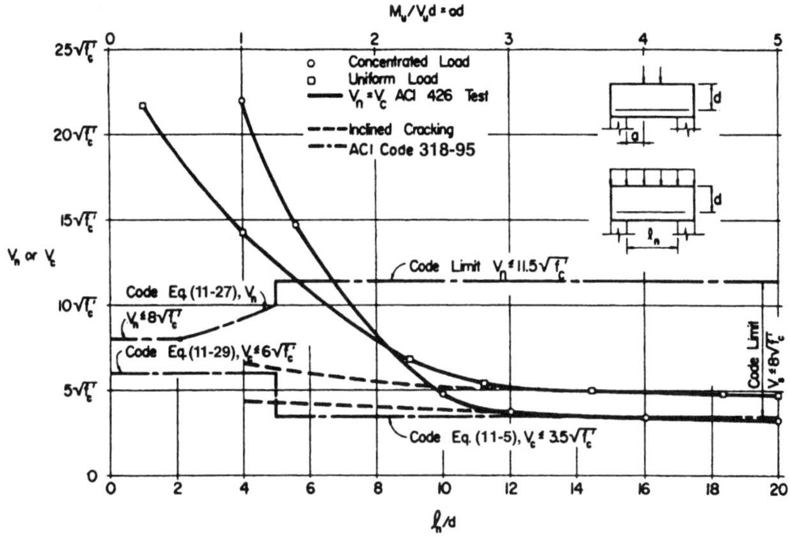

Figure 12-8 Comparison of Code limits and test results.

able shear on concrete as a function of the angle of the potential shear crack to the vertical when this angle varies from 45° to 0°. Shear strength increases rapidly from "diagonal tension" at 45° to "pure shear" at 0°.* See Figs. 12-9, and 12-10. The formulas are rational and possibly conservative as they utilize d instead of h, and w instead of a, where w = distance from face of the load (column) to *center* of the reaction (pile). Flexural reinforcement in deep pile caps where single lines of high-capacity piles are located on opposite sides of the column must be anchored for development of full tension at the centerline of the piles to prevent failure in a "tied arch" mode.†

For one-way shear at face of column (see Fig. 12-9):

$$v_c = k_1\sqrt{f'_c}$$

$$= \left(\frac{d}{w}\right)\left[3.5 - 2.5\frac{M_u}{V_u}\right]\left[1.9\sqrt{f'_c} + 0.1\sqrt{f'_c}\,\frac{V_u d}{M_u}\right] \leqslant 10\sqrt{f'_c}$$

* (1) Joint ASCE-ACI Com. 426 report; *ASCE ST-Jour.*, Part I, June 1973 and Part II, August 1974; (2) "Design of Thick Pile Caps," by A. B. Gogate and G. M. Sabnis, *ACI Journal, Proceedings* V. 77 No. 1, Jan.-Feb. 1980, pp. 18-24; (3) "Investigation of Thick Slab (Pile Cap) Behavior," by A. B. Gogate and G. M. Sabnis, *ACI Journal, Proceedings.* V. 81, No. 1, Jan.-Feb. 1984, pp. 35-39.

† "Shear Strength of Deep Reinforced Concrete Continuous Beams," by J. G. MacGregor and D. M. Ragowsky, Structural Engineering Report No. 110, University of Alberta, Nov. 1983.

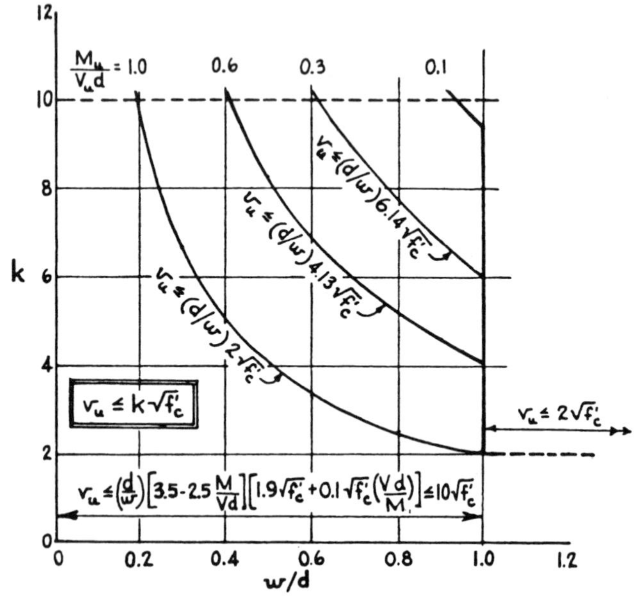

Figure 12-9 Derivation of allowable shear, v_c, for one-way deep beam sections, $w/d < 1.0$.

which reduces to $k_1 = 2$ at $(d/w) = 1$ when the product of bracketed expressions $= 2$ and thus gives the same value as the Code.

For two-way shear at perimeter of column (see Fig. 12-10):

$$v_c = k_2\sqrt{f_c'} \text{ ksi}$$

$$= 2\left(\frac{d}{w}\right)\left(1 + \frac{d}{c}\right)\sqrt{f_c'} \text{ ksi}$$

which reduces to

$$k_2 = 4\left(1 + \frac{d}{c}\right) \text{ at } \left(\frac{d}{w}\right) = 2.$$

When $(w/d) = 0.5$, and the critical section is taken at $d/2$ (Section 11.12.2), $b_o = 4(c + d)$ and $v_c = 4\sqrt{f_c'}$. Converting v_c in terms of critical section at perimeter of the column, $b_o = 4c$:

$$\text{At } \left(\frac{d}{w}\right) = 2;\ c_1 = c_2;\ \text{and } d = c$$

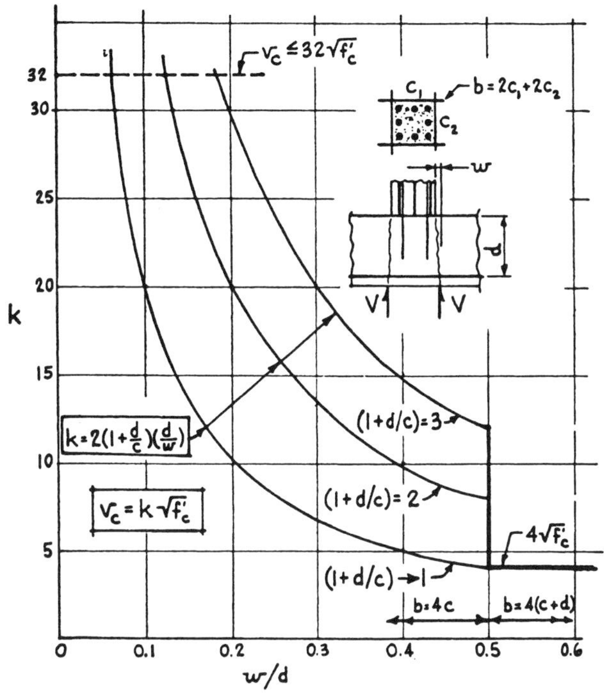

Figure 12-10 Derivation of allowable shear, v_c, for two-way deep corbels, $w/d < 0.5$.

ACI (Section 11.12.2.1-c), ACI 318-95:

$$V_c = 4\sqrt{f'_c}\,(b_o d) = 4\sqrt{f'_c}\,d(4)(c + d) \text{ kips}$$

$$V_c = 32\sqrt{f'_c}\,(d)^2 \text{ kips}$$

CRSI Handbook, 1996:

$$V_c = 4\left(1 + \frac{d}{c}\right)\sqrt{f'_c}\,(4c)d \text{ kips}$$

$$V_c = 32\sqrt{f'_c}\,(d)^2 \text{ kips}$$

13 Splices and Details of Reinforcement

GENERAL

Structurally sound details and proper arrangements of reinforcing bars are vital to the satisfactory performance of reinforced concrete members and structures. Concurrently, such details and bar arrangements should be practical and buildable, as well as cost-effective. Hence details, development, and splices of reinforcement—the topics of this chapter—are a crucial aspect of the design process. Of the ten more important items of information that must be included in the design drawings, typical details, or project specifications, two of the items cover development and splices of reinforcement (Sections 1.2.1-h and 1.2.1-i, and Commentary R1.2.1). These items are anchorage (embedment) length of reinforcement and location and length of lap splices; and the type and location of mechanical and welded splices. Code Chapter 12 also states: splices of reinforcement shall be made only as required or permitted on design drawings, or in project specifications, or as authorized by the engineer (Section 12.14.1).

Chapter 7 of the 1995 Code is titled "Details of Reinforcement," but the formulas and most of the basic design requirements controlling lap splice lengths, anchorage, hook effectiveness, and stirrup anchorage are located in Chapter 12. Special design requirements for details in shear and torsion are given in Chapter 11; for crack control in Chapter 10; for two-way slabs in Chapter 13; walls in Chapters 10 and 14; footings in Chapter 15; and seismic design in Chapter 21. In this Guide, the design requirements of Chapter 12 and applications as in Chapter 7 are presented in this chapter, together with cross references to other chapters herein on various different elements of design. The Code's requirements for structural integrity (Section 7.13.2) were cited in this Guide's chapters on one-way joist systems and beams. Seismic details are outside the scope of this Guide.

BOND

The bond of concrete to reinforcement is a complex phenomenon. It involves adherence and friction on smooth surfaces and bearing against the deformations (ribs) of rebars. Both effects produce tensile secondary stresses at right angles to the rebars which may cause the concrete to split. The state of stress at right angles to the bars due to external loads is also an important factor; confinement by external stress or lateral enclosing reinforcement also deters splitting.*

Most of the research on bond has involved at least several of these effects, making it impossible to separate them and to write a rational set of formulas from pure theory. All codes, including the 1995 Code, deal with bond as an average value over the length of a bar in question. Earlier codes differentiated between two kinds of bond: "flexural bond" as a function of shear, and "anchorage bond" or end embedment as a function of the bar stress, using the same average bond value for each kind. Research has indicated that the "flexural bond" is unimportant to the ultimate strength, generally that sufficient embedment, "development length" on either side of a point of peak stress will ensure against failure. The 1995 Code does not, therefore, use the term, "bond," but does frame all requirements in terms of development length. The only remnant of earlier Codes on flexural bond is a limitation on the *net* development length after deduction of anchorage length for positive moment tension bars (without standard end hooks) at simple supports and points of inflection (Section 12.11.3). *Net* development length for the bar size selected may not exceed M_n/V_u.

LAP SPLICES

Lap splices are treated as a special condition of stress transfer. A natural discontinuity develops at each cutoff end, somewhat helpful in a compression lap splice and a definitely detrimental stress raiser in tension lap splices. The latter is especially detrimental for fatigue resistance to repetitive stress reversals. The splitting effect of transferring stress through shear to the concrete is aggravated in a lap splice, as two opposing tensile bar forces are transferred (see Fig. 13-1).

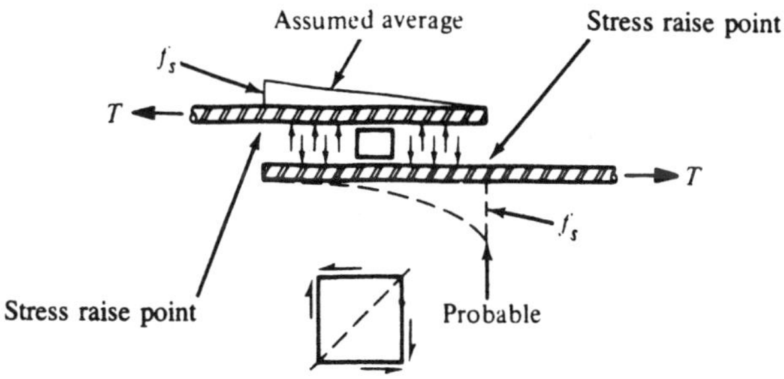

Figure 13-1 Two-dimensional splitting at tension lap splices.

*"Small Bar Spacing or Cover—A Bond Problem for the Designer," Phil M. Ferguson, *ACI Journal*, September 1977, Proc. Vol. 74.

EMBEDMENT

The term "embedment" is used, more or less interchangeably in practice, with "anchorage." Both terms represent the length of bar past the point of maximum stress to the end of the bar, in which the total stress in the bar is transferred to or from the concrete. The term "anchorage," often more specifically described as "end anchorage," connotes a transfer of tensile stress to the concrete. Ordinarily, both terms imply a bar length in a region of no stress or decreasing stress in concrete parallel to the bar, as in end anchorage or embedment past the face of a support. Embedment can refer to a tensile or compressive stress transfer at any point in a structure, and is, therefore, the more general term. Older codes called the stress transfer process "bond"; the 1995 Code does not use this outmoded term. The stress transfer is called "development"; and the length in which this transfer occurs is "development length," ℓ_d.

Compression Embedment

Compression embedment involves a length of bar in which compressive stress is transferred to the concrete. The Code term, compression development length, is that length required to develop from, or transfer to, the concrete the full compression capacity of a given size bar. Compression development length requirements in the Code include an allowance for the transfer of compressive stress by simple end bearing on the concrete (regardless of end condition, squared or sheared). Tests* have shown as much as 15,000 psi so transferred where splitting is somewhat restrained. The Code specifically forbids consideration of the bar length in an end hook as effective for transfer of compression for this reason (Sections 12.1.1 and 12.5.5). The basic Code requirement is that the stress at any section of all bars shall be developed on either side of that section. Compression development is commonly required in (1) compression lap splices, and (2) bars carrying compression to a zone of no computed parallel stress such as column dowels to footings, and extensions of bottom flexural bars into an interior support.

For single bars or 2-bar bundles, Grade 60 bar sizes #3 through #18, the compression development length required for fully stressed bars is stated quite simply as (1) $1200\ d_b/\sqrt{f'_c}$, or (2) $18\ d_b$, whichever is greater (Section 12.3.2), but not less than 8 in. (Section 12.3.1). For bundled bars, these development lengths must be increased by 20 percent for bars in a 3-bar bundle and by 33 percent for bars in a 4-bar bundle (Section 12.4). For dowels from column to footing, only the compressive stress in the column verticals *in excess* of that permitted to be transferred by bearing of the gross concrete column section area need be transferred by compression dowels (Section 15.8.1.2). See Table 13-1 for minimum

---

* "Column Tests with Offset Lap Splices of Zero Lap," PCA.

TABLE 13-1 Minimum Compression Dowel Embedment (Inches)

(Compression Development Lengths, ℓ_d; Grade 60 Bars)

For $f_c' \geqslant 3{,}000$ psi, Recommended Embedment for Compression Dowels = 22 bar dia. ($22d_b$)

Bar Size	f_c' (Normal Weight Concrete)			
	3,000 psi	3,500 psi	4,000 psi	Over 4,444 psi
# 3	9	8	8	8
# 4	11	11	10	9
# 5	14	13	12	11
# 6	17	16	14	14
# 7	19	18	17	16
# 8	22	21	19	18
# 9	25	23	22	20
#10	28	26	24	23
#11	31	29	27	25
#14	37	35	32	31
#18	50	46	43	41

For $f_c' > 4{,}444$ psi, minimum embedment = $18\,d_b$.

NOTE: For embedments enclosed by spirals or ties use 0.75 length shown but not less than 8 in. (Section 12.3.3.2).

compression dowel embedment in inches (without spiral or tie enclosure) in the usual range of strengths with normal weight concrete. Note that Table 13-1 gives embedment length into *footings* for full development of Grade 60 bars, or extension of bottom bars in flexural members past the face of a support for full development. Note also that the simplified expression "22 bar dia." ($22\,d_b$) for $f_c' \geqslant 3{,}000$ psi can be conservatively used for routine design.*

Footing Dowels

For footing dowels, the size of the dowel and the lap length for the splice in the column depend upon the factored load to be transferred from the column to the footing. The design bearing strength, ϕP_{nb}, of the footing and the column need to be determined. ϕP_{nb} is calculated as $\phi(0.85 f_c' A_1)$ (Section 10.17.1). When the supporting surface is wider on all sides than the loaded area, ϕP_{nb} can be increased by the factor $\sqrt{A_2/A_1} \leqslant 2$; where A_1 is the loaded area and A_2 is the area of the supporting surface.

Consider the columns described as Cases A and B, pictured in Figure 13-2. Assume the area of the footing > 4 times the area of the column so that $\sqrt{A_2/A_1} = 2$. In A, ϕP_{nb} at the top of the footing (the supporting surface) = $0.70(0.85)(3)(20 \times 20)(2) = 1{,}428$ kips > $\phi P_{n(\max)}$ of the column (= 1,147 kips). For the column with $\sqrt{A_2/A_1} = 1$, $\phi P_{nb} = 0.70(0.85)(5)(20 \times 20) = 1{,}190$ kips,

* Recommended for general use by CRSI.

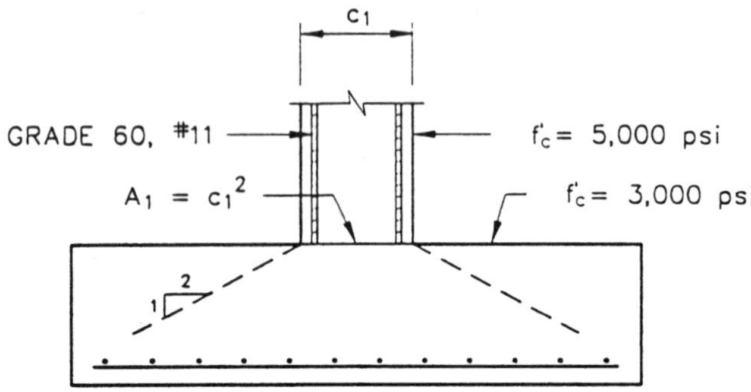

Case A: Column is reinforced with 4-#11 bars;
A_s = 6.24 in.^{2} or 1.56%; $\phi P_{n\,(max)} = 1{,}147$ kips.

Case B: Column is reinforced with 8-#11 bars;
A_s = 12.48 in.^{2} or 3.12%; $\phi P_{n\,(max)} = 1{,}342$ kips.

Figure 13-2 Dowels in compression.

which is also > $\phi P_{n(\max)}$ of the column. For this condition, the Code requires minimum dowels with total area $\geqslant$ 0.005 times the area of the column (Section 15.8.2.1). Required area of dowels $= 0.005(20 \times 20) = 2.0$ in.^{2} The authors would recommend at least four dowels, embedded at least $22d_b$ into the footing (Table 13-1), and at least #5 in size, to resist overturning in case of minor construction collisions, and to assist bar placers in locating and fastening down the column reinforcement cage. Four #7 dowels providing $A_s = 2.40$ in.^{2} and embedded 19 in. (Table 13-1) into the footing would suffice. The #7 dowels must be lap spliced with the #11 column bars. Lap splice length required is the larger of the compression ℓ_d for the #11 bars or the compression lap splice length for the #7 dowels (Section 12.16.2). Compression ℓ_d for #11 bars is 25 in. (Table 13-1). Compression lap splice length of the #7 dowels is 27 in. (Table 13-2). Use 27 in. lap splice length.

In Case B, $\phi P_{n(\max)}$ of the column $= 1{,}342$ kips which is less than ϕP_{nb} at the top of the footing (1,428 kips), but exceeds ϕP_{nb} of the column (1,190 kips). The excess compression for transfer by dowels $= 1{,}342 - 1{,}190 = 152$ kips; required area of dowels $= 152/60 = 2.53$ in.^{2}, which is greater than the minimum area (2.0 in.^{2}). Six #6 dowels with $A_s = 2.64$ in.^{2} and embedded 17 in. (Table 13-1) into the footing would be adequate. Lap splice length of the #6 dowels with the #11 column bars; compression ℓ_d for the #11 bars is 25 in. (Table 13-1); compression lap splice length for the #6 dowels is 23 in. (Table 13-2). Use 25 in. lap splice length.

TABLE 13-2 Compression Lap Splices for Grade 60 Bars (Inches)

Bar Size	MINIMUM LAP LENGTH ALL CONCRETES WITH $f'_c \geqslant$ 3,000 psi	
	Standard Lap (1)	Within Column Spirals (2)
# 3	12	12
# 4	15	12
# 5	19	15
# 6	23	17
# 7	27	20
# 8	30	23
# 9	34	26
#10	38	29
#11	43	32
#14 & #18	← Lap Splices not Permitted (3) →	

For $f'_c \geqslant$ 3,000 psi, Recommended Compression Lap Splice Lengths* = 30 bar dia. (30 d_b)

* Recommended for general use by CRSI.

(1) Lap length is minimum for use in all members, including tied columns, except as provided below.

(2) For use in spirally-reinforced columns with spirals conforming to Sections 7.10.4 and 10.9.3.

(3) Except as column-to-footing dowels, or to #11 and smaller bars.

Tension Embedment

Tension embedment is a length of bar in which the stress transferred to the concrete is tensile. Tension development length, ℓ_d, is the length of bar required to develop, or to transfer to the concrete, the full tensile capacity of the bar. The length, ℓ_d, of an uncoated bar embedded in normal weight concrete is made a function of yield strength of the bar, f_y; the square root of the compressive strength of the concrete, $\sqrt{f'_c}$; the diameter of the bar; depth of concrete below a horizontal bar (causing water rise); bar spacings and concrete cover; and lateral confinement, stirrups or ties (Section 12.2). The Code's provisions also account for epoxy-coated bars, and embedment of the bar in lightweight aggregate concrete. ℓ_d can be reduced when more reinforcement is provided than the amount required by analysis (Section 12.2.5).

The Code provides the designer with a choice of approaches for determining tension ℓ_d—a direct short-cut procedure (Section 12.2.2); or a more rigorous approach which is applicable to all conditions of bar spacings, concrete cover, and transverse reinforcement (Section 12.2.3).

Expressions are given in Table 13-3 for calculating ℓ_d under the short-cut procedure (Section 12.2.2). The minimum ℓ_d is 12 in.

Under the more rigorous approach in Section 12.2.3, tension development length is calculated:

$$\ell_d = \frac{0.075 f_y \alpha\beta\gamma\lambda(d_b)}{\sqrt{f'_c}\left(\frac{c + K_{tr}}{d_b}\right)} \geqslant 12 \text{ in.} \qquad \text{(Eq. 12-1)}$$

where

$\gamma = 0.8$ for bar sizes #3–#6 $\qquad \gamma = 1.0$ for bar sizes #7–#18

The term $\frac{c + K_{tr}}{d_b}$ is limited to a value of 2.5.

c = the smaller of: (1) one-half of the center-to-center spacing of the bars; or (2) the concrete cover to the center of the bar

$$K_{tr} = \frac{A_{tr} f_{yt}}{1500\, sn}$$

A_{tr} = total area of all transverse reinforcement within the spacing s, which crosses the potential plane of splitting through the bars being developed

f_{yt} = specified yield strength of transverse reinforcement

s = maximum c.-c. spacing of transverse reinforcement within ℓ_d

n = number of bars being developed along the plane of splitting

α, β, and λ are defined in Table 13-3.

Increased ℓ_d is required for bundled bars: in 3-bar bundles, 20 percent; in 4-bar bundles, 33 percent (Section 12.4; same as for compression development). See Chapter 10 in this Guide for applications to bundles.

Application of all the various interdependent tension development length requirements to each structural element in design would be extremely tedious and could be a confusing waste of design time. For practical design, the authors recommend instead that the Code user check the actual dimensions available for tension development in the connection (or from a cutoff point established as a fraction of the span on his typical details), compare to a table of development lengths required for each bar size, and select the bar size allowable. Such a table can be constructed for any special set of conditions. Table 13-4, which is based on Section 12.2.2, presents values of tension ℓ_d for each size bar for normal weight concrete with compressive strengths of 3,000, 4,000, and 5,000 psi. Note that separate values are tabulated for "top bars" and "other bars, " and that top bars are defined as horizontal bars with more than 12 in. of concrete cast below them (Section 12.2.4). This definition *excludes* from the "top" bars, the top layer of bars in a slab reinforced top and bottom, if thickness is less than the sum (12 in. + d_b + cover).

In Code Eq. 12-1, the term K_{tr} accounts for the beneficial effects of transverse reinforcement. For simplification, K_{tr} can be taken as zero even if transverse reinforcement is present (Section 12.2.4). Values of tension ℓ_d, based on

TABLE 13-3 Tension Development Length—Section 12.2.2 in ACI 318-95

Conditions	Bar Sizes #3–#6	Bar Sizes #7–#18
Clear spacing of bars being developed or spliced not less than d_b, clear cover not less than d_b, and stirrups or ties throughout ℓ_d not less than the code minimum or Clear spacing of bars being developed or spliced not less than $2\,d_b$ and clear cover not less than d_b	$\ell_d = \left(\dfrac{0.04 f_y \alpha\beta\lambda}{\sqrt{f'_c}}\right) d_b$	$\ell_d = \left(\dfrac{0.05 f_y \alpha\beta\lambda}{\sqrt{f'_c}}\right) d_b$
Other cases	$\ell_d = \left(\dfrac{0.06 f_y \alpha\beta\lambda}{\sqrt{f'_c}}\right) d_b$	$\ell_d = \left(\dfrac{0.075 f_y \alpha\beta\lambda}{\sqrt{f'_c}}\right) d_b$

Where:

$\alpha = 1.3$ for "top" bars $\quad \alpha = 1.0$ for "other" bars
$\beta = 1.0$ for uncoated bars
$\beta = 1.5$ for epoxy-coated bars with cover $<3\,d_b$, or clear spacing $<6\,d_b$
$\beta = 1.2$ for other cover and clear spacing conditions of epoxy-coated bars
The product of $\alpha\beta$ need not be taken greater than 1.7.
$\lambda =$ factor for lightweight aggregate concrete
$\lambda = 1.0$ for normal weight concrete

Section 12.2.3, can be tabulated for walls and slabs. The scope or use of such tables would be restricted to conditions where c in the quantity, $(c + K_{tr})/d_b$, in Code Eq. 12-1 is governed by concrete cover rather than by one-half of the center-to-center spacing of the bars.*

EXAMPLES OF DEVELOPMENT LENGTH. Some simplified applications of the Code requirements for tension development length are shown in Figure 13-3.

CANTILEVERED SLABS AND FOOTINGS

In Figure 13-3 (*a*), the simplest possible application of the tension development length concept is shown—end anchorage for reinforcing steel in a uniformly loaded cantilever slab. In Fig. 13-3 (*b*), note that the dimension available for development is ($\ell - 3$ in.) where ℓ is the overall clear span from the critical section at the face of the wall or column, and 3 in. is the cover required. The numbers in parentheses indicate Code sections regulating items so marked.

One-Way Slabs

Figure 13-4 shows the tension development length requirements for the end anchorage of positive moment reinforcing steel in slabs. The maximum bar size

*Tables based on Section 12.2.3 are included in *Reinforcement Anchorages and Splices*, 4th ed., CRSI, 1997.

TABLE 13-4 Tension Development Lengths, ℓ_d, for Grade 60 Uncoated Bars (Inches)

	$f'_c = 3{,}000$ psi				$f'_c = 4{,}000$ psi				$f'_c = 5{,}000$ psi			
	Top Bars		Other Bars		Top Bars		Other Bars		Top Bars		Other Bars	
Bar Size	Case 1	Case 2	Case 1	Case 2	Case 1	Case 2	Case 1	Case 2	Case 1	Case 2	Case 1	Case 2
# 3	22	32	17	25	19	28	15	22	17	25	13	19
# 4	29	43	22	33	25	37	19	29	22	33	17	26
# 5	36	54	28	41	31	47	24	36	28	42	22	32
# 6	43	64	33	50	37	56	29	43	33	50	26	38
# 7	63	94	48	72	54	81	42	63	49	73	37	56
# 8	72	107	55	82	62*	93	48	71	55	83	43	64
# 9	81	121	62	93	70	105	54	81	63	94	48	72
#10	91	136	70	105	79	118	61	91	70	105	54	81
#11	101	151	78	116	87	131	67	101	78	117	60	90
#14	121	181	93	139	105	157	81	121	94	140	72	108
#18	161	241	124	186	139	209	107	161	125	187	96	144

NOTES:
1. Values are based on Section 12.2.2.
2. Case 1 and Case 2 are defined:

	Structural Element	
Case	***Beams and Columns***	***Other Elements***
1	Concrete cover $\geqslant d_b$, c.-c. bar spacing $\geqslant 2\,d_b$ and with stirrups or ties throughout ℓ_d not less than Code minimum	Concrete cover $\geqslant d_b$ and c.-c. bar spacing $\geqslant 3\,d_b$
2	Concrete cover $< d_b$ or c.-c. bar spacing $< 2\,d_b$	Concrete cover $< d_b$ or c.-c. bar spacing $< 3\,d_b$

3. Values are for normal weight concrete.
4. Standard 90° or 180° end hooks may be used to replace part of the required development length. See Table 13-5.

* Sample Calculation: For Case 1, bar size #8, from Table 13-3 $\ell_d = (0.05 f_y \alpha \beta \lambda / \sqrt{f'_c}) d_b$ where $f_y = 60{,}000$ psi; $\alpha = 1.3$ for "top" bars; $\beta = 1.0$ for uncoated bars; $\lambda = 1.0$ for normal weight concrete; $f'_c = 4{,}000$ psi; and $d_b = 1.0$ in. Thus, $\ell_d = (0.05 \times 60{,}000 \times 1.3 \times 1.0 \times 1.0/\sqrt{4{,}000})(1.0) = 61.7$ or 62 in.

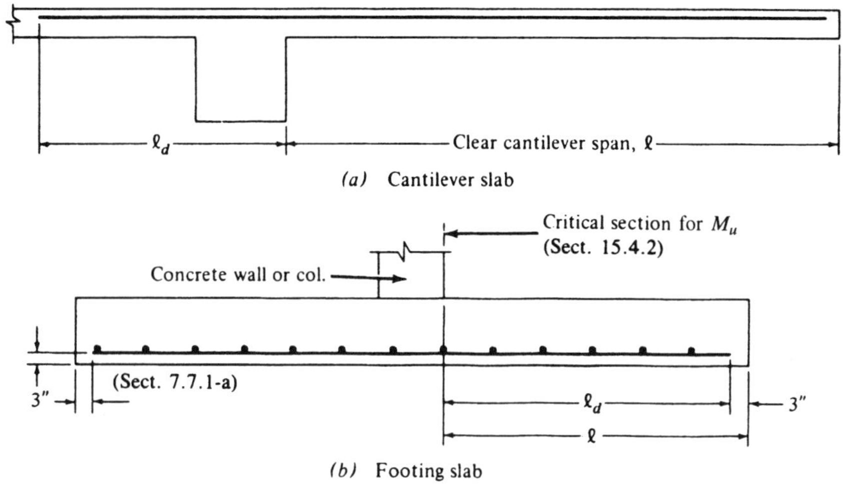

Figure 13-3 Tension development lengths.

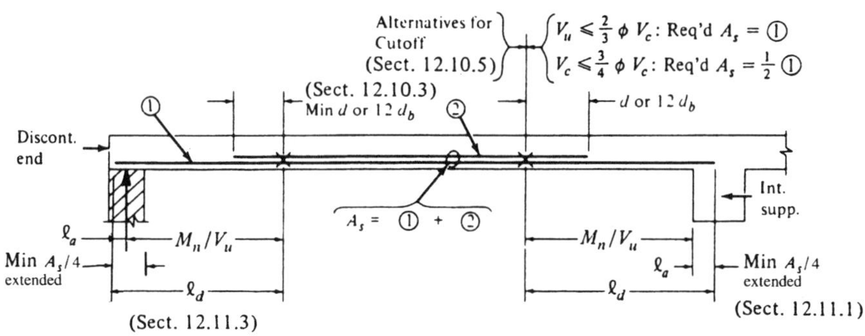

Figure 13-4 Tension development lengths for positive moment reinforcing steel in one-way slabs.

is established here by a computed dimension involving the moment/shear ratio taken at the point where the reinforcing steel is to be fully stressed. For the cantilever slab in Fig. 13-3 (*a*), uniformly loaded, the ratio (M_n/V_u) becomes 0.5ℓ where ℓ = clear cantilever span. For the simple end support in Fig. 13-4, the ratio for uniform loading (M_n/V_u) becomes 0.250ℓ where ℓ = clear span.

If some of the bars are to be cut off for savings in reinforcing steel, note how the cutoff point locations must be computed. For practical design within the ordinary range of spans, the authors suggest that no attempt be made to use varying lengths of bars. The savings in material will often be less than the additional costs of added design time, detailing, fabricating, placing, and inspection required. For longer spans or where a great number of identical spans are repeated, the savings in material will be worthwhile, but the variations in shear from concentrated or partial span loading would have to be considered (Sections 12.1, 12.10, and 12.11).

PRIMARY LATERAL LOAD RESISTING FRAMES

When the flexural member is a part of the "primary lateral load resisting system," the same percentages of positive reinforcement extended into the support as shown in Fig. 13-4 must be anchored for full development of f_y, in tension at the face of the support (Section 12.11.2) (see Fig. 13-5). "A primary lateral load resisting system" is usually defined as the beam-column frames without shear walls or in combination with shear walls designed for only a part of the lateral load resistance. Note that the standard end hook, if required, will become an effective part of the total anchorage for tension (see Table 13-5). When this detail is used, only the straight embedment from the face of the support to the tangent of the 90° bend can be considered effective for compression when the lateral loads reverse in direction (see Table 13-1).

TENSION DEVELOPMENT FOR NEGATIVE MOMENT REINFORCING STEEL

A similar detailing problem is shown in Figure 13-6. In this case note that full development in tension is required for the top bars in both directions from the face of the support (Section 12.12). The additional requirements are measured from the point of inflection (Sections 12.1 and 12.10.3). The requirements for anchorage of negative moment reinforcement apply to all connections, and are not limited only to those which are part of primary lateral load resisting systems. Where the connection involves principal framing elements, such as beams and columns, enclosure (to prevent splitting) must be provided for splices of continuing reinforcement and end anchorage of reinforcement terminating in the connection (Section 7.9). Such enclosure may consist of external concrete, where other members frame into the connection at right angles, or internal closure as provided by ties, stirrups, or spirals. See "Hooks" and Table 13-5. Figure 13-7 shows the development requirements for negative moment reinforcing

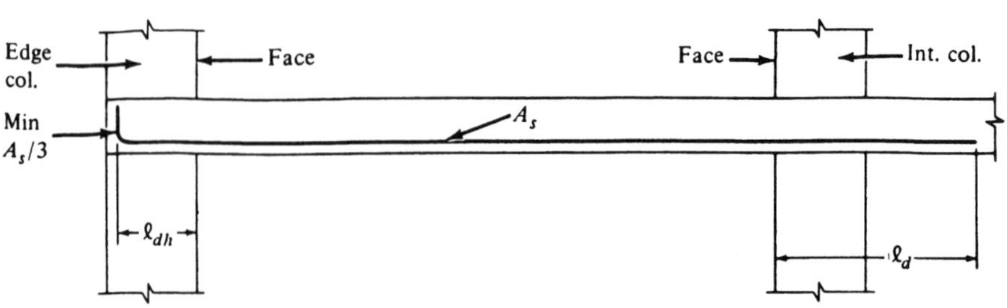

Figure 13-5 Tensile anchorage of positive moment reinforcing steel in beams (when beams are part of the primary lateral load-resisting system).

TABLE 13-5(a) Minimum Tension Embedment Lengths, ℓ_{dh}, for Standard End Hooks on Grade 60 Uncoated Reinforcing Bars in Normal Weight Concrete (Inches)

GENERAL USE (Nonseismic)

1. Side cover $\geq 2\frac{1}{2}$ in. 2. End cover (90° Hooks) $\geq$ 2 in.

Bar Size	Normal Weight Concrete, f'_c (psi)					
	3,000	4,000	5,000	6,000	7,000	8,000
# 3	6	6	6	6	6	6
# 4	8	7	6*	6 *	6*	6 *
# 5	10	9	8	7	7	6 *
# 6	12	10	9	8	8	7
# 7	14	12	11	10	9	9
# 8	16	14 ‡	12	11	10	10
# 9	18	15	14	13	12	11 *
#10	20	17	15	14	13*	12 *
#11	22	19	17	16	14	14 *
#14	37	32	29	27	25	23
#18	50	43	39	35	33	31

‡ Sample Calculation:

For bar size #8, d_b = 1.0 in.; f'_c = 4,000 psi

$\ell_{hb} = 1200\, d_b/\sqrt{f'_c}$ (Section 12.5.2)

$= 1200(1.0)/\sqrt{4{,}000} = 19.0$ in.

$\ell_{dh} = \ell_{hb} \times 0.7$ (Section 12.5.3.2)

$\ell_{dh} = 19.0 \times 0.7 = 13.3$ or 14 in.

SPECIAL CONFINEMENT (Nonseismic)

1. Side cover $\geq 2\frac{1}{2}$ in. 2. End cover (90° Hooks) $\geq$ 2 in. 3. Confining reinf., $s \leq 3\, d_b$

Bar Size	Normal Weight Concrete, f'_c (psi)					
	3,000	4,000	5,000	6,000	7,000	8,000
# 3	6	6	6	6	6	6
# 4	6 *	6 *	6 *	6 *	6 *	6 *
# 5	8	7	6 *	6 *	6 *	6 *
# 6	10	8	7	7 *	6 *	6 *
# 7	11	10	9	8 *	7 *	7 *
# 8	13	11	10	9 *	8 *	8 *
# 9	14	12	11 *	10 *	9 *	9 *
#10	16	14	12 *	11 *	11 *	10 *
#11	18	15	14 *	13 *	12 *	12 *

* For 180° hooks at right angles to exposed surfaces, see minimum lengths to provide 2 in. minimum cover to tail. See Table 13-5(c).

TABLE 13-5(b) Minimum Tension Embedment Lengths, ℓ_{dh}, for Standard End Hooks on Grade 60 Uncoated Reinforcing Bars in Lightweight Aggregate Concrete (Inches)

GENERAL USE (Nonseismic)
1. Side cover $\geqslant 2\frac{1}{2}$ in. 2. End cover (90° Hooks) $\geqslant$ 2 in.

Bar Size	Lightweight Aggregate Concrete, f'_c (psi)			
	3,000	4,000	5,000	6,000
# 3	8	7	6	6
# 4	10	9	8	8
# 5	13	11	10	9
# 6	15	13	12	11
# 7	18	16	14	13
# 8	20	18	16	15
# 9	23	20	18	16
#10	26	22	20	18
#11	29	25	22	20
#14	49	42	38	35
#18	65	56	50	46

SPECIAL CONFINEMENT (Nonseismic)
1. Side cover $\geqslant 2\frac{1}{2}$ in. 2. End cover (90° Hooks) $\geqslant$ 2 in. 3. Confining reinf., s $\leqslant 3 d_b$

Bar Size	Lightweight Aggregate Concrete, f'_c (psi)			
	3,000	4,000	5,000	6,000
# 3	6	6	6	6
# 4	8	7	7	6 *
# 5	10	9	8	8
# 6	12	11	10	9
# 7	14	13	11	10
# 8	16	14	13	12
# 9	18	16	14	13
#10	21	18	16	15
#11	23	20	18	16

* For 180° hooks at right angles to exposed surfaces, see minimum lengths to provide 2 in. cover to tail. See Table 13-5(c).

steel at interior supports, applicable to both slabs and beams. Note that use of varying bar lengths requires that the designer satisfy a number of minimum development length requirements, whichever is larger controlling.

END ANCHORAGES IN LIMITED DISTANCES

The end anchorage requirements for top bars at discontinuous ends of any principal member (beam) can present a practical problem of fitting into available

TABLE 13-5(c) Minimum Embedment Lengths to Provide 2 in. Cover to Tail of Standard 180° End Hook

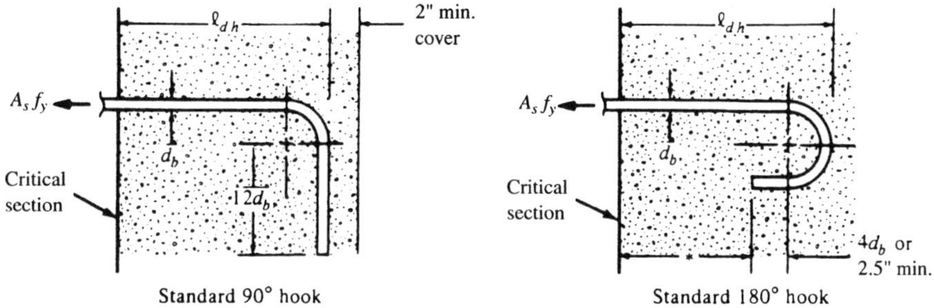

Bar size	#3	#4	#5	#6	#7	#8	#9	#10	#11	#14	#18
Embedment length (in.)	6	7	7	8	9	10	12	14	15	20	25

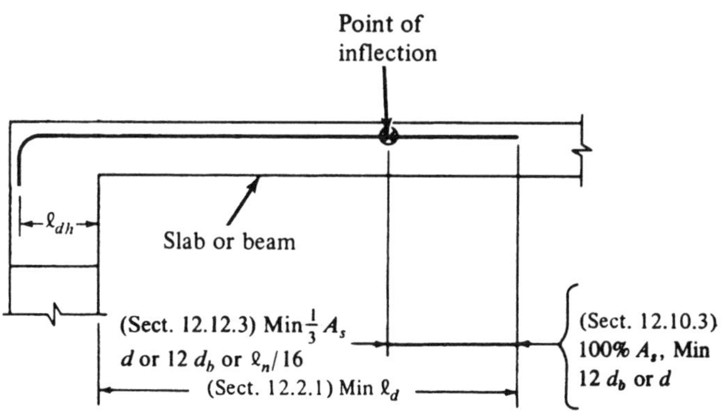

Figure 13-6 Tension development lengths for negative moment reinforcing steel at discontinuous ends.

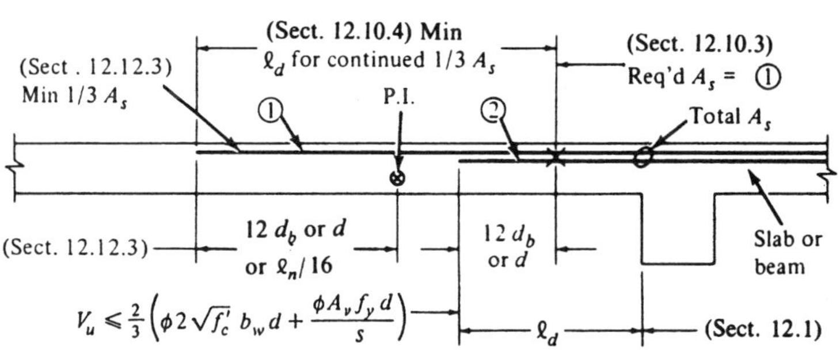

$$V_u \leqslant \frac{2}{3}\left(\phi 2\sqrt{f_c'}\, b_w d + \frac{\phi A_v f_y d}{s}\right)$$

Figure 13-7 Tension development lengths for negative moment reinforcing steel at interior supports.

form space. The same problem arises for the bottom bars in a primary lateral load resisting system. The problem becomes critical when the column depth is limited and an end hook is required, but the beam is too shallow for the standard size vertical end hook. To provide access for placing concrete in the column, the floor system reinforcing steel is frequently placed after the column is cast. For the end anchorage shown in Figure 13-8, one solution is to employ L-shaped dowels with one end set into the column and the other spliced to the main beam bars. If set before concreting, these bars interfere with the entry of concrete; if placed into the plastic concrete after concreting the column, they may easily be overlooked. Several alternative details will satisfy the Code and will in some cases facilitate bar placing and concrete placing. Since confining concrete is not present on all sides of the column, confining closed ties are required (Section 7.9). It will at times be possible to combine the dowel requirement for the bottom bars with a partial U-shaped tie with wide beams (see Figure 13-8). This tie (maximum size for usual columns #5) will be fully developed at the face of the column and can be combined with the development possible by straight extension of, and end hooks on, the bottom bars (see Table 13-6 for tension lap splice requirements, Class B). Since the Code requires anchorage sufficient to develop 33 percent of the bottom reinforcing steel fully, the same total tension can be developed by providing $1/2\ell_d$ for 66 percent. Although this connection would not develop the same ductility, if the anchorage consisted of some straight embedment plus a standard 90° end hook,

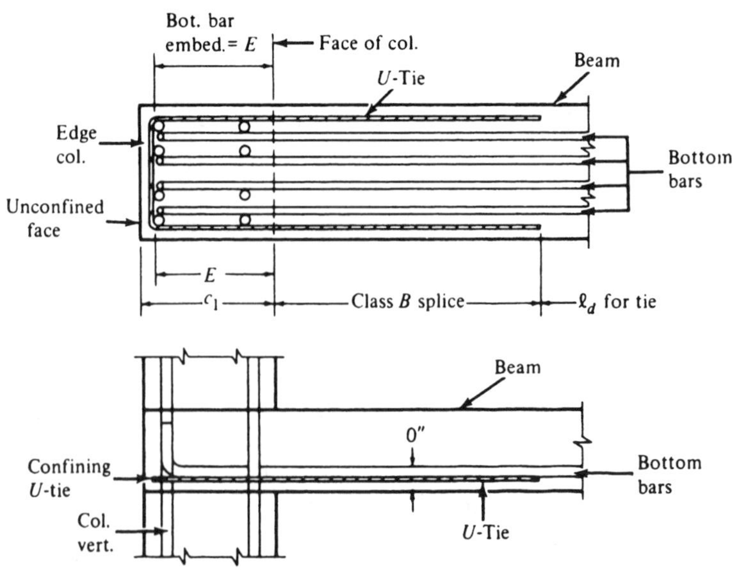

Figure 13-8 Anchorage in limited space for bottom bars in a primary load-resisting system.

TABLE 13-6 Tension Lap Splice Lengths for Grade 60 Uncoated Bars (Inches)

Bar Size	Lap Class	$f'_c = 3{,}000$ psi				$f'_c = 4{,}000$ psi				$f'_c = 5{,}000$ psi			
		Top Bars		Other Bars		Top Bars		Other Bars		Top Bars		Other Bars	
		Case 1	Case 2	Case 1	Case 2	Case 1	Case 2	Case 1	Case 2	Case 1	Case 2	Case 1	Case 2
# 3	A	22	32	17	25	19	28	15	22	17	25	13	19
	B	28	42	22	32	24	36	19	28	22	33	17	25
# 4	A	29	43	22	33	25	37	19	29	22	33	17	26
	B	37	56	29	43	32	48	25	37	29	43	22	33
# 5	A	36	54	28	41	31	47	24	36	28	42	22	32
	B	47	70	36	54	40	60	31	47	36	54	28	42
# 6	A	43	64	33	50	37	56	29	43	33	50	26	38
	B	56	84	43	64	48	72	37	56	43	65	33	50
# 7	A	63	94	48	72	54	81	42	63	49	73	37	56
	B	81	122	63	94	70	106	54	81	63	94	49	73
# 8	A	72	107	55	82	62	93	48	71	55	83	43	64
	B	93	139	72	107	80*	121	62	93	72	108	55	83
#9	A	81	121	62	93	70	105	54	81	63	94	48	72
	B	105	157	81	121	91	136	70	105	81	122	63	94
#10	A	91	136	70	105	79	118	61	91	70	105	54	81
	B	118	177	91	136	102	153	79	118	91	137	70	105
#11	A	101	151	78	116	87	131	67	101	78	117	60	90
	B	131	196	101	151	113	170	87	131	101	152	78	117

NOTES:

1. Values are based on Sections 12.2.2 and 12.15.
2. See notes under Table 13-4 for definitions of Case 1 and Case 2.
3. Values are for normal weight concrete.

* Sample Calculation:

From Sample Calculation under Table 13-4, for Case 1, bar size #8, top bars $\ell_d = 61.7$ in.

For Class B tension lap splice,

Lap length $= 1.3\,\ell_d$

$= 1.3\,(61.7)$

$= 80.2$ or 80 in.

tests show that the full f_y will be developed, but at larger slip. If the modified tie of Fig. 13-8 is used and a #5 bar capacity is inadequate, and bending radii for a larger size bar will not fit, a 2-bar bundle of #5 bars, one atop the other, can be used.

EXAMPLE. BOTTOM BAR ANCHORAGE. (Figure 13-8) $f'_c = 4{,}000$ psi; assume four #9 Grade 60 bottom bars with $A_s = 4.00$ in.^{2} are the required one-third midspan bottom A_s to be fully developed at the face of the column. $c_1 = 15$ in.; $E = 12.5$ in.; required $\ell_{dh} = 15$ in. (Table 13-5(a)); for the wide beam, assume c.-c. spacing of the #9 bars $\geqslant 2d_b$; cover to the #9 bars $\geqslant d_b$; and minimum stirrups will be provided. From Table 13-4, Case 1 applies and total ℓ_d for #9 bottom bars $= 54$ in.; $\ell_{dh} - E = 2.5$ in. The additional stress to be developed in the bar is $\frac{2.5}{54}$ (60) $= 2.8$ ksi; total additional force $=$ (4)(1.00)(2.8) $= 11.2$ kips. Use one U-tie, #4. Force $=$ (2)(0.20)(60) $= 24$ kips.

TOP BARS IN LIMITED ANCHORAGE DISTANCE

When space is limited, the top bar anchorages are even more difficult to achieve since the Code requires full development of 100 percent of the top reinforcing steel required for flexure. Under some conditions where the column is wider than the beam framing into it or where a deep spandrel beam is available for extension of top steel, the top beam bars may be lap spliced to a larger number of smaller bars spread into the adjacent slab flanges of the beam (see Fig. 13-9(*a*)).

Under very restricted conditions, such as with the important top bars in short span corbels, the designer often has no choice other than to weld the top bars,

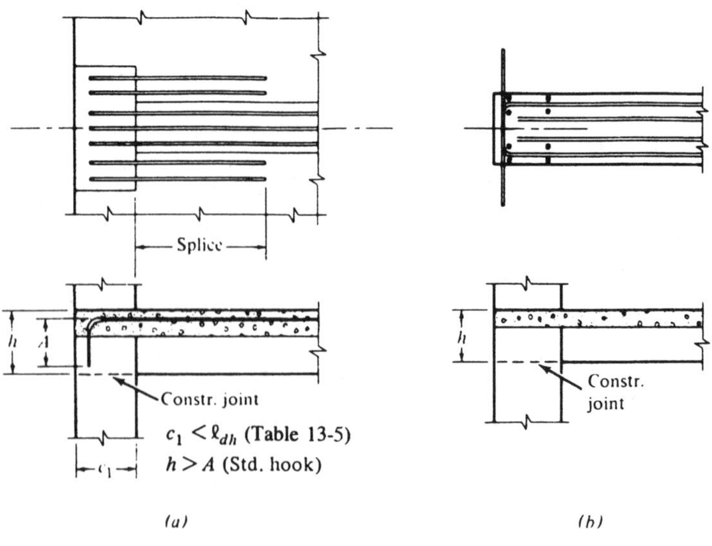

Figure 13-9 Top bar anchorages in limited dimensions.

preferably to a bar or plate at right angles. Fig. 13-9 (*a*) and (*b*) shows two alternatives that may sometimes be used to supplement the development required sufficiently to permit all floor system reinforcing steel to be placed after the lower portion of the column is cast. Another option would be to use mechanical devices or anchorages as a means to develop and anchor bars (Sections 12.1.1 and 12.6). Headed reinforcing bars, which have been used in the construction of offshore oil platforms, are a new available technology. Headed reinforcing bars can also relieve congestion of reinforcement to enhance constructability. At press time, ASTM had issued a new Specification A970 for headed reinforcing bars.

OTHER EXAMPLES. Other examples showing practical applications of development requirements to particular structural elements will be found in other chapters; one-way slabs, Figs. 3-3 and 3-4; one-way joist construction, Fig. 4-4; two-way flat plates, Figs. 5-18 and 5-19; waffle flat slabs, Fig. 7-12; footings and dowels, Figs. 12-4 and 12-5.

HOOKS

End hooks are a vital part of tension anchorage. When space is limited so that bars cannot be fully developed by straight embedment, hooks are effective in providing the required end anchorage (Fig. 13-10). Provisions are given in Section 12.5 for determining the anchorage capacity of deformed reinforcing bars with standard end hooks. End hooks can only be considered effective in developing bars in tension, and not in compression; see Sections 12.1.1 and 12.5.5. Only standard end hooks (Section 7.1) are considered in Section 12.5. The anchorage capacity of end hooks with bend diameters larger than those of standard hooks cannot be determined by the provisions in Section 12.5.

Application of the hook development provisions is to first calculate the basic development length of the hooked bar, ℓ_{hb}. For Grade 60 bars, $\ell_{hb} = 1200d_b/\sqrt{f_c'}$ (Section 12.5.2). Then the basic development length must be multiplied by the applicable modification factor or factors (Section 12.5.3) to determine the development length of the hook, i.e., $\ell_{dh} = \ell_{hb} \times$ applicable

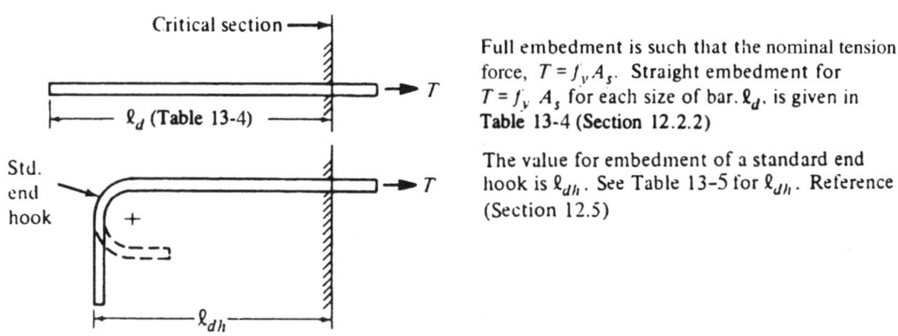

Figure 13-10 End anchorages for full development of bars.

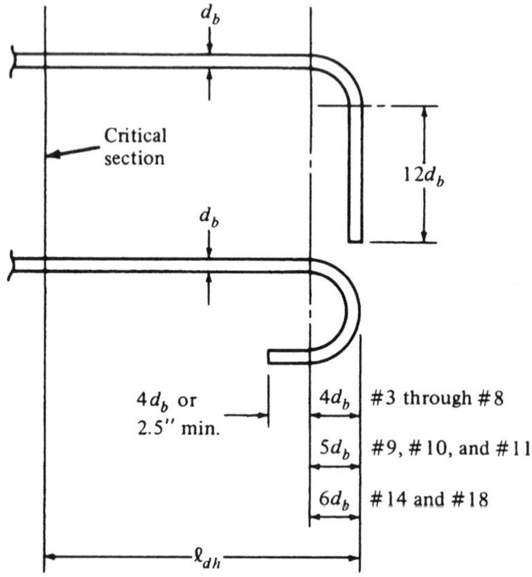

Figure 13-11 Development length ℓ_{dh} of standard hooks.

modification factors. The development length ℓ_{dh} is measured from the critical section to the outside end of the standard hook, i.e., the straight embedment length between the critical section and the start of the hook, plus the radius of bend of the hook, plus one bar diameter. Figure 13-11 shows the measurement of ℓ_{dh} from the critical section to the outside end of the standard hooks (Section 7.1) for all the bar sizes.

The ℓ_{hb} modification factors listed in Section 12.5.3 account for:

- Bar yield strength if f_y is not 60,000 psi.
- Favorable confinement conditions provided by increased concrete cover, or by transverse ties or stirrups to resist splitting of the concrete.
- More flexural reinforcement provided than that required by analysis.
- Lightweight aggregate concrete.
- Epoxy coating.

After multiplying the basic development length ℓ_{hb} by the applicable modification factor or factors, the resulting development length ℓ_{dh} must not be less than $8d_b$ nor 6 in.

Section 12.5.4 is a special provision for hooked bars terminating at discontinuous ends of members, such as at the ends of simply supported beams, at the free ends of cantilevers, and at the ends of members framing into a joint where the member does not extend beyond the joint. If the full strength of a hooked bar must be developed, and both side cover and top (or bottom) cover over the hook is less than $2\frac{1}{2}$ in., Section 12.5.4 requires the hook to be enclosed within ties or stirrup-ties over the development length, ℓ_{dh}. Spacing of the ties or stirrup-ties must not exceed $3d_b$, where d_b is the nominal diameter of the hooked bar. In addition, the modification factor of 0.8 for confinement provided by ties or stirrups (Section 12.5.3.3) does not apply to the special conditions covered by Section 12.5.4. At discontinuous ends of slabs with concrete confinement provided by the slab continuous on both sides normal to the plane of the hook, the provisions of Section 12.5.4 do not apply.

Table 13-5(a) gives the minimum tension development (embedment) lengths ℓ_{dh} for standard end hooks on Grade 60 uncoated reinforcing bars in normal weight concrete. Embedment lengths in Table 13-5(a) (general use) apply for end hooks with minimum side cover normal to the plane of the hook of $2\frac{1}{2}$ in., and minimum end cover (90° hooks only) of 2 in. Thus, for #11 and smaller hooked bars, the 0.7 factor of Section 12.5.3.2 applies, and ℓ_{dh} = $(0.7)1200d_b/\sqrt{f'_c}$, but not less than $8d_b$ nor 6 in. Table 13-5(a) (special confinement) includes the additional 0.8 reduction factor of Section 12.5.3.3 for confining ties or stirrup-ties, i.e., for #11 and smaller hooked bars, ℓ_{dh} = $(0.7)(0.8)1200d_b/\sqrt{f'_c}$, but not less than $8d_b$ nor 6 in.

Minimum tension embedment lengths for standard end hooks on Grade 60 uncoated reinforcing bars in lightweight aggregate concrete are given in Table 13-5(b). These tabulated values of ℓ_{dh} include the modification factor of 1.3 from Section 12.5.3.5.

For 180° hooks at right angles to exposed concrete surfaces, the authors recommend that a minimum 2 in. of concrete cover be provided to the tail of the hook (Section 7.7). The minimum tension embedment lengths which will provide the recommended 2 in. of cover to the tail of the hook are given in Table 13-5 (c). Values of ℓ_{dh} which are affected by the 2 in. cover recommendation are indicated by an asterisk in Tables 13-5(a) and 13-5(b).

The tension embedment lengths in Tables 13-5(a) and 13-5(b) are the dimensions required on the design drawings.

Pertinent dimensions of hooked bars for fitting into the concrete outline of each element as design progresses are shown in Figure 13-12. These standard hook dimensions are based upon the requirements in Sections 7.1 and 7.2 (minimum finished inside bend diameters). The dimensions for bar length *A* (or *G*) for 180° hooks are based upon the length of the center fiber of the bar as bent. Dimensions for fit into concrete, *A* (or *G*) for 90° hooks, *H* for 135° hooks, and *J* for 180° hooks are measured from out-to-out. The dimension *A* (or *G*) for 90° hooks is also used for estimating bar length. In standard practice, the length *A* or *G* is merely added to the detailing dimension of the straight portion as though the bar were bent square (as it is usually, sometimes unfortunately, drawn).

COMPRESSION LAP SPLICES

Compression lap splices for all values of $f'_c \geqslant 3{,}000$ psi are specified to vary as a direct function of the bar diameter, d_b, and the yield strength of the reinforcing steel, f_y (Section 12.16) (see Table 13-2). Two different lap lengths in terms

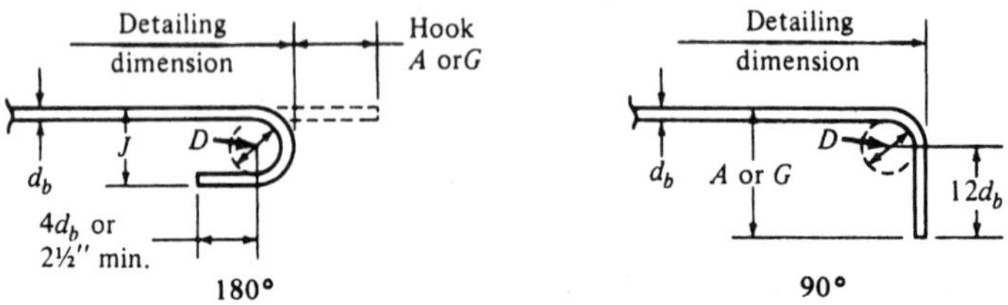

Bar size	Finished bend diameter D	END HOOKS, ALL GRADES OF STEEL: 180° hooks A or G	180° hooks J	90° hooks A or G
#3	2¼	5	3	6
#4	3	6	4	8
#5	3¾	7	5	10
#6	4½	8	6	12
#7	5¼	10	7	14
#8	6	11	8	16
#9	9½	15	11¾	19
#10	10¾	17	13¼	22
#11	12	19	14¾	24
#14	18¼	27	21¾	31
#18	24	36	28½	41

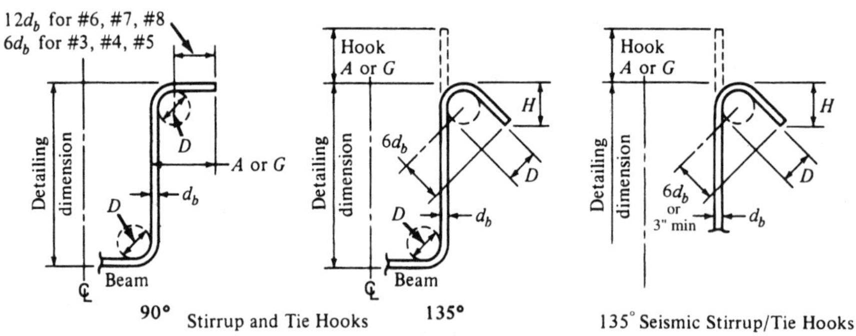

Bar size	D	Stirrup and tie hook dimensions: 90° hook A or G	135° hook A or G	135° hook H, approx.	135° stirrup-tie hook dimensions: 135° hook A or G	135° hook H, approx.
#3	1½	4	4	2½	4¼	3
#4	2	4½	4½	3	4½	3
#5	2½	6	5½	3¾	5½	3¾
#6	4½	12	8	4½	8	4½
#7	5¼	14	9	5¼	9	5¼
#8	6	16	10½	6	10½	6

Figure 13-12 Standard hooks*—all dimensions in inches.

*The Code limits use of "stirrup and tie hooks" to bar sizes #3-#8 (Section 7.1.3). Bars larger than #5 must be bent at the same minimum diameter as for other standard hooks (Section 7.2.2).

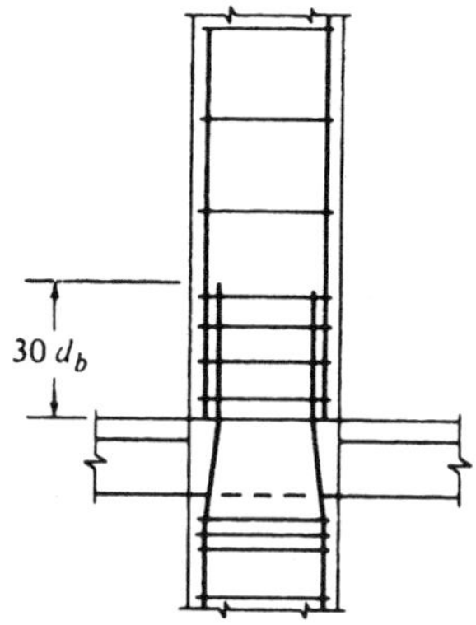

Figure 13-13 Typical interior column.

of bar diameters for compression splices are specified for full development of Grade 60 bars under common conditions of use:

1. 30 $d_b \geqslant$ 12 in., for general use without regard to confinement (Section 12.16.1) (See Fig. 13-13).
2. 25 $d_b \geqslant$ 12 in., for use when confined by ties or stirrups with an effective area of at least 0.0015 hs (Section 12.17.2.4) where h is the width of the member at right angles to the legs of the ties and s is the spacing of the ties. The authors' recommendation is not to fool around with this reduction. It is not cost-effective in terms of design time.
3. 22.5 $d_b \geqslant$ 12 in., for use only when confined within spirals (Section 12.17.2.5). Standard column spirals (Section 7.10.4) will satisfy this requirement.

Compression lap splices are commonly used for the bottom bars at the interior supports of continuous flexural members or for the vertical bars in multistory columns. In slabs and joists, it is not generally practicable nor required that confining ties or stirrups be provided, and so, if required, compression lap splice lengths of 30 d_b are used. For compression reinforcement in beams and girders, enclosing stirrups are required (Section 7.11.1); these must meet the minimum size and maximum spacing limits for column ties (Section 7.10.5).

Special Requirements for Compression Lap Splices

OTHER STRENGTHS OF MATERIALS There are a number of additional provisions for special conditions (Section 12.16). For concrete strengths, $f'_c < 3{,}000$ psi, all required lap lengths must be increased by one-third. For $f_y > 60{,}000$ psi, all required lap lengths must be increased by the ratio $(0.9f_y - 24)/30$, where f_y is in ksi (Section 12.16.1). For lower yield strength (grade) bars, the required lap length is reduced directly in the ratio $f_y/60$, where f_y is in ksi.

BUNDLED BARS For bars in 3-bar bundles, the above lap lengths must be increased 20 percent; in 4-bar bundles, 33 percent (Section 12.4). Note that these reductions in effectiveness prescribed for bundled bars are based on the exposed bar surface areas and, therefore, are intended to include any added dowels or "splice bars" as part of the number in the bundle. A bundle consisting of three through-bars with an added splice bar past butted cutoff points becomes a 4-bar bundle at the splice point for the purpose of dimensioning the length of the splice bar (see Chapter 10 of this Guide for examples).

LARGE BARS Lap splices are not permitted for bars larger than #11 (Section 12.14.2.1) except that: (1) #14 and #18 bars in compression may be lap spliced to #11 and smaller bars (Section 12.16.2); and (2) #14 and #18 column verticals lap spliced to smaller size bars used as dowels for column-to-footing connections where the vertical bars are required for compression only (Section 15.8.2.3) (see Chapter 12 for examples).

COLUMN VERTICALS For compression lap splices of column verticals, a number of special detailing requirements apply (Section 7.8). Where the column faces are offset 3 in. or more, separate dowels must be used (Section 7.8.1.5). Lap splices of column verticals must also meet additional requirements (Section 12.17.2). Each face of a column is required to have a minimum tension capacity even where structural analysis indicates compression only for all load combinations. Compression lap splices are deemed to provide the required minimum tension capacity.

Mechanical Connections in Compression Splicing

In addition to compression lap splices, the Code permits the use of mechanical connections or welding for compression splices of all bar sizes and requires same for #14 and #18.

Compression splices for bars required to transmit compressive stress only may consist of end bearing on square cut ends (Section 12.16.4). When conditions permit their use, end-bearing mechanical splices will be the most economical type for #14 and # 18 bars, since material costs, field time, and labor, and inspection requirements are minimal. Lateral confining reinforcement in the region of end-bearing splices must be provided (Section 12.16.4.3). The tolerance on squareness of square-cut ends, 1-1/2°, prescribed, is based upon full-size tests in which access for the concrete grout to the bearing area was not prevented (Commentary R12.16.4). Proprietary clamping devices are available which will hold the square-cut ends of the bars in close concentric contact even for changing bar sizes and also provide openings for inspection of fit that permit access by the grout. NOTE: when specifying end-bearing mechanical splices, it is usual practice to specify "saw-cut" to the required tolerance for maximum economy.

End-bearing mechanical splices of column verticals must also meet additional requirements (Section 12.17.4). Splices must be staggered or additional bars must be provided at the splice locations. See Fig. 13-14. The continuing bars in each face of the column must have a minimum tensile strength equal to $0.25f_y$ times the area of the vertical bars in that face.

TENSION SPLICES

Types

Tension splices for all #11 or less in size may consist of (1) lap splices, (2) full mechanical connections, (3) full welded splices, or (4) mechanical connections or welded splices designed for lesser values than "full" tension. NOTE: "full" tension design requires a splice capacity 125 percent of the minimum specified f_y of the bar being spliced, (Section 12.14.3).

Classes of Tension Lap Splices

Tension lap splices are specified as Class A or B (Section 12.15.1). The classes represent multiplies of tension development length, i.e., Class A = 1.0 ℓ_d and Class B = 1.3 ℓ_d. See Table 13-6 for the lap splice lengths required with the commonly used range of concrete compressive strengths of 3,000, 4,000 and

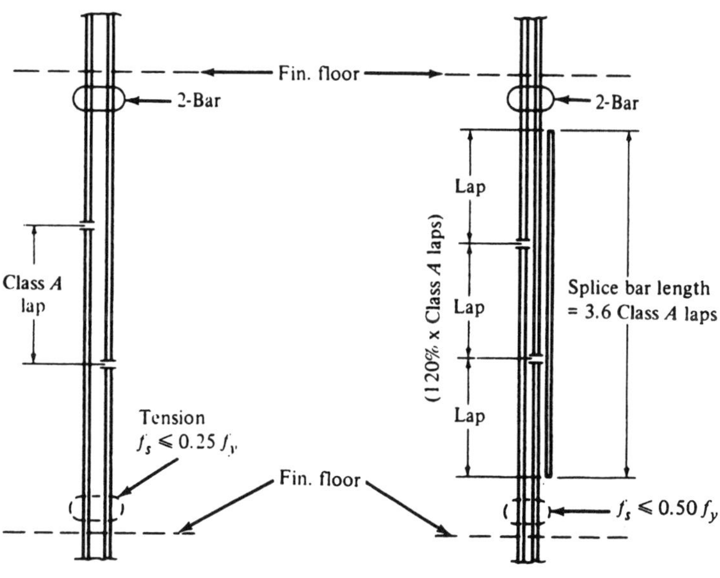

Figure 13-14 End-bearing mechanical splices and lap splices in bundled column verticals.

5,000 psi. Since tension lap splice lengths required are specified as a function of the tension development length, ℓ_d, the same variables which affect ℓ_d also influence tension lap splice length (see discussion of "Tension Embedment" on page 359). (Section 12.2). For tension development lengths, see Table 13-4. The location of the splice, tensile stress in the bar due to factored loads, confinement available, and the percentage of the total steel area spliced at the one point determine which class of lap splice may be used (Section 12.15.2). The Code requires Class B tension lap splices except Class A splices are permitted at locations or points where prescribed conditions are met. These conditions are: where A_s (provided) $\geqslant$ 2 times A_s (required) and not more than 50 percent of the total A_s is spliced within one Class A lap length. In terms of unit stresses, the conditions which permit the use of Class A tension lap splices are: where tensile stress in all bars due to factored loads $\leqslant 0.5f_y$ and no more than 50 percent of the total A_s is spliced within one Class A lap length. The end result of the Code's provisions for Class A splices is to encourage locating tension lap splices away from points of high tensile stress.

Examples of tension lap splices are shown in Figs. 13-15, 13-16, and 13-17.

Except for compression only to #11 and smaller bars (Section 12.16.2), and at footings (Section 15.8.2.3), mechanical connections or welded splices must be used for #14 and #18 bars (Section 12.14.2.1) and may be used for the smaller bars (Section 12.14.3.1). A "full" mechanical connection must be capable of developing in tension or compression or both, as required, 125 percent of the (specified) f_y of the bar being spliced; a "full" welded splice is a splice capa-

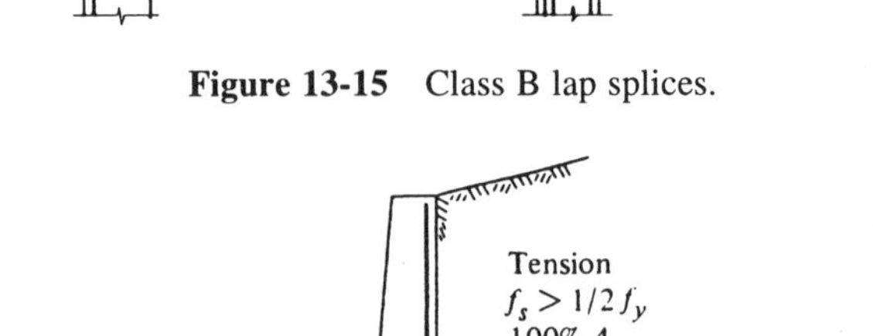

Figure 13-15 Class B lap splices.

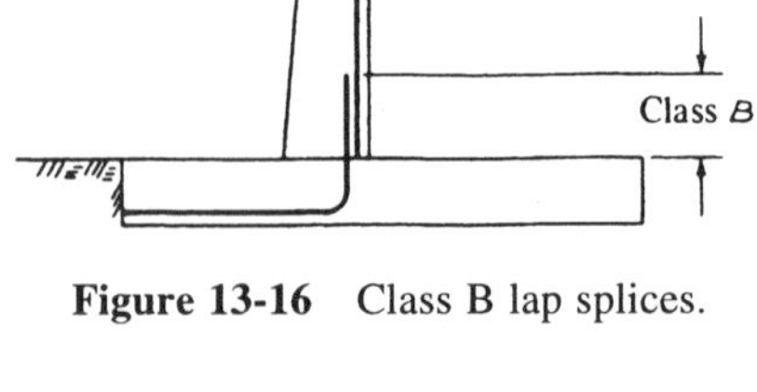

Figure 13-16 Class B lap splices.

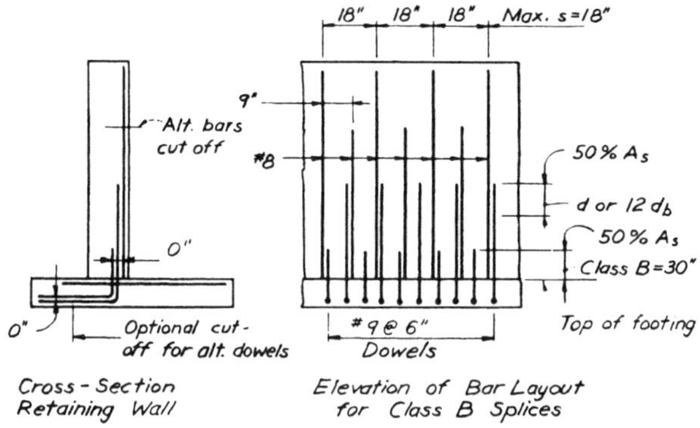

Figure 13-17 Class B lap splices.

ble of developing in tension 125 percent of f_y (Section 12.14.3). Lesser value mechanical connections and lesser value welded splices, including lap welds or groove welds may be used at points of known stress less than maximum tensile stress, where splices are staggered at least 24 in. and where, together with unspliced bars, at least twice the maximum calculated tensile capacity is provided, but not less than 20,000 psi on the total steel area through the section containing splices (Section 12.15.4.1).

WELDED WIRE FABRIC (WWF)

Plain WWF Wires

(Wire, ASTM A82; WWF, ASTM A185). Embedment of plain wire WWF is dependent entirely upon the anchorage of the welds to the cross-wires. Full development (embedment for full strength) requires two cross-wires embedded, with the closer wire at least 2 in. beyond the critical section. The minimum development length, ℓ_d, measured from the critical section to the outermost embedded cross-wire (Section 12.8) must be:

$$\ell_d \geqslant 0.27\left(\frac{A_w}{s_w}\right)\left(\frac{f_y}{\sqrt{f_c'}}\right)\lambda \geqslant 6 \text{ in.}$$

where A_w is the area of the wires being developed, s_w is the spacing of the wires being developed, and λ is a factor for lightweight aggregate concrete. See Figs. 13-18 and 13-19 for embedment applications.

Lap splice lengths are similarly established by the cross-wire embedment. Except for temperature and shrinkage reinforcement, lap splices of plain WWF at points of maximum stress should be avoided. Where calculated maximum tensile stress due to factored loads exceeds $1/2f_y$, the lap length required is one spacing of cross-wires plus 2 in. measured from the end cross-wires but not less than 1.5 times the calculated ℓ_d nor 6 in. (Section 12.19.1). For splices where

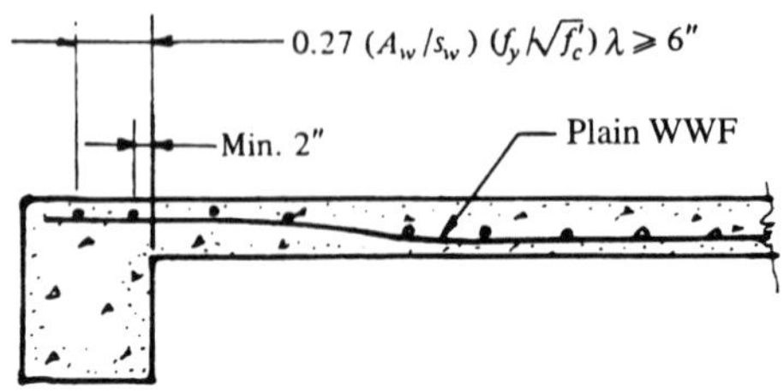

Figure 13-18 Plain WWF—Full Strength Embedment—straight.

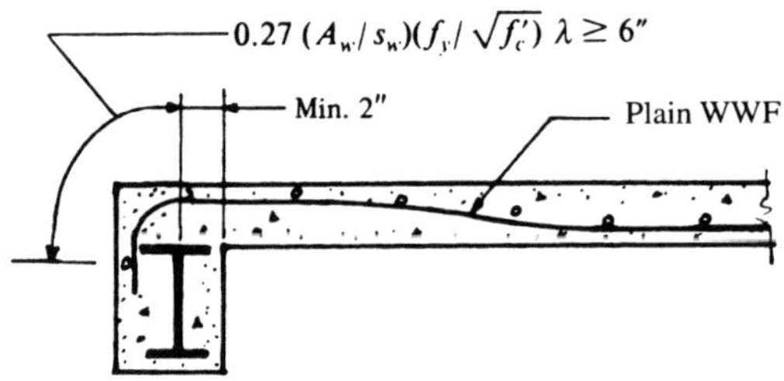

Figure 13-19 Plain WWF—Full Strength Embedment—hooked.

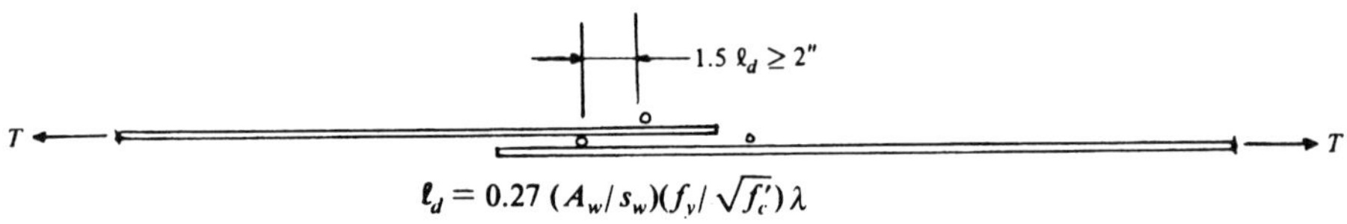

Figure 13-20 Plain WWF—Lap splice for A_s(actual) $\geq 2 \times A_s$(req'd).

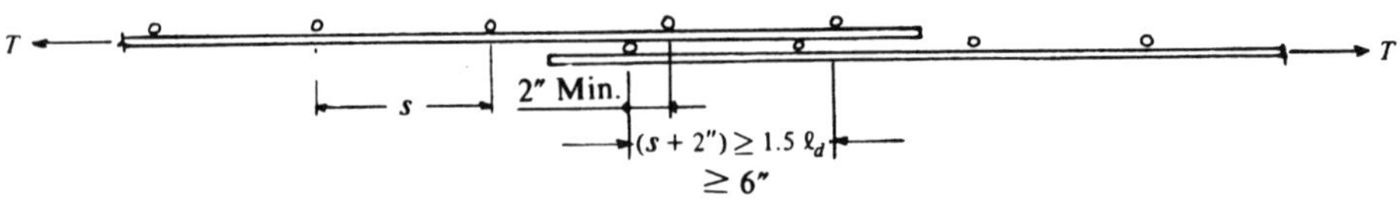

Figure 13-21 Plain WWF—Lap splice for A_s(actual) $< 2 \times A_s$(req'd).

the maximum calculated tensile stress is always less than $1/2f_y$, the outermost cross-wires must overlap 2 in. but not less than 1.5 times the calculated ℓ_d (Section 12.19.2) (see Figs. 13-20 and 13-21).

Deformed Wires and Deformed WWF

(Wire, ASTM A496; WWF, ASTM A497). Deformed wire has a development capacity recognized in the Code. The tension development length, ℓ_d, is calculated similar to ℓ_d for reinforcing bars (Section 12.2.2 or 12.2.3).

Deformed WWF is developed by both the deformations along the wire and the welded cross-wire anchorages. The development length of deformed WWF is computed as the ℓ_d of deformed wire (Section 12.2.2 or 12.2.3) times a wire fabric factor from Section 12.7.2 or 12.7.3. With at least one cross-wire present within the development length and located not less than 2 in. from the critical section, the wire fabric factor is the larger of:

$$\frac{f_y - 35{,}000}{f_y} \quad \text{or} \quad \frac{5d_b}{s_w}$$

but the factor is not required to exceed one (Section 12.7.2). If there are no cross-wires present within the development length or if there is a single wire located less than 2 in. from the critical section, the wire fabric factor can be taken as one and the development length is the same as that for deformed wire (Section 12.7.3). In the above expression, f_y is in psi, d_b is the nominal diameter of the wire, and s_w is the spacing of the wires being developed. Development length can be reduced by $A_{s(\text{required})}/A_{s(\text{provided})}$. The calculated ℓ_d cannot be less than 8 in. except for determining lap splice lengths. Figure 13-22 illustrates embedment for deformed WWF.

According to ASTM A497, deformed WWF can be made from a combination of deformed wires and plain wires. When deformed WWF has plain wires in the direction of ℓ_d, the development length requirements for plain WWF must be used (Section 12.7.4).

Lap splices for deformed wire follow the same requirements as those for deformed bars (Section 12.15). Lap splices of deformed WWF with no cross-wires within the lap splice length must meet the requirements for deformed wire (Section 12.18.2). If cross-wires are present and the overlap of the outermost cross-wires is at least 2 in., the lap splice length must be $1.3\ \ell_d \geqslant 8$ in. (Section 12.18.1). See Fig. 13-23. If the deformed WWF has plain wires in the direction of the lap splice, or when deformed WWF is spliced to plain WWF, the lap splice requirements for plain fabric must be used (Section 12.18.3).

It will be noted that all splice and embedment requirements for wire are based upon the full f_y. In most building elements using WWF as the principal reinforcement, of necessity the splices occur at the edges of the fabric, whether in

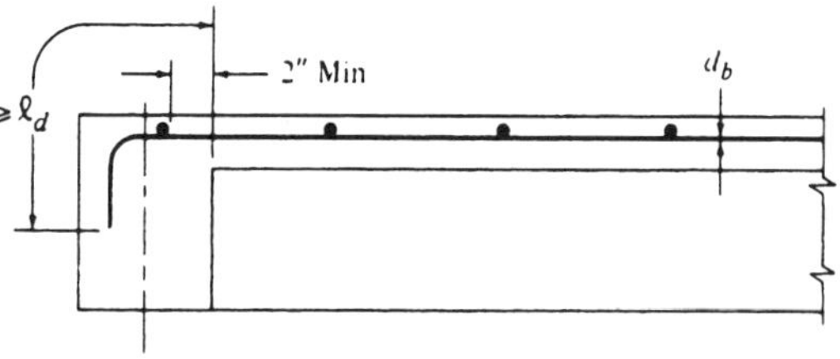

Figure 13-22 Embedment for deformed WWF.

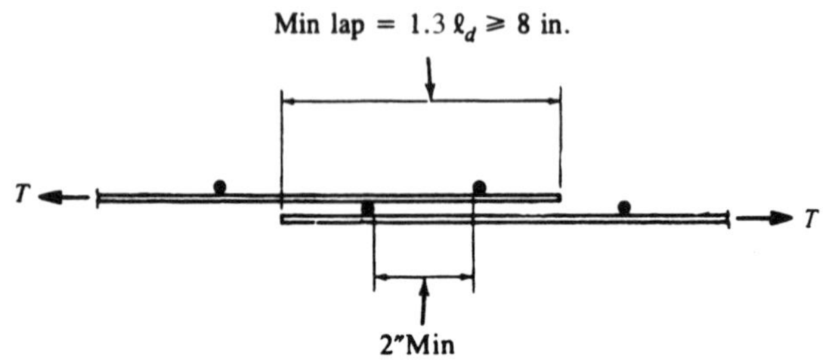

Figure 13-23 Lap splices for deformed WWF.

flat sheets or in rolls, and at ends of the rolls; it is seldom feasible to plan the location of all splices away from points of maximum stress. Also, it should be noted that temperature and shrinkage reinforcement in rigid concrete buildings is designed as fully stressed from end anchorage to end anchorage at each end of the element or structure, unlike temperature and shrinkage reinforcement in slabs on ground where stress is a function of the friction and weight of slab between the critical section and nearest free joint.

SPACING OF REINFORCEMENT

One of the most significant Code provisions regarding the spacing of reinforcement concerns bundled bars and noncontact lap splices. Where spacing and cover limitations are based on the bar size, a bundle of bars is treated as a single bar of a diameter derived from the equivalent total area of the bundle (Section 7.6.6). Contact lap splices are encouraged, and noncontact lap splices are limited so that the maximum transverse spacing between the single bars is not to exceed one-fifth of the required lap length or 6 in. (Section 12.14.2.3) (see Fig. 13-24).

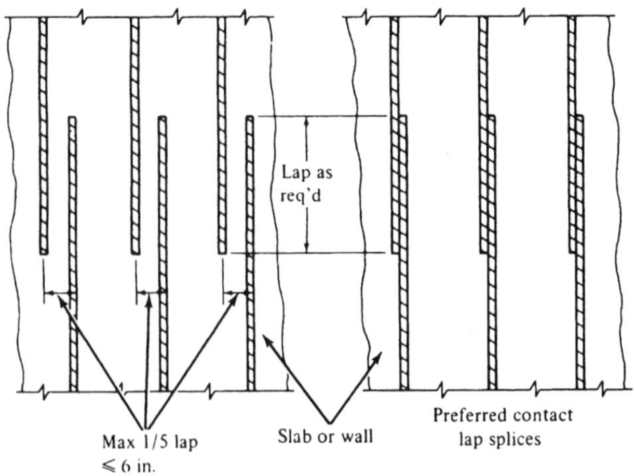

Figure 13-24 Spaced and contact lap splices.

In flexural members, the minimum clear spacing between bars is one bar diameter, but not less than 1 in. For parallel layers, as in beams, bars in the upper layer must be positioned directly above those in the layer below, but not less than 1 in. clear distance above (Section 7.6.2). In walls and one-way slabs, the maximum bar spacing is $3h$ (where h = thickness of wall or slab) but not more than 18 in. (Section 7.6.5). For two-way slabs, the maximum spacing of bars is $2h \leqslant 18$ in. (Section 13.3.2). For temperature and shrinkage reinforcement only, maximum spacing is $5h \leqslant 18$ in. (Section 7.12.2). Temperature and shrinkage reinforcement at right angles to the ribs, in the slabs of one-way slab-joist systems, must be supplied (Section 8.11.5) and the maximum spacing is $5h \leqq 18$ in., where h = thickness of top slab. See "Flexural Notes, " Chapter 9, for applications of the minimum spacing requirements to shrinkage and temperature reinforcement and minimum flexural reinforcement. In columns, vertical bars (or bundles) must have a clear spacing not less than $1\frac{1}{2}$ bar diameters or $1\frac{1}{2}$ in., whichever is larger (Section 7.6.3).

It will be noted that all of the above minimum bar spacings are contingent upon the use of concrete with a nominal maximum size of coarse aggregate not larger than three-fourths the minimum clear spacing (Section 3.3.2). Normal practice for design and detailing of reinforcement is to use the minimum bar spacings with the regulation of the aggregate size prescribed in the concrete section of the project specifications to suit the bar spacings. The engineer can waive the maximum limits on aggregate size if proper workability and consolidation of the concrete is achieved (Section 3.3.2). If not, adjustments in the concrete are required in any event (Section 5.2.1).

CONCRETE COVER

The Code requirements for concrete cover, intended to provide protection to the reinforcement for ordinary exposure conditions, are presented in Section 7.7. Except with earth forms, 2 in. is the maximum cover required. With earth forms a 3 in. cover is required to allow for a tolerance of -1 in. at each side of the excavation. The specified clear cover is measured to the outermost reinforcement (Section 7.7.1), be it ties, stirrups, or spirals if present; or to main reinforcing bars if not. For precast concrete (Section 7.7.2) or prestressed concrete (Section 7.7.3) separate requirements, applicable simultaneously, either of which may control, are given.

In the design of precast beams, girders, or columns, it is not sufficient to allow the minimum cover to the outermost reinforcement plus the diameter of same as cover to the main reinforcement. For example, if a precast column contains #8 vertical bars and #4 ties, cover to the vertical bars will control. Minimum cover to the #8 bar equals one bar diameter, 1 in.; minimum cover to the #4 tie equals 3/8 in., but 1/2 in. must be furnished in order to provide the 1 in. cover to the vertical bars (Section 7.7.2).

It will be noted that the required cover for precast or prestressed concrete is less than that for nonprestressed cast-in-place concrete. These reductions recognize the greater *accuracy* possible in the positioning of reinforcement under plant control when precast; and the greater *effectiveness* of cover when prestressed to keep cracks closed under dead load (tension $\leq$ zero) (Sections 7.7.3.2 and 7.7.3.3)

It is intended that the minimum covers prescribed be increased (1) where local building codes require additional fire protection, or (2) where members are exposed to highly corrosive environments (Sections 7.7.5 and 7.7.7). Such increases are not expected to be proportional to all specified minimum cover limits given. Beyond about 2 in., increases in cover do not provide proportional increases in protection against penetration of corrosive liquids or gases and improvement in the quality of the concrete itself or coated bars may be required (Section 7.7.5). As an example, for sewage retention, slab and wall cover prescribed for cast-in-place concrete (3/4 and $1\frac{1}{2}$ in.) might be increased to 2 in. together with use of a strength of concrete higher than that required structurally, use of sulfate-resisting cement (Section 4.3.1), and air entrainment (Section 4.2.1). Proprietary "shrinkage-controlled" cements have produced some construction with marked reduction in random shrinkage cracking. The use of the concrete made with such cements is not specifically regulated by the Code, but it is acceptable where performance requirements govern (Section 1.4).

JOINTS AND CORNERS

The Code has few prescriptive requirements for design of joints and corners. In this respect, at least, the Code requirements approach a pure "performance" style. Joints and corners are expected to be detailed to permit the structure to perform as assumed for analysis and design. The few general Code provisions require: enclosure within connections of "principal framing elements" (Section 7.9) and define enclosure as external concrete or internal closed transverse reinforcement; enclosing reinforcement for bars in compression, subject to stress reversals, or torsion in flexural members (Section 7.11); enclosure at end-bearing mechanical splices for compression applications (Section 12.16.6.3); and seismic connection details (Chapter 21) which are outside the scope of this book.

This book, however, would be incomplete without some further guidance to the user. For over-all study, a report on the "state of the art" in general has been published.* In addition, an interesting series of tests became available, which indicate the relative efficiency of "standard" corner and joint details, and will be helpful.** The latter will be summarized briefly.

---

* See "Selected Reference" No. 3.

** See "Selected Reference" No. 1.

The principal difficulty in design of corner connections details arises with moments tending to *open* the corner. Most standard design details focus on restraining the obvious tensile moment crack on the diagonal beginning from the inside corner. As the compression forces converge at right angles at the outside corner, induced tensile forces on the opposite diagonal develop and will, if unrestrained, pop off the outside corner destroying the compression side and causing premature joint failure. See Fig. 13-25. In designing very large rigid frame corners, a radial arrangement of stirrups is often used. See Fig. 13-26. For the usual wall corner or retaining wall to footing slab corner, this remedy is impracticable. Inefficient details for anchoring tensile reinforcement from the two directions are shown in Fig. 13-27. (Only the reinforcing steel in the tension faces is shown, together with approximate efficiencies, $E = M_{test}/M_{calc.}$). Efficient arrangements for the T-shaped joint between cantilever retaining wall and footing are shown in Fig. 13-28. Note that these examples are for moments in one direction only, and also note the improvement in relative crack width, *w*, with use of diagonal dowels. Finally, see Fig. 13-29, for corner details with satisfactory efficiencies. The details as tested have been modified slightly here to permit splicing corner reinforcing steel to long horizontal wall bars outside the joints so as to provide double-acting joints to resist either opening or closing moments.

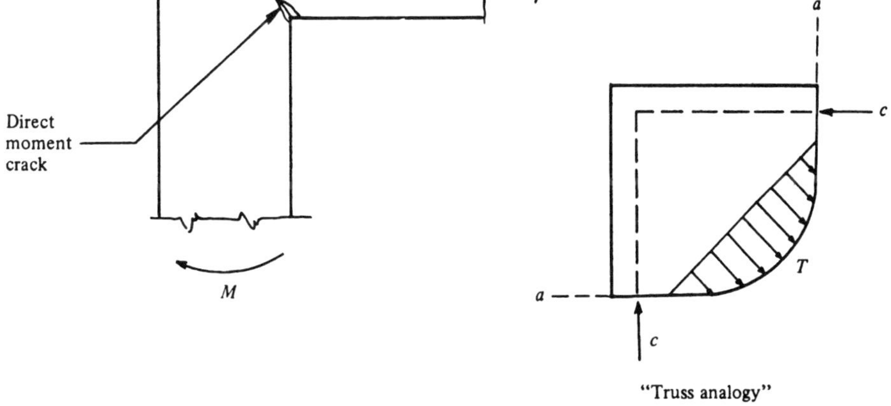

Figure 13-25 Corner joint with moments opening joint and potential cracking mechanism.

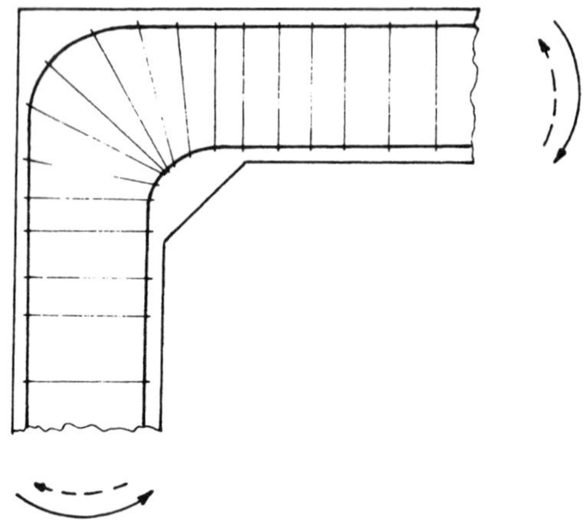

Figure 13-26

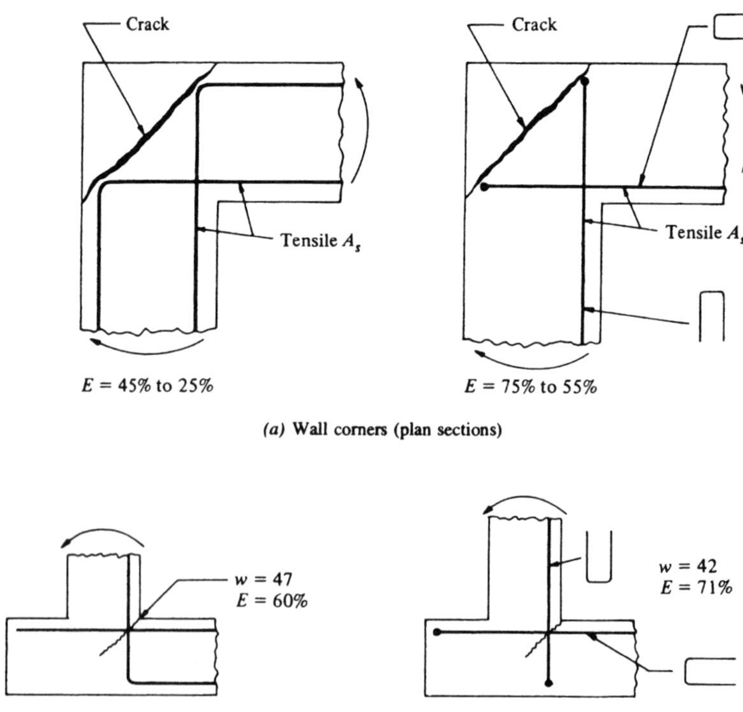

Figure 13-27 Inefficient design details at corners.

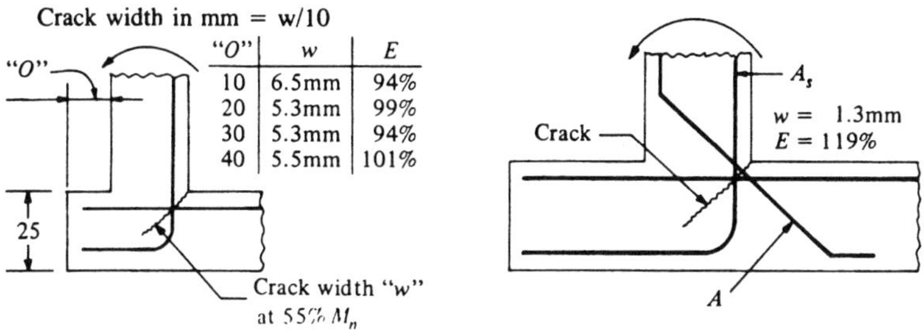

Figure 13-28 Satisfactory details for T-shaped joints of cantilever retaining wall-to-footing.

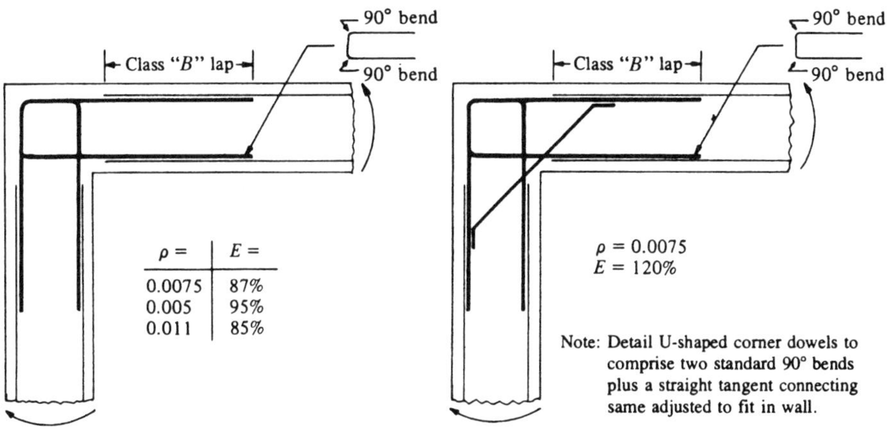

Figure 13-29 Wall corner design details—reinforcement to restrain both direct and induced potential cracks.

SELECTED REFERENCES

1. "Reinforced Concrete Corners and Joints Subjected to Bending Moment," I.H.E. Nilsson and A. Losberg, ASCE ST-6, June 1976.
2. "Detailing for Torsion," D. Mitchell and M. P. Collins, *ACI Journal,* Sept. 1976.
3. "Recommendations for Design of Beam-Column Joints in Monolithic Reinforced Concrete Structures, (ACI 352R-89)," *Manual of Concrete Practice*, Part 3.
4. "Mechanical Connections of Reinforcing Bars, (ACI 439.3R-91)," *Manual of Concrete Practice*, Part 3.

14 Prestressed Concrete

GENERAL

Prestressed concrete is reinforced concrete in which there have been introduced internal stresses of such magnitude and distribution that the stresses resulting from loads are counteracted to a desired degree. For design under the Code, the internal stresses can be introduced by pretensioning (a method of prestressing in which the tendons are tensioned before the concrete is placed) or by post-tensioning (a method of prestressing in which the tendons are tensioned after the concrete has hardened) (Section 2.1).

Although limited in no way in the Code, the provisions of the latter for prestressed concrete were developed primarily for slabs, beams, and columns, and contain no specific provisions for application to nonlinear prestressing for elements such as pipes and circular tanks (Commentary R18.1.1). Code provisions apply only to structural members prestressed with high-strength steel tendons made from wire, seven-wire strands, or from high-strength alloy steel bars (Section 18.1.1).

EXCEPTIONS TO THE GENERAL CODE

Prestressed concrete is subject to the provisions of the entire Code, except the following specifically excluded portions:

(Section 8.4)	Redistribution of moments on a limit design basis for prestressed concrete is covered in Section 18.10.4.
(Sections 8.10.2, 8.10.3 and 8.10.4)	The width and thickness of the flange of prestressed concrete T-beams is left to the judgment of the engineer. The empirical provisions for conventional reinforced concrete, if applied to prestressed concrete, would exclude many prestressed products in satisfactory use (Commentary Section R18.1.3).

(Section 8.11)	Concrete joists. Empircal limits are justified for con ventional reinforced concrete joist construction but not for prestressed concrete joist construction.
(Sections 10.3.2 and 10.3.3)	Limiting reinforcement ratio. For prestressed concrete, separate limits are prescribed (Section 18.8).
(Section 10.5)	Minimum reinforcement requirements. Prestressed concrete minimums are prescribed separately (Section 18.8).
(Section 10.9.1)	Limits for reinforcement in compression members. Compression members with an average prestress of 225 psi or higher are excluded from these limits. Compression members with an average prestress less than 225 psi for purposes of these requirements are not considered as prestressed (Section 18.11.2.1).
(Chapter 13)	Two-way slabs. Prestressed two-way slab design is separate (Section 18.12) .
(Sections 14.3, 14.5 and 14.6)	Walls. Empirical requirements for walls are not suitable for prestressed concrete.

HIGH-STRENGTH PRESTRESSING STEEL

High-strength steel tendons in prestressed concrete must conform to the "Standard Specification for Steel Strand, Uncoated Seven-Wire for Prestresed Concrete" (ASTM A416) or the "Standard Specification for Uncoated Stress-Relieved Wire for Prestressed Concrete" (ASTM A421) or the "Standard Specification for Uncoated High-Strength Steel Bar for Prestressing Concrete" (ASTM A722). Low-relaxation wire, conforming to the Supplement of ASTM A421, is also listed (Section 3.5.5.1). Other steel strands or wire or bars may be used providing that they conform to the minimum requirements of A416 or A421 or A722 and have no properties that make them less satisfactory (Section 3.5.5.2).

The prestressing strand is made by stranding six wires to form helices around a straight seventh wire. The straight wire has a slightly larger diameter to ensure that the outer wires, when under stress, will grip the center one. The properties of Grades 250 and 270 seven-wire uncoated stress-relieved steel strands conforming to ASTM A416, intended for use in prestressed concrete construction, are shown in Table 14-1. Note that the Grade 270 strand has a larger area, weight, and breaking strength than Grade 250. The minimum yield strength of ASTM A416 strand based on 1 percent total strain is $0.85f_{pu}$, as shown in Figure 14-1. Low-relaxation strand is regarded as the standard type.

The properties of uncoated stress-relieved wire conforming to ASTM A421 are shown in Table 14-2. The minimum yield strength for ASTM A421 wire

TABLE 14-1 Properties of ASTM A416 High-Strength Steel Strand

Strand Nominal Diameter (in.)	Nominal Strand Area (in.^{2})		Nominal Strand Weight (lb/t)		Strand Breaking Strength (kips)	
	Grade 250	Grade 270	Grade 250	Grade 270	Grade 250	Grade 270
1/4 (0.250)	0.036	—	0.122	—	9.0	—
5/16 (0.313)	0.058	—	0.197	—	14.5	—
3/8 (0.375)	0.080	0.085	0.272	0.290	20.0	23.0
7/16 (0.438)	0.108	0.115	0.367	0.390	27.0	31.0
1/2 (0.500)	0.144	0.153	0.490	0.520	36.0	41.3
(0.600)	0.216	0.217	0.737	0.740	54.0	58.6

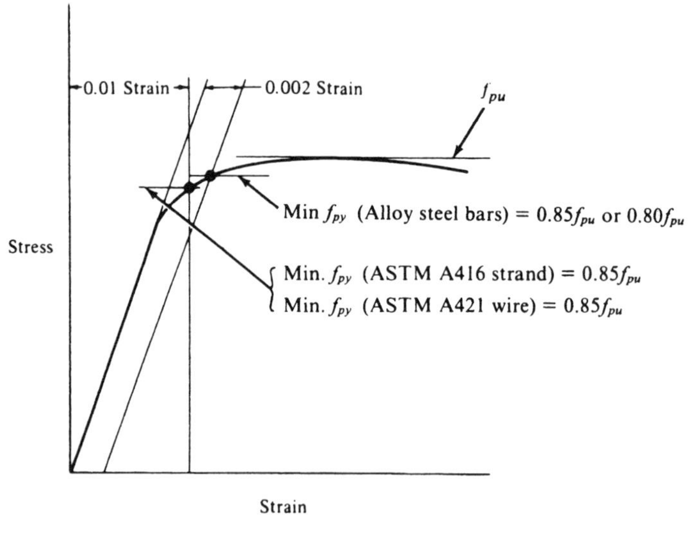

Figure 14-1 High-strength prestressing steel.

based on 1 percent, total strain is equal to 0.85 f_{pu}, as shown in Figure 14-1. Since the cessation of nuclear power plant construction the use of ASTM A421 wire is virtually nonexistent in new construction.*

High-strength alloy steel bars (Section 3.5.5.1-e) are used principally for post-tensioned prestressed concrete construction. They are available as plain

* "Steel Reinforcement—Physical Properties and U.S. Availability (ACI 439.4R-89)," *Manual of Concrete Practice*, Part 3, 1996.

TABLE 14-2 Properties of ASTM A421 High-Strength Steel Wire

Nominal Wire Diameter (in.)	Nominal Wire Area (in.^{2})	Minimum Wire Tensile Strength f_{pu} (ksi)	
		Button Anchorage *BA*	Wedge Anchorage *WA*
0.192	0.0290	—	250.0
0.196	0.0302	240.0	250.0
0.250	0.0491	240.0	240.0
0.276	0.0598	235.0	235.0

(Type I) or deformed (Type II) in diameters ranging from 5/8 in. (Type I) or 3/4 in. (Type II) to 1-3/8 in. These bars are required by ASTM A722, as shown in Fig. 14-1, to have a yield strength, f_{py}, not less than 0.85 f_{pu} for Type I and not less than 0.80 f_{pu} for Type II, and an elongation at rupture (0.2 percent offset) not less than 4 percent. High-strength alloy steel bars are produced with a minimum tensile strength as high as 150,000 psi.

Large-diameter high-strength wire strand made of 7, 19, 37, or more galvanized or uncoated hard-drawn wires, spirally wound, is also available for prestressing steel, but must meet the minimum requirements of ASTM A416 to conform to Code requirements (Section 3.5.5.2).

The modulus of elasticity of high-strength prestressing steel strand is usually less than wire and it should be determined by tests or supplied by the manufacturer (Section 8.5.3).

Cover

The minimum concrete cover for prestressed or nonprestressed reinforcement, ducts, and end fittings of cast-in-place and precast prestressed concrete members varies with exposure and manufacturing conditions. If members exposed to an earth, weather, or corrosive environment and members requiring greater cover for fire protection (Sections 7.7.3.2, 7.7.5, and 7.7.7) are excluded, the minimum concrete cover in inches for reinforcement, ducts, and end fittings in prestressed members, other than in members manufactured under plant control where cover can be the same as that for precast concrete (Section 7.7.2), is as follows:

	Tension in precompressed tension zone less than $6\sqrt{f'_c}$	Tension in precompressed tension zone greater than $6\sqrt{f'_c}$
Members cast against and permanently exposed to earth	3″	$4\frac{1}{2}$″
Members exposed to earth or weather:		
Wall panels, slabs, and joists	1″	$1\frac{1}{2}$″
Other members	$1\frac{1}{2}$″	$2\frac{1}{4}$″
Members not exposed to weather or in contact with ground:		
Slabs, walls, joists	$\frac{3}{4}$″	$\frac{3}{4}$″
Beams, girders, columns: Principal reinforcement	$1\frac{1}{2}$″	$1\frac{1}{2}$″
Beams, girders, columns: Ties, stirrups, spirals	1″	1″
Shells and folded plate members:		
Reinforcement $\frac{5}{8}$ in. and smaller	$\frac{3}{8}$″	$\frac{3}{8}$″
Other reinforcement	d_b but not less than $\frac{3}{4}$″	

When prestressed concrete members are subject to corrosive atmospheres, the amount of concrete cover must be suitably increased or other protection must be provided (Section 7.7.5).

Loss of Prestress

The amount of prestress in a prestressed concrete member varies with time. The initial prestressing force can be measured with reasonable accuracy by reading a pressure gauge on the jack and checked by measuring tendon elongation. The final prestressing force is more difficult to determine and can only be estimated because it varies with time and is affected by slip at anchorage, elastic shortening of concrete, creep of concrete, shrinkage of concrete, and relaxation of steel stress, and includes frictional losses due to intended or unintended curvature in the tendons. The difference between the initial prestressing force and the effective prestressing force at any later time is defined herein as the loss of prestress.

The loss of prestress is difficult to predict and can only be estimated. T. Y. Lin and N. H. Burns,* suggest that for average steel and concrete properties and curing conditions encountered with prestressed concrete members that the average percentage of loss of prestress for losses other than frictional losses can be taken as follows:

* *Design of Prestressed Concrete Structures*, 3rd. ed. (John Wiley & Sons, Inc., New York, 1981).

	Pretensioning, %	Post-tensioning, %
Elastic shortening and bending of concrete	4	1
Creep of concrete	6	5
Skinkage of concrete	7	6
Steel relaxation	8	8
Total loss	25	20

Pre-1983 editions of the Code Commentary contained lump sum values of prestress losses, e.g., when specific loss data are lacking, the loss in prestress—not including friction loss—can be assumed as 35,000 psi for pretensioning and 25,000 psi for post-tensioning. However, such lump sum values of prestress losses are now regarded as obsolete, and they have been deleted from the current Code Commentary. The Commentary suggests calculating losses in accordance with the recommendations in a report which was prepared under the direction of ACI-ASCE Committee 423.*

Frictional losses can be calculated from experimentally determined wobble and curvature coefficients which must be controlled during stressing operations by initial overstressing, measured elongation, and, if necessary, jacking from each end (Section 18.6.2). The Code Commentary sets forth a range of values for wobble and curvature coefficients for grouted tendons in metal sheathing and unbonded mastic coated and pregreased tendons.

The calculated nominal moment strength of a prestressed concrete member with bonded tendons is little affected by the magnitude of the loss of prestress, but any error in the provisions for the loss of prestress could result in a false prediction of the behavior of the member under service loads and cause calculations for stresses, cracking load, and deflection or camber to be in error.

Flexural Strength

The flexural strength of prestressed concrete members must be computed by the strength design methods of the Code (Section 18.7). In designing for strength, the assumptions provided for nonprestressed members apply (Section 10.2), except the assumption of "flat yielding" (Section 10.2.4) which applies only for nonprestressed reinforcing steel that conforms to Section 3.5.3 of the Code (Section 18.3.1).

In the calculation of the design moment strength, ϕM_n (Section 9.3.1) the Code requires that f_{ps}, stress in the prestressing steel at nominal strength be substituted for f_y, specified yield strength of nonprestressed reinforcement, (Section

* "Estimating Prestress Losses*" by P. Zia, H. K. Preston, N. L. Scott, and E. B. Workman; *ACI Concrete International: Design & Construction*, V. 1, No. 6, June 1979, pp. 32-38.

18.7). Different types of high-strength prestressing steels have different stress-strain characteristics and each lacks a sharp and distinct yield point (Fig. 14-1). It is therefore difficult to predict actual stress at nominal strength closely unless the stress-strain curve for the steel being used is known.

When the information to determine f_{ps} is not available, the Code allows approximate values to be used, provided that the effective prestressing force after losses at service loads is not less than 50 percent of the specifed tensile strength f_{pu} (Section 18.7.2).

For members with unbonded tendons and with a span-to-depth ratio of 35 or less, the approximate value given for f_{ps} is

$$f_{ps} = f_{se} + 10{,}000 + \frac{f'_c}{100\,\rho_p} \quad \text{(Section 18.7.2-b, Eq. 18-4)}$$

where f_{ps} must not be greater than f_{py} nor greater than f_{se} + 60,000.

The flexural design strength of members with unbonded tendons is generally less than, and occurs at, a lower steel stress than in members with equivalent bonded tendons. Large variations in the steel stress of unbonded tendons at design moment strength have been reported by different investigators for different members and are due to differences in these members such as profile of prestressing steel, moment diagram, length:depth ratio, friction coefficient between prestressing steel and duct and amount of bonded nonprestressed reinforcement. The rational method for prediction of design moment strength of unbonded members requires: (1) an accurate estimate of losses and (2) an accurate deflection analysis to predict increase in prestress from service load to factored load based on known stress-strain properties of the steel.

For members with bonded tendons, the approximate value given for f_{ps} is (Section 18.7.2-a, Eq. 18-3):

$$f_{ps} = f_{pu}\left(1 - \frac{\gamma_p}{\beta_1}\left[\rho_p \frac{f_{pu}}{f'_c} + \frac{d}{d_p}(\omega - \omega')\right]\right).$$

For members without nonprestressed tension reinforcement or compression reinforcement, Eq. 18-3 reduces to

$$f_{ps} = f_{pu}\left(1 - \frac{\gamma_p}{\beta_1}\left[\rho_p \frac{f_{pu}}{f'_c}\right]\right).$$

The expression for f_{ps} for members with bonded tendons and no nonprestressed tension reinforcement or compression reinforcement is shown graphically in Fig. 14-2, where high-strength prestressing steels (not low-relaxation, i.e., γ_p = 0.40) with f_{pu} = 250 and 270 ksi are used in combination with f'_c = 4, 5, and 6 ksi.

Note in Fig. 14-2 that values for f_{ps} for bonded tendons are discontinued at steel ratios where $\omega_p = \rho_p f_{ps}/f'_c$ becomes greater than 0.36 β_1, which is the maximum value permitted to avoid the condition of an overreinforced beam which will fail in compression and whose design moment strength will be controlled

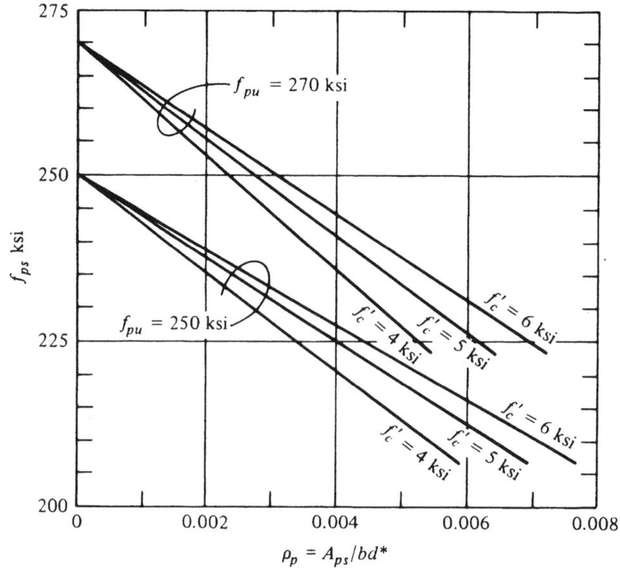

*For flanged sections, where the neutral axis or the depth of the equivalent rectangular stress block at design strength falls below the flange, ρ is calculated using the width of the web and the steel area is taken as only that amount that is required to develop the design compressive strength of the web. (Section 18.0).

Figure 14-2 Approximate values of f_{ps} for bonded tendons.

by the strength of the concrete (Section 18.8.1). For values of ω_p less than 0.36 β_1, a flexural member will fail in tension and the design moment strength will be controlled by the strength of the prestressing steel.

For rectangular sections without nonprestressed tension reinforcement or compression reinforcement, the design moment strength, ϕM_n, for flexural members reinforced only with high-strength prestressing steel can be derived as shown in Fig. 14-3.

When $\rho_p f_{ps}/f_c' < 0.36\, \beta_1$, tension controls and the flexural member is under-reinforced and the design moment strength is

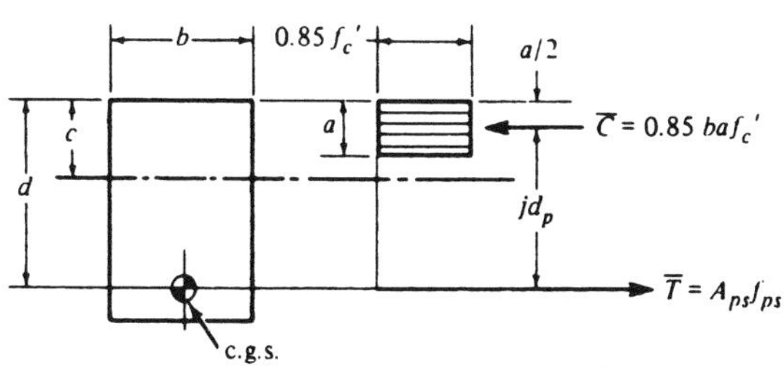

Figure 14-3 ϕM_n for rectangular prestressed sections.

$$a = \frac{A_{ps} f_{ps}}{0.85 f'_c\, b} = \frac{\rho_p f_{ps}\, d_p}{0.85 f'_c}$$

$$\begin{aligned}\phi M_n &= \phi\, T j\, d_p = \phi\, A_{ps} f_{ps}\, j d_p \\ &= \phi\, A_{ps} f_{ps} \left(d_p - \frac{a}{2}\right) \\ &= \phi\, A_{ps} f_{ps} \left(d_p - \frac{0.59\, \rho_p f_{ps}\, d_p}{f'_c}\right) \\ &= \phi\, A_{ps} f_{ps}\, d_p\, (1 - 0.59\omega_p)\end{aligned}$$

When $\rho_p f_{ps}/f'_c = 0.36\, \beta_1$, compression starts to control and the flexural member starts to become overreinforced and the design moment strength approaches

$$\begin{aligned}\phi M_n &= \phi\, A_{ps} f_{ps}\, d_p\, (1 - 0.59\omega_p) \\ &= \phi\, \rho_p\, b d_p\, f_{ps}\, d_p\, (1 - 0.59\omega_p) \\ &= \phi\, \rho_p \frac{f_{ps}}{f'_c} f'_c\, b d_p^2\, (1 - 0.59\omega_p) \\ &= \phi\, \omega_p f'_c\, b d_p^2\, (1 - 0.59\omega_p) \\ &= \phi\, (0.36\, \beta_1 f'_c\, b d_p^2\,)[1 - (0.59)\,(0.36\beta_1)] \\ &= \phi\, [f'_c\, b d_p^2\, (0.36\beta_1 - 0.08\beta_1^2)]\end{aligned}$$

The authors recommend the use of a ϕ-factor of 0.70 for this brittle type of concrete compression failure.

For flanged sections such as T-beams, where the neutral axis and the depth of the equivalent rectangular stress block fall below the flange, the design moment strength, ϕM_n, for flexural members reinforced only with high-strength prestressing steel can be derived as shown in Fig. 14-4.

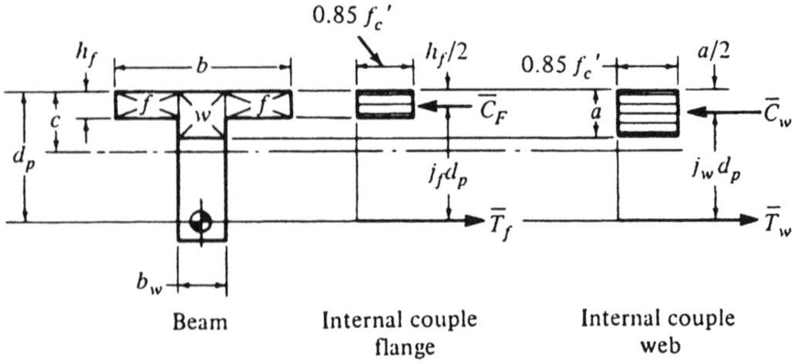

Figure 14-4 ϕM_n for T-beam prestressed sections.

Let $A_{psw} = A_{ps} - A_{psf}$ = the steel area required to develop the ultimate compressive strength of the web of a flanged section. When $A_{psw} f_{ps} / b_w \, d f_c' < 0.36 \, \beta_1$, tension controls and the flexural member is underreinforced and the design moment strength is

$$\overline{T}_f = \overline{C}_f = A_{psf} f_{ps} = 0.85 f_c' \, (b - b_w) \, h_f$$

$$\phi \, M_n = \phi \, (\overline{T}_f j_f \, d_p + T_w \, j_w \, d_p)$$

$$= \phi \left[A_{psf} f_{ps} \left(d_p - \frac{h_f}{2} \right) + A_{psw} \, f_{ps} \left(d_p - \frac{a}{2} \right) \right]$$

$$= \phi \left[0.85 f_c' \, (b - b_w) \, h_f \left(d_p - \frac{h_f}{2} \right) + A_{psw} \, f_{ps} \left(d_p - \frac{a}{2} \right) \right]$$

When $A_{psw} f_{ps} / b_w d_p f_c' = 0.36 \, \beta_1$, compression starts to control and the flexural member starts to become overreinforced and the design moment strength approaches

$$a = \frac{A_{psw} \, f_{ps}}{0.85 \, f_c' \, b_w} = \frac{0.36 \, \beta_1 \, b_w \, d_p f_c'}{0.85 f_c' \, b_w} = 0.423 \, \beta_1 \, d_p$$

$$\phi \, M_n = \phi \left[0.85 f_c' \, (b - b_w) \, h_f \left(d_p - \frac{h_f}{2} \right) + 0.36 \, \beta_1 \, b_w \, d_p f_c' \left(d_p - \frac{a}{2} \right) \right]$$

$$= \phi \left[0.85 f_c' \, (b - b_w) \, h_f \left(d_p - \frac{h_f}{2} \right) + 0.36 \beta_1 b_w d_p f_c' \left(d_p - \frac{0.423 \beta_1 d_p}{2} \right) \right]$$

$$= \phi \left[0.85 f_c' \, (b - b_w) \, h_f \left(d_p - \frac{h_f}{2} \right) + f_c' \, b_w \, d_p^2 \, (0.36 \, \beta_1 - 0.08 \, \beta_1^2) \right]$$

Again the authors recommend the use of a ϕ-factor of 0.70 here for a brittle concrete compression failure.

As stated before, the fact that bonded reinforcement is prestressed does not affect the calculated flexural strength of the member and nonprestressed reinforcement, in combination with prestressed steel, can be considered to contribute to the tension force in a member at factored load an amount equal to its area times its yield strength (Section 18.7.3).

To avoid the sudden flexural failure of an underreinforced prestressed concrete member due to an overload, such as when the flexural capacity is reached immediately after cracking, the total amount of prestressed and nonprestressed reinforcement cannot be less than that required to develop a factored load in flexure at least 20 percent greater than the cracking load (Section 18.8.3), except for flexural members which have shear and flexural strength equal to at least twice that required by Section 9.2.

Behavior at Service Conditions

In addition to having adequate flexural strength, prestressed concrete members must behave properly at service conditions at all loads that the member will be subjected to from the time the prestress is applied (Section 18.2.2). Service conditions requiring proper behavior include transfer of the prestressing force (when the tensile force in the prestressed steel is transferred to the concrete usually combined with all or part of the dead load, and compression stress in the concrete is usually high relative to ultimate concrete compressive strength). Service dead and live loads after long-time volume changes have occurred represent a condition that must also be considered. Other service conditions that must be considered for proper behavior include: stress concentrations due to prestressing (Section 18.2.3); effects on the adjoining structure of elastic and plastic deformations, deflections, changes in length, and rotations caused by the prestressing (Section 18.2.4); temperature and shrinkage (Section 18.2.4); possiblity of buckling in a member between points where the concrete and the prestressing steel are in contact and of buckling in their webs and flanges (Section 18.2.5); and deflections (Section 9.5.4).

Permissible stresses in the concrete and the prestressing steel at service loads are given to ensure serviceability (Commentary Section R18.4). They do not automatically guarantee adequate flexural strength which must satisfy factored load combinations (Section 9.2). Stresses in the concrete and the prestressing steel can be calculated using the basic assumptions that strains vary linearly with depth through the entire load range, and that concrete resists no tension at cracked sections (Section 18.3.2).

Permissible flexural stresses in the concrete at service loads must be within prescribed limits shown in Table 14-3, unless it can be shown experimentally or analytically that performance will not be impaired (Sections 18.4.1, 18.4.2, and 18.4.3).

When the calculated flexural tension stress after transfer but before losses exceeds $3\sqrt{f_c'}$, reinforcement must be provided to resist the total tension force in the concrete computed on the assumption of an uncracked section (Section 18.4.1).

Permissible flexural stresses in the high-strength prestressing steel at service loads must be within prescribed limits (Section 18.5) shown in Table 14-4.

Shear

Concrete is strong in *pure* shear. The Code limits the shear strength of normal weight concrete where it is inappropriate to use it as a measure of diagonal tension to a maximum value of $0.2\ f_c'$ but not more than 800 psi (Section 11.9.3.2.1) for members in one-way (beam) shear.

For design of prestressed concrete flexural members, the average or nominal shear stress, $V_u/\phi\, b_w\, d$, is used as a measure of the diagonal tension. In this expression the distance, d, is taken as that from the extreme compression fiber

TABLE 14-3 Permissible Stresses in Concrete at Service Loads

	f'_{ci} = 4 ksi	f'_{ci} = 5 ksi	f'_{ci} = 6 ksi
After transfer but before losses			
Compression, $f_c \leqslant 0.6 f'_{ci}$	2,400 psi	3,000 psi	3,600 psi
Tension in members without prestressed or nonprestressed bonded auxiliary reinforcement in the tension zone, f_t, $\leqslant 3\sqrt{f'_{ci}}$ except at ends of simply-supported members	190 psi	212 psi	232 psi
Tension at ends of simply-supported members, $f_t \leqslant 6\sqrt{f'_{ci}}$	379 psi	424 psi	465 psi
	f'_c = 4 ksi	f'_c = 5 ksi	f'_c = 6 ksi
Service load stresses after losses			
Compression, $f_c \leqslant 0.45 f'_c$ for sustained service live load	1,800 psi	2,250 psi	2,700 psi
Compression, $f_c \leqslant 0.60 f'_c$ for transient or temporary service live load	2,400 psi	3,000 psi	3,600 psi
Tension in precompressed tensile zone,	379 psi	424 psi	465 psi
(except two-way slab systems) where immediate and long-term deflection calculated on the basis of the transformed cracked section and on bilinear moment deflection comply with limits (Section 9.5), $f_t \leqslant 12\sqrt{f'_c}$	758 psi	848 psi	930 psi

to the centroid of the tension reinforcement, but not less than 0.8 h (Section 11.4.2.3). Sections located less than a distance $h/2$ from the face of the support can be designed for the shear, V_u, at $h/2$ (Section 11.1.3.2), when the reaction in the direction of the applied shear introduces compression into the end region of the member, and no concentrated load occurs between the face of the support and the section.

There are two types of diagonal tension cracks that occur in flexural members; web-shear cracks caused by principal tensile stresses that exceed the tensile strength of the concrete and flexural-shear cracks initiated by flexural cracks. When flexural cracks occur, the shear stresses above the crack increase resulting in increased principal tensile stresses that cause flexural-shear cracks when the principal tensile stresses exceed the tensile strength of the concrete.

To restrain the growth of both types of diagonal tension cracks, and to increase ductility (Commentary Section R11.5.5), the Code requires that a minimum area of shear reinforcement be provided in all prestressed concrete flexural members, except: (1) where V_u is less than one-half of ϕV_c: or (2) where the depth of the members (h) is less than 10 in., 2.5 times the thickness of the compression flange, or one-half the thickness of the web; or (3) where tests show that the required nominal flexural and shear strengths can be developed with the shear reinforcement omitted (Section 11.5.5.2).

TABLE 14-4 Permissible Stresses in Prestressing Steel at Service Loads

	ASTM A416 STRAND		ASTM A421 WIRE			ASTM A722 BAR	
f_{pu}	250 ksi	270 ksi	235 ksi	240 ksi	250 ksi	150 ksi	150 ksi
f_{py}	212.5 ksi	229.5 ksi	199.75 ksi	204 ksi	212.5 ksi	127.5 ksi (TYPE I)	120 ksi TYPE II)
Tension due to jacking $\leqslant 0.94 f_{py}$, but not greater than $0.85 f_{pu}$	199.7 ksi	215.7 ksi	187.7 ksi	191.7 ksi	199.7 ksi	119.8 ksi	112.8 ksi
Tension in pretensioning tendons, immediately after transfer, or in post-tensioning tendons immediately after anchoring $\leqslant 0.82 f_{py}$, but not greater than $0.74 f_{pu}$	174.2 ksi	188.1 ksi	163.7 ksi	167.2 ksi	174.2 ksi	104.5 ksi	98.4 ksi
Tension in post-tensioning tendons at anchorages and couplers immediately after tendon anchorage $\leqslant 0.7 f_{pu}$	175.0 ksi	189.0 ksi	164.5 ksi	168.0 ksi	175.0 ksi	105.0 ksi	105.0 ksi

Where shear reinforcement is required for prestressed concrete members the minimum area, A_v, must be equal to or greater than

$$A_v = 50\frac{b_w s}{f_y} \quad \text{(Section 11.5.5.3, Code Eq. 11-13)}$$

Alternately, a minimum area

$$A_v = \frac{A_{ps} f_{pu} s}{80 f_y d}\sqrt{\frac{d}{b_w}} \quad \text{(Section 11.5.5.4, Code Eq. 11-14)}$$

can be used if the effective prestress force is at least equal to 40 percent of the tensile strength of the flexural reinforcement (Section 11.5.5.4).

The design yield strength of the shear reinforcement for prestressed members may not be assumed greater than 60,000 psi. The limit on the design yield strength of deformed welded wire fabric is 80,000 psi (Section 11.5.2).

Shear reinforcement is required to consist of stirrups placed perpendicular to the axis of the member (Section 11.5.1.1-a). Such stirrups may consist of welded wire fabric with wires placed perpendicular to the axis of the member (Section 11.5.1.1-b). Vertical legs of shear reinforcement cannot be spaced further apart than 0.75 h or 24 in. (Section 11.5.4.1).

Figure 14-5 shows in graphical form the minimum area of shear reinforcement for prestressed concrete members of two different widths and varying depths (Section 11.5.5).

Minimum shear reinforcement required (Code Eq. 11-13) is shown in Fig. 14-5 for flexural members with minimum d ($d = 0.8\ h$), $f_y = 60{,}000$ psi and a

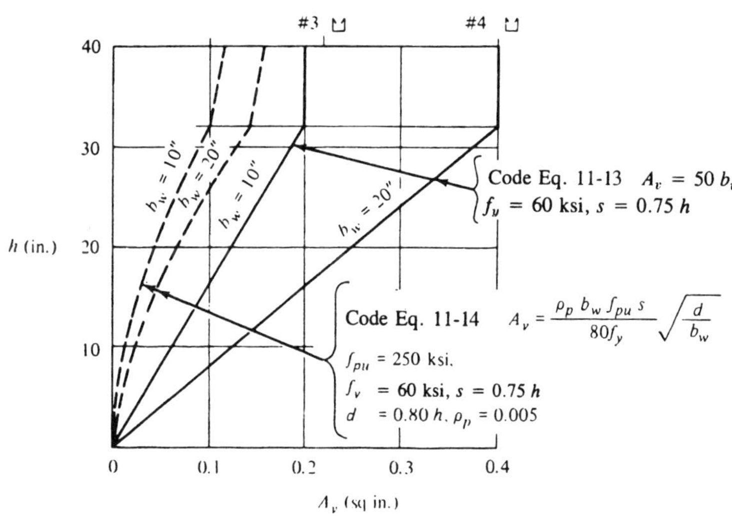

Figure 14-5 Minimum shear reinforcement for prestressed concrete members.

stirrup spacing equal to 0.75 h, but not greater than 24 in. The alternative minimum shear reinforcement required (Code Eq. 11-14) is also shown in Fig. 14-5 for $f_{pu} = 250{,}000$ psi, an effective prestressing force equal to or greater than 40 percent of the tensile strength of the flexural reinforcement, minimum d ($d = 0.8\,h$), $f_y = 60{,}000$ psi and a stirrup spacing equal to 0.75 h, but not greater than 24 in. It is obvious from Fig. 14-5 that, when minimum stirrups only are required, Code Eq. 11-14 will require less shear reinforcement.

In prestressed concrete members without shear reinforcement, the shear strength provided by the concrete is assumed equal to the shear causing significant inclined cracking and must not exceed

$$V_c = \left(0.6\sqrt{f_c'} + \frac{700\,V_u\,d}{M_u}\right) b_w\,d \quad \text{(Section 11.4.1, Eq. 11-9)}$$

unless a more detailed analysis is made (Section 11.4.1). Shear strength provided by the concrete, V_c, need not be less than $2\sqrt{f_c'}\,b_w\,d$ and cannot be greater than $5\sqrt{f_c'}\,b_w\,d$ (Section 11.4.1). The factored moment, M_u, occurs simultaneously with shear, V_u, at a particular section and d is the distance from the extreme compression fiber to the centroid of the prestressing tendons at this section. Note that d is not limited to a minimum value of 0.80 h. $V_u d/M_u$ cannot be taken greater than 1.0.

For a simple span beam, supporting a uniform load w, the quantity $V_u d/M_u$ in Code Eq. 11-9 at a distance x from the support can be expressed as follows (Commentary Section R11.4.1):

$$\frac{V_u\,d}{M_u} = \frac{w\,(\ell/2 - x)\,d}{(w/2)\,(\ell - x)\,x} = \frac{d\,(\ell - 2x)}{x\,(\ell - x)}$$

Using this expression to calculate $V_c/b_w d$

$$V_c/b_w\,d = 0.6\sqrt{f_c'} + \frac{700\,d\,(\ell - 2x)}{x\,(\ell - x)}$$

which is shown graphically in Fig. 14-6 for concretes with $f_c' = 4{,}000$ and 6,000 psi and for d/ℓ ratios between 0.01 and 0.06.

There is less than 4 percent difference between values of v_c for $f_c' = 4{,}000$ psi and for $f_c' = 6{,}000$ psi, except where $V_c/b_w d$ approaches a maximum permitted value of $5\sqrt{f_c'}$ near the support or a minimum value of $2\sqrt{f_c'}$ near the center two-thirds of the span.

When it is desired to use a more accurate method to determine the shear carried by the concrete or when the effective prestress force is less than 40 percent of the tensile strength of the flexural reinforcement, it will be necessary to use the lesser of the shear causing inclined flexure-shear cracking, V_{ci}, or web-shear cracking, V_{cw}.

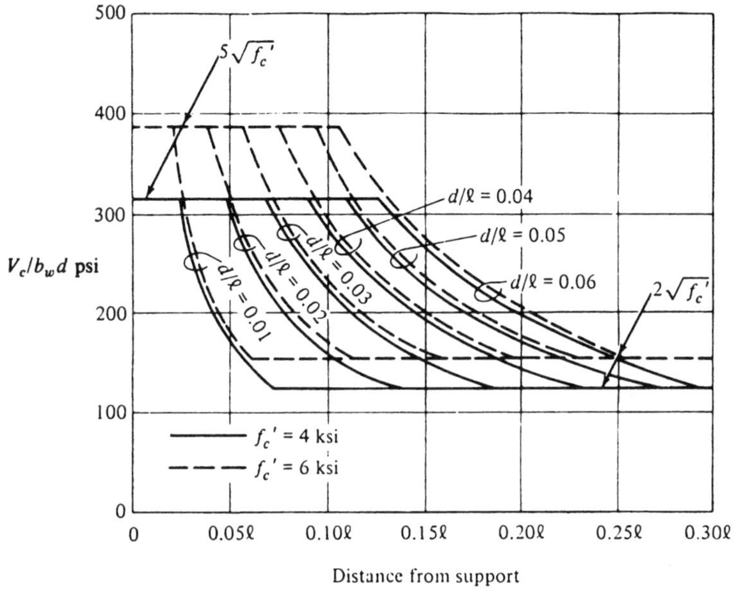

Figure 14-6 v_c for simple beam with uniform load.

$$V_{ci}/b_w d = 0.6\sqrt{f_c'} + \frac{V_d + (V_i M_{cr}/M_{max})}{b_w d} \qquad \text{(Section 11.4.2.1, Eq. 11-10)}$$

$$V_{cw}/b_w d = 3.5\sqrt{f_c'} + 0.3 f_{pc} + V_p/b_w d \qquad \text{(Section 11.4.2.2, Eq. 11-12)}$$

Web-shear cracking is predicted if the shear, V_{cw}, causes a principal tension stress approximately $4\sqrt{f_c'}$ at the centroidal axis of the cross section (Commentary Section R11.4). In general, $V_{cw}/b_w d$ will be the maximum value for the shear carried by the concrete similar to the maximum stress of $5\sqrt{f_c'}$ in the approximate equation. The computed shear stress carried by the concrete need not be less than $1.7\sqrt{f_c'}$, which is smaller than the minimum value of $2\sqrt{f_c'}$ for an approximate equation due to the possibility of a smaller prestress force.

For uniformly loaded noncomposite beams, the equation for v_{ci} reduces to the following expression (Commentary Section R11.4):

$$V_{ci}/b_w d = 0.6\sqrt{f_c'} + \frac{V_u M_{cr}}{M_u b_w d}$$

It is important to note that with equations for V_{ci} and V_{cw}, d is the distance from the extreme compression fiber to the centroid of the prestressing tendons, and need not be taken less than 0.8 h (Section 11.4.2.3). The approximate equation for V_c requires that d be the actual distance from the extreme compression fiber to the centroid of the prestressing steel with no minimum limit (Section 11.4.1).

When the factored shear force, V_u is greater than the shear strength, ϕV_c, of the concrete (Sections 11.4.1 and 11.4.2) shear reinforcement must be provided (Section 11.5.6.1). Proper shear reinforcement will ensure that the prestressed concrete member will develop its design flexural strength without serious shear distress.

Shear reinforcement must be placed perpendicular to the axis of the member and spaced not farther apart than 0.75 h, or not more than 24 in. (Section 11.5.4.1). When the nominal shear strength provided by the reinforcement, V_s is greater than $4\sqrt{f'_c}\, b_w d$ the maximum spacing of shear reinforcement must be reduced to 0.375 h but not more than 12 in. (Section 11.5.4.3).

The maximum shear strength provided by the reinforcement, V_s must be equal to or less than $8\sqrt{f'_c}\, b_w d$ (Section 11.5.6.8).

The shear reinforcement required to carry the shear in excess of that carried by the concrete can be determined as

$$A_v = \frac{(V_u - \phi V_c)\, s}{\phi f_y d} \qquad \text{(Commentary R11.5.6)}$$

The design yield strength of the shear reinforcement, f_y, except for deformed WWF, must not be greater than 60,000 psi. Design yield strength of deformed WWF is limited to 80,000 psi (Section 11.5.2). These limits provide a control on diagonal crack width (Commentary Section R11.5.2).

Shear reinforcement can also be determined from design aids such as are shown in Figs. 14-7 and 14-8 which give the shear stress capacity $(v_u - v_c)$ of #3 and #4 stirrups, respectively, for beam web widths from 2.5 to 30 in. where $v_u/\phi = V_u/\phi\, b_w d$ and $v_c = V_c/b_w d$. As an example of the use of these design aids, let it be required to select stirrups for a 10 in. wide by 20 in. deep prestressed concrete beam with a value $(v_u/\phi - v_c)$ equal to 300 psi at distance of $h/2$ from the support. From Figs. 14-7 and 14-8 it is evident that #3 stirrups at a 4 in. spacing or #4 stirrups at an 8 in. spacing are satisfactory.

The spacing of the #3 stirrups could also be obtained from Fig. 14-8 for #4 stirrups by simple proportion of the stirrup area as follows:

$$\text{(Spacing of \#3}\rceil\!\sqcup\!\lceil) = \frac{\text{(Spacing of \#4}\rceil\!\sqcup\!\lceil)\ \text{(Area of \#3}\rceil\!\sqcup\!\lceil)}{\text{(Area of \#4}\rceil\!\sqcup\!\lceil)}$$

$$= \frac{(8.00)\,(0.22)}{(0.40)}$$

$$= 4.4 \text{ in.}$$

The spacing of #5 stirrups or other sizes can be determined from Figs. 14-7 or 14-8 in a similar manner or from the figures in Chapter 9 of this Guide.

The design example at the close of this chapter shows in detail some of the problems in design for shear. In particular note Fig. 14-14.

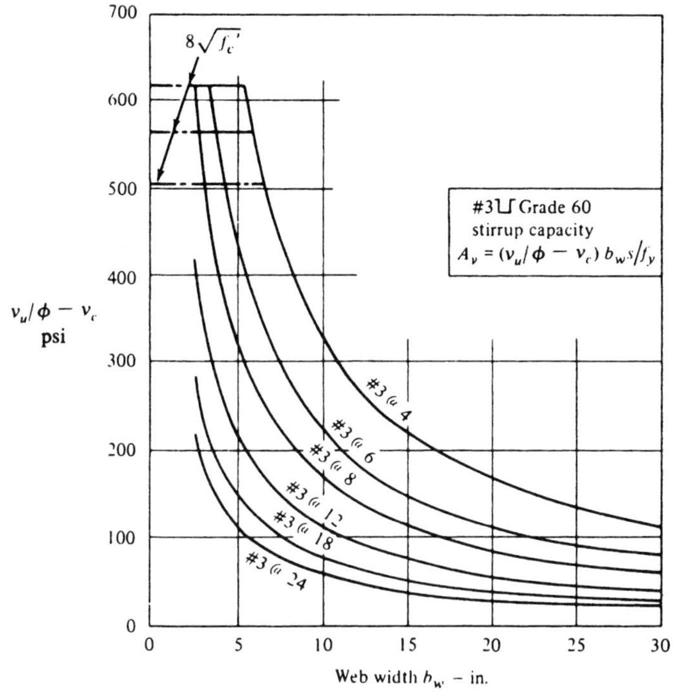

Figure 14-7 #3 stirrup capacity.

Design criteria for combined shear and torsion in prestressed concrete members were introduced in the 1995 Code (Section 11.6).

When the prestressing steel is unbonded, some bonded reinforcement must be provided in the precompressed tensile zone of flexural members and distributed uniformly over the tension zone near the extreme tensile fiber (Section 18.9.2.1).

The amount of bonded reinforcement provided for beams and one-way slabs must be equal to or greater than

$$A_s = 0.004\,A \quad \text{(Section 18.9.2, Eq. 18-6)}$$

where "*A*" is equal to the area of that part of the cross section between the flexural tension face and the center of gravity of the gross section.

Flat Plates

For two-way flat plates, defined as solid slabs of uniform thickness (Section 18.9.3), minimum bonded reinforcement is required in both positive and negative moment regions.

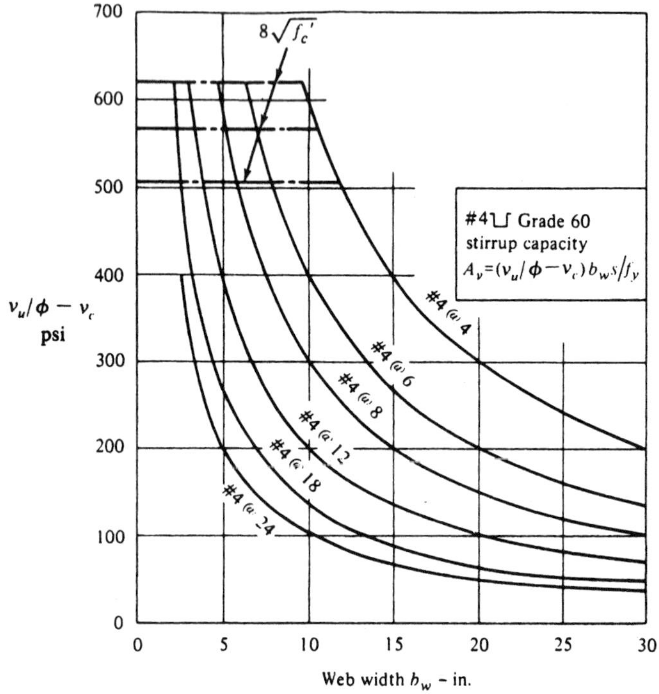

Figure 14-8 #4 stirrup capacity.

In positive moment areas where the tensile stress in the concrete at service loads (after allowance for all prestress losses) exceeds $2\sqrt{f_c'}$, the minimum area of bonded reinforcement must be equal to or greater than

$$A_s = \frac{N_c}{0.5 f_y} \quad \text{(Section 18.9.3.2, Eq. 18-7)}$$

where f_y is the yield strength of the bonded reinforcement which must not be greater than 60,000 psi (Section 18.9.3.2) and N_c = tensile force in the concrete under loads consisting of effective prestress (after losses) plus unfactored dead load (D) plus live load (L).

In negative moment areas at column supports, the minimum area of bonded reinforcement in each direction must be equal to or greater than

$$A_s = 0.00075 h \ell \quad \text{(Section 18.9.3.3, Eq. 18-8)}$$

where ℓ is the span parallel to the direction of the reinforcement. This area of bonded reinforcement must be placed within a slab width between lines that are 1.5 h outside opposite faces of the column support (Section 18.9.3.3).

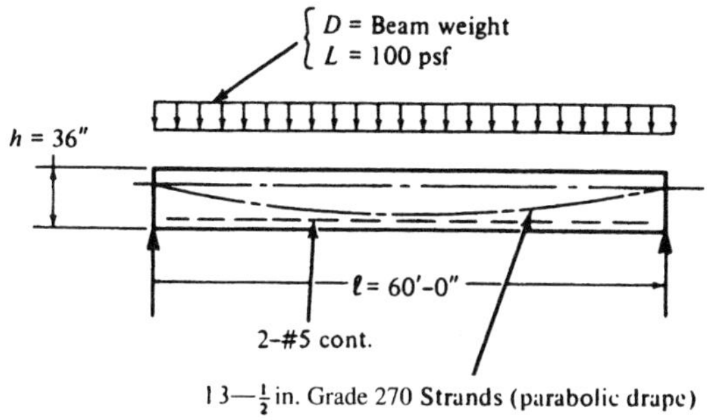

Figure 14-9 Prestressed concrete beam elevation and tendon profile.

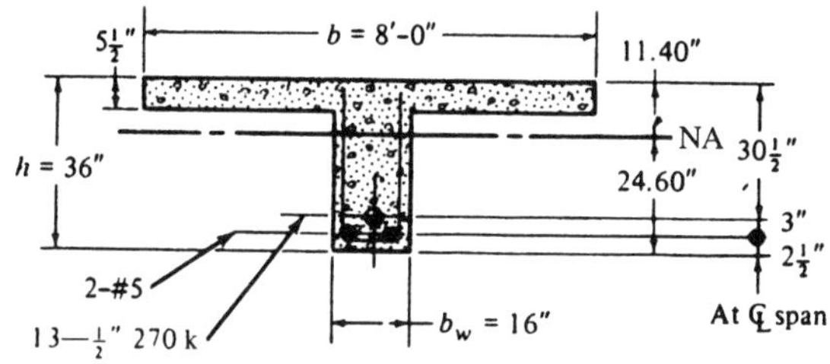

Figure 14-10 Beam cross section.

DESIGN EXAMPLE The final design of a cast-in-place prestressed concrete post-tensioned 60′-0″ simple span T-beam will be used to illustrate some of the design decisions required by the Code. An elevation of the beam is shown in Fig. 14-9 and the dimensions of the cross section are shown in Fig. 14-10. A preliminary design has been made and it indicates that 13—$\frac{1}{2}$ in. diameter Grade 270 parabolically draped bonded prestressing tendons will meet the Code requirements for strength. Two #5 reinforcing bars (f_y = 60,000 psi) are placed in the bottom of the beam in the corners of the stirrups for anchorage of web reinforcement (Section 12.13.3). A final design check will now be made to determine if the nonprestressed reinforcement used in combination with the prestressed steel will provide the necessary design moment strength for factored load (Section 18.7.3) and also meet the service load stress limitations (Section 18.4.2).

The final design check will be made for concrete with an ultimate compressive strength of 4,000 psi at the time the prestressing force is applied to the concrete (transfer) and 6,000 psi at factored load (1.4D + 1.7L).

The geometrical properties of the section are as follows:

Area and centroid of section

$$
\begin{aligned}
(80.0)\,(5.5) &= \quad 440 \times 33.25 = 14{,}630 \\
(16.0)\,(36.0) &= \quad \underline{576} \times 18.00 = \underline{10{,}368} \\
\Sigma A_c &= 1{,}016 \qquad \Sigma A_y = 24{,}998
\end{aligned}
$$

$$y_t = \frac{\Sigma A_y}{\Sigma A_c} = 24.60 \text{ in.}$$

Moment of inertia

$$\begin{aligned}
(16)(24.60)^3/3 &= 79{,}397 \\
(96)(11.4)^3/3 &= \underline{47{,}409} \\
& 126{,}806 \\
-(80)(5.9)^3/3 &= \underline{-5{,}476} \\
I_g &= 121{,}330 \text{ in.}^4
\end{aligned}$$

The beam loads, moments, and shear are as follows:

Service loads, moments, and shears

$$\begin{aligned}
D = (1016)(150)/(144)(1000) &= 1.058 \text{ kips/ft} \\
L = (100)(8)/(1000) &= \underline{0.800} \text{ kips/ft} \\
D + L &= 1.858 \text{ kips/ft}
\end{aligned}$$

$$V_{(D+L)} \text{ at support} = (1.858)(30) = 55.74 \text{ kips}$$

$$M_D \text{ at center of span} = (1.058)(60)^2/8 = 476 \text{ ft-kips}$$

$$M_{(D+L)} \text{ at center of span} = (1.858)(60)^2/8 = 836 \text{ ft-kips}$$

Factored loads, moments, and shears

$$\begin{aligned}
w_d &= (1.058)(1.4) = 1.481 \text{ kips/ft} \\
w_\ell &= (0.800)(1.7) = \underline{1.360} \text{ kips/ft} \\
w_d + w_\ell &= 2.841 \text{ kips/ft}
\end{aligned}$$

$$V_u = (\text{at support}) = (2.841)(30) = 85.23 \text{ kips}$$

$$M_u = (\text{at center of span}) = (2.841)(60)^2/8 = 1278 \text{ ft-kips}$$

The behavior of the beam under service load conditions will be determined by calculating the flexural stresses due to service dead load and the prestressing force that will be acting immediately after transfer and before prestress losses. This condition will produce the maximum compression stress in the concrete in the bottom of the beam and the maximum tension stress in the concrete in the top of the beam, if any. The flexural stresses due to service dead plus service live load and the prestressing force that will be acting after all prestress losses will also be calculated to determine the maximum compression stress in the concrete in the top of the beam and the maximum tension stress in the concrete in the bottom of the beam. The stress in the prestressing steel at transfer will be assumed equal to the maximum value of $0.7\,f_{pu}$ (Section 18.5.1) and the effective prestress force after losses will be based on an effective stress of $0.6\,f_{pu}$.

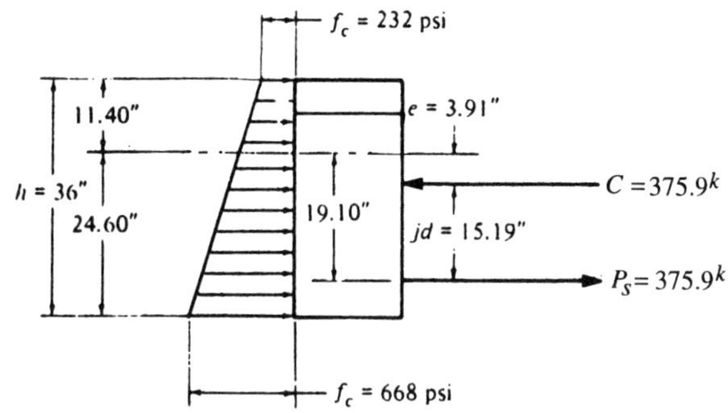

Figure 14-11 Service load stresses at transfer.

The locations of the prestressing strands and the reinforcing bars are shown in Figs. 14-9 and 14-10.

Behavior at Service Loads

Stresses at jacking with service dead load (refer to Fig. 14-11):

$$f'_{ci} = 4{,}000 \text{ psi}$$

$$f_{pu} = 270 \text{ ksi}$$

$$M_D = 476 \text{ ft-kips}$$

$$A_{ps} = (13)\ 1/2 \text{ in. Grade 270 strands} = (13)(0.153) = 1.989 \text{ in.}^2$$

$$P_s = 0.70\, f_{pu} A_{ps} = (0.70)(270)(1.989) = 375.9 \text{ kips}$$

where P_s = the prestressing force at jacking end. Summing moments about the centroid of the prestressing force

$$Cjd = M_D$$

and

$$jd = M_D/C = (476)(12)/375.9 = 15.19 \text{ in.}$$

letting e equal eccentricity of the compression force about the centroid of the gross concrete section,

$$e = 19.10 - 15.19 = 3.91 \text{ in. and}$$

$$f_c \text{ (top fiber)} = -\frac{P_s}{A_c} + \frac{P_s ey}{I_g}$$

$$= -\frac{375.9}{1016} + \frac{(375.9)(3.91)(11.40)}{121{,}330}$$

$$= -0.370 + 0.138$$

$$= -0.232 \text{ ksi compression OK} < 0.60 f'_{ci} = 2.40 \text{ ksi}$$

$$f_c \text{ (bottom fiber)} = -\frac{P_s}{A_c} + \frac{P_s e y_t}{I_g}$$

$$= -\frac{375.9}{1016} - \frac{(375.9)(3.91)(24.60)}{121{,}330}$$

$$= -0.370 - 0.298$$

$$= -0.668 \text{ ksi compression OK} < -2.40 \text{ ksi}$$

Service load stresses after losses (refer to Fig. 14-12):

$$f'_c = 6{,}000 \text{ psi}$$

$$f_{pu} = 270 \text{ ksi}$$

$$M_{(D+L)} = 836 \text{ ft-kips}$$

$$A_{ps} = 1.989 \text{ in.}^2$$

$$P_s = 0.60 f_{pu} A_{ps} = (0.60)(270)(1.989) = 322.2 \text{ kips}$$

Summing moments about of the centroid of the prestressing force

$$Cjd = M_{(D+L)}$$

and

$$jd = \frac{M_{(D+L)}}{C} = \frac{(836)(12)}{322.2} = 31.14 \text{ in.}$$

letting e equal the eccentricity of the compression force about the centroid of the gross concrete section

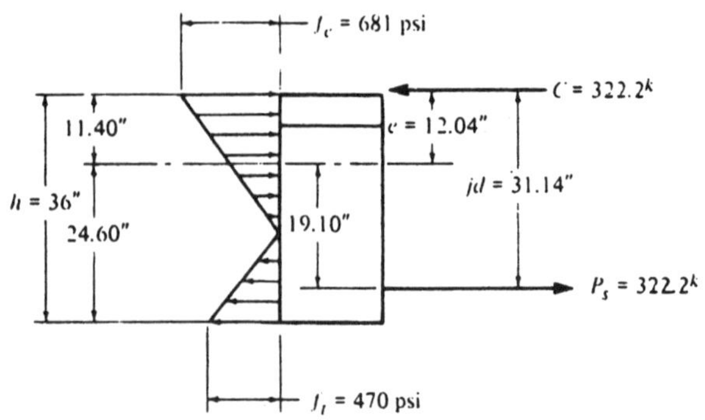

Figure 14-12 Service load stresses after losses.

$$e = 31.14 - 19.10 = 12.04 \text{ in.}$$

and

$$f_c \text{ (top fiber)} = -\frac{P_s}{A_c} - \frac{P_s ey}{I_g}$$

$$= -\frac{322.2}{1016} - \frac{(322.2)(12.04)(11.40)}{121{,}330}$$

$$= -0.317 - 0.364$$

$$= -0.681 \text{ ksi compression OK} < 0.45 f_c' = -2.70 \text{ ksi}$$

When a large portion of the total service load is made up of transient or temporary live load, the limit on f_c can be taken as 0.60 f_c' rather than 0.45 f_c'. Since f_c (top fiber) is only 681 psi, which meets the 0.45 f_c' limit with margin to spare, it is not necessary in this example to examine the nature of the live load as to whether it is sustained or transient.

$$f_t \text{ (bottom fiber)} = -\frac{P_s}{A_c} + \frac{P_s ey_t}{I_g}$$

$$= -\frac{322.2}{1016} + \frac{(322.2)(12.04)(24.60)}{121{,}330}$$

$$= -0.317 + 0.787$$

$$= +0.470 \text{ ksi tension OK} \approx 6\sqrt{f_c'} = 0.465 \text{ ksi}$$

It is obvious from an examination of Fig. 14-12 that the tension stress of 470 psi in the concrete in the bottom of the beam under service dead load and service live load after losses is less than the modulus of rupture of $7.5\sqrt{f_c'} = 7.5\sqrt{6{,}000} = 581$ psi (Section 9.5.2.3, Eq. 9-9). This means that the beam will not crack under service dead and live loads. A service load equal to dead load plus 115 percent service live load will be required to produce a tension stress equal to the modulus of rupture and cause cracking due to flexure. The absence of cracks in any beam will result in a member with less deflection and one that is more corrosion resistant than if cracks were present.

If the tension stress in the concrete due to flexure caused by service loads does not exceed $6\sqrt{f_c'}$, the instantaneous deflection of a prestressed concrete member can be based on the modulus of elasticity of the concrete (Section 8.5.1) and on the gross moment of inertia of the concrete (Section 9.5.4.1). A tension stress greater than $6\sqrt{f_c'}$ but less than $12\sqrt{f_c'}$ is permitted only if it can be shown that the immediate and long-term deflections based on the transformed cracked moment of inertia and on bilinear moment deflection relationships will conform to Code requirements (Section 18.4.2).

The beam in this example will not crack under service loads because the maximum tension stress is less than the modulus of rupture (470 psi < 581 psi) and the immediate and long-term deflection can be approximated as follows:

$$E_c = 57{,}000\sqrt{f'_c} \quad \text{(Section 8.5.1)}$$

$$= 57{,}000\sqrt{6{,}000} = 4{,}415{,}220 \text{ psi}$$

$$\Delta\ \text{(Gravity loads)} = \frac{5w\ell^3}{384E_cI_g}$$

$$\Delta\ \text{(Prestress force)} = \frac{5P_se\ell^2}{48E_cI_g}$$

$$\Delta\ \text{(Service dead load only)} = \frac{(5)(1058)(60)(720)^3}{(384)(4{,}415{,}220)(121{,}330)} = 0.576 \text{ in.} \downarrow$$

$$\Delta\ \text{(Prestress before losses)} = \frac{(5)(375{,}900)(19.1)(720)^2}{(48)(4{,}415{,}220)(121{,}330)} = 0.724 \text{ in.} \uparrow$$

$$\Delta\ \text{(Inst. service dead load with prestress before losses)} = 0.148 \text{ in.} \uparrow$$

$$\Delta\ \text{(Prestress losses at 15 percent)} = (0.15)(0.724) = 0.109 \text{ in.} \downarrow$$

$$\Delta\ \text{(Service dead load with prestress after losses)} = 0.039 \text{ in.} \uparrow$$

$$\Delta\ \text{(Creep and shrinkage of concrete)} = (2)(0.39) = 0.078 \text{ in.} \uparrow$$

$$\Delta\ \text{(Long-term service dead load with prestress after losses)} = 0.117 \text{ in.}^* \uparrow$$

$$\Delta\ \text{(Inst. service live load)} = (0.576)(800)/(1058) = 0.436 \text{ in.} \downarrow$$

The example beam is a floor member that is not connected to and does not support any nonstructural elements that would be damaged by large deflections. The maximum instantaneous service live load deflection permitted for such a member (Code Table 9.5(b); Section 9.5.2.6) is 1/480 of the span or 1.50 in. This is much greater than the calculated instantaneous live load deflection of 0.436 in. The example beam is therefore satisfactory under the criteria of the Code for behavior under service loads.

Flexural Strength

The flexural strength necessary for the factored load (1.4 D + 1.7 L) must be computed in accordance with the strength design method (Sections 10.2 and 18.7). The nominal moment strength as shown in Fig. 14-13 is equal to the inter-

* The additional long-term deflection due to creep and shrinkage (Section 9.5.2.5) has been assumed equal to twice the instantaneous deflection due to service dead load with prestress after losses.

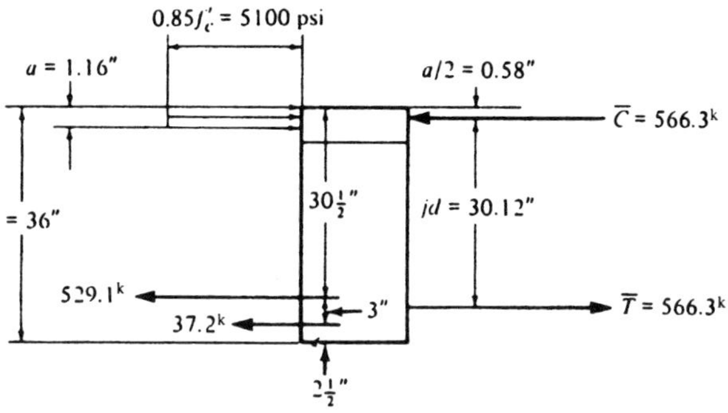

Figure 14-13 Nominal moment strength.

nal couple $\overline{T}jd$, where d is the distance to the combined centroid of the prestressing steel and the nonprestressed steel and $\overline{T}$ is equal to the tension force in the prestressed and nonprestressed reinforcement. This nominal tension force is equal to the area of the prestressing steel multiplied by the stress in the prestressing steel at nominal strength, f_{ps}, plus the area of the nonprestressed reinforcement times its yield strength.

$$\overline{T} = A_{ps}f_{ps} + A_s f_y$$

In order to determine $\overline{T}$ and then jd the following steps can be taken:

Step 1. Assume neutral axis will fall in flange.
Step 2. Assume $f_{ps} = 266$ ksi (Fig. 14-2).
Step 3. Determine $\overline{T}$ and d.

13—1/2 in. diameter strands

$$\begin{aligned} &(1.989)(266) = 529.1 \times 30.5 = 16{,}140 \text{ kip-in.} \\ \text{two \#5 bars}\quad &(0.62)(\ 60) = \underline{\ 37.2} \times 33.5 = \underline{\ 1246\ } \\ &\overline{T} = 566.3 \text{ kips} \qquad 17{,}386 \text{ kip-in.} \end{aligned}$$

$$d = (17{,}386)/566.3 = 30.70 \text{ in.}$$

Step 4. Determine jd

$$\overline{C} = \overline{T} = 0.85f'_c ba$$

$$a = \frac{\overline{T}}{0.85f'_c b} = \frac{566{,}300}{(0.85)(6000)(96)}$$

$$= 1.16 \text{ in.}$$

$$jd = d - a/2$$
$$= 30.70 - 1.16/2 = 30.78 - 0.58$$
$$= 30.12 \text{ in.}$$

Step 5. Determine ratio of prestressing steel

$$\rho_p = \frac{A_{ps}}{bd} = \frac{1.989}{(96)(30.70)}$$
$$= 0.000675$$

Step 6. Determine f_{ps}

$$f_{ps} = f_{pu}\left(1 - \frac{\gamma_p}{\beta_1\,[\rho_p f_{pu}/f'_c]}\right), \text{ neglecting the 2-\#5 bars}$$
$$= 270\left(1 - \frac{0.40}{0.75[(0.000675)(270)/(6)]}\right)$$
$$= (270)\,(0.984)$$
$$= 266 \text{ ksi which checks the assumed value in Step 2.}$$

Step 7. Calculate the design moment strength

$$\phi M_n = \phi \bar{T} jd = \frac{(0.90)(566.3)(30.12)}{12}$$
$$= 1279 \text{ ft-kips OK} > \text{factored moment of } 1278 \text{ ft-kips}$$

Shear Reinforcement

The maximum value of the factored shear to be carried by the shear reinforcement for a uniformly loaded simple span beam is limited to a maximum value of $8\sqrt{f'_c}\, b_w\, d$ (Section 11.5.6.8). It will usually occur between one-eighth and one-quarter of the span from the support for a uniformly loaded simple span beam.

Sample calculations for the shear stresses will be made at one-eighth of the span or 7′-6″ from the support. The results of these calculations and similar calculations for other points are shown in Fig. 14-14.

Calculations for the shear stresses at 7'-6" from the support are as follows:

$$V_u = 85.2 - (2.841)\,(7.5) = 85.2 - 21.3 = 63.9 \text{ kips}$$

$$\frac{v_u}{\phi} = \frac{V_u}{\phi\, b_w\, d} = \frac{63.9}{(0.85)\,(16)\,(0.80)\,(36)} = 163 \text{ psi}$$

(In calculating v_u, d need not be taken less than $0.80h$.)

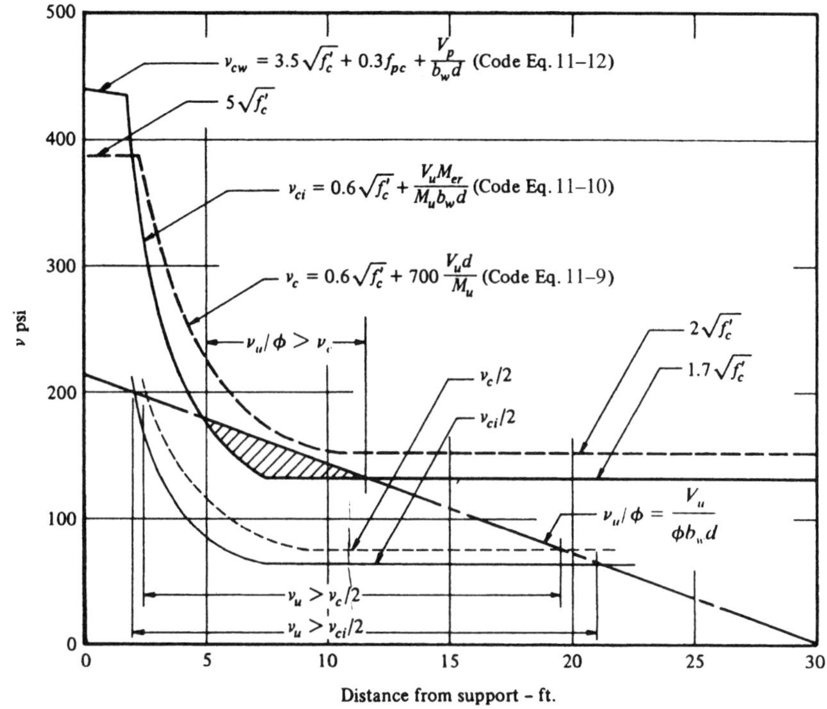

Figure 14-14 Shear stresses.

$$v_c \quad (\text{Code Eq. 11-9}) = 0.6\sqrt{f_c'} + 700\frac{V_u d}{M_u}$$

$$v_c = 0.6\sqrt{6000} + \frac{(700)\,(63.9)\,(19.75)}{(560)\,(12)}$$

$$= 46 + (700)\,(0.188) \text{ OK, } V_u d/M_u < 1.0 \quad (\text{Section 11.4.1})$$

$$= 46 + 132$$

$$= 178 \text{ psi}$$

In calculating v_c, d is taken as the distance from the extreme compression fiber to the centroid of the prestressing tendons (Section 11.4.1).

It is obvious that v_c, the shear stress carried by the concrete is greater than v_u/ϕ, the average or nominal shear stress. However, v_u/ϕ is greater than $v_c/2$ (Section 11.5.5.1), and minimum stirrups are required. The stirrups can be calculated by the less conservative requirement (Code Eq. 11-14) (see Fig. 14-5).

$$A_v(\text{min}) = \frac{A_{ps} f_{pu} s \sqrt{d}}{80 f_y d \sqrt{b_w}} = \frac{(1.989)(270)(24)\sqrt{(0.80)\,(36)}}{(80)(60)(0.80)(36)(\sqrt{16})}$$

$$= 0.125 \text{ in.}^2$$

#3 ⅂_┌ stirrups at a maximum spacing of 24 in. are more than adequate for minimum shear reinforcement area. Fig. 14-14 indicates that minimum stirrups are required from 2′-3″ from the support to a location 19′-9″ from the support (Code Eq. 11-9). #3 ⅂_┌ stirrups spaced one @12 in. and nine @24 in. from each support will meet this requirement.

More than minimum stirrups would have been required for the length of the span shown shaded in Fig. 14-14 (5′-0″ from the support to 11′-9″ from the support), if v_c had been calculated as the lesser value v_{ci} or v_{cw} (Section 11.4.2). Calculations for v_{ci} and v_{cw} at 7'-6" from the support are as follows:

e (eccentricity of prestressing force about the centroid of the concrete section at a distance 7′-6″ from the support) = 8.35 in.

f_{pe} (compressive force in the concrete at the bottom of the beam due only to the prestress force after losses)

$$f_{pe} = \frac{P_s}{A_c} + \frac{P_s e y_t}{I_g} = \frac{322{,}200}{1016} + \frac{(322{,}000)(8.35)(24.60)}{121{,}330}$$

$$= 317 + 545$$

$$= 862 \text{ psi}$$

f_d (tension stress in bottom of beam due to dead load)

$$f_d = \frac{M_D\, y_t}{I_g} = \frac{(209{,}000)(12)(24.60)}{121{,}330}$$

$$= 510 \text{ psi}$$

$$M_{cr} = (\text{cracking moment}) = \frac{I_g}{y_t}[6\sqrt{f_c'} + f_{pe} - f_d] \qquad \text{(Section 11.4.2)}$$

$$M_{cr} = \frac{(121{,}330)}{(24.60)}\left[\frac{(6\sqrt{6000})}{(12{,}000)} + 862 - 510\right]$$

$$= 336 \text{ ft-kips}$$

v_{ci} = (shear stress at diagonal cracking due to factored loads when such cracking is the result of combined shear and moment)

v_{ci} = (Code equation 11-10)

$$= 0.6\sqrt{f_c'} + \frac{V_d + V_i M_{cr}/M_{max}}{b_w d}$$

$$= 0.6\sqrt{f_c'} + \frac{V_u M_{cr}}{M_u b_w d}$$

$$= 0.6\sqrt{6{,}000} + \frac{(63.9)(336{,}000)}{(560)(16)(0.80)(36)}$$

$$= 47 + 83$$

$$= 130 \text{ psi.}$$

f_{pc} = (compressive stress in concrete at neutral axis due to dead and live load after prestress losses)

$$f_{pc} = \frac{P_s}{A_c} = \frac{322{,}200}{1016}$$

$$= 317 \text{ psi}$$

V_p = (vertical component of prestress force after losses)

$V_p = P_s$ (Slope of tendon profile)

$$= (322{,}200)\frac{(19.1)(2)(22.5)}{(360)(30)}$$

$$= (322{,}200)(0.0796)$$

$$= 25{,}600 \text{ lb}$$

v_{cw} = (shear stress at diagonal cracking due to factored loads, when such cracking is the result of principal tensile stress in the web)

v_{cw} = (Code Eq. 11-12)

$$v_{cw} = 3.5\sqrt{f_c'} + 0.3 f_{pc} + V_p/b_w d$$

$$= 3.5\sqrt{6000} + (0.3)\,(317) + \frac{25{,}600}{(16)(0.80)(36)}$$

$$= 271 + 95 + 56$$

$$= 422 \text{ psi}$$

v_c = the smaller of v_{ci} or v_{cw}

$= v_{ci}$

$= 130$ psi (refer to Fig. 14-14).

It is evident from the calculations that the shear carried by the concrete as v_{ci} or v_{cw} (Section 11.4.2) is conservative for the example beam. Stirrups based on v_c are more economical and utilize simpler manual calculation methods for the example beam (Section 11.4.1). From Fig. 14-14 it is also evident for the example beam that more than minimum stirrups are required when v_c is calculated as v_{ci} or v_{cw} (Section 11.4.2).

15 Structural Lightweight Aggregate Concrete

Unless special exceptions are made, the entire Code is applicable to lightweight aggregate concrete. The principal exceptions and special provisions for lightweight aggregate concrete are discussed and illustrated in the following section.

Lightweight concrete is defined for purposes of this book (and implicitly in the Code) as *structural* concrete containing lightweight aggregate and weighing less than 115 pcf (Section 2.1). Lightweight aggregate is defined and limited to aggregates conforming to ASTM C330 (Section 3.3.1), and having a dry, loose weight of 70 pcf or less (Section 2.1). Foamed lightweight concrete or lightweight concrete containing styrofoam aggregate suitable principally for insulation would not conform to the Code definition of structural lightweight aggregate concrete and would require special approval for proposed use under the Code as reinforced concrete (Section 1.4). Lightweight concrete in which only the coarse aggregate is lightweight is termed "sand-lightweight" concrete; where both coarse and fine aggregates are lightweight, "all-lightweight" concrete (Section 2.1).

Maximum water-cementitious material ratios, by weight, for lightweight aggregate concrete are not given for special exposure conditions, because the amount of water absorbed by these aggregates is uncertain, making the water-cementitious ratios uncertain (Section 4.4.2). Lightweight concrete intended to have low permeability when exposed to water, must be proportioned for a specified strength, f'_c, at least 4,000 psi (Section 4.4.2, Table 4.2.2); concrete exposed to freezing and thawing in a moist condition or to deicing chemicals, f'_c = 4,500 psi; and for corrosion protection of reinforcement in concrete exposed to chlorides from deicing chemicals, salt, salt water, brackish water, seawater, or spray from these sources, f'_c = 5,000 psi. New requirements of the 1995 Code for concrete exposed to sulfate solutions that are moderate must be proportioned for a specified strength of, f'_c = 4,000 psi (Section 4.3.1, Table 4.3.1); and for sulfate solutions that are severe and very severe, f'_c = 4,500 psi.

It will have been noted that shear resistance and development (bond) are functions of $\sqrt{f_c'}$ to which the pure tensile strength of concrete is considered proportional. Since the tensile strength of lightweight concrete utilizing certain acceptable lightweight aggregates is substantially below that for normal weight concretes of equal compressive strength, reductions for shear (Section 11.2) and development (Section 12.2.4.3) are provided. Alternate procedures permit higher values to be used, up to the value for normal weight concretes when the "splitting strength," f_{ct} of the lightweight aggregate concrete to be used is known (Section 5.1.4). Such tests are *not* intended to be used as routine acceptance tests on field concrete (Section 5.1.5).

Aside from the simply handled lesser weight and lower tensile strengths, the most important property of lightweight aggregate concrete differentiating its design from normal weight concrete is its lower modulus of elasticity (Section 8.5.1). Note that E_c is proportional to $w^{1.5}$, where w is the weight in pcf. The modular ratio, $n = E_s/E_c$ by definition. For design with unity load and strength-reduction factors using the straight-line theory of stress and strain in flexure, the Code provides that, "except in calculations for deflection ... n for lightweight concrete shall be assumed to be the same as (that) for normal weight concrete of the same strength" (Appendix A.5.4).

Under design utilizing load factors and strength-reduction (ϕ) factors, the term n is not used (Section 10.2, Assumptions) in strength design. The computation of deflections is required to follow usual formulas for elastic deflections, using the modulus of elasticity as defined for both normal and lightweight concretes (Section 9.5.2.3). $E_c = w^{1.5}\, 33\sqrt{f_c'}$ in psi; (Section 8.5.1). The effective moment of inertia for a reinforced concrete section is required to be taken as

$$I_e = \left(\frac{M_{cr}}{M_a}\right)^3 I_g + \left[1 - \left(\frac{M_{cr}}{M_a}\right)^3\right] I_{cr} \leq I_g \quad \text{(Code Eq. 9-7)}$$

where

$$M_{cr} = \frac{f_r I_g}{y_t} \quad \text{(Section 9.5.2.3)}$$

in which when f_{ct} is not specified

$$f_r = 5.63\sqrt{f_c'} \text{ for all-lightweight concrete}$$

$$f_r = 6.39\sqrt{f_c'} \text{ for sand-lightweight concrete}$$

Due to the use of the lower value for E_c, lightweight concrete will have a higher moment of inertia than normal weight concrete for the cracked section transformed to concrete, I_{cr} (see Fig. 15-1 comparing I_{cr} for lightweight to normal weight concrete). Thus, for low ratios of M_{cr}/M_a (heavily reinforced sections), I_{cr} will be larger for lightweight concrete, offsetting considerably the effect of its lower value of E_c in the deflection equations.

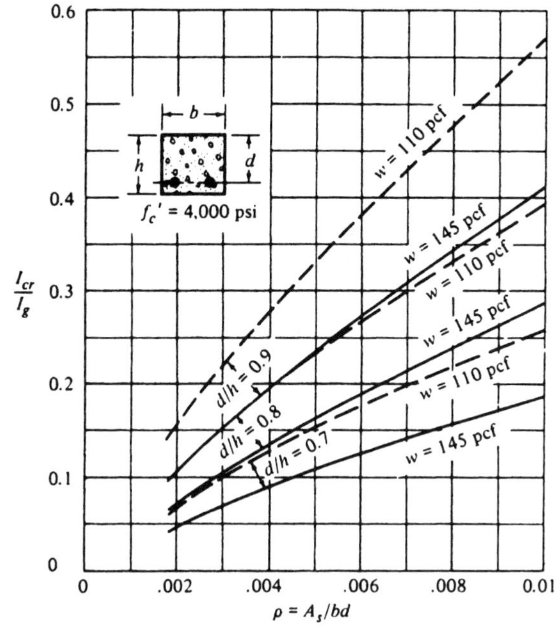

Figure 15-1 I_{cr} for normal vs lightweight concrete.

EXAMPLE. Compute relative deflections for a reinforced concrete slab in which $f_c' = 4{,}000$ psi; $f_y = 60{,}000$ psi; $h = 5$ in.; $d = 4$ in.; $\rho = 1.0\%$; $d/h = 0.8$; (1) for a normal weight concrete, $w = 145$ pcf and (2) for a lightweight aggregate concrete $w = 110$ pcf.

$$(1)\ E_c = 3.644 \times 10^6 \text{ psi} \quad n = 8 \quad I_g = \frac{(12)(5)^3}{12} = 125 \text{ in.}^4/\text{ft}$$

From Fig. 15-1, $I_{cr}/I_g = 0.293$. $I_{cr} = (0.293)(125) = 36.6$ in.^{4}/ft

$$M_{cr} = \frac{(7.5)(63.2)(125)}{(2.5)} = 23{,}700 \text{ in.-lb/ft}$$

$$\phi M_n = \phi f_y A_s (d - a/2) \qquad a = \frac{(60)}{(0.85)(4)(12)} = 1.47\,A_s$$

$$= (0.90)(60)(0.48)(4 - 0.35) \qquad = 1.47 \times 0.48 = 0.7 \text{ in.}$$

$$= 94{,}500 \text{ in.-lb/ft} \qquad 1/2a = 0.35 \text{ in.}$$

Use average load factor for $D + L = 1.6$

$$M_a = \frac{94{,}500}{1.6} = 59{,}000 \text{ in.-lb/ft}$$

$$\left(\frac{M_{cr}}{M_a}\right)^3 = \left(\frac{23.7}{59.0}\right)^3 = (0.4)^3 = 0.064$$

$$I_e = (0.064)(125) + (0.936)(36.6) = 42.2 \text{ in.}^4/\text{ft}$$

Immediate live load deflection,

$$w = 145 \text{ pcf}; \Delta_L \approx \frac{100}{E_c I_e} = \frac{100}{(3.644)(42.2)} = 0.65$$

$$(2)\ w = 110 \text{ pcf}, E_c = 2.407 \times 10^6 \text{ psi}$$

from Fig. 15-1, $I_{cr}/I_g = 0.396$. $I_{cr} = (0.396)(125) = 49.5$ in.^{4}/ft

Since $f_r = 0.75 f_r$ for normal weight concrete, $\dfrac{M_{cr}}{M_a} = (0.75)(0.4)$

$$\left(\frac{M_{cr}}{M_a}\right)^3 = (0.3)^3 = 0.027$$

$$I_e = (0.027)(125) + (0.973)(49.5) = 51.6 \text{ in.}^4/\text{ft}$$

Immediate live load deflection,

$$w = 110 \text{ pcf} \quad \Delta_L \approx \frac{100}{E_c I_e} = \frac{100}{(2.407)(51.6)} = 0.81$$

To compare relative total deflections, let the ratio of live to dead loads, $L/D = 1/2$ for the lightweight concrete slab, $L = 1.0$; $D = 2.0$; $L + D = 3$; the total immediate deflection is proportional to $(L + D)/E_c I_e$. Total deflection, $\Delta \approx (0.81)(3) = 2.43$. For the normal weight concrete, the dead load, $D = (2)(145/110) = 2.64$ $D + L = 3.64$. Total deflection, $\Delta \approx (0.65)(3.64) = 2.37$.

For the usual lightly reinforced slab, where the ratio of the cracking moment, M_{cr}, to the maximum applied (service load) moment, M_a, approaches or exceeds unity, I_e will become I_g. In this case the relative deflections of the lightweight and normal weight concretes will be computed as proportional to their respective moduli of elasticity. When Code Eq. 9-7 gives a value for I_e larger than that of I_g, the requirement that I_e shall not be taken as larger than I_g controls (Section 9.5.2.3).

16 Structural Plain Concrete

When the title of the 1989 Code (Revised 1992) was changed from "Building Code Requirements for Reinforced Concrete" to *Building Code Requirements for Structural Concrete (ACI 318-95)*, Chapter 22 on structural plain concrete was added. Structural plain concrete is defined as either unreinforced or with less reinforcement than specified in the Code for reinforced concrete (Section 2.1). Chapter 22 provides minimum requirements for design and construction of structural plain concrete members (cast-in-place or precast).

LIMITATIONS

Structural plain concrete basement walls are exempted from the requirements for special exposure conditions listed in Code Table 4.2.2, because of the successful use of large amounts of concrete with 28-day compressive strengths of 2500 and 3000 psi in basement walls of residences and minor structures (Section 22.1.1.1)

The design and construction of soil-supported slabs, such as sidewalks and slabs on grade are not covered unless they transmit vertical loads from other parts of the structure to the soil (Section 22.1.1.2).

The use of structural plain concrete is limited to members that are continuously supported by soil or supported by other structural members capable of providing continuous vertical support, members for which arch action provides compression under all conditions of loading, or walls and pedestals. The use of structural plain concrete columns is not permitted (Section 22.2.2).

Code Chapter 22 does not cover the design and installation of plain concrete piles and piers (Section 22.2.3) because these requirements are specified in the general building codes.

Structural plain concrete must have a minimum specified compressive strength of 2500 psi (Section 22.2.4).

JOINTS

Contraction joints or isolation joints must be provided to limit the length of structural plain concrete members in order to resist the large buildup of tension stresses due to creep, shrinkage and temperature effects (Section 22.3.1). A minimum 25 percent reduction of member thickness is considered sufficient for contraction joints to be effective (Commentary R22.3). Transverse contraction joints or isolation joints are not required in continuous wall footing members because random cracks are usually not objectionable and do not affect the wall footing's structural integrity (Commentary R22.3).

STRENGTH DESIGN

Strength design for flexure and axial loads is based on a linear stress–strain relationship in both tension and compression. Tension stresses are limited to a maximum of $5\sqrt{f_c'}$ (Section 22.5.1 Eq. 22-2) provided that construction, contraction, or isolation joints are provided to relieve the restraint and resulting tensile stresses due to creep, temperature, and shrinkage effects (Section 22.4.4 and Commentary R22.4.4).

In computing strength in flexure, combined flexure and axial load, and shear, the entire cross section of a member can be considered in design, except for concrete cast against soil where overall thickness h must be taken as 2 in. less than the actual thickness because of possible unevenness of soil excavation and contamination of the concrete adjacent to the soil (Section 22.4.8).

The design of cross sections for flexure, compression, bearing, and shear is based on factored moments and loads and their nominal strengths. The strength-reduction factor, ϕ, is 0.65 for all strength conditions (Section 9.3.5).

Moment strength $\phi M_n \geqslant M_u$ (Section 22.5.1)

$$(0.65)(5)\sqrt{f_c'}\, S \geqslant M_u$$

where S is the elastic section modulus.

Compressive strength $\phi P_n \geqslant P_u$ (Section 22.5.2)

$$(0.65)(0.60)f_c'\left[1 - \left(\frac{\ell_c}{32h}\right)^2\right]A_1 \geqslant P_u$$

where A_1 is the loaded area.

Beam shear strength $\phi V_n \geqslant V_u$ (Section 22.5.4)

$$(0.65)\left(\frac{4}{3}\right)\sqrt{f_c'}\, bh \geqslant V_u.$$

Two-way shear strength as in spread footings $\phi V_n \geqslant V_u$

$$(0.65)\left(\frac{4}{3} + \frac{8}{3\beta_c}\right)(\sqrt{f_c'}\, b_o h) \geqslant V_u$$

except that $\phi V_n \leqslant (0.65)(2.66)\sqrt{f_c'}\, b_o h$ where b_o is the perimeter of the critical section for shear.

Bearing strength except when the supporting surface is wider on all sides than the loaded area: $\phi B_n \geqslant P_u$ (Section 22.5.5)

$$(0.65)(0.85)f'_c A_1 \geqslant P_u$$

Bearing strength when the supporting surface is wider on all sides than the loaded area:

$$\phi B_n \geqslant (0.65)(0.85)f'_c \sqrt{(A_1/A_2)}$$

where A_2 is the area of the lower base of the largest frustum of a pyramid, cone or tapered wedge contained wholly within the support and having for its upper base the loaded area, and having side slopes of 1 vertical to 2 horizontal (Commentary R10.17 Fig. R10.17). The expression $\sqrt{(A_1/A_2)}$ must not be more than 2.

The design of cross sections for combined flexure and axial load in compression is based on limiting stresses and proportioning the members such that on the compression face (Section 22.5.3):

$$\frac{P_u}{\phi P_n} + \frac{M_u}{\phi M_n} \leqslant 1$$

and on the tension face:

$$\frac{M_u}{S} - \frac{P_u}{A_g} \leqslant 5\phi\sqrt{f'_c}$$

where A_g is the gross area of the cross section.

WALLS

Structural plain concrete walls must be continuously supported by soil, footings, foundation walls, grade beams, or other structural members (Section 22.6.1). Walls must be designed for vertical, lateral, and any other loads they are subjected to (Section 22.6.2). The Code requires that they be designed for minimum factored moment of $P_u(0.10h)$ (Section 22.6.3). If the resultant axial load is within the kern of the section (within the middle third of a rectangular cross section), walls can be designed by the empirical design method (Section 22.6.5). It is the authors' opinion that this method is no simpler than the strength design method of Code Section 22.5.3.

Limitations on wall design;

Thickness of bearing walls, other than exterior basement walls, must not be less than 1/24 of the unsupported height or length whichever is shorter, nor less than $5\frac{1}{2}$ in. (Section 22.6.6.2).

The thickness of exterior basement walls must not be less than $7\frac{1}{2}$ in. (Section 22.6.6.3).

The only reinforcement required for plain concrete occurs for walls (Section 22.6.6.5). A minimum of 2-#5 bars are required around all window and door openings and they must extend at least 24 in. beyond the corners of openings.

FOOTINGS

The area of footings must be determined from unfactored loads (service loads) by allowable soil pressures selected through principles of soil mechanics (Section 22.7.2). However, plain concrete footings must be designed for strength using factored loads (Section 22.7.1). Plain concrete must not be used for footings on piles (Section 22.7.3). The thickness of plain concrete footings must not be less than 8 in. (Section 22.7.4).

Critical sections for maximum factored moments in footings supported on soil (Section 22.7.5):

1. At the face of a column, pedestal, or wall, for a footing supporting a concrete column, pedestal, or wall.
2. Halfway between the middle and edge of a wall, for a footing supporting a masonry wall.
3. Halfway between the face of a column and edge of a steel base plate, for a footing supporting a column with a steel base plate.

Critical sections for maximum shear forces in footings supported on soil (Section 22.7.6) must be measured from the face of a column, pedestal, or wall for a footing supporting a column, pedestal, or wall. For a footing supporting a column with a steel base plate, it must be measured from a location halfway between the face of the column and the edge of the base plate. Critical sections for maximum shear forces in footings supported on soil are as follows:

1. For beam action, with a critical section extending in a plane extending across the entire footing width, and located at a distance h from the face of a column pedestal or wall.
2. For two-way action, on a perimeter around a column or wall at a distance $h/2$ away from the face of the column, pedestal, or wall.
3. For two-way action, on a perimeter around a column supported by a steel base plate at a distance $h/2$ from a location halfway between the face of the column and the edge of the steel base plate.

PEDESTALS

Pedestals must be designed for vertical, lateral, or any other loads to which they may be subjected. The ratio of their unsupported height to least lateral dimension should not exceed 3 (Section 22.8.2).

EXAMPLE NO. 1

Using $f'_c = 3000$ psi determine the thickness required for a plain concrete exterior basement wall of a residence that supports a factored axial load of 2,000 plf with an eccentricity of 1.20 in. and resists a lateral factored soil pressure equal to an equivalent fluid pressure of $50 \times 1.7 = 85$ psf. Assume a wall thickness of $h = 12$ in. and check the strength of the wall. See Fig. 16-1.

$$A_g = bh = (12)(12) = 144 \text{ in.}^2$$

$$S = \frac{bh^2}{6} = \frac{(12)(12)^2}{6} = 288 \text{ in.}^3$$

$$P_u = 2{,}000 \text{ lb}$$

$$M_u = \left(\frac{2}{3}\right)(5.20)(1148) + 84 \text{ (see Fig. 16-1)}$$

$$= 3{,}980 + 84 + 4{,}064 \text{ ft-lb}$$

$$V_u = 1{,}572 + 22 = 1{,}592 \text{ lb}$$

$$\phi P_n = \phi 0.60 f'_c \left[1 - \left(\frac{\ell_c}{32h}\right)^2\right] A_1$$

$$= (0.65)(0.60)(3000)\left\{1 - \left[\frac{108}{(32)(12)}\right]^2\right\} 144$$

$$= 155{,}000 \text{ lb}$$

$$\phi V_n = \phi\left(\frac{4}{3}\right) bh\sqrt{f'_c} = (0.65)\left(\frac{4}{3}\right)(12)(12)(\sqrt{3000}) = 6{,}836 \text{ lb}$$

$$\phi M_n = \phi 5\sqrt{f'_c}\, S = \frac{(0.65)(5)(\sqrt{3000})(288)}{12} = 4{,}272 \text{ ft-lb}$$

On the compression side of the wall, check if

$$\frac{P_u}{\phi P_n} + \frac{M_u}{\phi M_n} \leqslant 1$$

$$\frac{2{,}000}{155{,}000} + \frac{3{,}980 + 84}{4{,}272} = 0.013 + 0.950$$

$$= 0.963 < 1 \text{ OK}$$

On the tension side of the wall, check if $M_u/S - P_u/A_g \leqslant \phi 5\sqrt{f'_c}$

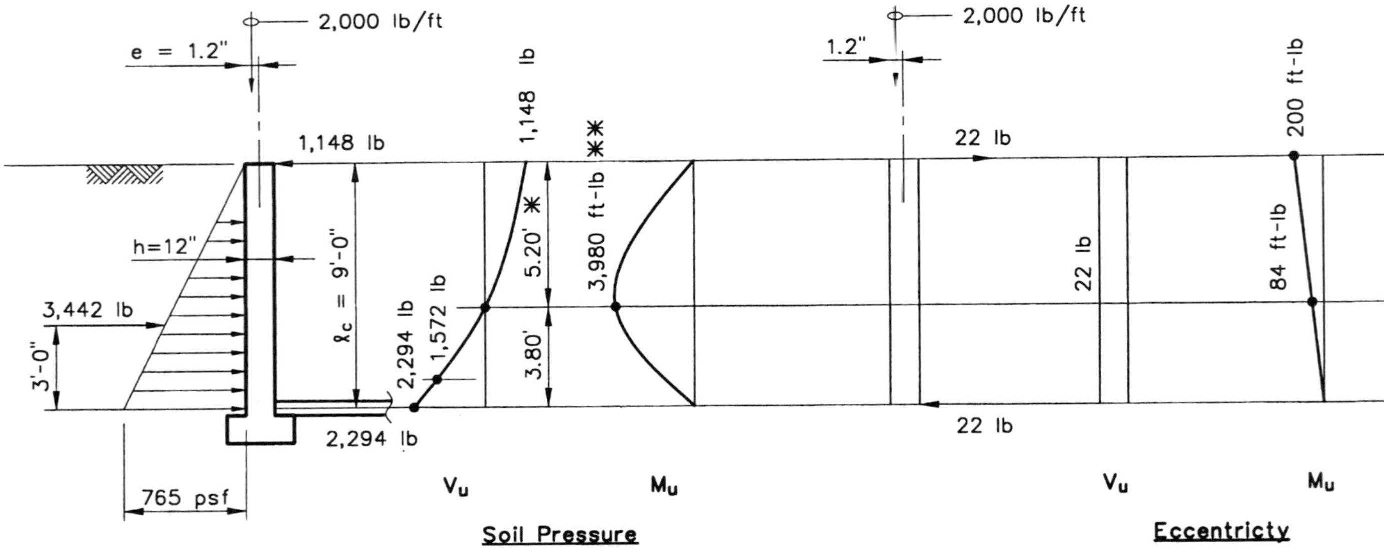

$$* \ \frac{85y^2}{2} = 1{,}148, \quad y = 5.20'$$

$$** \ M_u = \frac{2}{3}(5.20)(1{,}148) = 3{,}980 \text{ ft-lb}$$

Figure 16-1 Plain concrete wall.

$$\frac{(3{,}980 + 84)(12)}{288} - \frac{2{,}000}{144} = 169 - 14$$

$$= 155 \text{ psi} < \phi 5\sqrt{f'_c}$$

$$= 155 < (0.65)(5)(\sqrt{3000})$$

$$= 155 < 178 \text{ psi} \quad \text{OK}$$

EXAMPLE NO. 2

Using $f'_c = 4000$ psi and an allowable unfactored soil pressure $q_a = 3{,}000$ psf design a square structural plain concrete footing to support an unfactored column dead load of 16,000 lbs and an unfactored live load of 20,000 lb. The dead load includes an allowance for the weight of the footing.

Determine A, the required footing area and the footing dimensions.

$$A = \frac{P}{q_s} = \frac{16{,}000 + 20{,}000}{3{,}000} = 12 \text{ ft}^2$$

$$\text{Footing plan dimensions} = \sqrt{12} = 3.46, \text{ say } 3'\text{-}6'' \text{ square}$$

Determine the factored load:

$$P_u = (16{,}000)(1.4) + (20{,}000)(1.7) = 56{,}400 \text{ lb}$$

Determine the factored soil pressure, q_s, for checking the strength of the footing:

$$q_s = \frac{P_u}{(3.5)^2} = \frac{56{,}400}{(3.50)^2} = 4{,}600 \text{ psf}$$

Assume a footing thickness of 14 in. and check the footing strength in flexure and shear, remembering to deduct 2 in. from the total thickness to allow for soil uneveness and concrete contamination (Section 22.4.8). See Fig. 16-2.

$$S = \frac{bh^2}{6} = \frac{(42)(14 - 2)^2}{6} = \frac{(42)(12)^2}{6} = 1{,}008 \text{ in.}^3$$

$$\phi M_n = \phi 5\sqrt{f'_c}\, S = \frac{(0.65)(5)(\sqrt{4000})(1{,}008)}{12} = 17{,}266 \text{ ft-lb}$$

$$M_u = \frac{[4{,}600 - (14/12)(150)(1.4)](3.5)(1.50)^2}{2}$$

$$= \frac{[4{,}600 - 245](3.5)(1.50)^2}{2} = 17{,}148 \text{ ft-lb OK}$$

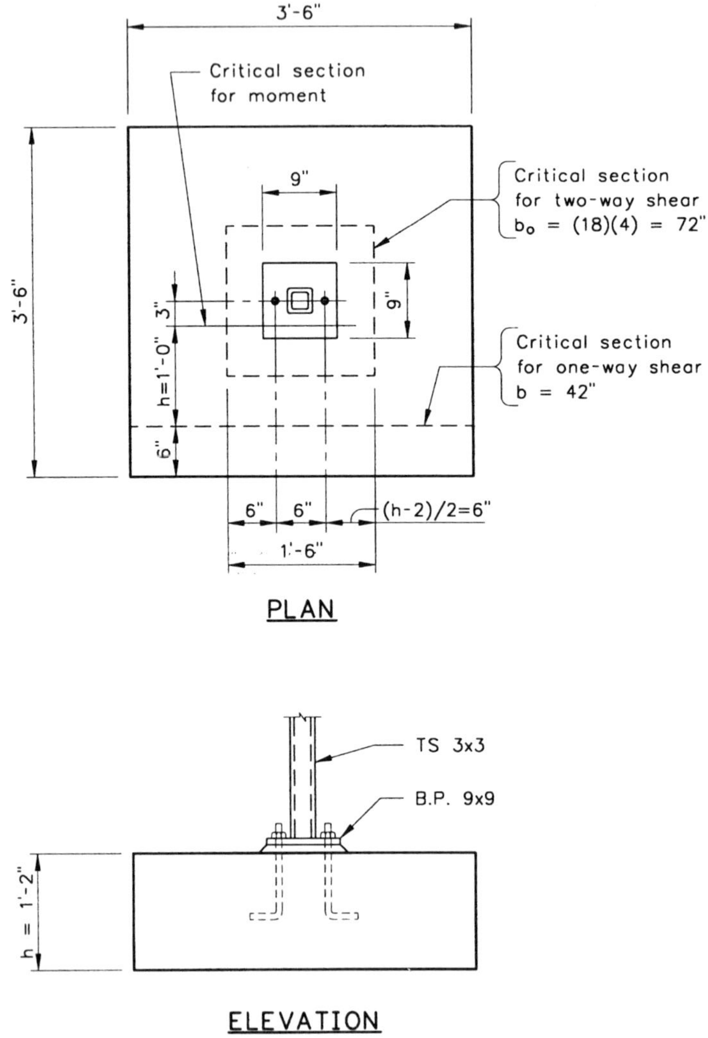

Figure 16-2 Plain concrete footing.

$$V_u(\text{one-way}) = \left[4{,}600 - \left(\frac{14}{12}\right)(150)(1.4)\right](3.5)(0.5)$$

$$= 7{,}621 \text{ lb}$$

$$\phi V_n(\text{one-way}) = \phi\left(\frac{4}{3}\right)\sqrt{f_c'}\, bh = (0.65)\left(\frac{4}{3}\right)(\sqrt{4{,}000})(42)(14 - 2)$$

$$= 27{,}626 \text{ lb}$$

$$> 7{,}621 \text{ lb OK}$$

$$V_u(\text{two-way}) = [4{,}600 - (14/12)(150)(1.4)][(3.5)^2 - (1.5)^2]$$

$$= 43{,}550 \text{ lb}$$

$$\phi V_n(\text{two-way}) = \phi\left(\frac{4}{3} + \frac{8}{3\beta_c}\right)(\sqrt{f_c'})b_o h \leqslant \phi(2.66)\sqrt{f_c'}\, b_o h$$

$$= (0.65)\left[\frac{4}{3} + \frac{8}{(3)(1)}\right]\sqrt{4000}\,(72)(14 - 2)$$

$$\leqslant (0.65)(2.66)(\sqrt{4000})(72)(12)$$

$$= (0.65)(4)(\sqrt{4000})(72)(12) \leqslant 94{,}479 \text{ lb}$$

$$= 142{,}075 \text{ lb} \leqslant 94{,}479 \text{ lb}$$

$$V_u = 43{,}550 \text{ lb} < 94{,}479 \text{ lb OK}$$

17 Field Inspection and Construction

GENERAL

The ACI Building Code is a document intended to become part of a law regulating all construction for the benefit of the public. Thus, there are frequent references outlining obligations (which could be undertaken by the Engineer, or Architect, representing the Owner or the Contractor) for securing approval of the Building Official. The Code covers the entire subject of field inspection with a few short minimum requirements (Section 1.3). It states that inspection of concrete construction shall be performed by an Engineer or Architect or competent representative thereof (Section 1.3.1). Minimum records which must be kept during construction and for 2 years thereafter by the inspecting Engineer or Architect (Section 1.3.4) are prescribed. Since the Code is intended to regulate concrete construction for small and large projects, no requirements for continuous inspection, frequency of intermittent inspection, special inspection or inspection procedures can be given.

A full treatment of field inspection or construction is outside the scope of this book. In this chapter, important or frequently overlooked Code requirements concerning inspection and construction have been collected and are presented with brief explanations and practical precautions. The best reference in detail to accepted inspection practice is the ACI *Manual of Concrete Inspection*, SP-2; the authors strongly recommend it to anyone undertaking concrete inspection. An excellent resource regarding project specifications for concrete construction is ACI 301-96, "Standard Specifications for Structural Concrete". The document is intended to be used by reference or incorporation in its entireity in the Project Specifications.

The actual design through construction is a complex procedure with many interdependent obligations among designer, material suppliers, contractors, subcontractors, inspectors acting for the Owner and for the Building Official. Very simply, each party concerned would like his responsibility to be limited; the design Engineer, to correct design drawings and project specifications; the Contractor and subcontractors, to correct execution of same; material suppliers, to material meeting specifications and design dimensions; and inspectors to following the contract documents. When the Code is part of the law regulating construction, and, as often the case, a part of the contract documents (with a statement that the term "Building Official" means "Engineer"), it protects the public only. It does not define separate duties of each party involved except in general terms.

If defective form materials were supplied and accepted, put together defectively according to defective shop drawings for formwork made by a subcontractor and approved by Contractor, design Engineer, and city inspector, and—after approvals by all for concreting—and forms failed, the Code does not assign responsibilities. If a completed structure is found unsafe for design loads, the public is protected by the provision that the Building Official may approve a lower load rating (Section 20.6) and further by standard general code provisions for condemning.

The structural Engineer, through project drawings and project specifications which become part of the "contract documents," establishes terms for material quality and construction procedures. With special provisions, he may require approvals prior to form design, reinforcing bar details and placing drawings, construction sequences for casting, schedules for form removal, curing, etc. In general, the Code merely states performance requirements.

MATERIALS

Structural inspection begins with material tests. Compliance with ASTM specifications for concrete materials, aggregate, water, and cement is required (Section 3.1.2). Code requirements supplementing those of ASTM should not be overlooked (see Chapter 1).

FORMWORK

The Code requires that formwork be properly designed and prescribes design considerations[1] (Section 6.1). The contract may or may not require submission of formwork design for approvals by the Engineer. Even when such approval is required and secured, ordinarily the approval is for overall conception only. Where the contract requires approval of the construction before concreting, the inspection is usually limited to dimensions, proper surface preparation, cleanout, etc. For protection of all concerned, the inspector should always check camber, whether specified or not, and record same. Since all formwork design is related to rate and method of casting as well as temperatures, all under control of the Contractor, the Engineer's approval of formwork is not intended to guarantee final safety.

REINFORCEMENT AND PLACING[2]

Ordinarily, the Engineer has required and approved details and placing drawings before shipment. Field inspection of this phase should ensure proper placing. Bar supports as shown on approved placing drawings should be adequate to carry the load of the bars and *necessary* foot traffic (Section 7.5.1). Usually auxiliary documents prescribing bar support spacings are specified.[3] Tack welding for assembly of reinforcement is prohibited, except by approval of the Engineer (Section 7.5.4). Ordinarily this approval will be given only for tack welds at points of low stress in each crossing bar. The Code provides tolerances applied simultaneously to cover or effective depth, d (Section 7.5.2). The dimension d is the most structurally important dimension in the inspection. Other important points are proper position of column verticals in plan; end anchorages as detailed; and splices. Dimensions locating bend-up points, cutoff points, spacing between slab, footing, or wall bars arc relatively unimportant so long as the correct total number of the correct bars are used. *A* tolerance of ± 2 in. is prescribed for bends and ends of bars, except at discontinuous ends of members (Section 7.5.2.2).

Coatings, other than rust, on bars generally should be removed (Section 7.4.1). Rust is generally harmless since "loose" rust possibly detrimental to bond is by definition knocked off in handling. Where rust is considered excessive, the Code provides a specific test to determine if it is excessive (Section 7.4.2). A specimen, any length, is to be wire brushed by hand and subsequently checked for weight and height of deformations. If either is less than the minimum required by the relevant ASTM specification, the inspector may (1) reject the bars, or (2) require cleaning and permit use with suffictent additional material to make up any design deficiencies.

As a practical matter, perhaps the most critical points for inspection of bar placing are (1) dowels for staggered butt splices in column verticals (see Chapter 10) and (2) the first floor reinforcement in multistory projects repeating same. In regard to correcting mislocated dowels, dowels out of plumb, etc., it should be noted that the Code requires bars to be bent cold, and forbids field bending of embedded bars, except as permitted by the Engineer (Section 7.3).

PLACING CONCRETE[4]

The Code requires clean forms, clear of ice or debris (Section 5.7.1). Concrete during placing should not be permitted to segregate (Sections 5.9.1, 5.9.2, and 5.10.1). The inspector should not only take (or supervise taking) samples for required cylinder tests, but should observe slump to avoid use of unnecessarily high slump concrete and avoid segregation. If the contract documents or accepted trial batch proportions are based on limited slump, these provisions, of course, should be enforced. Honeycomb is a structural deficiency due to segregation in casting or form leakage with high slump concrete. To prevent it at the base of deep (wall or column) forms, placing mortar to a depth of at least 1 in. is encouraged (Commentary R5.10). This mortar consists simply of the same mix used for concrete with the coarse aggregate omitted and water reduced if necessary.

The location of planned construction joints establishes extent of a single casting. Acceptable locations for joints not indicated on the project drawings can be specified. Otherwise, casting should be a continuous operation till completion of a panel, section, or element. Generally, joints may be allowed near the center of flexural members, except that joints in a girder should be offset two beam widths from a beam at the center (Section 6.4.4). Unless a construction joint is located at the top of walls or columns (best practice), a delay in casting (usually 1 to 4 hours) until the concrete in the wall or column takes an initial set is required (Section 6.4.5). Note that column capitals, brackets, beams, girders, haunches, etc. are to be "considered part of the floor system and shall be placed monolithically therewith" (Section 6.4.6).

CURING

Minimum requirements only are given (Section 5.11). For all cast-in-place concrete, additional curing is to be expected, since curing is not a formal process ending in 28 days. The reference to high early strength *concrete* (not defined) is usually interpreted as concrete made with high-early strength cement, or equivalent concrete made so by accelerating admixtures or additional cement, generally resulting in achievement of 28-day f_c' in 7 days under laboratory curing. Extremes of temperatures, hot or cold, require special precautions (Sections 5.12 and 5.13) including but not limited to curing.[5,6,7]

FORM REMOVALS

A number of studies have shown that, in many types of concrete structures lightly loaded in use, the most critical combinations of load-strength occur during construction. Two contradictory considerations are involved. The Contractor is held responsible for demonstrating that the structure is adequately *strong* to support construction loads (Section 6.2.1). Adequate safety (strength) does not ensure adequate *stiffness* to prevent permanent deflections considered excessive by the Engineer (or owner). Many Engineers specify arbitrary minimum periods before form removal or reshoring for an arbitrary number of floors or days. The Code prescribes performance requirements only, and not any arbitrary time limits which add to cost, except that vertical forms and horizontal forms separate from shoring must remain in place until concrete is strong enough not to be damaged by form removal (Section 6.2.1).

One suggestion that may be helpful to the Engineer in his decision to permit construction loads and in assessing the behavior of proposed shoring or reshoring[1] is to require modulus of elasticity tests (stress-strain) curves for job cured test specimens, beams or cylinders, as well as strength tests. Such tests will be found also especially valuable for long span shell or folded plate structures where large deflections are anticipated (Section 19.5.1).[1]

REFERENCES

1. ACI 347R-94: *Guide to Formwork for Concrete*; also *Formwork for Concrete*, ACI SP-4 (95).
2. *Placing Reinforcing Bars*, 7th Ed., 1997, CRSI.
3. *ACI Detailing Manual-1994* (SP-66)
4. ACI 304R-89: *Guide for Measuring, Mixing, and Placing Concrete.*
5. ACI 305R-91: *Hot Weather Concreting.*
6. ACI 306R-88: *Cold Weather Concreting.*
7. ACI 306.1-90: *Standard Specifications for Cold Weather Concreting.*

18 Strength Evaluation of Existing Structures

GENERAL

Chapter 20 of the 1995 Code has been extensively revised to provide more specific guidance regarding the strength evaluation of existing structures. If doubt exists that a part or all of a structure meets the safety requirements of the ACI Code, it is mandatory that a strength evaluation be carried out. Existing structures include any completed or partially completed construction the safety of which is in doubt, as well as old and/or deteriorated structures for which the present safety is unknown. In this chapter, the practical application of Code provisions is briefly explained with some practical precautions, but for an in-depth study, the ACI Committee 437 report* is recommended.

Strength evaluation can be performed by an analysis based on the member dimensions and material properties determined for the completed structure. In the alternative, a load test may be used to evaluate the strength of the structure. Some strength evaluations may be performed as a combination of both analysis and load testing. The minimum age of construction for testing and terms of testing are specified (Section 20.3.3). Analytical evaluations are recommended for structures where the effect of the strength deficiency is well understood as related to the effects of flexure, axial load, and combined flexure and axial load. (Section 20.1.2). If the effect of the strength deficiency is not well understood then the strength evaluation must be based on a load test. Structures with suspected deficiencies in shear or bond, and structures which cannot be adequately field measured or cannot have adequate evaluation of material strengths are best evaluated by a means of a load test (Section 20.1.3). It is noted that the 1995 Code requires a load test of structures to evaluate the effects of shear and bond, whereas the 1992 Code stated that analysis was the preferable method of eval-

* "Strength Evaluation of Existing Concrete Buildings," ACI Committee 437, 1991.

uation for such structures due to the lack of well-established criteria to evaluate the results of load tests. Although the 1995 Code has not provided well-defined acceptance criteria for the effects of shear and bond, it requires an evaluation of test observations (Sections 20.5.3, 20.5.4, 20.5.5) and has given general guidance for the basis of the evaluation in the Commentary Sections (R20.5.3, R20.5.4, R20.5.5).

The Code specifically permits the establishment of a *lower* load rating, based upon evaluation procedures outlined (Section 20.6). Code Chapter 20 is *not* intended for use as an acceptance test procedure for new materials or new design or construction methods (Commentary Section R20.1). Conversely, the provisions for approval of special systems of design or construction contained in Section 1.4 are not to be used for the strength evaluation of existing structures (Commentary Section R1.4).

PROCEDURE

When the safe load rating is doubtful, a five-step procedure is implied:

1. Identify the most doubtful elements or areas.
2. Determine the stress condition of cross-sections of members having doubtful strength or stiffness.
3. Determine the combination of design live loads and the pattern of loading required in design which will produce the maximum critical stress.
4. Compute the actual amount and extent of test load required, taking into account the live load reductions permitted by both the ACI Code and the general code and taking care to avoid overload of the elements supporting the element to be tested.
5. Execution of the load test (or analysis) and determination of the safe load rating.

DOUBTFUL CONSTRUCTION

Doubtful strength of completed construction may concern the concrete or the reinforcement. The quality of the concrete may be questioned for one or more of the following reasons, among others: (1) low 28-day acceptance test strengths, (2) unanticipated curing and temperature conditions, including freezing or plastic shrinkage (drying) cracking, (3) visibly poor consolidation (honeycomb), and (4) any failure to harden as expected, as evidenced by surface abrasion or impact tests. The effectiveness of the reinforcing steel, tensile or compressive capacity, can be questioned because of one or more of the following reasons among others: (1) low tests on field samples, (2) concrete strengths less than f_c' (3) use of a reinforcing steel grade *lower* than specified (higher grade is no problem, obviously), (4) misplaced reinforcing bars due to wrong size bar supports, top bars trodden down during concreting traffic, etc., and (5) discovery of overlooked bundles of main reinforcing bar or stirrups marked for incorporation into elements cast.

The elements or area of elements containing suspected low-strength concrete or reduced reinforcing steel capacity can usually be located from construction records. At this point all possible methods of nondestructive testing should be employed. These test results may dispel the doubt of quality, or verify the doubt and aid in locating the most doubtful portion, and will provide data on actual conditions for necessary analytical evaluations.

ANALYSIS

Even though a load test may be required and its results will then comprise the final acceptance or the basis for a lower load rating, an analysis is required before the testing. The analysis will determine whether the cause of doubtful strength will critically affect crack widths; capacity in shear, negative moment, positive moment; or deflection (low E_c). Further analysis is necessary to determine the critical required load combination (Section 9.2, Eq. 9-1, 9-2, and 9-3) and the required pattern, or patterns, of loading to produce the maximum values of the effects (such as deflection, rotation, or stress) necessary to demonstrate the adequacy of the structure (Section 20.3.1).

Field measurements of members and strengths of materials from tests are used in the strength evaluation by analytical methods. Since actual values of completed construction are available, there is less uncertainty regarding their values than would be assumed in the original design calculations. Consequently, values of the strength-reduction factors given in Section 9.3 can be increased. Maximum values of the strength-reduction factors to be used for strength evaluation of existing structures are given in Section 20.2.5.

TEST LOADS

At this point it is necessary to consider different factored loads used for slabs or joist-slabs and the supporting elements, beams or girders, where the loads and spans are such that the live load reductions of the general code are applicable. For example, if no live load reductions are allowed for slabs supported on beams, and slab reinforcement capacity is in question, the test load for the slabs must be applied so as not to "overtest" the beams. This may require shores to be placed beneath the beams during the slab test.

Test Load Patterns, Measurements, and Criteria

If negative moment capacity or shear capacity of a flexural member is considered the critical property to be established by the load test, the Engineer, acting as the representative of the owner, cannot accept the Code limit on deflection only as a criterion of acceptability. The deflection acceptance criteria (Section

20.5.2) are intended to represent satisfactory performance under load tests for only the positive moment capacity for end or continuous spans. For cantilevers the span of the member to be used in Eq. 20-1 (Section 20.5.2) is defined as twice the distance from support to the cantilever end (Section 20.0). An obvious interpretation (which has been misapplied even in published load test reports) for the term "maximum deflection" (Section 20.5.2) is that it means "maximum deflection at the point which should deflect most under the loading pattern causing maximum deflection at that point." The term "maximum deflection" and the numerical criteria provided in the Code for same are significant only when so interpreted.

It is possible and practically desirable to utilize the deflection measurements during a load test intended to develop critical shear or negative moment (Section 20.4.3). In cases where the Code requires that a load test be used to evaluate the strength of an existing structure (Section 20.1.3), it is desirable to support the results of a load test by analysis (Commentary Section R20.1.3). Analysis should also be used to determine an appropriate reduction factor to be applied to the maximum numerical limits for deflection (Section 20.5.2) when the test load pattern is established to test the shear or negative moment capacity in continuous flexural members. See Fig. 18-1 for loading patterns on one-way flexural members.

For two-way construction the ideal loading pattern to test positive moment capacity for direct application of the deflection criteria extends at least two panels in each direction from the panel in which the test measurements are taken (see Fig. 18-2). Where practical considerations make such an extensive loading pattern impracticable, the maximum allowable deflection criterion should be reduced appropriately.

Critical negative moment capacity and shear capacity in two-way construction is developed by a simpler loading pattern (see Fig. 18-3), but again the maximum allowable deflection criteria, if utilized, must be reduced appropriately.

EXAMPLE 1. For the construction shown in Fig. 18-4, assume that all tests indicate concrete strength in the panel *C-D*-3-4 substantially below f_c'. The live load prescribed in the general code for the floor is 100 psf. Compute the test loads applicable for load testing (1) slabs, (2) beams *B*-5, and (3) girder *G*-5. No dead loads other than the weight of the concrete floor are involved.

(1) *Slab.* Live load reduction allowed, $R = 0$. Test load = the total load (Section 20.3.2) minus the dead load already in place. Test load, $w_{test} = (0.85)(1.4D + 1.7L) - 1.0D = 0.19D + 1.45L$

(2) *Beam.* Live load reduction allowed by general codes, $R = 0$. Test load, $w_{test} = (0.85)[(1.4D) + (1.7L)] - 1.0D = 0.19D + 1.45L$

(3) *Girder.* Live load reduction allowed. *Test load,* $w_{test} = (0.85)[(1.4D) + (1.7)(0.86L)] - 1.0D = 0.19D + 0.73L$

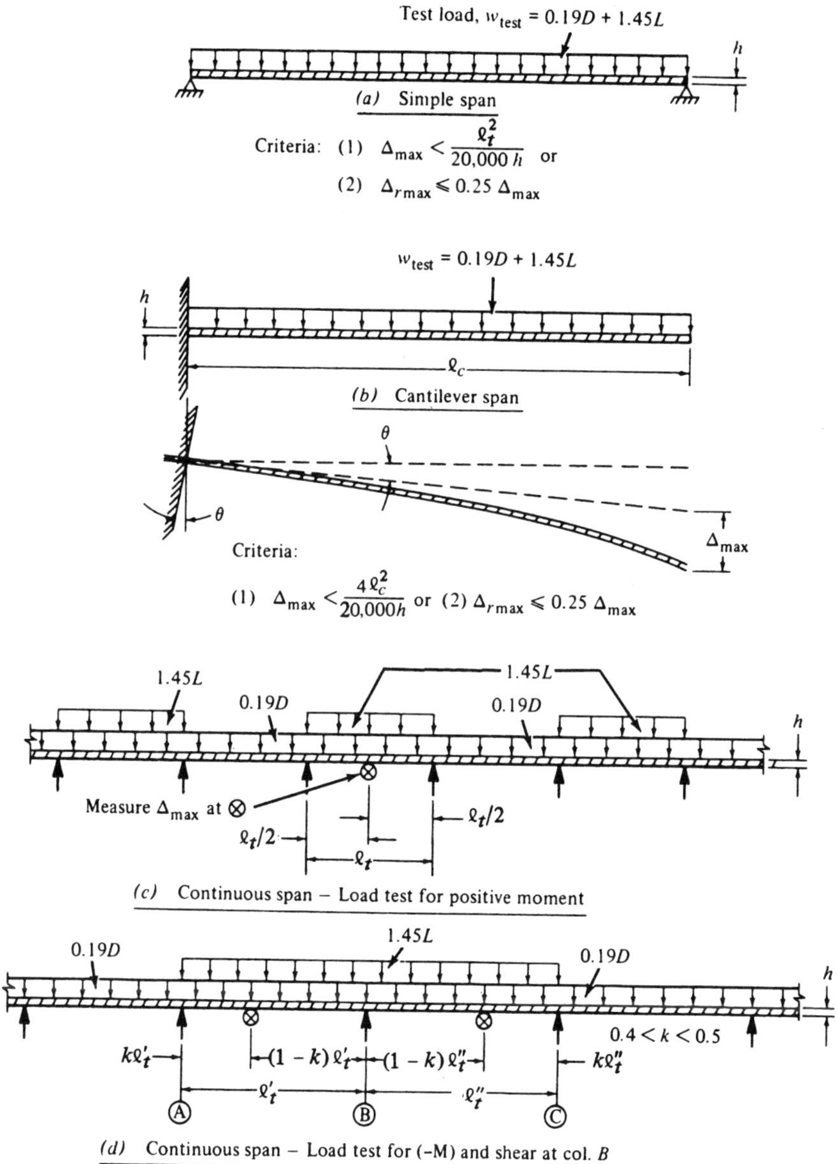

Figure 18-1 One-way slabs—load tests.

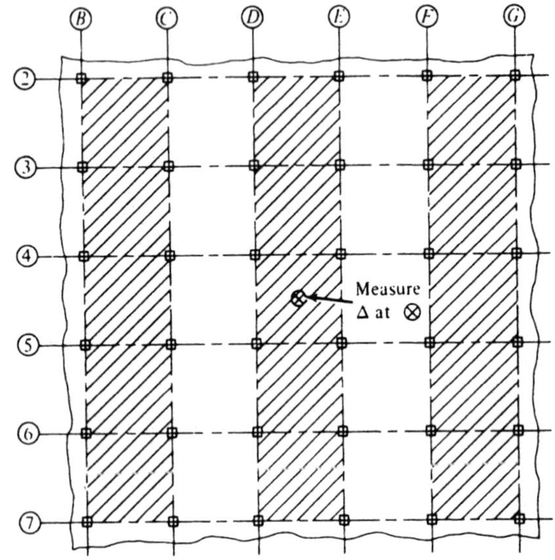

$$\Delta_{max} = k \frac{w\ell^4}{E_c h^3}\ ^*$$ where units are consistent

For square panels: k = 0.103 alternate bays loaded
k = 0.054 checkerboard loading
k = 0.063 all panels loaded

*S. Timoshenko, Theory of Plates and Shells

Figure 18-2 Two-way slabs—loading for Δ_{max} and $(+\ M_{E-W})$.

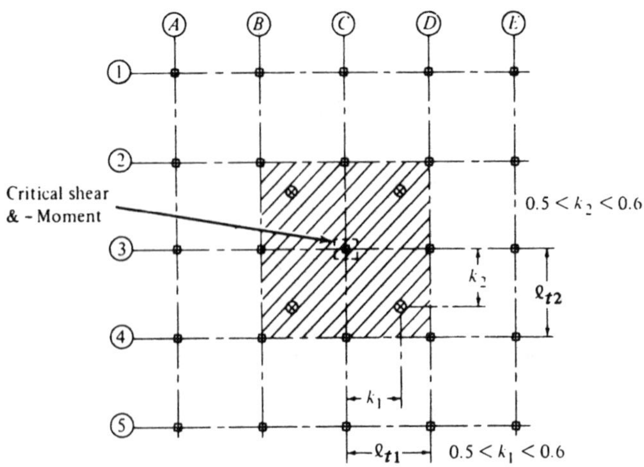

Figure 18-3 Two-way slabs—loading for critical shear and—moment at col. C3.

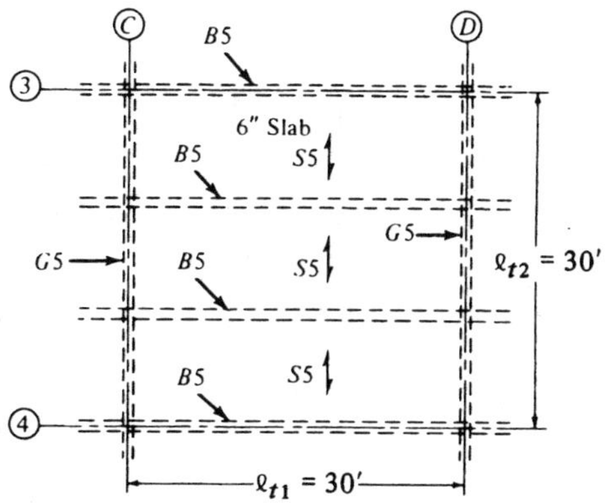

Figure 18-4 One-way slab, beam, and girder.

Appendix A

Metrication

Since 1983, the American Concrete Institute has issued separate metric editions of the 318 building code. The companion metric code and commentary to the 1995 inch-pound code and commentary is "Building Code Requirements for Structural Concrete (ACI 318M-95) and Commentary—ACI 318RM-95." The style, format, and content of ACI 318M-95 and ACI 318RM-95 closely follow the 1995 inch-pound code. Each section in the metric code and commentary matches up with the corresponding section in the inch-pound code and commentary. Except for the necessary modications of numerical values and units, and the expressions which are impacted by the units of parameters, such as f'_c or f_y, the provisions in ACI 318M-95 are identical to the corresponding provisions in ACI 318RM-95.

At press time, federal government-funded construction projects were required to be metric, i.e., designed in metric units and built with metric materials. Private sector projects, however, were not under any such mandate regarding units. An aspect of the federal government's metric initiative which impacts non-metric construction is soft metric reinforcing bars. In early 1997, several steel producing mills began to phase in the production of soft metric reinforcing bars. It is anticipated that virtually all production will soon be soft metric bars. If the mills follow through with shifting their production to only soft metric bars, it is expected that the production of inch-pound bars will be discontinued. As a consequence of the steel mills' initiative to produce only soft metric reinforcing bars, it is likely then that soft metric bars will be furnished to non-metric construction projects. Since the body of this Guide is concerned with the inch-pound version of the 1995 ACI Building Code, the authors believe it is essential to include basic information regarding soft metric reinforcing bars in this appendix. Thus, the purpose of the following discussion is to acquaint the reader with the impact of soft metric reinforcing bars on non-metric construction projects.

ASTM has metricated several of the steel reinforcement specifications which are referenced in ACI 318M-95. The ASTM specifications for steel reinforcement, which have been metricated, are presented in a "dual-units" format. A dual-units format means that a single document, e.g., Specification

A615/A615M-96a, contains inch-pound units and SI* metric units, and the specification prescribes the requirements for both metric and inch-pound billet-steel bars. Thus, in ACI 318M-95, billet-steel reinforcing bars are required to conform to ASTM A615M-96a. Similarly, low-alloy steel reinforcing bars are required to meet the requirements of ASTM A706M-96a. The other two ASTM Specifications are A616/A616M-96a for rail-steel and A617/A617M-96a for axle-steel reinforcing bars. The metric bar sizes in these two specifications have also been soft converted. These specifications are not discussed here, because rail-steel and axle-steel rebars are not generally available except in a few areas of the country.

The main differences in the metric provisions in the ASTM specifications for reinforcing bars, as compared with the requirements for inch-pound bars, are concerned with:

- Designations of the bar sizes
- Minimum yield strengths or grades
- Bar marking
- Substitution of metric bars for inch-pound bars

DESIGNATIONS OF BAR SIZES. Soft metric conversions of the eleven inch-pound sizes are now the standard sizes of metric bars. The one-to-one correspondence of bar sizes is shown in Table A-1. Each metric bar size has the same physical characteristics as its corresponding inch-pound bar size. For example, consider a #13 metric bar. The corresponding inch-pound bar size is

TABLE A-1 ASTM Standard Metric Reinforcing Bars

Bar Size[a]	Nominal Dimensions[b]		
	Diameter mm [in.]	Cross-Sectional Area, mm^2 [in.^{2}]	Weight kg/m [lbs/ft]
#10 [#3]	9.5 [0.375]	71 [0.11]	0.560 [0.376]
#13 [#4]	12.7 [0.500]	129 [0.20]	0.994 [0.668]
#16 [#5]	15.9 [0.625]	199 [0.31]	1.552 [1.043]
#19 [#6]	19.1 [0.750]	284 [0.44]	2.235 [1.502]
#22 [#7]	22.2 [0.875]	387 [0.60]	3.042 [2.044]
#25 [#8]	25.4 [1.000]	510 [0.79]	3.973 [2.670]
#29 [#9]	28.7 [1.128]	645 [1.00]	5.060 [3.400]
#32 [#10]	32.3 [1.270]	819 [1.27]	6.404 [4.303]
#36 [#11]	35.8 [1.410]	1006 [1.56]	7.907 [5.313]
#43 [#14]	43.0 [1.693]	1452 [2.25]	11.38 [7.65]
#57 [#18]	57.3 [2.257]	2581 [4.00]	20.24 [13.60]

[a] Equivalent inch-pound bar sizes are the designations enclosed within brackets.
[b] The equivalent nominal dimensions of inch-pound bars are the values enclosed within brackets.

* SI stands for International System of Units.

a #4. The #13 metric bar's nominal diameter, cross-sectional area, and weight per unit of length—except for being expressed in metric units and its weight more properly called "mass" in metric terminology—are exactly the same as the nominal diameter, area and weight of its counterpart #4 inch-pound bar. In addition, the height and spacing of the #13 metric bar's deformations are the same as the #4 bar's deformations. The end result of the soft conversion of the inch-pound bar sizes is that the soft metric bars are physically the same as the inch-pound bars.

YIELD STRENGTHS OR GRADES. Minimum yield strengths in metric units are 300, 420 and 520 MPa (megapascals), which are equal to 43,000, 60,900 and 75,400 psi, respectively. Metric Grade 420 is the counterpart of standard Grade 60.

BAR MARKING. Soft metric rebars are required to be identified or marked distinctively for size and minimum yield strength or grade. For example, consider the marking requirements for a #25, Grade 420 metric bar. The ASTM Specifications require the number "25" to be rolled onto the surface of the metric bar to indicate its size. For identifying or designating the minimum yield strength or grade, the ASTM Specifications provide an option. A mill can choose to roll a "4" (the first digit in the grade number) onto the bar, or roll on a single longitudinal rib or grade line to indicate Grade 420. Marking requirements for metric rebars are shown in Figure A-1.

SUBSTITUTION. The ASTM Specifications contain a provision for substitution of metric bars for corresponding inch-pound size bars and strength level. For example, Section 20.3.5 is the substitution provision in the A615/A615M-96a Specification for billet-steel rebars. It states:

> "20.3.5 It shall be permissible to substitute: a metric size bar of Grade 300 for the corresponding inch-pound size bar of Grade 40, a metric size bar of Grade 420 for the corresponding inch-pound size bar of Grade 60, and a metric size bar of Grade 520 for the corresponding inch-pound size bar of Grade 75."

The substitution clause in the A706/A706M-96B Specification for low-alloy steel rebars states:

> "16.3.5 It shall be permissible to substitute a metric size bar of Grade 420 for the corresponding inch-pound size bar of Grade 60."

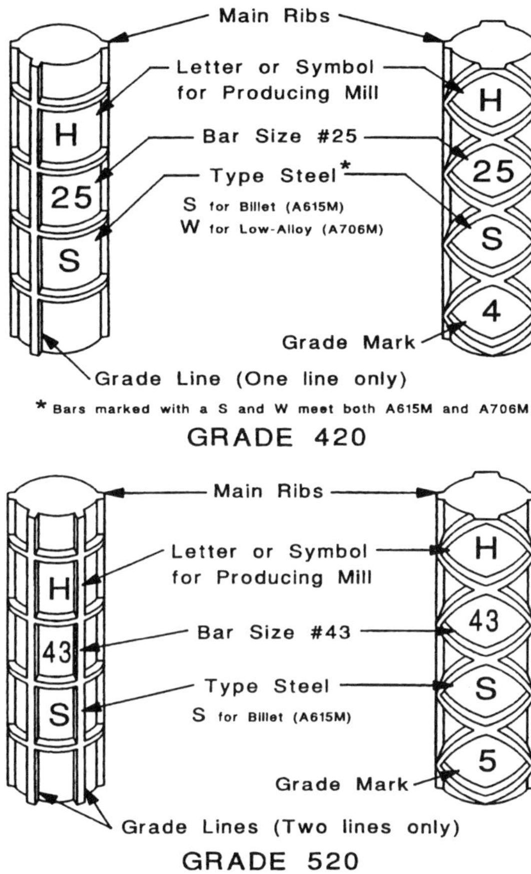

Figure A-1 Marking requirements for metric reinforcing bars

MECHANICAL PROPERTIES. Requirements for tensile and bending properties in the current ASTM specifications for reinforcing bars are presented in Table A-2. The metric bar sizes and criteria for tensile and bending properties in metric units comprise the requirements in ASTM A615M-96a and A706M-96b. Similarly, the data for inch-pound bars, which are enclosed within brackets in Table A-2, reflect the requirements in the inch-pound parts of the two specifications.

The tabulated data in Table A-2 serve to further illustrate soft conversion. Since the tensile property requirement "minimum percentage of elongation" is non-dimensional and related to bar size, the stipulated values are exactly the same for soft metric bars and corresponding inch-pound bar sizes. Note the gage length of 8 in. has also been soft-converted to $(8)(25.4) = 203.2$ mm. The minimum tensile strengths for Grade 60 and Grade 75 have been rounded

TABLE A-2 Bar sizes, grades, and tensile and bending properties

Type of Steel and ASTM Specification	Bar Sizes	Grade	Minimum Yield Strength MPa [psi]	Minimum Tensile Strength MPa [psi]	Minimum Percentage of Elongation in 203.2 mm [8 in.]	Cold Bend Test Pin Diameter (d = nominal diameter of bar)
	#10–#19 [#3–#6]	300 [40]	300 [40,000]	500 [70,000]	#10 [#3] . . . . .11 #13, #16, #19 [#4, #5, #6] 12	#10, #13, #16 [#3, #4, #5] . . . . .3.5d #19 [#6] 5d
Billet-Steel A615/A615M-96a	#10–#36, #43, #57 [#3–#11, #14, #18]	420 [60]	420 [60,000]	620 [90,000]	#10, #13, #16, #19 [#3, #4, #5, #6] . . . . .9 #22, #25, [#7, #8] . . . . . . .8 #29, #32, #36, #43, #57 [#9, #10, #11, #14, #18] . .7	#10, #13, #16 [#3, #4, #5] . . . . .3.5d #19, #22, #25 [#6, #7, #8] . . . . .5d #29, #32, #36 [#9, #10, #11] . . . . .7d #43, #57 (90°) [#14, #18, (90°)] 9d
	#19–#36, #43, #57 [#6–#11, #14, #18]	520 [75]	520 [75,000]	690 [100,000]	#19, #22, #25 [#6, #7, #8] . . . . .7 #29, #32, #36, #43, #57 [#9, #10, #11, #14, #18] . .6	#19, #22, #25 [#6, #7, #8] . . . . .5d #29, #32, #36 [#9, #10, #11] . . . . .7d #43, #57 (90°) [#14, #18 (90°)] 9d
Low-Alloy Steel A706/A706M-96b	#10–#36, #43, #57 [#3–#11, #14, #18]	420 [60]	420 [60,000]	550 [80,000]	#10, #13, #16, #19 [#3, #4, #5, #6] . . . . .14 #22, #25, #29, #32, #36 [#7, #8, #9, #10, #11] . . .12 #43, #57 [#14, #18] 10	#10, #13 #16 [#3, #4, #5] . . . . .3d #19, #22, #25 [#6, #7, #8] . . . . .4d #29, #32, #36 [#9, #10, #11] . . . . .6d #43, #57 [#14, #18] 8d

Bar sizes and other data enclosed within brackets are for inch-pound bars.
For low-alloy steel rebars, ASTM A706/A706M-96b prescribes a maximum yield strength of 540 MPa [78,000 psi] and tensile strength must be 1.25 times the actual yield strength.
Bend tests are 180° except ASTM A615/A615M-95a permits 90° for bar sizes #43 and #57 [#14 and #18].

slightly: 90,000 psi = 620.7 MPa, rounded to 620 MPa; 100,000 psi = 689.7 MPa, rounded to 690 MPa; and 80,000 psi = 551.7 MPa, rounded to 550 MPa. Since bend test requirements are a function of bar size, the pin diameters required for soft metric rebars are exactly the same as those for corresponding inch-pound bar sizes.

DESIGN ASPECTS

Soft metric bar sizes are identical to corresponding inch-pound bar sizes with respect to cross-sectional area (Table A-1). The minimum yield strength, f_y, of metric bars is slightly larger than the corresponding f_y of inch-pound bars. Thus, in all cases with respect to the reinforcement, the calculated design strength of a structural member, which is reinforced with soft metric bars, will be slightly larger than the design strength of the same member with inch-pound bars of equivalent size and yield strength.

For example, compare the design strength of a structural member reinforced with 4–#36 Grade 420 bars versus the design strength of the same member reinforced with 4–#11 Grade 60 bars. From Table A-1, the area of a #36 bar is 1,006 mm^2; the area of a #11 bar is 1.56 $in.^2$ The design strength of the member would be equal to the product of the strength-reduction factor, ϕ, and the nominal strength of the reinforcement, A_sf_y, where $\phi = 0.90$ for flexure, $\phi = 0.70$ for compression in a tied column, and so forth. Since A_sf_y for the metric rebars exceeds that of the inch-pound rebars, consequently ϕA_sf_y of the member reinforced with the metric bars is greater than ϕA_sf_y of the member reinforced with the inch-pound bars. See Table A-3.

This straightforward example can also serve as a basis to discuss substitution of soft metric bars for inch-pound bars in a non-metric construction project. Suppose a tied column in a non-metric designed building is reinforced with 4–#11, Grade 60 bars; ties are #4. If 4–#36 Grade 420 bars and #13 ties are substituted, determine the effect on the design strength of the column. With respect to the reinforcement, the design strength, ϕA_sf_y of the soft metric bars is (0.70)(380 kips) = 266 kips. The design strength of the inch-pound bars is (0.70)(374.4) = 262 kips. Substitution of the soft metric bars does not have an adverse effect on the design strength of the column. Instead the soft metric bars provide a slight increase in design strength.

DEVELOPMENT AND LAP SPLICES. ACI 318M-95, similar to the inch-pound code, requires anchorage or embedment length, the location and length of lap splices, and the type and location of mechanical splices and welded splices to be included on design drawings or in project specifications (Sections 1.2.1-h and 1.2.1-l in either Code). Consider an example concerning substitution of soft metric reinforcing bars for inch-pound bars. Suppose the embedment

TABLE A-3 Comparison of design strengths, metric bars vs. inch-pound bars

	4–#36 Grade 420	4–#11 Grade 60
Area of bars, A_s	$4(1006) =$ 4,024 mm^2 or 6.24 in.^{2}	$4(1.56) =$ 6.24 in.^{2} or 4,024 mm^2
Minimum yield strength, f_y	420 MPa[(a)] or 60.9 ksi	60 ksi or 413.7 MPa[(b)]
Nominal strength of the bars, $A_s f_y$	$(4{,}024)(420) =$ 1.69×10^6 Newtons or 380 kips	$(6.24)(60) =$ 374.4 kips or 1.66×10^6 Newtons

[(a)] Expressing strength or stress values in MPa is inconvenient in the authors' opinion. For convenience in design calculations, the minimum yield strength value of 420 MPa can be taken as 420 Newtons per mm^2 or 420 N/mm^2. To convert Newtons to kips, divide the value in Newtons by 4.45×10^3. In this case, $(1.69 \times 10^6) \div (4.45 \times 10^3) = 380$ kips.
[(b)] To convert ksi to Megapascals, multiply the value in ksi by 6.895×10^6 to get Pascals. Then divide the result by 10^6 to get Megapascals. In this case, $(60)(6.895 \times 10^6) \div 10^6 =$ 413.7 MPa.

and lap splice requirements for Grade 60 reinforcing bars are specified on the design drawings or in the project specifications for an inch-pound construction project. If soft metric Grade 420 bars are substituted for the inch-pound bars, will the embedment and lap splice lengths specified for the Grade 60 bars be suitable for the metric Grade 420 bars? Development lengths and lap splice lengths of reinforcing bars, whether tension or compression, are directly proportional to the specified yield strength, f_y, of the bar. The ratio of the minimum yield strengths, metric vs. inch-pound, is $(60.9/60) = 1.015$ or a variance of 1.5 percent. Because of the nearly identical minimum yield strengths, the anchorage or embedment lengths and lap splice lengths which the designer has specified for the Grade 60 rebars will be suitable for the substitute Grade 420 rebars.

FABRICATION AND DETAILS OF REINFORCEMENT

Since soft metric rebars are physically the same as corresponding sizes of inch-pound bars, all aspects of fabricating and detailing soft metric rebars will be identical to the fabricating and detailing practices used for inch-pound bars. CRSI has prepared and published Hook Details for ASTM Standard Metric Rebars—dimensions of end hooks, stirrup hooks, and seismic stirrup/tie hooks—soft-converted values of the standard hook dimensions used for inch-pound bars. See Figure A-2.

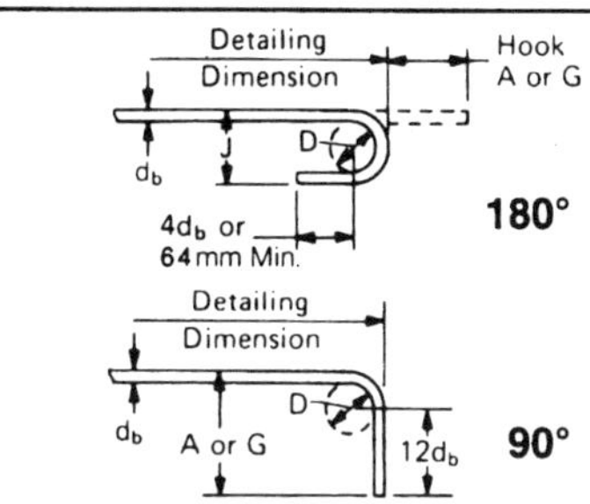

RECOMMENDED END HOOKS, ALL GRADES OF STEEL

BAR SIZE	D	180° HOOKS		90° HOOKS
		A or G	J	A or G
#10	60	125	80	150
#13	80	150	105	200
#16	95	175	130	250
#19	115	200	155	300
#22	135	250	180	375
#25	155	275	205	425
#29	240	375	300	475
#32	275	425	335	550
#36	305	475	375	600
#43	465	675	550	775
#57	610	925	725	1050

NOTE: All dimensions are in millimeters (mm).

ALL GRADES OF STEEL

D = Finished inside bend diameter

d_b = Nominal bar diameter

Min. D = 6 d_b for #10 through #25

Min. D = 8 d_b for #29, #32 and #36

Min. D = 10 d_b for #43 and #57

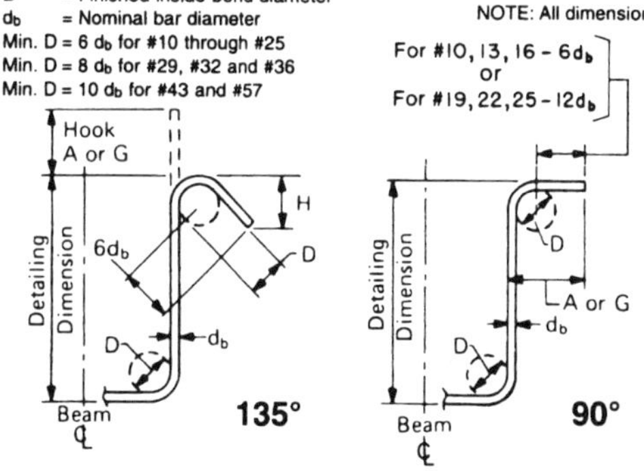

STIRRUP HOOKS (Tie Bends Similar)

BAR SIZE	D	90°	135°	
		A or G	A or G	H*
#10	40	105	105	65
#13	50	115	115	80
#16	65	155	140	95
#19	115	305	205	115
#22	135	355	230	135
#25	155	410	270	155

*H dimension is approximate.

NOTE: All dimensions are in millimeters (mm).

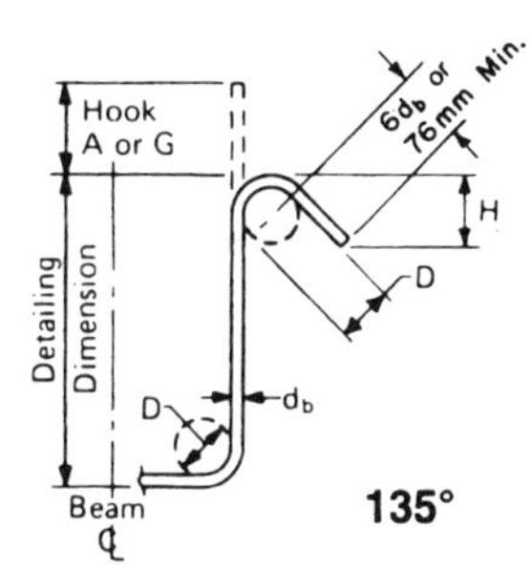

SEISMIC STIRRUP / TIE

BAR SIZE	135° SEISMIC HOOK		
	D	A or G	H*
#10	40	110	80
#13	50	115	80
#16	65	140	95
#19	115	205	115
#22	135	230	135
#25	155	270	155

*H dimension is approximate.

NOTE: All dimensions are in millimeters (mm).

Figure A-2 Hook details for ASTM standard metric reinforcing bars (courtesy of Concrete Reinforcing Steel Institute)

Subject Index

Commentary Section Index

Code Section Index